CHILTON'S GUIDE TO
SMALL ENGINE
REPAIR—UP TO 20 HP

Repair, maintenance and service for gasoline engines up to and including 20 horsepower.

Vice President & General Manager	John P. Kushnerick
Editor-In-Chief	Kerry A. Freeman, S.A.E.
Managing Editor	Dean F. Morgantini, S.A.E.
Senior Editor	Richard J. Rivele, S.A.E.
Senior Editor	W. Calvin Settle Jr., S.A.E.

CHILTON BOOK COMPANY

ONE OF THE **ABC PUBLISHING COMPANIES,**
A PART OF **CAPITAL CITIES/ABC, INC.**

Manufactured in USA
© 1983 Chilton Book Company
Chilton Way, Radnor, PA 19089
ISBN 0–8019–7899–8
Library of Congress Card Catalog No. 88–43168
1234567890   8765432109

SAFETY NOTICE

Proper service and repair procedures are vital to the safe, reliable operation of all motor vehicles, as well as the personal safety of those performing repairs. This manual outlines procedures for servicing and repairing vehicles using safe, effective methods. The procedures contain many NOTES, CAUTIONS and WARNINGS which should be followed along with standard safety procedures to eliminate the possibilty of personal injury or improper service which could damage the vehicle or compromise its safety.

It is important to note that the repair procedures and techniques, tools and parts for servicng motor vehicles, as well as the skill and experience of the individual performing the work vary widely. It is not possible to anticipate all of the conceivable ways or conditions under which vehicles may be serviced, or to provide cautions as to all of the possible hazards that may result. Standard and accepted safety precautions and equipment should be used when handling toxic or flammable fluids, and safety goggles or other protection should be used during cutting, grinding, chiseling, prying, or any other process that can cause material removal or projectiles.

Some procedures require the use of tools specially designed for a specific purpose. Before substituting another tool or procedure, you must be completely satisfied that neither your personal safety, nor the performance of the vehicle will be endangered

PART NUMBERS

Part numbers listed in this reference are not recomendations by Chilton for any product by brand name. They are references that can be used with interchange manuals and aftermarket supplier catalogs to locate each brand supplier's discrete part number.

ACKNOWLEDGEMENTS

Chilton Book Company expresses appreciation to Briggs & Stratton Corp., Milwaukee, Wis.; Onan Corp., Minneapolis, Minn., Kohler Co., Kohler, Wis.; Wisconsin Engines, Teledyne Wisconsin Motor Co., Milwaukee, Wis., Clinton Engines Corp., Maquoketa, Iowa, and Tecumseh Products Co., Grafton, Wis. for their generous assistance in the preperation of this book.

No part of this publication may be reproduced, transmitted or stored in any form or by any means, electronic or mechanical, including photocopy, recording, or by information storage or retrieval system withiout prior written permission from the publisher.

CONTENTS

Section 1   Engine Operation and General Information

Section 2   Routine Care, Maintenance, Storage and Operating Precautions

Section 3   Troubleshooting the Small Engine

Section 4   Briggs and Stratton Engines 0-6 hp

CONTENTS

CONTENTS

Section 6   Clinton Engines 0-6 hp

CONTENTS

CONTENTS

CONTENTS

Section 12 Kohler 4-Stroke Engines 6-20 hp

Section 13 Onan Engines 0-6 hp

CONTENTS

CONTENTS

Section 15 Tecumseh-Lauson Light Frame 4-Stroke Engines 0-6 hp

Section 16 Tecumseh-Lauson Light Frame 4-Stroke Engines 6-20 hp

CONTENTS

Section 17 Tecumseh-Lauson 2-Stroke Engines 0-6 hp

CONTENTS

Section 18 Tecumseh-Lauson
2-Stroke Engines 6-20 hp

CONTENTS

Section 19 Wisconsin Robin Engines 0-6 hp

Section 20 Wisconsin Engines 6-20 hp

CONTENTS

Section 1

Engine Operations & General Information

HOW AN INTERNAL COMBUSTION ENGINE DEVELOPS POWER

The energy source that runs an internal combustion engine is heat created by the combustion of an air/fuel mixture. The combustion process takes place within a sealed cylinder containing a piston which is able to move up and down in the cylinder. The piston is connected to a crankshaft by a connecting rod. The lower end of the rod is connected to the crankshaft at a point which is offset from the centerline of the crankshaft, allowing it to turn a large circle. This is why the piston moves up and down, and why pressure on the top of the piston eventually becomes torque, or turning force, on the crankshaft.

The sealed cylinder, or "combustion chamber" traps the air/fuel mixture. When burning takes place in a confined space such as this, the heat it produces becomes pressure which can be used mechanically to produce power. As the fuel burns and the piston goes down, the chamber becomes larger and larger, allowing for continued use of this pressure. The fact that the size of the chamber changes with the position of the piston also allows the air/fuel to be compressed, or packed into a confined space before it is burned, which has the effect of greatly increasing the amount of pressure made by the heat of burning, and makes the engine produce more power on less fuel. The variable size of the chamber also permits the engine to do its own breathing-to expel burnt gases and pull in fuel and fresh air (see the description of the four events in the operating cycle of a four-stroke engine below).

As you can see from how the engine produces power, leakage from the combustion chamber will have a tremendous effect on operating efficiency. One of the most important aspects of engine overhaul work involves repair or replacement of parts so that the combustion chamber will be as tightly sealed as possible.

Four-Stroke Engines

The entire series of four events that occur in order for an engine to operate may take place in one revolution of the crankshaft or it may take two revolutions of the crankshaft. The former is termed a two-cycle engine and the latter a four-cycle engine.

The four events that must occur in order for any internal combustion engine to operate are: intake, compression, expansion or power, and exhaust. When all of these take place in succession, this is considered one cycle.

In a four-cycle engine, the intake portion of the cycle takes place when the piston is traveling downward, creating a vacuum within the cylinder. Just as the piston starts to travel downward, a mechanically operated valve opens, allowing the fuel/air mixture to be drawn into the cylinder.

As the piston begins to travel upward, the valve closes and the fuel/air mixture becomes trapped in the cylinder. The piston travels upward and compresses the air/fuel mixture. This is the compression part of the cycle.

Intake stroke of a four-stroke engine

Compression stroke of a four-stroke engine

Power stroke of a four-stroke engine

Exhaust stroke of a four-stroke engine

Just as the piston reaches the top of its stroke and starts back down the cylinder, an electric spark ignites the air/fuel mixture and the resulting explosion and rapid expansion of the gases forces the piston downward in the cylinder. This is the expansion or power stroke of the cycle.

Just as the piston reaches the end of its downward travel on the compression stroke, another mechanically operated valve is opened. The next upward stroke of the piston forces the burned gases out the opened valve. This is the exhaust stroke.

When the piston reaches the top of the cylinder, thus ending the exhaust stroke, the exhaust valve is closed and the intake valve opened. The next downward stroke of the piston is the intake stroke that the whole series of events began with.

Two-Stroke Engines

In a two-stroke engine, intake, compression, power, and exhaust take place in one downward stroke and one upward stroke of the piston. The spark plug fires every time the piston reaches the top of each stroke, not every other stroke as in a four-stroke engine. (On some four-stroke engines, the magneto is operated by the crankshaft so that the spark plug actually fires once in every engine revolution. However, since the plug fires only into already burnt gases, this has no appreciable effect on engine operation.)

The piston in a two-cycle engine is used as a sliding valve for the cylinder intake and exhaust ports. The crankcase is used as a pump in order to slightly compress new fuel/air mixture and force it through the cylinder. In some designs, the piston also opens and closes a third port connecting the intake tube and carburetor to the crankcase. The intake and exhaust ports are both

The compression stroke of a two-stroke engine; the intake port is open and the air/fuel mixture is entering the crankcase

The power stroke of a two-stroke engine; the intake port is closed and, as the piston is being forced down by the expanding gases above, the air/fuel mixture in the crankcase is being compressed

LOOP SCAVENGE WITH 3RD PORT

Some engines use a third port, which is opened and closed by the bottom of the piston, to control the admission of air/fuel mixture into the crankcase

The exhaust stroke of a two-stroke engine; the piston travels past the exhaust port, thus opening it, then past the intake port, opening that. As the exhaust gases flow out, the air/fuel mixture flows in due to being under pressure in the crankcase. The next stroke of the piston is the compression stroke and the series of events starts over again

open when the piston is at the end of its downward stroke, which is called bottom dead center or BDC. Since the exhaust port is opened to the outside atmospheric pressure, the exhaust gases, which are under a much higher pressure due to combustion, will escape to the outside through the exhaust port. After the pressure of the exhaust gases has been somewhat relieved, the piston uncovers the intake port, and a fresh charge of air/fuel is pumped through the intake port, forcing the exhaust gases out in front of it. We say that the air/fuel mixture is being pumped into the cylinder because it is under pressure caused by the downward movement of the piston. In the most common type of two-stroke engine, the air/fuel mixture is drawn through a one-way valve, known as a reed valve, and into the crankcase by the vacuum caused by upward movement of the piston. When the piston reaches top dead center or TDC and starts back down, the one-way valve in the crankcase is closed by its natural spring action and the building pressure caused by the downward movement of the piston. In the three port type of engine, the upward movement of the piston creates a vacuum in the crankcase. When the skirt (bottom) of the piston uncovers the third port as the piston nears the top of its travel, the vacuum in the crankcase draws in air/fuel mixture. As the piston descends, the third port is again covered by the piston skirt, and the crankcase is sealed for compression of the mixture. The air/fuel mixture is compressed until the piston moves past the intake port, thus allowing the compressed mixture to enter the cylinder. The piston starts its upward stroke, closing off the intake port and then the exhaust port, thus sealing the cylinder. The air/fuel mixture in the cylinder is compressed and at the same time, a fresh charge

of air/fuel mixture is being drawn into the crankcase. When the piston reaches TDC, a spark ignites the compressed air/fuel mixture and the resulting expansion of the gases forces the piston back down the cylinder. The piston passes the exhaust port first, allowing the exhaust gases to begin to escape. The piston travels down the cylinder a little farther and past the intake port. The fresh air/fuel mixture in the crankcase, which was compressed by the downward stroke of the piston, is forced through the intake port and the whole cycle starts over.

FUEL SYSTEMS

The fuel system in a small engine consists of a fuel supply or storage vessel, a fuel pump, various fuel lines, and the carburetor. Fuel is stored in the fuel tank and pumped from the tank into the carburetor. Most small engines do not have an actual fuel pump. The carburetor receives gasoline by gravity feed or the fuel is drawn into the carburetor by venturi vacuum. The function of the carburetor is to mix the fuel with air in the proper proportions.

Types of Carburetors

Small engine carburetors are categorized by the way in which the fuel is delivered to the carburetor fuel inlet passage in the venturi (fuel nozzle).

PLAIN TUBE CARBURETORS

Carburetors used on engines that run at a constant speed, carrying the same load all of the time, can be relatively simple in design because they are only required to mix fuel and air at a constant ratio. Such is the case with plain tube carburetors.

All carburetors operate on the principle that a gas will flow from a large or wide volume passage through a smaller or narrower volume passage at an increased speed and a decreased pressure over that of the larger passage. This is called the venturi principle. The narrower passage in the carburetor where this acceleration takes place is therefore called the venturi.

A fuel inlet passage is placed at the carburetor venturi. Because of the reduced pressure (vacuum) and the high speed of the air rushing past at that point in the carburetor, the fuel is drawn out of the passage and atomized with the air.

The fuel inlet passage, in most carburetors, can regulate how much fuel is allowed to pass. This is done by a needle valve which consists of a tapered rod (needle) inserted in the opening (seat), partially blocking the flow of fuel out of the opening. The needle is movable in and out of the opening and, since it is tapered at that end, it can regulate the flow of fuel.

All carburetors operate on the venturi principle. The numbers represent hypothetical ratios of air speed and vacuum in relation to the venturi. Zero vacuum at the fuel nozzle is atmospheric pressure

The choke of a plain tube carburetor is located before the venturi or on the atmospheric side. The purpose of the choke is to increase the vacuum within the carburetor during low cranking or starting speeds of the engine. During cranking speeds, there is not enough vacuum present to draw the fuel out through the inlet and into the carburetor to become mixed with the air. When the choke plate closes off the end of the carburetor open to the atmosphere, the vacuum condition in the venturi increases greatly, thus enabling the fuel to be drawn out of the opening.

In most cases the choke is manufactured with a small hole in it so that not all air is blocked off. When the engine starts, the choke is opened slightly to allow more air to pass. The choke usually is not opened all the way until the engine is warmed up and can operate on the leaner fuel/air mixture that comes into the engine when the choke is completely opened and not restricting the air flow at all. Thus, the choke also provides the richer mixtures (more fuel for the amount of air) required when the engine is cold.

Many of the chokes used on small engine carburetors are entirely automatic. In many cases, the choke is required to provide just sufficient vacuum to get the fuel moving and give rich mixture during cranking. Once the engine is running, the choke can be opened in just a few seconds without causing it to stall. It will

run fairly well on a normal, lean mixture because the carburetor is close to the engine and puddling of fuel that occurs in engines with long manifolds while they are cold is no problem.

This type of choke is held shut by the pressure of a spring, and opened by the action of a diaphragm. Manifold vacuum is fed to the side of the diaphragm on which the spring is located through a small orifice. There is no vacuum when the engine is stopped, but upon starting, manifold vacuum gradually draws air out of the diaphragm chamber, and draws the diaphragm downward, against spring pressure. A rod linking the diaphragm to the choke then pulls it open.

This type of choke is also capable of keeping the mixture adequately rich during sudden increases in throttle opening; vacuum normally gets very low under these circumstances, and the choke tends to close slightly as the throttle opens, thus aiding fuel flow.

The throttle plate in a carburetor is installed on the engine side of the venturi. The purpose of the throttle plate is to regulate the flow of air/fuel mixture going into the engine. Thus the throttle plate regulates the speed and power output of an engine. By restricting the flow of air/fuel mixture going into the engine, the throttle plate is also restricting the combustion explosion and the energy created by the explosion.

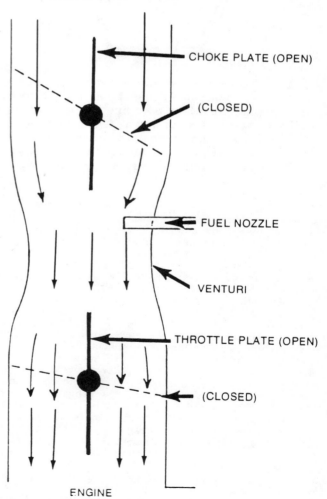

The positioning of the choke and throttle plates in relation to the position of the venturi

These are the basic components required for any carburetor to operate on an engine. It is possible for such a carburetor to be installed on an engine and work. However, most engines operate at various speeds, under various load conditions, at different altitudes, and in a variety of temperatures. All of these variations require that the fuel/air mixture can be changed on command. In other words, the simple plain tube carburetor is not sufficient in most cases.

SUCTION CARBURETOR

Next to the plain tube carburetor, this is the simplest in design. With this type of carburetor, the fuel supply is located directly below the carburetor in a fuel tank. In fact, the carburetor and fuel tank are considered one assembly in most cases because a pipe extends from the carburetor venturi down into the fuel tank. When the engine is running, the partial vacuum in the venturi, and the relatively higher atmospheric pressure in the fuel tank, force the fuel up through the fuel pipe and into the carburetor venturi. There is a check ball located in the bottom of the pipe that prevents the fuel in the pipe from draining back into the fuel tank when the engine is shut down. In most engines there is a screen located at the end of the pipe to prevent dirt from entering and blocking the fuel nozzle.

Some suction carburetors are designed with an extra fuel inlet passage for when the engine is idling. This extra passage would be located on the engine side of the throttle plate. Since the vacuum condition in front of an idling engine might not be enough for the fuel to be drawn out at that point, the extra passage is installed behind the throttle plate where the vacuum is great enough to draw the fuel out. This extra fuel inlet passage would also have a regulating needle as does the main inlet.

FLOAT TYPE CARBURETORS

The fuel tank used with float type carburetors is usually located on top of, or at least above, the level of the carburetor. Fuel is fed to the carburetor by gravity. If the fuel tank is located below the level of the carburetor, a fuel pump is employed to pump fuel to the carburetor. Fuel enters the carburetor through a valve and into a float bowl and, as it fills the bowl, a float rises on the surface of the fuel. The float is connected to the inlet valve and, as the level rises, a needle is inserted into the inlet valve and the flow coming into the float bowl is checked. As the fuel is drawn into the main fuel nozzle in the venturi and the level in the bowl drops, the float drops and the needle opens the valve allowing more fuel to run into the bowl.

Although there are many different float carburetors, all operate in this manner.

DIAPHRAGM TYPE CARBURETORS

Fuel is delivered to the diaphragm type carburetor in the same manner as to the float type carburetor.

A flexible diaphragm operates the fuel inlet valve, hence the description "diaphragm carburetor."

Atmospheric pressure is maintained on the under side of the diaphragm by a vent hole in the bottom of the carburetor. The other side of the diaphragm is acted upon by the varying vacuum conditions in the carburetor.

When the vacuum condition in the carburetor is increased by the opening of the throttle plate, and the demand for an increased flow of fuel, the center diaphragm is bellowed upward by the increased vacuum. The center of the diaphragm is connected by a lever to the fuel inlet valve needle. The needle is dropped down away from the inlet valve and fuel is allowed to drop in. When the throttle plate is closed and the need for fuel is reduced, the vacuum condition also decreases and the diaphragm is returned by a spring located on the atmospheric side to its normal flattened position. This closes the fuel inlet valve by raising the needle up into position against its seat.

This type of carburetor also has an idle orifice positioned behind the throttle plate to compensate for the low vacuum condition in front of the throttle plate during idle speeds.

Some diaphragm type carburetors incorporate an integral fuel pump. It consists of a second diaphragm having fuel on one

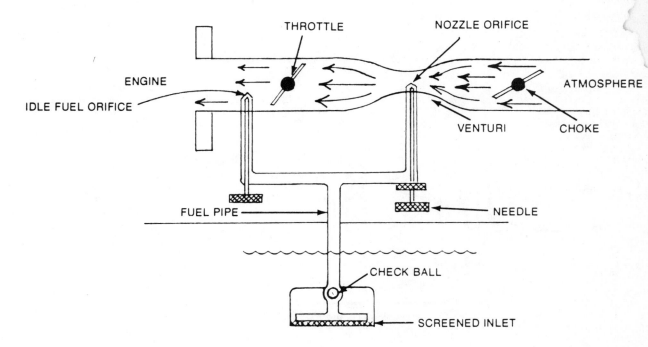

Diagram of a suction type carburetor

side, and exposed to the pressure fluctuations of the crankcase or the intake manifold on the other side. The vibration of this diaphragm, in conjunction with the action of check valves in the passages leading to and from the fuel side of the diaphragm chamber work together to pump fuel, under pressure, to the fuel inlet valve.

REED VALVES

Reed valves are used on those two stroke engines in which the intake tube leading from the carburetor empties right into the crankcase. The reed serves as a kind of check valve, so that air/fuel cannot be forced back out of the crankcase. When the

Diagram of a float type carburetor

1. Choke
2. Diaphragm
3. Fuel chamber
4. Fuel inlet
5. Inlet valve needle
6. Lever
7. Nozzle
8. Orifice
9. Pivot pin
10. Spring
11. Throttle
12. Vent
13. Valve seat

Diagram of a diaphragm type carburetor

piston begins to rise, a slight vacuum is created under it, sucking the reed open against its spring pressure, as shown in the illustration. When the piston reaches the top of its travel, and the pressure in the crankcase gets near to outside (atmospheric) pressure, the valve's spring action pulls it flat against the intake opening (the reed is really a kind of flat spring). Then, as the piston descends, the build-up in pressure in the crankcase, in combination with the slight amount of oil that gets onto the reed, seals the intake very tightly, forcing practically the full charge into the cylinder.

IGNITION SYSTEMS

The function of an ignition system is to provide the electrical spark that ignites the air/fuel mixture in the cylinder at precisely the correct time. The two types of ignition systems used on small engines are either a battery ignition or a magneto ignition. The difference between the two types are where they get their initial electrical charge from. The battery ignition, as the name implies, gets its initial electrical charge from a storage battery. The magneto ignition actually generates its own electricity, thus eliminating the need for a battery as far as ignition is concerned.

Battery and magneto ignition systems are similar in one respect. Both systems take a relatively small amount of voltage, such as 12 volts in the case of a battery ignition, then step up that to an extremely high voltage, in some systems as high as 20,000 volts.

Basic Battery Ignition System

Battery ignition systems are found mostly on larger one cylinder engines, such as those installed on lawn tractors, larger pumps and generators.

As previously stated, the initial electrical charge comes from a storage battery. The entire ignition system is grounded so that current will flow from the battery throughout the primary circuit.

The ignition system is composed of two segments, the primary circuit and the secondary circuit. Current from the battery flows through the primary circuit and the current that fires the

The reed valve in the open position (piston moving upward)

spark plug flows through the secondary circuit. The current in the primary circuit actually creates the secondary current magnetically.

When the engine is running, current flows from the battery, through the breaker points, and on to the ignition coil. The current then flows through the primary windings of the coil which are wrapped around a soft iron core and grounded. The primary current flowing through these primary windings causes a magnetic field to be created around the soft iron core of the coil. At precisely the right time, the breaker points open and break the primary circuit, cutting off the flow of current coming from the battery. This causes the magnetic field in the coil to collapse toward the center of the soft iron core. As the field collapses, the lines of force of the magnetic field must pass through the secondary windings of the coil. These windings are also wrapped

Diagram of a battery type ignition for a single cylinder engine

around the soft iron core of the coil, except that there are many more windings and the wire is thinner than those of the primary windings. When the magnetic field collapses and passes through the secondary windings, current flow is induced in the secondary circuit which is grounded at the spark plug. Because the wire of the secondary windings is smaller and there are many more coils around the soft iron core, the current created or induced in the secondary side of the ignition is of much greater voltage than the primary side. This current in the secondary circuit flows toward the grounded end of the circuit which is the spark plug. The current flows down the center of the spark plug to the center electrode. In order to reach the ground, it must jump across a gap to the grounded electrode. When it does, a spark is created and it is this spark that ignites the air/fuel mixture in the cylinder.

Magneto Type Ignition Systems

Magneto ignition systems create the initial primary circuit current, thereby eliminating the need for ignition batteries.

Flywheel Type Magnetos

On this type of magneto, the flywheel of the engine carries the permanent magnets that are used to create the primary current.

The magnets are arranged so that about ⅓ of the area enclosed by the flywheel is a magnetic field. In the center of the flywheel is a three pronged coil, the three prongs representing the core. The center prong has the primary and secondary windings and is set up just like a battery ignition coil.

As the magnets in the flywheel pass the two outside prongs, the primary current is induced in the coil and a magnetic field is set up around the center prong. At the point when the primary

current is at its strongest which is also the time when the secondary current is needed at the spark plug, the breaker points interrupt the primary current, causing the magnetic field to collapse through the secondary windings. This induces the secondary current which flows to the grounded spark plug.

Unit Type Magnetos

Unit type magnetos operate in the same way as flywheel magnetos.

Permanent magnets are rotated through a mechanical connection to the crankshaft of the engine. The rotating magnets create the primary current which is routed through the breaker points and on to windings around a soft iron core which is also wrapped by the secondary windings. At precisely the right moment, the points interrupt the primary current, causing the magnetic field to collapse through the secondary windings and inducing the secondary current to the spark plug.

When the breaker points are opened, the induced magnetic field collapses. The collapsing of the magnetic field induces the secondary current into secondary windings leading to the spark plug

P—PERMANENT MAGNET
N—NORTH POLE SHOE
S—SOUTH POLE SHOE
F—FLYWHEEL
C—LAMINATED COIL CORE

A cutaway view of a flywheel used in a flywheel magneto. The magnets are arranged so that there is a magnetic field covering about ⅓ of the area enclosed by the flywheel. On flywheel magnetos where the coil and core are mounted on the outside of the flywheel, the magnets would be arranged on the outside of the flywheel

Because the magnets in the unit magneto are driven by the crankshaft of the engine through a gear mechanism, starting is a problem. At starting speeds, the magnets in the magneto cannot be rotated fast enough to create a primary current. To overcome this difficulty, unit magnetos have an impulse coupling on their shaft which drives the magnets. When the engine is turned over at starting speed, a catch engages a coil spring that is wound up in much the same manner as those that propel wind-up toys. As the shaft rotates further, it releases the spring. The spring unwinds rapidly, spinning the magnets fast enough to cause the primary current to be induced in the primary circuit. When the engine starts to run, centrifugal force keeps the catch in the impulse coupling from engaging the wind-up spring.

LUBRICATION SYSTEMS
Four-Stroke Engines

Most small four stroke engines are lubricated by the splash system. All vital moving parts are splashed with lubricating oil that

The three legs of the core pass through the magnetic field created by the rotating magnets. In this position, the lines of force are concentrated in the left and center core legs and are interlocking the coil windings

The flywheel has turned to the point where the lines of force of the permanent magnets are being withdrawn from the left and center cores and are being attracted by the center and right cores. Since the center core is both drawing and attracting the lines of force, the lines are cutting up one side and down the other (indicated by the heavy black arrows). The breaker points are now closed and a current is now induced in the primary circuit by the lines of force cutting up and down the center core leg

With the induced current in the primary windings, a magnetic field is created around the center core leg (coil). When the field is created around the center leg, the points are opened and the condenser begins to absorb the reverse flow of current

Diagram for a unit type magneto for a single cylinder engine

is stored in the crankcase. The connecting rod bearing gap usually has an arm extending down into the area in which the oil lies. As the crankshaft turns, the arm, commonly called a dipper, splashes oil up onto the cylinder walls, crankshaft bearings and camshaft bearings. In some cases, the connecting rod bearing cap screws are locked in place by lock plates which double as oil dippers. When the lock plate's tabs are bent up beside the cap screws, they extend a little past the tops of the screws and, when they enter the oil, they provide a sufficient splash to lubricate the engine.

In larger four stroke engines, a pressure system is used to lubricate the engine. This type of lubrication system pumps the oil through special passages in the block and to the components to be lubricated. In some cases, the camshaft and crankshaft have their center drilled so that oil can be pumped through the center and to the bearing journals. Most oil pumps in small engines are gear types.

In addition to lubricating the engine, oil in four stroke engines has other important functions. One of those additional functions is to help cool the engine. The oil actually absorbs heat from high temperature areas and dissipates it throughout other parts of the engine that are not directly exposed to the very high temperatures of combustion.

Oil in the crankcase also functions as a sealer. It helps to seal off the combustion chamber (top of the piston) from the crankcase, thus maintaining compression which is vital to satisfactory engine operation, while at the same time lubricating the cylinder walls and piston rings.

Engine oil also keeps harmful bits of dust, metal, carbon, or any other material that might be present in the crankcase, in suspension. This keeps potentially harmful abrasives away from vital moving parts.

Most manufacturers recommend a good grade, medium weight detergent oil for their engines. The detergent properties of engine oil do not mean that the oil is capable of cleaning away dirt or sludge deposits already present in the engine. It means that the oil will help fight the formation of such deposits. In other words, the oil keeps the dirt in suspension.

A certain amount of combustion leaks past the piston rings and into the crankcase. Raw gas, carbon, products of combustion, and other undesirable material still manage to make their way into the lubricating oil. The oil becomes diluted by the gas and loses its cooling properties. When it is filled with abrasives, it loses its detergent properties due to chemical reactions, constant heat, and pollutants. The oil can then actually become harmful to the engine. Thus, one can see the need for keeping oil as clean as possible by changing it frequently.

Two-Stroke Engines

Since the two-stroke engine uses its crankcase to compress the air/fuel mixture so that it can be forced up into the combustion chamber, it cannot also be used as an oil sump. The oil would be splashed around and forced up into the combustion chamber, resulting in the loss of great amounts of lubricating oil while at the same time "contaminating" the air/fuel mixture. In fact the air/fuel mixture would be so filled with oil that it would not ignite.

A two stroke engine is lubricated by mixing the lubricating oil in with the fuel. A mixture of fuel and lubricating oil is sucked into the crankcase and, while it is being compressed by the downward movement of the piston, enough oil attaches itself to the moving parts of the engine to sufficiently lubricate the engine. The rest of the oil is burned along with the fuel/air mixture. As can well be imagined, the mixture ratio of fuel and oil is critical to a two stroke engine. Too much oil will foul the spark plug and too little oil will not lubricate the engine sufficiently, causing excessive wear and possibly engine seizure. Follow the manufacturer's recommendations closely. Most recommend that detergent oils not be used in two-stroke engines.

Lubrication of a four-stroke engine

GOVERNORS

The governor is a control that opens and closes the engine throttle to maintain its speed when the load on the engine changes. Consider an engine which is operating a generator. The generator must operate within a very narrow speed range in order to provide the stable output current electrical appliances require. Yet, as appliances in the circuit are turned on and off, the load on the generator changes, which tends to either allow the engine to speed up or force it to slow down.

The governor measures engine speed and responds accordingly. As the speed drops, the throttle is opened. As speed increases the throttle is closed. The governor has a "speed droop," which means that it does not try to keep the engine at exactly the same speed, but allows a gradual drop in speed as load increases. If a governor has a speed droop of 100 rpm, and is set so that the engine runs at 3650 rpm with no load, loading the engine will cause the throttle to be opened to about the half-way point at 3600 rpm, and all the way at 3550 rpm.

Lubrication of a two-stroke engine

Operation of an air vane governor

The governor is basically just a spring that pulls the throttle open while some other force, which varies with engine speed, tries to close it. The simplest method of providing the closing force is the air vane. The vane is hinged at the leading edge. When it turns on the hinge, it closes the throttle through a simple wire link. The vane is housed inside the cooling system, directly in the path of cooling air. When engine speed increases, air pressure on the vane increases. At a certain speed, the force on the vane is great enough to overcome the tension of the spring and begin stretching the spring, closing the throttle. The point at which the spring has stretched enough to nearly close the throttle, and keep the engine speed steady without any load, is somewhat above the point where the vane just begins to stretch the spring. This range is the speed droop. It accounts for the fact that as load is applied to an engine, it finds a slightly lower speed at which it operates.

The flyweight type governor works very similarly. Here, the increasing closing force comes from hinged weights which spin in a circle, driven through gearing from the engine's crankshaft. As engine speed increases, centrifugal force makes the weights turn outward on their hinges and force the throttle closed via a collar which the bottoms of the weights work against.

Operation of a flyweight governor

The spring and flyweights or wind vane are connected via a lever which contains various holes in which the end of the spring can be installed. If the governor is too sensitive (speed droop too small), it will hunt back and forth, causing the throttle to jump violently from full open to idle position and back. If the governor is not sensitive enough (speed drop too great), the engine will slow excessively before it adjusts to the load. The spring is moved from hole to hole in order to change this sensitivity. If the spring is hooked near the shaft the lever turns on, a small change in force from the flyweights or air vane will stretch the spring a great deal, making the governor very sensitive. If the spring is hooked to a hole farther away from the pivot point of the lever, a large change in engine speed will be required to stretch it, making the governor slow to respond. The governor speed setting is changed by simply stretching the spring—pulling the end opposite the point where it fits into the lever away from the lever, utilizing any of various kinds of mechanical linkages.

COOLING SYSTEMS

Most small engines are cooled directly by the outside air because this type of system is much less expensive to manufacture, is more easily maintained and repaired, and; if the engine is operating correctly, is more reliable. This type of cooling also saves engine weight, and this is an important side benefit, as many small engines are used to operate hand carried or hand propelled machinery.

TEMPERATURE ON CYLINDER BORE

The illustration shows how, as the one-third of energy handled by the engine passes through the cylinder walls and fins, the temperature of the metal drops from 350°F on the cylinder wall to only 100°F at the tips of the fins

Air is a much less efficient coolant than water. It carries much less heat, so that a given volume of air will rise to a very high temperature without actually carrying away very much heat. And, it readily forms stagnant areas around solid objects. Water carries a lot of heat without much temperature rise; it grips tightly to whatever metal object it is cooling, and can easily be kept in uniform motion. For these reasons, air cooled engines are not as tolerant of inefficient, heat producing operation. They must be kept running at peak efficiency if maximum component life is to be realized.

The biggest key to making air cooling work is finning. The parts of the cylinder and cylinder head which are exposed to burning gases inside are cast in such a way as to very greatly increase the metal surface exposed to air. At a number of points, the metal of a cylinder, for example, is forced outward from the main structure into a thin sheet or fin that may increase the surface available to air for cooling as much as ten times. While the fins at first might seem to represent excess weight, they actually form an integral portion of the structure of the engine, increasing its rigidity, and decreasing the thickness required in the main structure. While the outer ends of the fins seem to be very far from the source of the heat they in fact carry lots of heat to the cooling air because metal parts are excellent heat conductors, moving energy from hot to cool portions in a process that is a little like the flow of electricity.

Since the air heats fast and can easily become stagnant, forming a blanket near the metal, the second secret of air cooling is high velocity. The air must be guided in a precise manner around every part of the engine coming into contact with the burning fuel. This is done by shrouding the outer edges of the fins where necessary with light metal ducts. Use of a very powerful, centrifugal fan which slings the air outward and through the engine (or draws the air through the engine at high speed) ensures that there will be enough air moving along all the hot parts to break up any stagnant pockets and prevent any of the air from being there long enough to reach excessively high temperatures.

Since air cooling is more sensitive than conventional water cooling systems, always observe the following precautions when operating any air-cooled engine:

1. You have to have a sharper eye with air cooled engines. Watch for sluggish operation and excessive oil consumption that may mean things are getting too hot.

2. Never overload the engine. If the engine runs full throttle below governed rpm, it will be running short of cooling air, and will usually overheat.

3. Use oil of the proper viscosity. Too low a viscosity (or even very dirty oil) will cause excess friction and overheating.

4. Keep all cooling system ducts, shrouds, or seals in top shape—replace any that become loose or even dented. Replace parts whose fins are broken off.

5. Keep fins clean of dust, oil and debris.

6. Keep the cooling air blower clean, and watch carefully for even partial clogging of the air intake screen. If any of the blower's blades break off, replace the blower or flywheel.

7. Some air cooled engines use thermostatically or throttle controlled air vane to restrict cooling air at low temperatures or under light loads; make sure it works freely and opens when it's supposed to.

8. Keep the engine in good tune. Lean fuel mixtures, late ignition timing, engine knock, or any other combustion problem will severely overheat the engine.

STARTING SYSTEMS
Electric Starters

Many small engines utilize direct current electric motors for starting purposes. These are run off a standard battery, and the entire system is quite similar to ordinary automotive equipment of this type, except for its smaller size. Direct current motors consist of a stationary coil of wire which is mounted on the inside of the housing, and a rotating coil, which is mounted on the shaft, and is surrounded by the "stator," or stationary coil. Direct current from the battery passes through both coils in series, being carried to the "armature" or rotating coil through brushes. The brushes are mounted inside the housing and rub against commutator rings, which spin with the rotor shaft. The brushes are sprung against the rings so they stay in constant contact.

Electric starter with Bendix type engagement

The most common problems with these motors occur when the brushes wear and lose their spring tension.

Some starter motors turn the engine over by engaging a pinion—a gear on the end of the shaft—with a toothed flywheel. Some starter pinions are moved forward into mesh with the flywheel through the action of a magnetic solenoid, while others employ the Bendix drive. The Bendix type pinion spins forward, against the tension of a spring, via the action of spiral cut grooves in the armature shaft. When the engine starts, the pinion is spun back toward the starter, and then held by the spring. In the solenoid design, an overrunning clutch allows the pinion to spin faster than the starter armature, thus keeping the engine from over-speeding it as it starts. When operating engines

Electric starter with solenoid type engagement

with this type of drive, allow the starter to disengage as soon as the engine fires to avoid overheating the overrunning clutch. In both types, but especially the Bendix type, poor engagement may result from poor lubrication of the spiral grooves which guide the pinion along the armature shaft.

Some electric starters drive through belts and double as generators when the engine starts. This design incorporates a relay that switches the polarity (direction of current flow) through the unit when the engine comes up to speed. Because of the large amount of torque required to turn the engine over, it is important to keep the drive belt under proper tension, and to replace it if it becomes glazed (which can cause it to slip).

Recoil Starters

Recoil starters employ a rope or cable wound around a pulley, but automatically rewind the rope when it is released. The rewind force is provided by a large clock type spring which connects the pulley to the engine housing. When the rope is pulled, a cam action engages dogs which cause the engine flywheel to turn with the pulley. When the engine starts, its motion disengages the dogs.

When working on recoil starters, it is absolutely necessary to completely release the tension of the spring before disassembling the unit. If spring tension is released suddenly or unexpectedly, the force released is tremendous, and is very likely to cause personal injury.

The illustration shows how the rewind spring is mounted inside the cover on one popular type of recoil starter

Section 2

Basic Care and Maintenance

OPERATING PRECAUTIONS

General

1. Before starting the engine, make sure it is full of the proper grade of lubricating oil. Some sort of record of the viscosity of the oil in use should be kept so that the proper grade may be added when necessary and the oil may be changed to keep the viscosity in conformance with recommendations for the outside air temperature range that is prevalent. If the engine is run in hot weather with an oil of too low a viscosity in the crankcase, very serious damage can result. The same sort of damage (from overheating) can occur if the oil level is too low. Remember that all engines must consume some oil to run properly and that, even if the engine is in continuous service, it should be stopped every few hours for an oil level check, and the crankcase refilled.

In the case of two-stroke engines, a recommended two-stroke oil must be used, and it must be mixed in the proportions recommended for that engine by the manufacturer. If several two-strokes using different fuel mixes are in use at the same location, each fuel container should be marked so that only the mix recommended for each engine is used in it. Do not simply use a pre-mixed batch of fuel — there is no standard mix. Make sure what you're using is in the exact proportion recommended by the manufacturer. Also, fuel and oil must be uniformly mixed. Usually, this is done by filling the container about 25% of the way with gas, pouring in the oil, and then completing the filling job. After this, the container should be shaken vigorously with the cap applied tightly to complete the mixing process. When using fuel that has been setting for several weeks, shake up the can before refilling the engine fuel tank to ensure uniform mixing.

2. Check that all cooling system air inlets are clear before starting the engine, and check occasionally to make sure nothing has clogged the air intake, especially if a screen is used. Check the condition of fins, fan blades, and thermostatic airflow controls as described above in the description of air cooling systems, periodically.

3. Watch the engine during operation to ensure it is running smoothly at governed speed. If operation is sluggish, stop it immediately and check for possible overloading, overheating, or lack of oil or two stroke lubrication. If the engine lacks power and overheats, check fuel air mixture adjustments (especially on two-stroke engines) and ignition timing.

4. Check for exhaust smoke during operation. If the engine smokes, check first for excessive oil consumption. If this is not due to running too hot or with the crankcase too full, the crankcase breather or engine must be repaired. Smoke can also occur due to a very rich mixture. If necessary, service the air cleaner and check for any restrictions in the air intake. Make sure the choke is fully opened if it is a manual design, and that it is opening properly if it is an automatic design. Adjust the carburetor mixture, if necessary. Smoke may also accompany severe engine knock due to very advanced ignition timing. Check the timing if there is knock.

5. Make sure to use only clean, fresh fuel. Fuel can begin deteriorating only a month after it is purchased. Fuel containing water or rust or other dirt should be descarded. Use leaded or unleaded fuel as per manufacturer's recommendations. Make sure the fuel also meets octane requirements.

6. Set the throttle only slightly above idle in starting (unless this conflicts with specific recommendations for starting), and idle the engine for several minutes before putting it to work. This will allow oil to work. This will allow oil to become thin enough to reach all moving parts of the engine before they begin carrying too much load. It's also a good idea to idle the engine for a minute or two before shutdown, to cool the hottest parts more gradually.

7. Make sure the engine remains tightly mounted, as vibration can severely damage the engine or other parts, or cause potentially dangerous mechanical damage.

Safety Precautions

1. Never refuel the engine unless it is stopped. Hot exhaust system parts or an electrical spark could ignite the fuel. Also, avoid spillage of fuel, especially when the engine is hot. If fuel spills, make sure it is completely removed before starting the engine. Check the engine occasionally for fuel leaks and repair them immediately. Also, keep ignition high tension wiring in top shape. Brittle insulation can crack, causing a spark which could ignite spilled fuel.

2. Keep all sources of ignition away from batteries, especially when they are being rapidly charged or caps are off. When installing jumper cables between batteries, remove caps and place a rag over open vent holes. Make positive connections first (make sure you connect positive to positive); make negative connections by first connecting the cable to the negative side of the

good battery, and making the final connection to a bare spot on the frame of the piece of equipment with the dead battery.

3. Be careful not to come in contact with output terminals on an electric generator, which are often located externally. Remember that metal tools are excellent conductors, and that they too must be kept away from electrical terminals. Even if you do not become exposed to electrical shock via tools, they can serve as a conductor and become hot enough to cause serious burns. This can happen if they touch positive and negative terminals of a battery.

Many electric generators are used only in case of a failure in commercial power. In these cases, they are often connected into a circuit through a transfer switch. Remember that the transfer switch is energized even when the generator is not running. Remember, too, that the output terminals of a generator are hot once the generator is running and electrically excited – it does not need to be carrying a load.

4. Make sure that exhaust gases are properly vented. If you are working near an engine which is enclosed, leaks in the exhaust system can prove dangerous even when the bulk of the exhaust is being carried to the outside air.

5. If your small engine is utilized in a marine engine compartment, remember that the compartment must be thoroughly aired out before starting the engine, or explosion of accumulated fumes may result.

6. Remember that many pieces of equipment driven by small engines are too heavy to be safely carried by a single person. Often, even though the total weight may be within reasonable limits, you may strain yourself because of the difficulty of handling the unit's bulk. Get help!

7. Keep the operating area clean. It should be wiped clean of fluids which could catch fire, and kept free of debris which might be drawn into a fan or blower and thrown out at high speed. This, of course, includes removing debris from a lawn which is to be moved with a rotary mower or blown clean by a rotary blower.

8. When working on any engine driven accessory, disconnect the spark plug wire to avoid possible accidental starting of the motor if it should be turned over.

9. Keep all safety guards tightly in place, and replace them should they become damaged. Always be fully aware of all moving parts and the possibility of coming into contact with them even if guards are in place.

10. Keep governors in good operating condition, and do not reset for a speed higher than the recommendations of the manufacturer of the equipment.

11. Periodically tighten mounting bolts of both the engine and any machinery driven off it, especially that which turns at crankshaft rpm, such as rotary blades.

Hot and Cold Weather Operation

There are a few things that must be done when a small engine is to be operated in hot or cold weather, that is, when temperatures are either below 30°F, or above 75°F.

Hot Weather

In hot weather, do the following:

1. Keep the cooling fins of the engine block clean and free of all obstructions. Remove all dirt, built up oil and grease, flaking paint and grass.

2. Air should be able to flow to and from the engine with no obstructions. Keep all fairing and cover openings free from obstructions.

3. During hot weather service, heavier weight oil should be used in the crankcase. Follow the manufacturer's recommendations as to the heaviest weight oil allowed in the crankcase.

4. Check the oil level each time the fuel tank is filled. An engine will use more oil in extremely hot weather.

5. Check the battery water level more frequently since, in hot weather, the water in the battery will evaporate more quickly.

6. Be on the lookout for vapor lock, which occurs within the carburetor.

7. Use regular grade gasoline rather than premium.

8. Use unleaded gasoline if possible.

9. The most important thing to remember is to keep the engine as clean as possible. Blow it off with compressed air or wash it as often as possible.

CAUTION

Wash the engine only after it has had sufficient time to cool down to ambient temperatures. Avoid getting water in or even near the carburetor intake opening.

Cold Weather

1. A lightweight oil should be installed in the crankcase when operating in cold weather. Consult the manufacturer's recommendations.

2. If the engine is filled with summer weight oil, the engine should be moved to a warm (above 60°F) location and allowed to reach ambient temperature before starting. This is because a heavy summer weight oil will be even thicker at cold temperatures. So thick, in fact, that it will be unable to sufficiently lubricate the engine when it is first started and running. Damage could occur due to lack of lubrication.

3. Change the oil only after the engine has been operated long enough for operating temperatures to have been reached. Change the oil while the engine is still hot.

4. Use fresh gasoline. Fill the gas tank daily to prevent the formation of condensation in the tank and fuel lines.

5. Keep the battery in a fully charged condition, since cold weather infringes upon a battery's maximum current output capabilities.

6. If the engine is run only for short periods of time, have the battery charged every so often to ensure maximum power output when it is needed most.

ROUTINE MAINTENANCE

Care of a small engine is divided into the following five categories: lubrication; filter service; tune-up; carburetor overhaul and fuel pump repair/replacement; combustion chamber deposit removal and valve repair; complete overhaul. Routine maintenance consists of the first three categories. These are described below.

Lubrication

This category includes simple replenishment of lost fluids, and, in part, is the responsibility of the operator. Operators must be aware of not only the need to run the engine only when it is adequately lubricated, but of the need to cease operation and perform required maintenance, even if the actual work is done by a mechanic.

Before starting the engine, fill the crankcase and the air cleaner with the proper oil and fill the gasoline tank. Never try to fill the fuel tank of an engine that is running, and if the engine is still hot from running, allow it to cool down before refueling it.

Use a good grade, clean, fresh, lead free or leaded regular grade automotive gasoline. The use of highly leaded gasoline (high octane) should be avoided, as it causes deposits on the valves and valve seats, spark plugs, and the cylinder head, thus shortening engine life.

Any high quality detergent oil having the American Petroleum Institute classification "For Service SC or SD or MS" can be used. Detergent oils keep the engine cleaner by retarding the formation of gum and varnish deposits. Do not use any oil additives. In the summer (above 40°F) use SAE 30 weight oil. If that is not available, use SAE 10W-30 or SAE 10W-40 weight oil. In the winter (under 40°F) use SAE 5W-20 or SAE 5W-30 weight oil. If neither of these is available, use SAE 10W or SAE 10W-30 weight oil. If the engine is operated in ambient temperatures that are below 0°F, use SAE 10W or SAE 10W-30 weight oil diluted 10% with kerosene.

The oil should be changed after each 25 hours of service or engine operation, and more often under dirty or dusty operating conditions, or as the manufacturer specifies. In normal running of any engine, small particles of metal from the cylinder walls, pistons and bearings will gradually work into the oil. Dust particles from the air also get into the oil. If the oil is not changed regularly, these foreign particles cause increased friction and a grinding action which shorten the life of the engine. Fresh oil also assists in cooling the engine, for old oil gradually becomes thick and cannot dissipate the heat fast enough. Oil oil will also gradually lose its lubrication properties.

In two-stroke engines, lubrication consists of ensuring the engine runs on a mix of fuel and oil which is in the proper proportion, and that the oil used meets the specifications of the manufacturer. Since running a two-stroke engine on straight gasoline or an improper mix is very much like operating a four stroke engine without oil, it must be seen that proper preparation of fuel/oil mix is literally a life and death matter for the engine — failure to provide the proper mix may result in immediate engine failure. Always observe the following points:

1. Use an oil specifically designed for two-stroke engines, and of the viscosity recommended by the manufacturer. The wrong oil may solidify in many different parts of the engine, may leave ash deposits in the combustion chamber, or foul the spark plug. Don't forget that the oil must not only lubricate well, but burn well.

2. Measure the oil accurately into the fuel container in the exact proportion recommended. Mix thoroughly according to the directions on the can (see "Two-Stroke Lubrication" above). Do not simply use a standard, pre-mixed fuel unless you can determine that it is in the correct proportion. Where engines requiring different mixes are used at a common site, label fuel cans with the fuel/oil mixture ratio contained.

3. Remember that available lubrication in a two-stroke also depends on fuel/air ration. Ensure that carburetors are properly adjusted, and that there are no air leaks so that sufficient lubrication will always be available. Watch, too, for clogged air cleaners, partially closed chokes, or too rich an adjustment, as these will lead to plug fouling. Correct immediately any conditions causing four-cycling or misfire, as gasoline may dilute oil lubing pistons and rings under these conditions.

Filter Service

Air filter service is usually performed at the time of oil change, but may be performed at a longer interval — check specific recommendations. The air cleaner must be serviced much more frequently if the engine is operated in dusty conditions. Check specific recommendations here, also.

Oil type air cleaners require draining of old oil; a thorough cleaning of oil bowl and element with solvent; oiling of the element; and refilling of bowl to the specified level with new oil of the type used in the engine.

Most dry element air cleaners require that the element be replaced — they usually cannot be cleaned with compressed air. Some also employ a swirl chamber to remove large dust particles before they reach the main element. This chamber must be thoroughly cleaned out and, in some cases, a dust catching bowl must be emptied and cleaned.

Other types of dry element type air cleaners may require cleaning in soap and water, thorough drying and, in some cases, oiling.

Fuel sediment bowls and strainers, or filters are used on many engines to ensure that the use of dirty fuel or the entrance of dirt into the gas tank will not cause dirt to get into the carbu-

retor. Since only dirt that enters with the fuel or works its way into the tank reaches the filter, it should be obvious that the first step in fuel filter maintenance is the use of clean gas, and the second, proper maintenance of the tank filler cap and gasket. The filter is usually serviced at the same time the oil is changed or at twice that interval. Fuel tank valves are turned off, and the bowl or filter housing is removed and cleaned. Strainers are cleaned in solvent and dried, and pleated paper type elements are replaced.

Oil filters are replaced at every oil change or every other change — consult specific recommendations. Throw-away type filters usually require the use of a strap wrench for ready removal. Wipe the filter base clean, lubricate the seal on the filter with clean oil, and tighten only by hand, or the amount specified on the filter. In the case of cartridge type filters, clean the housing with solvent and dry. Make sure to replace seals both at the filter base and around the mounting bolt, as applicable.

Filters deserve the same consistency of attention to recommended service intervals as oil changes. Oil change intervals are determined by the ability of the filter to prolong the life of the oil directly in mind. If the filter is allowed to accumulate dirt to the point where it bypasses due to loss of oil pressure, the oil will be subjected to a much greater than normal amount of material to keep in suspension. This shortens the potential life of the oil drastically, greatly increases the changes of clogging engine oil passages, and may allow abrasive particles large enough to be trapped between moving parts to circulate with the oil. Remember, too, that operation in dusty areas can cause a filter to become clogged and by-passed very quickly. Follow manufacturer's recommendations for more frequent changes under these conditions.

Tune-Up

The following list of procedures is rather extensive for a simple tune-up. Normally one would just check the condition of the spark plug, points, condenser, and wiring, make the necessary adjustments to these components and the carburetor, maybe change the oil, and service the carburetor if needed.

However if the following is performed, you will either be sure that the engine is functioning properly or you will know what major repairs should be made. In other words the engine is going to run well or you will find the cause of any problems.

1. Remove the air cleaner and check for the proper servicing.

2. Check the oil level and drain the crankcase. Clean the fuel tank and lines if separate from the carburetor.

3. Remove the blower housing and inspect the rope, rewind assembly and starter clutch of the starter mechanism. Thoroughly clean the cooling fins with compressed air, if possible, and check that all control flaps operate freely.

4. Spin the flywheel to check compression. It should be spun in the direction opposite to normal rotation, and as rapidly as possible. A sharp rebound indicates good compression. Four-stroke engines may also be checked with a compression gauge in place of the spark plug — consult manufacturer's specifications in the individual repair section.

5. Remove the carburetor and disassemble and inspect it for wear and damage. Wash it in solvent, replace parts as necessary, and assemble. Set the initial adjustments.

6. Inspect the crossover tube or the intake elbow for damaged gaskets.

7. Check the governor blade, linkage, and spring for damage or wear; if it is mechanical, check the linkage adjustment.

8. Remove the flywheel and check for seal leakage, both on the flywheel and power take off sides. Check the flywheel key for wear and damage.

9. Remove the breaker cover and check for proper sealing.

10. Inspect the breaker points and condenser. Replace or clean and adjust them. Check the plunger or the cam. Lubricate the cam follower.

11. Check the coil and inspect all wires for breaks or damaged insulation. Be sure the lead wires do not touch the flywheel. Check the stop switch and the lead.

12. Replace the breaker cover, using sealer where the wires enter.

13. Install the flywheel and time the ignition if necessary. Set the air gap and check for ignition spark.

14. Remove the cylinder head, check the gasket, remove the spark plug, clean off the carbon, and inspect the valves for proper seating.

15. Replace the cylinder head, using a new gasket, torque it to the proper specification, and set the spark plug gap or replace the plug if necessary.

16. Replace the oil and fuel and check the muffler for restrictions or damage.

17. Adjust the remote control linkage and cable, if used, for correct operation.

18. Service the air cleaner and check the gaskets and element for damage.

19. Run the engine and adjust the idle mixture and high speed mixture of the carburetor.

STORAGE

If an engine is to be out of service for more than 30 days, the following steps should be performed:

1. Run the engine for 5–10 minutes until it is thoroughly warmed up to normal operating temperatures.

2. Turn off the fuel supply while the engine is still running, and continue running it until the engine stops from lack of fuel. This procedure removes all fuel from the carburetor.

3. Drain the oil from the crankcase while the engine is still warm.

4. Fill the crankcase with clean oil and tag the engine to indicate what weight oil was installed.

5. Remove the spark plug and squirt about an ounce of oil into the cylinder. Turn the engine over a few times to coat the cylinder walls, the top of the piston, and the head with a protective coating of oil. Reinstall the spark plug and tighten it to the proper torque.

6. Clean or replace the air cleaner. Refer to the manufacturer's recommendations.

7. Clean the governor linkage, making sure that it is in good working order and oiling all joints.

8. Plug the exhaust outlet and the fuel inlet openings. Use clean, lintless rags.

9. Remove the battery and store it in a cool place where there is no danger of freezing. Do not store any wet cell battery directly in contact with the ground or cement floor, as it will establish a ground and discharge itself. A completely discharged battery can never be brought back to its original output capacity. Store the battery on a work bench or on blocks of wood on the floor.

10. Wipe off or wash the engine. Wash only after the engine has had time to cool down to ambient temperature and avoid getting water in the carburetor intake port.

11. Coat all parts that might rust with a light coating of oil. Paint all non-operating parts with a rust inhibiting paint.

12. Provide the entire unit with a suitable covering. Plastic is good where the application and removal of sunlight will not promote the formation of condensation under the plastic covering. If this is the case, use a covering that is able to "breathe," such as a canvas tarpaulin.

Section 3

Troubleshooting

HOW TO GO ABOUT IT

Start with the simplest, most obvious causes first — many engine mechanics and operators have difficulty identifying trouble because they start out assuming everything that is obvious has already been checked. Check to see that there is fuel in the tank, and that it is clean, that the tank is properly vented, and that the fuel filter or sediment bowl is not full of dirt. Check to see that the spark plug wire is connected and that the spark plug is not fouled. If the cause of the trouble is not immediately obvious, use your basic knowledge of how the engine works. For example, if the engine runs fine but is very hard to start, you might conclude that the choke does not close, since its function is confined, mainly, to engine starting.

The guide below will point out many possible causes of the most basic problems. Find the "PROBLEM" which matches the engine's behavior, and then check out the possibilities listed under "CAUSES AND REMEDIES." Refer to the manufacturer's section which pertains to your engine, if necessary, in making repairs.

Troubleshooting Guide

PROBLEM: The engine does not start or is hard to start.
CAUSES AND REMEDIES:
1. The fuel tank is empty.
2. The fuel shut-off valve is closed; open it.
3. The fuel line is clogged. Remove the fuel line and clean it. Clean the carburetor, if necessary.
4. The fuel tank is not vented properly. Check the fuel tank cap vent to see if it is open.
5. There is water in the fuel supply. Drain the tank, clean the fuel lines and the carburetor, and dry the spark plug. Fill the tank with fresh fuel. Check the fuel supply before pouring it into the engine's fuel tank. Chances are it might be the source of the water.
6. The engine is overchoked. Open the choke and throttle wide on manual choke engines. On engines with automatic chokes, close the throttle. Then, turn the engine over with several pulls of the starter rope. If engine does not start, set throttle to just above idle, close choke again, and again attempt to start the engine. If one or two pulls does not make engine fire, try cranking with the choke closed only half way. If engine still fails to start, remove the spark plug and dry it, and spin the engine over several times to clean excess fuel out of the engine. Replace the spark plug and perform the normal starting procedure. Overchoking is most often due to continued cranking with the choke fully shut.
7. The carburetor is improperly adjusted; adjust it to the standard recommended preliminary settings. See the carburetor section.
8. Magneto wiring is loose or defective. Check the magneto wiring for shorts or grounds and repair it, if necessary.
9. No spark. Check for spark, and if there is none, check and, if necessary, replace the contact points, and set contact gap and timing. If there is still no spark, replace further magneto parts (especially coil and high tension wire) as necessary.
10. The spark plug is fouled. Remove, clean, and regap the spark plug.
11. The spark plug is damaged (cracked procelain, bent electrodes etc.). Replace the spark plug.
12. Compression is poor. The head is loose or the gasket is leaking. Sticking or burned valves or worn piston rings could also be the cause. In any case, the engine will have to be disassembled and the cause of the problem corrected.

PROBLEM: The engine misses under load (if a two-stroke, it may "four-cycle.")

CAUSES AND REMEDIES:
1. The spark plug is fouled. Remove, clean, and regap the spark plug.
2. The spark plug is damaged. Replace the spark plug.
3. The spark plug is improperly gapped. Regap the spark plug to the proper gap.
4. The breaker points are pitted or improperly gapped. Replace the points, or set the gap.
5. The breaker point's breaker arm is sluggish. Clean and lubricate it.
6. The condenser is faulty. Replace it.
7. The carburetor is not adjusted properly. Adjust it.
8. The fuel system is partly clogged, or the fuel shut-off valve is partly closed. Open the valve and check the fuel filter/strainer, tank, lines, and carburetor for dirt. Clean all parts as necessary.
9. If the engine is a two-stroke, the exhaust ports may be clogged. Remove the exhaust manifold and inspect the ports. If they are clogged with carbon, clean them with a soft tool such as a wooden stick. Check also for bad crankshaft seals.

10. The valves are not adjusted properly. Adjust the valve clearance.

11. The valve springs are weak. Replace them.

PROBLEM: The engine knocks.
CAUSES AND REMEDIES:

1. The magneto is not timed correctly. Time the magneto.

2. The carburetor is not properly adjusted (may be too lean). Adjust the carburetor for best mixture.

3. The engine has overheated. Stop the engine and find the cause of overheating.

4. Carbon has built up in the combustion chamber, resulting in retention of excess heat and an increase in compression which causes pre-ignition. Remove the cylinder head, and remove the carbon from the head and the top of the piston.

5. The connecting rod is loose or worn. Replace it.

6. The flywheel is loose. Check the flywheel key and keyway and the end of the crankshaft. Replace any worn parts. Tighten the flywheel nut to the specified torque.

7. The cylinder is worn. Rebuild/replace parts as necessary.

PROBLEM: The engine vibrates excessively.
CAUSES AND REMEDIES:

1. The engine is not mounted securely to the equipment that it operates. Tighten any loose mounting bolts.

2. The equipment that the engine operates is not balanced. Check the equipment.

3. The crankshaft is bent. Replace the crankshaft.

4. The counter balance shaft is improperly timed (recent reassembly) or broken. Disassemble the crankcase, inspect, and replace or repair parts as necessary.

PROBLEM: The engine lacks power.
CAUSES AND REMEDIES:

1. The choke is partially closed. Open the choke.

2. The carburetor is not adjusted correctly. Adjust it.

3. The ignition is not timed correctly. Time the ignition.

4. There is a lack of lubrication or not enough oil in the crankcase. Fill the crankcase to the correct level.

5. The air cleaner is fouled. Clean it.

6. The valves are not sealing. Do a valve job.

7. Ring seal is poor. Repair/replace rings, piston, or cylinder/cylinder liner.

8. If the engine is a two stroke, the exhaust ports may be clogged with carbon. Remove the exhaust manifold and inspect. Clean with a soft instrument such as a wooden stick, if dirty. Ports may clog frequently if the carburetor mixture is adjusted too rich, or if there is excessive oil or oil of the wrong type in the fuel.

PROBLEM: The engine operates erratically, surges, and runs unevenly.
CAUSES AND REMEDIES:

1. The fuel line is clogged. Unclog it.

2. The fuel tank cap vent is clogged. Open the vent hole.

3. There is water in the fuel. Drain the tank, the carburetor, and the fuel lines and refill with fresh gasoline.

4. The fuel pump is faulty. Check the operation of the fuel pump if so equipped.

5. The governor is improperly set or parts are sticking or binding. Set the governor and check for binding parts and correct them.

6. The carburetor is not adjusted properly. Adjust it.

PROBLEM: Engine overheats.
CAUSES AND REMEDIES:

1. The ignition is not timed properly. Time the engine's ignition.

2. The fuel mixture is too lean. Adjust the carburetor.

3. The air intake screen or cooling fins are clogged. Clean away any obstructions.

4. The engine is being operated without the blower housing or shrouds in place. Install the blower housing and shrouds.

5. The engine is operating under an excessive load. Reduce the load and check associated equipment.

6. The oil level is too high. Check the oil level and drain some out if necessary.

7. There is not enough oil in the crankcase. Check the oil level and adjust accordingly.

8. The oil in the crankcase is of too low a viscosity or is excessively contaminated with fuel (four stroke). If the engine is a two-stroke, check for adequate fuel/oil mix — oil must be mixed with the fuel in proper proportions and be fully mixed. Check condition of crankcase oil (four-stroke) and if it appears very dirty, or there is doubt about proper viscosity, replace it.

9. The valve tappet clearance is too close. Adjust the valves to the proper specification.

10. Carbon has built up in the combustion chamber. Remove the cylinder and clean the head and piston of all carbon.

11. An improper amount of oil is mixed with the fuel (two stroke engines only). Drain the fuel tank and fill with correct mixture.

PROBLEM: The crankcase breather is passing oil (four stroke engines only).
CAUSES AND REMEDIES:

1. The crankcase is substantially overfilled with oil. Check oil level several minutes after engine has stopped. Wipe the dipstick clean before checking the level. If the crankcase is too full, drain oil as necessary until oil level is at or slightly below the upper mark.

2. The engine is being operated at two high rpm. Slow it down by adjusting the governor.

3. The oil fill cap or gasket is missing or damaged. Install a new cap and gasket and tighten it securely.

4. The breather mechanism is damaged. Replace the reed plate assembly.

5. The breather mechanism is dirty. Remove, clean, and replace it.

6. The drain hole in the breather is clogged. Clean the breather assembly and open the hole.

7. The piston ring gaps are aligned. Disassemble the engine and offset the ring gaps 90 degrees from each other.

8. The breather is loose or the gaskets are leaking. Tighten the breather to the crankcase.

9. The rings are not seated properly or they are worn. Install new rings.

PROBLEM: The engine backfires.
CAUSES AND REMEDIES:

1. The carburetor is adjusted so the air/fuel mixture is too lean. Adjust the carburetor.

2. The ignition is not timed correctly. Time the engine.

3. The valves are sticking. Do a valve job.

Section 4

Briggs and Stratton

Small Engines — 0-6 hp

ENGINE IDENTIFICATION

The Briggs and Stratton model designation system consists of up to a six digit number. It is possible to determine most of the important mechanical features of the engine by merely knowing the model number. An explanation of what each number means is given below.

1. The first one or two digits indicate the cubic inch displacement (cid).

2. The first digit after the displacement indicates the basic design series, relating to cylinder construction, ignition and general configuration.

3. The second digit after the displacement indicates the position of the crankshaft and the type of carburetor the engine has.

4. The third digit after the displacement indicates the type of bearings and whether or not the engine is equipped with a reduction gear or auxiliary drive.

5. The last digit indicates the type of starter. The model identification plate is usually located on the air baffle surrounding the cylinder.

MAINTENANCE

Air Cleaners

A properly serviced air cleaner protects the engine from dust particles that are in the air. kWhen servicing an air cleaner, check the air cleaner mounting and gaskets for worn or damaged mating surfaces. Replace any worn or damaged parts to prevent dirt and dust from entering the engine through openings caused by improper sealing. Straighten or replace any bent mounting studs.

SERVICING

Oil Foam Air Cleaners

Clean and re-oil the air cleaner element every 25 hours of operation under normal operating conditions. The capacity of the oil/foam air cleaner is adequate for a full season's use without cleaning. Under very dusty conditions, clean the air cleaner every few hours of operation.

The oil/foam air cleaner is serviced in the following manner:

1. Remove the screw that holds the halves of the air cleaner shell together and retains it to the carburetor.

Oil foam air cleaner

2. Remove the air cleaner carefully to prevent dirt from entering the carburetor.

3. Take the air cleaner apart (split the two halves).

4. Wash the foam in kerosene or liquid detergent and water to remove the dirt.

5. Wrap the foam in a clean cloth and squeeze it dry.

6. Saturate the foam in clean engine oil and squeeze it to remove the excess oil.

Oil foam air cleaner

7. Assemble the air cleaner and fasten it to the carburetor with the attaching screw.

Oil Bath Air Cleaner

Pour the old oil out of the bowl. Wash the element thoroughly in solvent and squeeze it dry. Clean the bowl and refill it with the same type of oil used in the crankcase.

Dry Element Air Cleaner

Remove the element of the air cleaner and tap (top and bottom) it on a flat surface or wash it in non-sudsing detergent and flush

Oil bath air cleaner

General Engine Specifications

Model	Bore Size (in.)	Horsepower
Aluminum Engines		
6B 60000	2.375	2
8B, 80000, 82000	2.375	3
92000	2.5625	3.5
100000	2.5	4
110000, 111000, 111,200	2.7812	4
130000	2.5625	5
Cast Iron Engines		
5, 6, N	2.000	2
8	2.250	3
9	2.250	3.5
14	2.625	5

Briggs and Stratton Model Numbering System

Cubic Inch Displacement	First Digit After Displacement — Basic Design Series	Second Digit After Displacement — Crankshaft, Carburetor Governor	Third Digit After Displacement — Bearings, Reduction Gears & Auxiliary Drives	Fourth Digit After Displacement — Type of Starter
6	0	0-	0-Plain Bearing	0-Without Starter
8	1	1-Horizontal Vacu-Jet	1-Flange Mounting Plain Bearing	1-Rope Starter
9	2	2-Horizontal Pulsa-Jet	2-Ball Bearing	2-Rewind Starter
10	3			
13	4			
14	5	3-Horizontal Flo-Jet (Pneumatic Governor)	3-Flange Mounting Ball Bearing	3-Electric-110 Volt, Gear Drive
17	6			
19	7	4-Horizontal Flo-Jet (Mechanical Governor)	4-	4-Elec. Starter-Generator-12 Volt, Belt Drive
20	8			
23	9			
24		5-Vertical Vacu-Jet	5-Gear Reduction (6 to 1)	5-Electric Starter Only-12 Volt, Gear Drive
30				
32				
		6-	6-Gear Reduction (6 to 1) Reverse Rotation	6-Wind-up Starter
		7-Vertical Flo-Jet	7-	7-Electric Starter, 12 Volt Gear Drive, with Alternator
		8-	8-Auxiliary Drive Perpendicular to Crankshaft	8-Vertical-pull Starter
		9-Vertical Pulsa-Jet	9-Auxiliary Drive Parallel to Crankshaft	

Dry element air cleaner

it from the inside until the water coming out is clear. After washing, air dry the element thoroughly before reinstalling it on the engine. NEVER OIL A DRY ELEMENT.

Heavy Duty Air Cleaner

Clean and re-oil the foam pre-cleaner at three month intervals or every 25 hours, whichever comes first.

Clean the paper element every year or 100 hours, whichever comes first. Use the dry element procedure for cleaning the paper element of the heavy duty air cleaner.

Use the oil foam cleaning procedure to clean the foam sleeve of the heavy duty air cleaner.

Briggs and Stratton heavy duty air cleaner

If the engine is operated under very dusty conditions, clean the air cleaner more often.

LUBRICATION

Oil and Fuel Recommendations

Briggs & Stratton recommends either a leaded regular grade gasoline or low lead or unleaded fuel. Low lead or unleaded fuel is preferable because of the reduction in deposits that results from its use, but its use is not required. Nor is it necessary to use leaded fuel occasionally in an engine which runs primarily on unleaded fuel, as is the case with some older automobile engines. Premium fuel is not required, as regular or unleaded will have sufficient knock resistance if the engine is in proper condition. The factory recommends that fuel be purchased in lots small enough to be used up in 30 days or less. When fuel is older than that, it can form gum and varnish, or may be improperly tailored to the prevailing temperature.

Oil Viscosity Recommendations

Winter
(under 40°F.)
SAE 5W-20, or
SAE 5W-30
If above are not
available:
SAE 10W or
SAE 10W-30
(under 0°F.)
Use SAE 10W or
SAE 10W-30 in
proportions of
90% motor oil/
10% kerosene

Summer
(above 40°F.)
SAE 30

If above is not
available:
SAE 10W-40 or
SAE 10W-30

Engine Oil Capacity Chart

Basic Model Series	Capacity Pints
Aluminum	
6, 8, 9, 11 Cu. in. Vert. Crankshaft	1¼
6, 8, 9 Cu. in. Horiz. Crankshaft	1¼
10, 13 Cu. in. Vert. Crankshaft	1¾
10, 13 Cu. in. Horiz. Crankshaft	1¼
Cast Iron	
9, 14, 19, 20 Cu. in. Horiz. Crank.	3

You should use a high quality detergent oil designated "For Service SC, SD, SE, or MS." Detergent oil is recommended because of its important ability to keep gum and varnish from clogging the lubrication system. Briggs & Stratton specifically recommends that no special oil additives be used.

Oil must be changed every 25 hours of operation. If the atmosphere in which the engine is operating is very dirty, oil changes should be made more frequently, as often as every 12 hours, if necessary. Oil should be changed after 5 hours of operation in the case of brand new engines. Drain engine oil when hot.

In hot weather, when under heavy load, or when brand new, engines may consume oil at a rate which will require you to refill the crankcase several times between oil changes. Check the oil level every hour or so until you can accurately estimate how

Location of oil fill and level check plugs, 6-1 gear reduction equipped engines

long the engine can go between refills. To check oil level, stop the engine and allow it to sit for a couple of minutes, then remove the dipstick or filler cap. Fill the crankcase to the top of the filler pipe when there is no dipstick, or wipe the dipstick clean, reinsert it, and add oil as necessary until the level reaches the upper mark.

On cast iron engines with a gear reduction unit, crankcase and reduction gears are lubricated by a common oil supply. When draining crankcase, also remove drain plug in reduction unit.

On aluminum engines with reduction gear, a separate oil supply lubricates the gears, although the same type of oil used in the crankcase is used in the reduction gear cover. On these engines, remove the drain plug every fourth oil change (100 hours), then install the plug and refill. The level in the reduction gear cover must be checked during the refill operation by removing the level plug from the side of the gearcase, removing the filler plug, and then filling the case through the filler plug hole until oil runs out the level plug hole. Then, install both plugs.

On 6–1 gear reduction engines (models 6, 8, 8000, 10000, and 13000), no changes are required for the oil in the reduction gear case, but level must be checked and the case refilled, as described in the paragraph above, every 100 hours. Make sure the oil level plug (with screwdriver slot and no vent) is installed in the hole on the side of the case.

TUNE-UP
Spark Plugs

Remove the spark plug with a ¾ in. (1½ in. plug) or a ¹³⁄₁₆ in. (2 in. plug) deep well socket wrench. Clean carbon deposits off the center and side electrodes with a sharp instrument. If possible, you should also attempt to remove deposits from the recess between the insulator and the threaded portion of the plug. If the electrodes are burned away or the insulator is cracked at any point, replace the plug. Using a wire type feeler gauge, adjust

Checking spark plug gap with a wire feeler gauge

the gap by bending the side electrode where it is curved until the gap is 0.030 in.

When installing the plug, make sure the threads of the plug and the threads in the cylinder head are clean. It is best to oil the plug threads very lightly. Be careful not to overtorque the plug, especially if the engine has an aluminum head. If you use a torque wrench, torque to about 15 ft. lbs.

Breaker Points

All Briggs and Stratton engines have magneto ignition systems. Three types are used: Flywheel Type—Internal Breaker, Flywheel Type—External Breaker, and Magna-Matic.

FLYWHEEL TYPE—INTERNAL BREAKER

This ignition system has the magneto located on the flywheel and the breaker points located under the flywheel.

The flywheel is located on the crankshaft with a soft metal key. It is held in place by a nut or starter clutch. The flywheel key must be in good condition to insure proper location of the flywheel for ignition timing. Do not use a steel key under any circumstances. Use only a soft metal key, as originally supplied.

The keyway in both flywheel and crankshaft should not be distorted. Flywheels are made of aluminum, zinc, or cast iron.

Tune-Up Specifications

Model	Plug Type	Plug Gap (In.)	Point Gap (In.)	Armature Gap 2 leg	Armature Gap 3 leg	Idle Speed
Aluminum Block						
6B, 6000, 8B	①	.030	.020	.006–.010	.012–.016	1750
80000, 82000, 92000, 110900	①	.030	.020	.006–.010	.012–.016	1750
100000, 130000	①	.030	.020	.010–.014	.016–.019	1750
Cast Iron Block						
5, 6, N, 8	①	.030	.020	—	.022–.026	1750
9	①	.030	.020	—	—	1200
14	①	.030	.020	—	—	1200

① Manufacturer's Code		Manufacturer
1½ in. plug	2 in. plug	
CJ-8	J-8	Champion
RCJ-8	RJ-8	Champion (resistor)
A-7NX	A-71	Autolite
AR-7N	AR-80	Autolite (resistor)
CS-45	GC-46	A.C.
—	R-46	A.C. (resistor)

Flywheel magneto ignition with internal breaker points and external armature

Removing the flywheel

Flywheel, Nut, and/or Starter Clutch

REMOVAL AND INSTALLATION

Place a block of wood under the flywheel fins to prevent the flywheel from turning while you are loosening the nut or starter clutch. Be careful not to bend the flywheel. lThere are special flywheel holders available for this purpose; Briggs & Stratton recommends their use on flywheels of 6¾ in. diameter or less.

On rope starter engines, the ½ in. flywheel nut has a lefthand thread and the ⅝ in. nut has a righthand thread. The starter clutch used on rewind or wind-up starters has a righthand thread.

Some flywheels have two holes provided for the use of a flywheel puller. Use a small gear puller or automotive steering wheel puller to remove the flywheel if a flywheel puller is not available. Be careful not to bend the flywheel if a gear puller is used. On rope starter engines leave the nut on for the puller to bear against. Small cast iron flywheels do not require a puller. Install the flywheel in the reverse order of removal after inspecting the key and keyway for damage or wear.

Breaker Point Removal and Installation

Remove the breaker cover. Care should be taken when removing the cover, to avoid damaging it. If the cover is bent or damaged, it should be replaced to insure a proper seal.

The breaker point gap on all models is 0.020 in. Check the points for contact and for signs of burning or pitting. Points that are set too wide will advance the spark timing and may cause kickback when starting. Points that are set too close will retard the spark timing and decrease engine power.

Removing the flywheel with a puller

Removing the breaker point assembly

On models that have a separate condenser, the point set is removed by first removing the condenser and armature wires from the breaker point clip. Loosen the adjusting lock screw and remove the breaker point assembly.

On models where the condenser is incorporated with the breaker points, loosen the screw which holds the post. The condenser/point assembly is removed by loosening the screw which holds the condenser clamp.

When installing a point set with the separate condenser, be sure that the small boss on the magneto plate enters the hole in the point bracket. Mount the point set to the magneto plate or the cylinder with a lock screw. Fasten the armature lead wire to the breaker points with the clip and screw. If these lead wires do

Removing the integral breaker point and condenser assembly

not have terminals, the bare end of the wires can be inserted into the clip and the screw tightened to make a good connection. Do not let the ends of the wire touch either the point bracket or the magneto plate, or the ignition will be grounded.

To install the integral condenser/point set, place the mounting post of the breaker arm into the recess in the cylinder so that the groove in the post fits the notch in the recess. Tighten the mounting screw securely. Use a ¼ in. wrench. Slip the open loop of the breaker arm spring through the two holes in the arm, then hook the closed loop of the spring over the small post protruding from the cylinder. Push the flat end of the breaker arm into the groove in the mounting post. This places tension on the spring and pulls the arm against the plunger. If the condenser post is threaded, attach the soil primary wire and the ground wire (if furnished) with the lock washer and nut. If the primary wire is fastened to the condenser with a spring fastener, compress the spring and slip the primary wire and ground wire into the hole in the condenser post. Release the spring. Lay the condenser in place and tighten the condenser clamp securely. Install the spring in the breaker arm.

Adjusting point gap on the integral point and condenser assembly

Point Gap Adjustment

Turn the crankshaft until the points are open to the widest gap. When adjusting a breaker point assembly with an integral condenser, move the condenser forward or backward with a screwdriver until the proper gap is obtained (0.020 in.). Point sets with a separate condenser are adjusted by moving the contact point bracket up and down after the lock screw has been loosened. The point gap is set to 0.020 in.

Breaker Point Plunger

If the breaker point plunger hole becomes excessively worn, oil will leak past the plunger and may get on the points, causing them to burn. To check the hole, loosen the breaker point mounting screw and move the breaker points out of the way. Remove the plunger. If the flat end of the #19055 plug gauge will enter the plunger hole for a distance of ¼ in. or more, the hole should be rebushed.

To install the bushing, it is necessary that the breaker points, armature, and crankshaft be removed. Use a #19056 reamer to ream out the old plunger hole. This should be done by hand. The reamer must be in alignment with the plunger hole. Drive the bushing, #23513, into the hole until the upper end of the bushing is flush with the top of the boss. Remove all metal chips and dirt from the engine.

If the breaker point plunger is worn to a length of 0.870 in. or less, it should be replaced. Plungers must be inserted with the

Adjusting the point gap

Replacing the breaker plunger bushing

groove at the top or oil will enter the breaker box. Insert the plunger into the hole in the cylinder.

Armature Air Gap Adjustment

Set the air gap between the flywheel and the armature as follows: With the armature up as far as possible and just one screw tightened, slip the proper gauge between the armature and flywheel. Turn the flywheel until the magnets are directly below the armature. Loosen the one mounting screw and the magnets should pull the armature down firmly against the thickness gauge. kTighten the mounting screws.

Adjustment of the armature gap

Variations in armature positioning

Flywheel magneto ignition with an external breaker assembly

FLYWHEEL TYPE – EXTERNAL BREAKER

Breaker Point Set Removal and Installation

Turn the crankshaft until the points open to their widest gap. This makes it easier to assemble and adjust the points later if the crankshaft is not removed. Remove the condenser and upper and lower mounting screws. Loosen the lock nut and back off the breaker point screw. Install the points in the reverse order of removal.

To avoid the possibility of oil leaking past the breaker point plunger or moisture entering the crankcase between the plunger and the bushing, a plunger seal is installed on the engine models using this type of ignition system. To install a new seal on the plunger, remove the breaker point assembly and condenser. Remove the retainer and eyelet, remove the old seal, and install the new one. Use extreme care when installing the seal on the plunger to avoid damaging the seal. Replace the eyelet and retainer and replace the points and condenser.

NOTE: Apply a small amount of sealer to the threads of both mounting screws and the adjustment screw. The sealer prevents oil from leaking into the breaker point area.

Point Gap Adjustment

Turn the crankshaft until the points open to their widest gap. Turn the breaker point adjusting screw until the points open to 0.020 in. and tighten the lock nut. When the cover is installed, seal the point where the primary wire passes under the cover. This area must be resealed to prevent the entry of dust and moisture.

Replacing Threaded Breaker Plunger and Bushing

Remove the breaker cover and the condenser and breaker point assembly.

Place a thick ³⁄₈ in. inside diameter washer over the end of the bushing and screw on the ³⁄₈–24 nut. Tighten the nut to pull the bushing out of the hole. After the bushing has been moved about ¹⁄₈ in., remove the nut and put on a second thick washer and repeat the procedure. A total stack of ³⁄₈ in. washers will be required to completely remove the bushing. Be sure the plunger does not fall out of the bushing as it is removed.

Removing a threaded plunger bushing

Breaker point gap adjustment

Place the new plunger in the bushing with the large end of the plunger opposite the threads on the bushing. Screw the ³⁄₈–24 in. nut onto the threads to protect them and insert the bushing into the cylinder. Place a piece of tubing the same diameter as the nut and, using a hammer, drive the bushing into the cylinder until the square shoulder on the bushing is flush with the face of the cylinder. Check to be sure that the plunger operates freely.

Replacing Unthreaded Breaker Plunger and Bushing

Pull the plunger out as far as possible and use a pair of pliers to break the plunger off as close as possible to the bushing. Use a ¹⁄₄–20 in. tap or a #93029 self threading screw to thread the hole in the bushing to a depth of about ¹⁄₂–⁵⁄₈ in. Use a ¹⁄₄–20x¹⁄₂ in. Hex. head screw and two spacer washers to pull the bushing out of the cylinder. The bushing will be free when it has been extracted ⁵⁄₁₆ in. Carefully remove the bushing and the remainder of the broken plunger. Do not allow the plunger or metal chips to drop into the crankcase.

Correctly insert the new plunger into the new bushing. Insert the plunger and the bushing into the cylinder. Use a hammer and the old bushing to drive the new bushing into the cylinder until the new bushing is flush with the face of the cylinder. Make sure that the plunger operates freely.

Removing an unthreaded plunger bushing

Plunger Seal

Later models with Flywheel Type—External Breaker Ignition feature a plunger seal. This seal keeps both oil and moisture from entering the breaker box. If the points have become contaminated on an engine manufactured without this feature, the seal may be installed. Parts, part numbers, and their locations are shown in the illustration. Install the seal onto the plunger very carefully to avoid fracturing it.

Plunger seal used on later model engines

Measuring the armature gap

MAGNA-MATIC IGNITION SYSTEM

Removing the Flywheel

Flywheels on engines with Magna-Matic ignition are removed with pullers similar to factory designs numbered #19068 and 19203. These pullers employ two bolts, which are screwed into holes tapped into the flywheel. The bolts are turned until the flywheel is forced off the crankshaft. Only this type of device should be used to pull these flywheels.

Armature Air Gap

The armature air gap on engines equipped with Magna-Matic ignition system is fixed and can change only if wear occurs on the crankshaft journal and/or main bearing. Check for wear by inserting a ½ in. wide feeler gauge at several points between the rotor and armature. Minimum feeler gauge thickness is 0.004

The Magna-Matic ignition system

in. Keep the feeler gauge away from the magnets on the rotor or you will have a false reading.

Rotor Removal and Installation

The rotor is held in place by a woodruff key and a clamp on later engines, and a woodruff key and set screw on older engines. The rotor clamp must always remain on the rotor, unless the rotor is in place on the crankshaft and within the armature, or loss of magnetism will occur.

Loosen the socket head screw in the rotor clamp which will allow the clamp to loosen. It may be necessary to use a puller to remove the rotor from the crankshaft. On older models, loosen the small lock screw, then the set screw.

To install the set screw type rotor, place the woodruff key in the keyway on the crankshaft, then slide the rotor onto the crankshaft until the set screw hole in the rotor and the crankshaft are aligned. Be sure the key remains in place. Tighten the set screw securely, then tighten the lock screw to prevent the set screw from loosening. The lock screw is self-threading and the hole does not require tapping.

Removing the rotor

To install the clamp type rotor, place the woodruff key in place in the crankshaft and align the keyway in the rotor with the woodruff key. If necessary, use a short length of pipe and a hammer to drive the rotor onto the shaft until a 0.025 in. feeler gauge can be inserted between the rotor and the bearing support. The split in the clamp must be between the slots in the rotor. Tighten the clamp screws to 60–70 inch lbs.

Rotor Timing Adjustment

The rotor and armature are correctly timed at the factory and require timing only if the armature has been removed from the engine, or if the cam gear or crankshaft has been replaced.

If it is necessary to adjust the rotor, proceed as follows: with the point gap set at 0.020 in., turn the crankshaft in the normal direction of rotation until the breaker points close and just start to open. Use a timing light or insert a piece of tissue paper between the breaker points to determine when the points begin to open. With the three armature mounting screws slightly loose, rotate the armature until the arrow on the armature lines up with the arrow on the rotor. Align with the corresponding number of engine models, for example, on Model 9, align with #9. Retighten the armature mounting screws.

Coil and/or Armature Replacement

Usually the coil and armature are not separated, but left assembled for convenience. However, if one or both need replacement, proceed as follows: the coil primary wire and the coil ground wire must be unfastened. Pry out the clips that hold the coil and coil core to the armature. The coil core is a slip fit in the coil and can be pushed out of the coil.

To reassemble, push the coil core into the coil with the rounded side toward the ignition cable. Place the coil and core on the armature with the coil retainer between the coil and the armature and with the rounded side toward the coil. Hook the lower end of the clips into the armature, then press the upper end onto the coil core.

Fasten the coil ground wire (bare double wires) to the armature support. Next, place the assembly against the cylinder and around the rotor and bearing support. Insert the three mounting screws together with the washer and lockwasher into the three long oval holes in the armature. Tighten them enough to hold the armature in place but loose enough so the armature can be moved for adjustment of the timing. Attach the primary wires from the coil and the breaker points to the terminal at the upper side of the backing plate. This terminal is insulated from the backing plate. Push the ignition cable through the louvered hole at the left side of the backing plate.

NOTE: On Model 9 engines, knot the ignition cable before inserting it through the backing plate. Be sure all wires are clean of the flywheel.

Breaker Point Removal and Installation

Turn the crankshaft until the points open to the widest gap. This makes it easier to assemble and adjust the points later if the crankshaft is not removed. With the terminal screw out, remove the spring screw. Loosen the breaker shaft nut until the nut is flush with the end of the shaft. Tap the nut to free the breaker arm from the tapered end of the breaker shaft. Remove the nut, lockwasher, and breaker arm. Remove the breaker plate screw, breaker plate, pivot, insulating plate, and eccentric. Pry out the breaker shaft seal with a sharp pointed tool.

To install the breaker points, press in the new oil seal with the metal side out. Put the new breaker plate on the top of the insulating plate, making sure that the detent in the breaker plate engages the hole in the insulating plate. Fasten the breaker plate screw enough to put a light tension on the plate. Adjust the eccentric so that the left edge of the insulating plate is parallel to the edge of the box and tighten the screw. This locates the breaker plate so that the proper gap adjustments may be made.

Adjustment of the timing

Turn the breaker shaft clockwise as far as possible and hold it in this position. Place the new breaker points on the shaft, then the lockwasher, and tighten the nut down on the lockwasher. Replace the spring screw and terminal screw.

Breaker Box Removal and Installation

Remove the two mounting screws, then remove the breaker box, turning it slightly to clean the arm at the inner end of the breaker shaft. The breaker points need not be removed to remove the breaker box.

To install, pull the primary wire through the hole at the lower left corner of the breaker box. See that the primary wire rests in the groove at the top end of the box, then tighten the two mounting screws to hold the box in place.

Breaker Shaft Removal and Installation

The breaker shaft can be removed, after the breaker points are removed, by turning the shaft one half turn to clear the retaining spur at the inside of the breaker box.

Install by inserting the breaker shaft with the arm upward so the arm will clear the retainer boss. Push the shaft all the way in, then turn the arm downward.

Breaker Point Adjustment

To adjust the breaker points, turn the crankshaft until the breaker points open to the widest gap. Loosen the breaker point plate screw slightly. Rotate the eccentric to obtain a point gap of 0.020 in. Tighten the breaker plate screw.

The breaker point assembly

Adjusting the breaker point gap

Mixture Adjustment

920000 ENGINES BUILT SINCE 1968 (AUTOMATIC CHOKE)

1. Start the engine and run it long enough to reach operating temperature. If the carburetor is so far out of adjustment that it will not start, close the needle valve by turning it clockwise. Then open the needle valve 1½ turns counterclockwise.

2. Move the control so that the engine runs at normal operating speed. Turn the needle valve clockwise until the engine starts to lose speed because of too lean a mixture. Then slowly turn the needle valve counterclockwise and out past the point of smoothest operation until the engine just begins to run unevenly because of too rich a mixture. Turn the needle back clockwise to the midpoint between the rich and lean mixture extremes. This should be where the engine operates smoothest. The final adjustment of the needle valve should be slightly on the rich side (counterclockwise) of the mid-point.

3. Move the engine control to the slow position and turn the idle adjusting screw until a fast idle of about 1750 rpm is obtained. If the engine idles at a speed lower than 1750 rpm, it may not accelerate properly. It is not practical to attempt to obtain acceleration from speeds below 1750 rpm, because the mixture which would be required would be too rich for normal operating speeds.

4. To check the idle adjustment, move the engine control from slow to fast speed. The engine should accelerate smoothly. If the engine tends to stall or die out, increase the idle speed or readjust the carburetor, usually to a slightly richer mixture.

Carburetor adjustment screws

Flooding can occur if the engine is tipped at an angle for a prolonged period of time, if the engine is cranked repeatedly with the spark plug wire disconnected, or if the carburetor mixture is too rich.

In case of flooding, move the governor control to the stop position and pull the starter rope at least six times.

When the control is placed in the stop position, the governor spring holds the throttle in a closed idle position. Cranking the engine with a closed throttle creates a higher vacuum which opens the choke rapidly, permitting the engine to clean itself of excess fuel.

Then move the control to the fast position and start the engine. If the engine continues to flood, lean the carburetor needle valve by about ⅛–¼ of a turn clockwise.

PULSA-JET AND VACU-JET (MODEL SERIES 82000, 92000 ONLY)

Models 82500 and 92500 have a Vacu-Jet carburetor and Models 82900 and 92900 have a Pulsa-Jet carburetor.

Adjust the carburetor with the air cleaner installed and the fuel tank half full.

Turn the needle valve clockwise to close it. Then open it about 1½ turns. This will permit the engine to be started and warmed up before making the final adjustment.

With the engine running at normal operating speed (about 3000 rpm without a load) turn the needle valve clockwise until the engine starts to lose speed because of a too lean mixture.

Then slowly turn the needle valve counterclockwise past the point of smoothest operation, until the engine just begins to run unevenly. This mixture will give the best performance under a load.

Hold the throttle in the idle position. Turn the idle speed adjusting screw until a fast idle is obtained (about 1750 rpm).

Test the engine under full load. If the engine tends to stall or die out, it usually indicates that the mixture is slightly lean and it may be necessary to open the needle valve slightly to provide a richer mixture. This slightly richer mixture may cause a slight unevenness in idling.

The breather tube and fuel intake tube thread into the cylinder on the model 82500 and 82900 engines. The fuel intake tube is bolted to the cylinder on the model 92500 and 92900 engines. Check for a good fit to prevent any air leaks or dirt entry. The fuel intake tube must not be distorted at the point where the carburetor O-ring fits or air leaks will occur.

TWO PIECE FLO-JET

1. Start the engine and run it at 3000 rpm until it warms up.

2. Turn the needle valve (flat handle) to both extremes of operation noting the location of the valve at both points. That is, turn the valve inward until the mixture becomes too lean and the engine starts to slow, then note the position of the valve. Turn it outward slowly until the mixture becomes too rich and the engine begins to slow. Turn the valve back inward to the mid-point between the two extremes.

3. Install a tachometer on the engine. Pull the throttle to the idle position and hold it there through the rest of this step. Adjust the idle speed screw until the engine idles at 1750 rpm if it's an aluminum engine, or 1200 rpm, if it's a cast iron engine. Then, turn the idle valve in and out to adjust mixture, as described in Step 2. If idle valve adjustment changes idle speed, adjust speed to specification.

4. Release the throttle and observe the engine's response. The engine should accelerate without hesitation. If response is poor, one of the mixture adjustments is too lean. Readjust either or both as necessary. If idle speed was changed after idle valve was adjusted, readjust the idle valve first.

Idle valve and needle valve locations on the One Piece Flo-Jet

ONE PIECE FLO-JET

Follow the instructions for adjusting the Two Piece Flo-Jet carburetor (above). On the large, One Piece Flo-Jet, the needle valve is located under the float bowl, and the idle valve on top of the venturi passage. On the small One Piece Flo-Jet, both valves are adjusted by screws located on top of the venturi passage. The needle valve is located on the air horn side, is centered above the float bowl, and uses a larger screw head.

Governor Adjustments

SETTING MAXIMUM GOVERNED SPEED WITH ROTARY LAWNMOWER BLADES

NOTE: Strict limits on engine rpm must be observed when setting top governed speed on rotary lawnmowers. This is done so that blade tip speeds will be kept to less than 19,000 feet per minute. Briggs & Stratton suggests setting the governor 200 rpm low to allow for possible error in the tachometer reading. These figures below, based on blade length, must be strictly adhered to, or a serious accident could result!

Blade Length (in.)	Max. Governed Speed (R.P.M.)
18	4032
19	3820
20	3629
21	3456
22	3299
23	3155
24	3024
25	2903
26	2791

Bending the spring anchor tang to get desired top speed

MODELS N, 6 AND 8

There is no adjustment between the governor lever and the governor crank on these models. However, governor action can be changed by inserting the governor link or spring in different holes of the governor and throttle levers. In general, the closer to the pivot end of the lever, the smaller the difference between load and no-load engine speed. The engine will begin to "hunt" if the spring is brought too close to the pivot point. The farther the spring is from the pivot end, the tendency to hunt will decrease, but the speed drop will be greater as the load increases. If the governor speed is lowered, the spring can usually be moved closer to the pivot. The standard setting is the 4th hole from the pivot point.

MODELS 6B, 8B, 60000 AND 80000

Loosen the screw which holds the governor lever to the governor shaft. Turn the governor lever counterclockwise until the carburetor throttle is wide open. With a screwdriver, turn the governor shaft counterclockwise as far as it will go. Tighten the screw which holds the governor lever to the governor shaft.

CAST IRON MODELS 9 AND 14

Loosen the screw which holds the governor lever to the governor shaft. Push the lever counterclockwise as far as it will go. Hold it in position and turn the governor shaft counterclockwise as far as it will go. This can be done with a screwdriver. Securely tighten the screw that holds the governor lever to the shaft.

You can make a tool like the one shown to adjust spring anchor tang

Obtaining closer governing on generator applications

ALUMINUM MODELS 100000 AND 130000

Vertical and horizontal shaft engine governors are adjusted by setting the control lever in the high speed position. Loosen the nut on the governor lever. Turn the governor shaft clockwise with a screwdriver to the end of its travel. Tighten the nut. The throttle must be wide open. Check to see if the throttle can be moved from idle to wide open without binding.

Adjusting Top No Load Speed

Set the control lever to the maximum speed position with the engine running. Bend the spring anchor tang to get the desired top speed.

Adjustment For Closer Governing
(Generator Applications Only)

1. Snap knob upward to release adjusting nut.
2. Pull knob out against stop.
3. Then, bend the spring anchor tang to get top no-load speed as described below, depending upon the application.
On models 100200 and 130200 with 3600 rpm generator, set the no load speed at 4,600 using the standard governor spring.
On models 100200 and 130200 with 1800 rpm generator, set the no load speed at 2800 rpm, and set throttle stop at 1600 rpm.
4. Snap knob back into its normal position.
5. Adjust the knob for the desired generator speed.

Choke Adjustment

CHOKE-A-MATIC, PULSA-JET AND VACU-JET CARBURETORS

To check the operation of the choke linkage, move the speed adjustment lever to the choke position. If the choke slide does not fully close, bend the choke link. The speed adjustment lever must make good contact against the top switch.

Install the carburetor and adjust it in the same manner as the Pulsa-Jet carburetor.

TWO-PIECE FLO-JET AUTOMATIC CHOKE

Hold the choke shaft so the thermostat lever is free. At room temperature (68°F), the screw in the thermostat collar should be in the center of the stops. If not, loosen the stop screw and adjust the screw.

Loosen the set screw on the lever of the thermostat assembly. Slide the lever to the right or left on the shaft to ensure free movement of the choke link in any position. Rotate the thermostat shaft clockwise until the stop screw strikes the tube. Hold it in position and set the lever on the thermostat shaft so that the choke valve will be held open about $\frac{1}{8}$ in. from a closed position. Then tighten the set screw in the lever.

Adjusting automatic choke—Two Piece Flo-Jet

Rotate the thermostat shaft counterclockwise until the stop screw strikes the opposite side of the tube. Then open the choke valve manually until it stops against the top of the choke link opening. The choke valve should now be open approximately $\frac{1}{8}$ in. as before.

Check the position of the counterweight lever. With the choke valve in a wide open position (horizontal) the counterweight lever should also be in a horizontal position with the free end toward the right.

Operate the choke manually to be sure that all parts are free to move without binding or rubbing in any position.

Compression Checking

You can check the compression in any Briggs and Stratton engine by performing the following simple procedure: spin the flywheel counterclockwise (flywheel side) against the compression stroke. A sharp rebound indicates that there is satisfactory compression. A slight or no rebound indicates poor compression. It has been determined that this test is an accurate indication of compression and is recommended by Briggs and Stratton. Briggs and Stratton does not supply compression pressures.

Loss of compression will usually be the result of one or a combination of the following:
1. The cylinder head gasket is blown or leaking.
2. The valves are sticking or not seating properly.
3. The piston rings are not sealing, which would also cause the engine to consume an excessive amount of oil.

Carbon deposits in the combustion chamber should be removed every 100 or 200 hours of use (more often when run at a steady load), or whenever the cylinder head is removed.

FUEL SYSTEM

Carburetors

There are three types of carburetors used on Briggs and Stratton engines. They are the Pulsa-Jet, Vacu-Jet and Flo-Jet. The first two types have three models each and the Flo-Jet has two versions.

Before removing any carburetor for repair, look for signs of air leakage or mounting gaskets that are loose, have deteriorated, or are otherwise damaged.

Note the position of the governor springs, governor link, remote control, or other attachments to facilitate reassembly. Be careful not to bend the links or stretch the springs.

AUTOMATIC CHOKE

All 92000 model engines built since August 1968 have an automatic choke system.

The automatic choke operates in conjunction with engine vacuum, similar to the Pulsa-Jet fuel pump.

A diaphragm under the carburetor is connected to the choke shaft by a link. A calibrated spring under the diaphragm holds the choke closed when the engine is not running. Upon starting, vacuum created during the intake stroke is routed to the bottom of the diaphragm through a calibrated passage, thereby opening the choke.

This system also has the ability to respond in the same manner as an accelerator pump. As speed decreases during heavy loads, the choke valve partially closes, enriching the air/fuel mixture, thereby improving low speed performance and lugging power.

To check the automatic choke, remove the air cleaner and replace the stud. Observe the position of the choke valve; it should be fully closed. Move the speed control to the stop position; the governor spring should be holding the throttle in a closed position. Give the starter rope several quick pulls. The choke valve should alternately open and close.

The two types of Flo-Jet carburetors

The three types of Vacu-Jet carburetors

If the choke valve does not react as stated in the previous paragraph, the carburetor will have to be disassembled to determine the problem. Before doing so, however, check the following items so you know what to look for:

ENGINE IS UNDERCHOKED

1. Carburetor is adjusted too lean.
2. The fuel pipe check valve is inoperative (Vacu-Jet only).
3. The air cleaner stud is bent.
4. The choke shaft is sticking due to dirt.
5. The choke spring is too short or damaged.
6. The diaphragm is not preloaded.

ENGINE IS OVERCHOKED

1. Carburetor is adjusted too rich.
2. The ari cleaner stud is bent.
3. The choke shaft is sticking due to dirt.
4. The diaphragm is ruptured.
5. The vacuum passage is restricted.
6. The choke spring is distorted or stretched.
7. There is gasoline or oil in the vacuum chamber.
8. There is a leak between the link and the diaphragm.
9. The diaphragm was folded during assembly, causing a vacuum leak.
10. The machined surface on the tank top is not flat.

Replacing the Automatic Choke

Inspect the automatic choke for free operation. Any sticking problems should be corrected as proper choke operation depends on freedom of the choke to travel as dictated by engine vacuum.

The three types of Pulsa-Jet carburetors

Automatic choke system

Installing the diaphragm and spring into the spring pocket

Assembling the diaphragm spring to the new diaphragm

Positioning the diaphragm on top of the fuel tank

Remove the carburetor and fuel tank assembly from the engine. The choke link cover may now be removed and the choke link disconnected from the choke shaft. Disassemble the carburetor from the tank top, being careful not to damage the diaphragm.

CHECKING THE DIAPHRAGM AND SPRING

The diaphragm can be reused, provided it has not developed wear spots or punctures. On the Pulsa-Jet models, make sure that the fuel pump valves are not damaged. Also check the choke spring length. The Pulsa-Jet spring minimum length is $1\frac{1}{8}$ in. and the maximum is $1\frac{7}{32}$ in. Vacu-Jet spring length minimum is $\frac{15}{16}$ in., maximum length 1 in. If the spring length is shorter or longer than specified, replace the diaphragm and the spring.

Checking the tank top for warpage

CHECKING THE TANK TOP

The machined surface on the top of the tank must be flat in order for the diaphragm to provide an adequate seal between the carburetor and the tank. If the machined surface on the tank is not flat, it is possible for gasoline to enter the vacuum chamber by passing between the machined surface and the diaphragm.

Once fuel has entered the vacuum chamber, it can move through the vacuum passage and into the carburetor. The flatness of the machined surface on the tank top can be checked by using a straightedge and a feeler gauge. The surface should not vary more than 0.002 in. Replace the tank if a 0.002 in. feeler gauge can be passed under the straightedge.

If a new diaphragm is installed, assemble the spring to the replacement diaphragm, taking care not to bend or distort the spring.

Place the diaphragm on the tank surface, positioning the spring in the spring pocket.

Place the carburetor on the diaphragm ensuring that the choke link and diaphragm are properly aligned between the carburetor and the tank top. On Pulsa-Jet models, place the pump spring and cap on the diaphragm over the recess or pump chamber in the fuel tank. Thread in the carburetor mounting screws to about two threads. Do not tighten them. Close the choke valve and insert the choke link into the choke shaft.

Remove the air cleaner gasket, if it is in place, before continuing. Insert a $\frac{3}{8}$ in. bolt or rod into the carburetor air horn. With the bolt in position, tighten the carburetor mounting screws in a staggered sequence. Please note that the insertion of the $\frac{3}{8}$ in. bolt opens the choke to an over-center position, which preloads the diaphragm.

Remove the $\frac{3}{8}$ in. bolt. The choke valve should now move to a fully closed position. If the choke valve is not fully closed, make sure that the choke spring is properly assembled to the diaphragm, and also properly inserted in its pocket in the tank top.

All carburetor adjustments should be made with the air cleaner on the engine. Adjustment is best made with the fuel tank half full. See the "Tune-Up" section.

Inserting the choke link into the choke shaft

Remove the O-ring and inspect the metering valve

Pre-loading the diaphragm to adjust the choke

PULSA-JET AND VACU-JET (MODEL SERIES 82000 AND 92000 ONLY)

Models 82500 and 92500 have a Vacu-Jet carburetor and Models 82900 and 92900 have a Pulsa-Jet carburetor.

Rebuilding

1. Remove the carburetor and fuel tank assembly from the engine by removing the two attaching bolts.

2. Disconnect the governor link at the throttle, leaving governor link and the governor spring hooked to the governor blade and control lever.

3. Slip the carburetor and tank assembly off of the engine.

4. Remove the carburetor from the tank. Always remove all nylon and rubber parts if the carburetor is soaked in solvent.

5. Remove the O-ring and discard it. Remove and inspect the needle valve, packing and seat.

6. Metering holes in the carburetor body should be cleaned with solvent and compressed air. Do not clean the holes with a pin or a length of wire because of the danger of alterning their size.

7. Remove the choke parts on models 82500 and 82900 by pulling the nylon choke shaft sideways to separate the choke shaft from the choke valve. On the 92500 and 92900, remove the choke parts by first disconnecting the choke return spring at the pin in the carburetor body. Then pull the nylon choke shaft sideways to separate the choke shaft from the choke valve.

8. If the choke valve is heat-sealed to the choke shaft, loosen it by sliding a sharp pointed tool along the edge of the choke shaft. Do not re-seal parts on assembly.

9. When replacing the choke valve and shaft, install the choke valve so the poppet valve spring is visible when the valve is in full choke position.

On these models, the nylon fuel pipe is threaded into the carburetor body. Use a socket to remove and replace it. Be careful not to overtighten it and do not use any sealer.

The Pulsa-Jet diaphragm also serves as a gasket between the carburetor and the tank. Inspect the diaphragm for punctures, wrinkles, and wear. Replace it if it is damaged in any way.

To assemble the carburetor to the tank, first position the diaphragm on the tank. Then place the spring cap and spring on

Removing the nylon fuel pipes

Removal and inspection of a Pulsa-Jet diaphragm

the diaphragm. Install the carburetor, tightening the mounting screws evenly to avoid distortion.

To install the carburetor and tank assembly onto the engine, make sure that the governor link is hooked to the governor blade. Connect the link to the throttle and slip the carburetor into place. Align the carburetor with the intake tube and breather tube grommet. Hold the choke lever in the open position so it does not catch on the control plate. Be sure the O-ring in the carburetor does not distort when fitting the carburetor to the intake tube. Install the mounting bolts. Adjust the carburetor as described in the Tune-Up section.

PULSA-JET CARBURETOR

Throttle Plate Removal

Cast throttle plates are removed by backing off the idle speed adjustment screw until the throttle clears the retaining lug on the carburetor housing.

Stamped throttles are removed by using a phillips screwdriver to remove the throttle valve screw. After removal of the valve, the throttle may be lifted out. Installation is the reverse of removal.

The two types of throttle shafts

Some carburetors may have a spiral in the carburetor bore. To remove it, fasten the carburetor in a vise about ½ in. below the top of the jaws. Grasp the spiral firmly with a pair of pliers. Place a screwdriver under the edge of the pliers. Using the edge of the vise, push down on the screwdriver to pry out the spiral. When installing the spiral, keep the top flush, or 1/32 in. below the carburetor flange, and parallel with the fuel tank mounting face.

Fuel Pipe

Check balls are not used in these fuel pipes. The screen housing or pipe must be replaced if the screen cannot be satisfactorily cleaned. The long pipe supplies fuel from the tank to the pump. The short pipe supplies fuel from the tank cup to the carburetor. Fuel pipes are nylon or brass. Nylon pipes are removed and installed by using a socket, or open-end wrench.

NOTE: Where brass pipes are used, replace only the screen housing. The housing is driven off the pipe with a screwdriver with the pipe held in a vise. The new housing is installed by lightly tapping it onto the pipe with a soft hammer.

Removing the cast throttle shafts

Removing the throttle plate

Needle Valve and Seat

Remove the needle valve to inspect it. If the carburetor is gummy or dirty, remove the seat to allow better cleaning of the metering holes. Do not insert pins or wires in the metering holes. Use solvent or compressed air.

Pump

Remove the fuel pump cover, diaphragm, spring, and cup. Inspect the diaphragm for punctures, cracks, and fatigue. Replace it if damaged. On early models, the spring cap is solid; on later models, the cap has a hole in it. The new style supersedes the old style. When installing the pump cover, tighten the screws evenly to insure a good seal.

Choke-A-Matic (Except 100900 Models)

To remove the choke link, remove the speed adjustment lever and stop switch insulator plate. Remove the speed adjustment

Removing and installing the spiral

Removal and inspection of the pump cover diaphragm from a Pulsa-Jet carburetor

lever from the choke link, then pull out the choke link through the hole in the choke slide.

Replace worn or damaged parts. To assemble, slip the washers and spring over the choke link. Hook the choke link through the hole in the choke slide. Place the other end of the choke link through the hole in the speed adjustment lever and mount the lever and stop switch insulator plate to the carburetor.

VACU-JET CARBURETORS

Vacu-Jet carburetors are removed from the engine together with the fuel tank as one unit. The throttle plates are removed and installed in the same manner as the throttles in the Pulsa-Jet carburetors.

Fuel Pipe

The fuel pipe contains a check ball and a fine mesh screen. To function properly, the screen must be clean and the check ball free. Replace the pipe if the screen and ball cannot be satisfactorily cleaned in carburetor cleaner.

NOTE: Do not leave the carburetor in the cleaner for more than $\frac{1}{2}$ hour without removing all nylon parts. Nylon fuel pipes are removed and replaced with a $\frac{9}{16}$ in. socket. Brass fuel pipes are removed by clamping the pipe in a vise and prying out the pipe with two screwdrivers.

To install the brass fuel pipes, remove the throttle, if necessary, and place the carburetor and pipe in a vise. Press the pipe into the carburetor until it projects $2\frac{9}{32}$–$2\frac{5}{16}$ in. from the carburetor face.

Needle Valve and Seat

Remove the needle valve assembly to inspect it. If the carburetor is gummy or dirty, remove the seat to allow better cleaning of the metering holes. Do not clean the metering holes with a pin or a length of wire.

Choke-A-Matic Linkage

To remove the choke link, remove the speed adjustment lever and the top switch insulator plate. Work the link out through the hole in the choke slide.

Adjustment of the Choke-a-Matic choke linkage

Fuel pipes

The two types of fuel pipes

Removal of brass fuel pipes

Cutaway view of a two piece Flo-Jet carburetor

Adjustment of a Choke-a-Matic choke linkage on a Vacu-Jet carburetor

Replace all worn or damaged parts. To assemble a carburetor using a choke slide, place the choke return spring and three washers on the choke link. Push the choke link through the hole in the carburetor body, turning the link to line up with the hole in the choke slide. The speed adjustment lever screw and the stop switch insulator plate should be installed as one assembly after placing the choke link through the end of the speed adjustment lever.

TWO PIECE FLO-JET CARBURETORS (LARGE AND SMALL LINE)

Checking the Upper Body for Warpage

With the carburetor assembled and the body gasket in place, try to insert a 0.002 in. feeler gauge between the upper and lower bodies at the air vent boss, just below the idle valve. If the gauge can be inserted, the upper body is warped and should be replaced.

Checking the Throttle Shaft and Bushings

Wear between the throttle shaft and bushings should not exceed 0.010 in. Check the wear by placing a short iron bar on the upper carburetor body so that it just fits under the throttle shaft. Measure the distance with a feeler gauge while holding the shaft down and then holding it up. If the difference is over 0.010 in., either the upper body should be rebushed, the throttle shaft replaced, or both. Wear on the throttle shaft can be checked by comparing the worn and unworn portions of the shaft. To replace the bushings, remove the throttle shaft using a thin punch to drive out the pin which holds the throttle stop to the shaft; remove the throttle valve, then pull out the shaft. Place a ¼ in. x 20 tap or an E-Z Out in a vise. Turn the carburetor body so as to thread the tap or E-Z Out into the bushings enough to pull the bushings out of the body. Press the new bushings into the carburetor body with a vise. Insert the throttle shaft to be sure it is

Checking throttle shaft wear with a feeler gauge

Two piece Flo-Jet carburetor

Removing the throttle shaft bushing

free in the bushings. If not, run a size $\frac{7}{32}$ in. drill through both bushings to act as a line reamer. Install the throttle shaft, valve, and stop.

Disassembly of the Carburetor

1. Remove the idle valve.
2. Loosen the needle valve packing nut.
3. Remove the packing nut and needle valve together. To remove the nozzle, use a narrow, blunt screwdriver so as not to damage the threads in the lower carburetor body. The nozzle projects diagonally into a recess in the upper body and must be removed before the upper body is separated from the lower body, or it may be damaged.
4. Remove the screws which hold the upper and lower bodies together. A pin holds the float in place.
5. Remove the pin to take out the float valve needle. Check the float for leakage. If it contains gasoline or is crushed, it must be replaced. Use a wide, proper fitting screwdriver to remove the float inlet seat.
6. Lift the venturi out of the lower body. Some carburetors have a welch plug. This should be removed only if necessary to remove the choke plate. Some carburetors have nylon choke shaft.

Repair

Use new parts where necessary. Always use new gaskets. Carburetor repair kits are available. Tighten the inlet seat with the gasket securely in place, if used. Some float valves have a spring clip to connect the float valve to the float tang. Others are nylon with a stirrup which fits over the float tang. Older float valves and engines with fuel pumps have neither a spring nor a stirrup.

A voton tip float valve is used in later models of the large, two-piece Flo-Jet carburetor. The seat is pressed into the upper body and does not need replacement unless it is damaged.

Replacing the Pressed-In Float Valve Seat

Clamp the head of a #93029 self threading screw in a vise. Turn the carburetor body to thread the screw into the seat. Continue turning the carburetor body, drawing out the seat. Leave the seat fastened to the screw. Insert the new seat #230996 into the carburetor body. kThe seat has a starting lead.

Replacing the float valve seat

NOTE: If the engine is equipped with a fuel pump, install a #231019 seat. Press the new seat flush with the body using the screw and old seat as a driver. Make sure that the seat is not pressed below the body surface or improper float-to-float valve contact will occur. Install the float valve.

Checking the Float Level

With the body gasket in place on the upper body and the float valve and float installed, the float should be parallel to the body mounting surface. If not, bend the tang on the float until they are parallel. Do not press on the flat to adjust it.

Assembly of the Carburetor

Assemble the venturi and the venturi gasket to the lower body. Be sure that the holes in the venturi and the venturi gasket are aligned. Some models do not have a removable venturi. Install the choke parts and welch plug if previously removed. Use a sealer around the welch plug to prevent entry of dirt.

Fasten the upper and lower bodies together with the mounting screws. Screw in the nozzle with a narrow, blunt screwdriver, making sure that the nozzle tip enters the recess in the upper body. Tighten the nozzle securely. Screw in the needle valve and idle valve until they just seat. Back off the needle valve $1\frac{1}{2}$ turns. Do not tighten the packing nut. Back off the idle valve $\frac{3}{4}$ of a turn. These settings are about correct. Final adjustment will be made when the engine is running. See the Tune-Up section for mixture and choke adjustments.

Small one piece Flo-Jet carburetor

ONE-PIECE FLO-JET CARBURETOR

The large, one-piece Flo-Jet carburetor has its high speed needle valve below the float bowl. All other repair procedures are similar to the small, one-piece Flo-Jet carburetor.

Disassembly

1. Remove the idle and needle valves.
2. Remove the carburetor bowl screw. A pin holds the float in place.
3. Remove the pin to take off the float and float valve needle. Check the float for leakage. If it contains gasoline or is crushed, it must be replaced. Use a screwdriver to remove the carburetor nozzle. Use a wide, heavy screwdriver to remove the float valve seat, if used.

Large one piece Flo-Jet carburetor

If it is necessary to remove the choke valve, venturi throttle shaft, or shaft bushings, proceed as follows:

1. Pry out the welch plug.
2. Remove the choke valve, then the shaft. The venturi will then be free to fall out after the choke valve and shaft have been removed.
3. Check the shaft for wear. (Refer to the "Two-Piece Flo-Jet Carburetor" section for checking wear and replacing bushings.)

Repair of the Carburetor

Use new parts where necessary. Always use new gaskets. Carburetor repair kits are available. If the venturi has been removed, install the venturi first, then the carburetor nozzle jets. The nozzle jet holds the venturi in place. Replace the choke shaft and valve. Install a new welch plug in the carburetor body. Use a sealer to prevent dirt from entering.

A viton tip float valve is used in the large, one-piece Flo-Jet carburetor. The seat is pressed in the upper carburetor body and does not need replacement unless it is damaged. Replace the seat in the same manner as for the two-piece Flo-Jet carburetor.

Checking the Float Level

With the body gasket in place on the upper body and float valve and the float installed, the float should be parallel to the body mounting surface. If not, bend the tang on the float until they are parallel. Do not press on the float.

Install the float bowl, idle valve, and needle valve. Turn in the needle valve and the idle valve until they just seat. Open the needle valve $2\frac{1}{2}$ turns and the idle valve $1\frac{1}{2}$ turns. On the large carburetors with the needle valve below the float bowl, open the needle valve and the idle valve $1\frac{1}{8}$ turns.

These settings will allow the engine to start. Final adjustment should be made when the engine is running and has warmed up to operating temperature. See the "Two-Piece Flo-Jet Carburetor" adjustment procedure.

Float Level Chart

Carburetor Number	Float Setting (In.)
2712-S	$\frac{19}{64}$
2713-S	$\frac{19}{64}$
2714-S	$\frac{1}{4}$
*2398-S	$\frac{1}{4}$
2336-S	$\frac{1}{4}$
2336-SA	$\frac{1}{4}$
2337-S	$\frac{1}{4}$
2337-SA	$\frac{1}{4}$
2230-S	$\frac{17}{64}$
2217-S	$\frac{11}{64}$

*When resilient seat is used, set float level at $\frac{9}{32} \pm \frac{1}{64}$.

Governors

The purpose of a governor is to maintain, within certain limits, a desired engine speed even though the load may vary.

AIR VANE GOVERNORS

The governor spring tends to open the throttle. Air pressure against the air vane tends to close the throttle. kThe engine speed at which these two forces balance is called the governed

Air vane governor installed on horizontal crankshaft engine

Air vane governor installed on vertical crankshaft engine

speed. The governed speed can be varied by changing the governor spring tension.

Worn linkage or damaged governor springs should be replaced to insure proper governor operation. No adjustment is necessary.

MECHANICAL GOVERNORS

The governor spring tends to pull the throttle open. The force of the counterweights, which are operated by centrifugal force, tends to close the throttle. The engine speed at which these two forces balance is called the governed speed. The governed speed can be varied by changing the governor spring tension.

GOVERNOR REPAIR

The procedures below describe disassembly and assembly of the various kinds of mechanical governors. Look for gears with worn or broken teeth, worn thrust washers, weight pins, cups, followers, etc. Replace parts that are worn and reassemble.

Models N, 6 and 8
DISASSEMBLY

1. Remove the two governor housing mounting screws, and remove the housing.
2. Pull the cup off the governor gear, and then slide the gear off the shaft.
3. Disassemble the governor crank by driving the roll pin out of the end of the governor lever and then remove the crank bushing. Pull the governor crank out of the housing.

ASSEMBLY

1. Push the governor crank, lever end first, into the housing.

Governor housing and gear assembly

Installing crank and lever

2. Slip the bushing onto the shaft, and then thread it into the housing and tighten securely.

3. Position the lever on the shaft with the lever and shaft pin holes lined up and the lever pointing away from the housing mounting flange. Push in the pin.

4. Push the governor gear onto the shaft in the engine block.

5. Position the gasket on the governor housing, put the housing into position on the block, and install the two housing mounting screws. Connect linkage.

Models 6B, 8B, 60000 and 80000

DISASSEMBLY

1. Loosen the governor lever mounting screw and pull the lever off the shaft.

2. Remove the two housing mounting screws. Carefully pull the housing off the block being careful to catch the governor gear, which will slip off the shaft. Pull the steel thrust washer off the shaft.

3. Remove the governor lever roll pin and washer. Unscrew the governor lever shaft by turning it clockwise and remove it.

ASSEMBLY

1. Push the governor lever shaft into the crankcase cover, threaded end first. Assemble the small washer onto the inner end of the shaft, and then screw the shaft into the governor crank follower by turning it counterclockwise. Tighten it securely.

2. Turn the shaft until the follower points down slightly, in a position where it would press against the cup when the housing is installed.

3. Place the washer on the outside end of the shaft. Install the rollpin, so the leading end just reaches the outside diameter of the shaft and the back end protrudes.

4. Install the thrust washer and the governor gear on the shaft in the housing (in that order).

5. Hold the crankcase cover in a vertical (the normal) position and install the housing with the gear in position so the point of

Mechanical governor exploded view

the steel cup on the gear contacts the follower. Install and tighten the housing mounting screws.

6. Install the lever on the shaft pointing downward at an angle of about 30 degrees. Adjust as described in the Tune-Up section.

Cast Iron Models

DISASSEMBLY

1. Remove the cotter key and washer from the outer end of the governor shaft. Remove the governor crank from inside the crankcase.

2. Slide the governor gear off the shaft.

Assembling the mechanical governor

ASSEMBLY

1. Install the governor gear onto the shaft inside the crankcase. Then, insert the governor shaft assembly through the bushing from inside the crankcase.

2. Install the governor lever to the shaft loosely, and then adjust it as described in the Tune-Up section.

Aluminum Models

DISASSEMBLY

On horizontal shaft models: Remove the governor assembly as a unit from the crankcase cover.

On vertical shaft models: Remove the entire assembly as part of the oil slinger (see the Overhaul Section).

ASSEMBLY

1. Assemble governors on horizontal crankshaft models with crankshaft in a horizontal position. The governor rides on a short stationary shaft which is integral with the crankcase cover. The governor keeps the governor from sliding off the shaft after the cover is installed. The governor shaft must hang straight down, or it may jam the governor assembly when the crankcase cover is installed, breaking it when the engine is started. The governor shaft adjustment should be made (see the Tune-Up section) as soon as the crankcase cover is in place so that the governor lever will be clamped in the proper position.

Assembling cover with governor and governor shaft in proper position (horizontal shaft engines)

Installing spring on camshaft (Models 100900 and 130900)

2. On both horizontal and vertical crankshaft models, the governor is held together through normal operating forces. For this reason, the governor link and all other external linkages must be in place and properly adjusted whenever the engine is operated.

3. On vertical shaft models 100900 and 130900, be sure the spring washer is in place on the camshaft after the governor is in position.

ENGINE OVERHAUL

Cylinder Head

REMOVAL AND INSTALLATION

Always note the position of the different cylinder head screws so that they can be properly reinstalled. If a screw is used in the wrong position, it may be too short and not engage enough threads. If it is too long, it may bottom on a fin, either breaking the fin, or leaving the cylinder head loose.

Remove the cylinder screws and then the cylinder head. Be sure to remove the gasket and all remaining gasket material from the cylinder head and the block.

Assemble the cylinder head with a new gasket, cylinder head shield, screws, and washers in their proper places. Graphite grease should be used on aluminum cylinder head screws.

Do not use a sealer of any kind on the head gasket. Tighten the screws down evenly by hand. Use a torque wrench and tighten the head bolts in the correct sequence.

Adjusting the breaker point gap

Cylinder head bolt tightening sequences

Cylinder Head Bolt Torque Specifications

Basic Model Series	In. lbs. Torque
Aluminum Cylinder	
6B, 60000, 8B, 80000 82000, 92000, 110000, 100000, 130000	140
Cast Iron Cylinder	
5, 6, N, 8, 9	140
14	165

Valves

REMOVAL AND INSTALLATION

Using a valve spring compressor, adjust the jaws so they touch the top and bottom of the valve chamber, and then place one of the jaws over the valve spring and the other underneath, between the spring and the valve chamber. This positioning of the valve spring compressor is for valves that have either pin or collar type retainers. Tighten the jaws to compress the spring. Remove the collars or pin and lift out the valve. Pull out the compressor and the spring.

To remove valves with ring type retainers, position the compressor with the upper jaw over the top of the valve chamber and the lower jaw between the spring and the retainer. Compress the spring, remove the retainer, and pull out the valve. Remove the compressor and spring.

Before installing the valves, check the thickness of the valve springs. Some engines use the same spring for the intake and exhaust side, while others use a heavier spring on the exhaust side. Compare the springs before installing them.

If the retainers are held by a pin or collars, place the valve spring and retainer and cup (Models 9–14–19–20–23–24–32) into the valve spring compressor. Compress the spring until it is solid. Insert the compressed spring and retainer into the valve chamber. Then drop the valve into place, pushing the stem through the retainer. Hold the spring up in the chamber, hold the valve down, and insert the retainer pin with needle nose pliers or place the collars in the groove in the valve stem. Loosen the spring until the retainer fits around the pin or collars, then pull out the spring compressor. Be sure the pin or collars are in place.

The three types of valve spring retainers

Removing the valve springs with the help of a valve spring compressor

To install valves with ring type retainers, compress the retainer and spring with the compressor. The large diameter of the retainer should be toward the front of the valve chamber. Insert the compressed spring and retainer into the valve chamber. Drop the valve stem through the larger area of the retainer slot and move the compressor so as to center the small area of the valve retainer slot onto the valve stem shoulder. Release the spring tension and remove the compressor.

Valve Guides

REMOVAL AND INSTALLATION

Aluminum Models

First check valve guide for wear with a plug gauge, Briggs & Stratton, Part #19122 or equivalent. If the flat end of valve guide plug gauge can be inserted into the valve guide for a distance of $5/16$, the valve guide is worn and should be rebushed in the following manner.

See the illustation.

Bushing the valve guide

Procure a reamer (part #19064) and a reamer guide bushing (part #19191). Equivalent tools available from other sources may be used. Lubricate the reamer with kerosene. Use reamer and reamer guide bushing to ream out the worn guide. Ream to only $1/16$ in. deeper than valve guide bushing #63709. BE CAREFUL NOT TO REAM THROUGH THE GUIDE.

Press in valve guide bushing #63709 until top end of bushing is flush with top end of valve guide. Use a soft metal driver (brass, copper, etc.) or driver #19065 so top end of bushing is not peened over.

Procure a reamer #19066 (or equivalent), lubricate it with kerosene, and finish reaming the bushing. A standard valve can now be used.

NOTE: It is usually not necessary to bush factory installed brass valve guides. However, if bushing is required, DO NOT REMOVE ORIGINAL BUSHING, but follow standard procedure outlined.

Cast Iron Models

First check valve guide for wear with a plug gauge, kBriggs & Stratton part #19151 or equivalent. If the flat end of the valve guide plug gauge can be inserted into the valve guide for a distance of $5/16$, the guide is worn and should be rebushed in the following manner.

Procure a reamer #19183 and reamer guide bushing #19192, and lubricate the reamer with kerosene. Then, use reamer and reamer guide bushing to ream out the worn guide. Ream to only $1/16$ deeper than valve guide bushing #230655. BE CAREFUL NOT TO REAM THROUGH THE GUIDE.

Bushing the valve guide

Press in valve guide bushing #230655 until top end of bushing is flush with top end of valve guide. Use a soft metal driver (brass, copper, etc.) so top end of bushing is not peened over.

The bushing #230655 is finish reamed to size at the factory, so no further reaming is necessary, and a standard valve can be used.

— CAUTION —
Valve seating should be checked after bushing the guide, and corrected if necessary by refacing the seat.

REFACING VALVES AND SEATS

Faces on valves and valve seats should be resurfaced with a valve grinder or cutter to an angle of 45 degrees.

NOTE: Some engines have a 30 degree intake valve and seat.

The valve and seat should then be lapped with a fine lapping compound to remove the grinding marks and ensure a good seat. The valve seat width should be $3/64$–$1/16$ in. If the seat is wider, a narrowing stone or cutter should be used. If either the seat or valve is badly burned, it should be replaced. Replace the valve if the edge thickness (margin) is less than $1/64$ in. after it has been resurfaced.

CHECK AND ADJUST TAPPET CLEARANCE

Insert the valves in their respective positions in the cylinder. Turn the crankshaft until one of the valves is at its highest position. Turn the crankshaft one revolution. Check the clearance with a feeler gauge. Repeat for the other valve. Grind off the end of the valve stem if necessary to obtain proper clearance.

NOTE: Check the valve tappet clearance with the engine cold.

Valve Seat Inserts

Cast iron cylinder engines are equipped with an exhaust valve insert which can be removed and replaced with a new insert. The intake side must be counterbored to allow the installation of an intake valve seat insert (see below). Aluminum alloy cylinder models are equipped with inserts on both the exhaust and intake valves.

REMOVAL AND INSTALLATION

Valve seat inserts are removed with a special puller.

NOTE: On aluminum alloy cylinder models, it may be necessary to grind the puller nut until the edge is $1/32$ in. thick in order to get the puller nut under the valve insert.

When installing the valve seat insert, make sure that the side with the chamfered outer edge goes down into the cylinder. Install the seat insert and drive it into place with a driver. The seat should then be ground lightly and the valves and seats lapped lightly with grinding compound.

NOTE: Aluminum alloy cylinder models use the old insert as a spacer between the driver and the new insert. Drive in the new insert until it bottoms. The top of the insert will be slightly below the cylinder head gasket surface. Peen around the insert using a punch and hammer.

NOTE: The intake valve seat on cast iron cylinder models has to be counterbored before installing the new valve seat insert.

COUNTERBORING CYLINDER FOR INTAKE VALVE SEAT ON CAST IRON MODELS

1. Select the proper seat insert, cutter shank, counter bore cutter, pilot and driver from the table. These numbers refer to Briggs & Stratton parts—you may get equivalent parts from other sources if available.

Valve Tappet Clearance Chart

Model Series	Intake		Exhaust	
	Max	Min	Max	Min
Aluminum Cylinder				
6B, 60000, 8B, 80000	.007	.005	.011	.009
82000, 92000, 100000, 110900	.007	.005	.011	.009
130000	.007	.005	.011	.009
Cast Iron Cylinder				
5, 6, 8, N, 9, 14	.009	.007	.016	.014

Valve Seat Inserts Chart

Basic Model Series	Intake Standard	Exhaust Standard	Exhaust Stellite	Insert # Puller Assembly	Puller Nut
Aluminum Cylinder					
6B, 8B	211291	211291	210452	19138	19140 Ex. 19182 In.
60000, 80000	210879*	211291	210452	19138	19140 Ex. 19182 In.
82000, 92000, 110000	210879	211291	210452	19138	19140 Ex. 19182 In.
100000, 130000	211158	211172	211436	19138	19182 Ex. 19139 In.
Cast Iron Cylinder					
5, 6, N	63838	21865		19138	19140
8	210135	21865		19138	19140
9	63007	63007		19138	19139
14	21880	21880	21612	19138	19141

*21191 used before serial #5810060—210808 used from serial #5810060—6012010
Includes puller and #19182, 19141, 19140 and 19139 nuts

2. With cylinder head resting on a flat surface, valve seats up, slide the pilot into the intake valve guide. Then, assemble the correct counterbore cutter to the shank with the cutting blades of the cutter downward.

3. Insert the cutter straight into the valve seat, over the pilot. Cut so as to avoid forcing the cutter to one side, and be sure to stop as soon as the stop on the cutter touches the cylinder head.

4. Blow out all cutting chips thoroughly.

Pistons, Piston Rings, and Connecting Rods

REMOVAL

To remove the piston and connecting rod from the engine, bend down the connecting rod lock. Remove the connecting rod cap. Remove any carbon or ridge at the top of the cylinder bore. This will prevent breaking the rings. Push the piston and rod out of the top of the cylinder.

Pistons used in sleeve bore, aluminum alloy engines are marked with an **L** on top of the piston. These pistons are tin plated and use an expander with the oil ring. This piston assembly is not interchangeable with the piston used in the aluminum bore engines (Kool bore).

Pistons used in aluminum bore (Kool bore) engines are not marked on the top.

To remove the connecting rod from the piston, remove the piston pin lock with thin nose pliers. One end of the pin is drilled to facilitate removal of the lock.

Remove the rings one at a time, slipping them over the ring lands. Use a ring expander to remove the rings.

INSPECTION

Check the piston ring fit. Use a feeler gauge to check the side clearance of the top ring. Make sure that you remove all carbon

Installing valve seat inserts

from the top ring groove. Use a new piston ring to check the side clearance. If the cylinder is to be resized, there is no reason to check the piston, since a new oversized piston assembly will be installed. If the side clearance is more than 0.007 in., the piston is excessively worn and should be replaced.

Check the piston ring end gap by cleaning all carbon from the ends of the rings and inserting them one at a time 1 in. down into the cylinder. Check the end gap with a feeler gauge. If the gap is larger than recommended, the ring should be replaced.

NOTE: When checking the ring gap, do not deglaze the cylinder walls by installing piston rings in aluminum cylinder engines.

Chrome ring sets are available fro all current aluminum and cast iron cylinder models. No honing or deglazing is required.

Removing the connecting rod cap

Replacing the piston rings

Removing the wrist pin and connecting rod from the piston

Measuring the piston ring gap

The cylinder bore can be a maximum of 0.005 in. oversize when using chrome rings.

If the crankpin bearing in the rod is scored, the rod must be replaced. 0.005 in. oversize piston pins are available in case the connecting rod and piston are worn at the piston pin bearing. If, however, the crankpin bearing in the connecting rod is worn, the rod should be replaced. Do not attempt to file or fit the rod.

If the piston pin is worn 0.0005 in. out of round or below the rejection sizes, it should be replaced.

INSTALLATION

The piston pin is a push fit into both the piston and the connecting rod. On models using a solid piston pin, one end is flat and the other end is recessed. Other models use a hollow piston pin. Place a pin lock in the groove at one side of the piston. From the opposite side of the piston, insert the piston pin, flat end first for solid pins; with hollow pins, insert either end first until it stops against the pin lock. Use thin nose pliers to assemble the pin lock in the recessed end of the piston. Be sure the locks are firmly set in the groove.

Install the rings on the pistons, using a piston ring expander. Make sure that they are installed in the proper position. The scraper groove on the center compression ring should always be down toward the piston skirt. Be sure the oil return holes are clean and all carbon is removed from the grooves.

NOTE: Install the expander under the oil ring in sleeve bore aluminum alloy engines.

Measuring piston ring side gap

Connecting Rod Bearing Specifications

Basic Model Series	Crank Pin Bearing	Piston Pin Bearing
Aluminum Cylinder		
6B, 60000	.876	.492
8B, 80000	1.001	.492
82000, 92000, 110000	1.001	.492
100000	1.001	.555
130000	1.001	.492
Cast Iron Cylinder		
5	.752	.492
6, 8, N	.751	.492
9	.876	.563
14	1.001	.674

Piston Ring Gap Specifications

Basic Model Series	Comp. Ring	Oil Ring
Aluminum Cylinder		
6B, 60000, 8B, 80000		
82000, 92000, 110000, 111000	.035	.045
Cast Iron Cylinder		
5, 6, 8, N, 9	.035	.035

Oil the rings and the piston skirt, then compress the rings with a ring compressor. On cast iron engines, install the compressor with the two projections downward; on aluminum engines, install the compressor with the two projections upward. These instructions refer to the piston in normal position—with skirt downward. Turn the piston and compressor upside down on the bench and push downward so the piston head and the edge of the compressor band are even, all the while tightening the compressor. Draw the compressor up tight to fully compress the rings, then loosen the compressor very slightly.

───────────── CAUTION ─────────────
Do not attempt to install the piston and ring assembly without using a ring compressor.
────────────────────────────────

Place the connecting rod and piston assembly, with the rings compressed, into the cylinder bore. Push the piston and rod down into the cylinder. Oil the crankpin of the crankshaft. Pull the connecting rod against the crankpin and assemble the rod cap so the assembly marks align.

NOTE: Some rods do not have assembly marks, as the rod and cap will fit together only in one position. Use care to ensure proper installation. On the 251000 engine, the piston has a notch on the top surface. The notch must face the flywheel side of the block when installed. On models 300000 and 320000, the piston has an identification mark "F" located mext to the piston pin bore. The mark must appear on the same side as the assembly mark on the rod. The assembly mark on the rod is also used to identify rod and cap alignment. Note, on these pistons, that the top ring has a beveled upper surface on the outside, while the center ring has a flat outer surface. The "F" mark or notch must face the flywheel when the piston is installed.

Wrist Pin Specifications

Basic Model Series	Piston Pin	Pin Bore
Aluminum Cylinder		
6B, 60000	.489	.491
8B, 80000	.489	.491
82000, 92000, 110000, 111000	.489	.491
100000	.552	.554
130000	.489	.491
Cast Iron Cylinder		
5, 6, 8, N	.489	.491
9	.561	.563
14	.671	.673

Cross-sectional views and positioning of the various types of piston rings used in Briggs and Stratton engines

Installing the piston and connecting rod assembly into the cylinder block

Connecting rod installation

Connecting Rod Capscrew Torque

Basic Model Series	Inch lbs Avg. Torque
Aluminum Cylinder	
6B, 60000	100
8B, 80000	100
82000, 92000, 110000, 111000	100
100000, 130000	100
Cast Iron Cylinder	
5, 6, N, 8	100
9	140
14	190

Where there are flat washers under the cap screws, remove and discard them prior to installing the rod. Assemble the cap screws and screw locks with the oil dippers (if used), and torque to the figure shown in the chart to avoid breakage or rod scoring later. Turn the crankshaft two revolutions to be sure the rod is correctly installed. If the rod strikes the camshaft, the connecting rod has been installed wrong or the cam gear is out of time. If the crankshaft operates freely, bend the cap screw locks against the screw heads. After tightening the rod screws, the rod should be able to move sideways on the crankpin of the shaft.

Crankshaft and Camshaft Gear

REMOVAL

Aluminum Cylinder Engines

To remove the crankshaft from aluminum alloy engines, remove any rust or burrs from the power take-off end of the crankshaft. Remove the crankcase cover or sump. If the sump or

Alignment of the camshaft and crankshaft timing marks

Alignment of the camshaft and crankshaft timing marks on engines equipped with ball bearings

cover sticks, tap it lightly with a soft hammer on alternate sides near the dowel. Turn the crankshaft to align the crankshaft and camshaft timing marks, lift out the cam gear, then remove the crankshaft. On models that have ball bearings on the crankshaft, the crankshaft and the camshaft must be removed together with the timing marks properly aligned – see illustration.

Cast Iron Cylinder Models

To remove the crankshaft from cast iron models, remove the crankcase cover. Revolve the crankshaft until the crankpin is pointing upward toward the breather at the rear of the engine (approximately a 45 degree angle). Pull the crankshaft out from the drive side, twisting it slightly if necessary. On models with ball bearings on the crankshaft, both the crankcase cover and bearing support should be removed.

On cast iron models with ball bearings on the drive side, first remove the magneto. Drive out the camshaft. Push the camshaft forward into the recess at the front of the engine. Then draw the crankshaft from the magneto side of the engine. Double thrust engines have cap screws inside the crankcase which hold the bearing in place. These must be removed before the crankshaft can be removed.

To remove the camshaft from all cast iron models, except the 300400 and 320400, use a long punch to drive the camshaft out toward the magneto side. Save the plug. Do not burr or peen the end of the shaft while driving it out. Hold the camshaft while driving it out. Hold the camshaft while removing the punch, so it will not drop and become damaged.

CHECKING THE CRANKSHAFT

Discard the crankshaft if it is worn beyond the allowable limit. Check the keyways for wear and make sure they are not spread. Remove all burrs from the keyway to prevent scratching the bearing. Check the three bearing journals, drive end, crankpin, and magneto end, for size and any wear or damage. Check the cam gear teeth for wear. They should not be worn at all. Check the threads at the magneto end for damage. Make sure that the crankshaft is straight.

Crankshaft Specifications

Basic Model Series	PTO Journal	Mag. Journal	C Crankpin
Aluminum Cylinder			
6B, 60000	.873	.873	.870
8B, 80000*	.873	.873	.996

Crankshaft Specifications

Basic Model Series	PTO Journal	Mag. Journal	C Crankpin
Aluminum Cylinder			
82000, 92000*, 110900*	.873	.873	.996
100000, 130000	.998	.873	.996
Cast Iron Cylinder			
5, 6, 8, N	.873	.873	.743
9	.983	.983	.873
14	1.179	1.179	.996

*Auxiliary drive models P.T.O. bearing reject size—1.003

Camshaft Specifications

Basic Model Series	Cam Gear or Shaft Journals	Cam Lobe
Aluminum Cylinder		
6B, 60000	.498	.883
8B, 80000*	.498	.883
82000, 92000	.498	.883
110900	.436 MAG. / .498 PTO.	.870
100000, 130000	.498	.950
Cast Iron Cylinder		
5, 6, 8, N	.372	.875
9	.372	1.124
14	.497	1.115

*Auxiliary drive models P.T.O. .751

NOTE: There are 0.020 in. undersize connecting rods available for use on reground crakpin bearings.

Removal and Installation of the Ball Bearings

The ball bearings are pressed onto the crankshaft. If either the bearing or the crankshaft is to be removed, use an arbor press to remove them.

To install, heat the bearing in hot oil (325°F maximum). Don't let the bearing rest on the bottom of the pan in which it is heated. Place the crankshaft in a vise with the bearing side up. When the bearing is quite hot, it will slip fit onto the bearing journal. Grasp the bearing, with the shield down, and thrust it down onto the crankshaft. The bearing will tighten on the shaft while cooling. Do not quench the bearing (throw water on it to cool it).

CHECKING THE CAMSHAFT GEAR

Inspect the teeth for wear and nicks. Check the size of the camshaft and camshaft gear bearing journals. Check the size of the cam lobes. If the cam is worn beyond tolerance, discard it.

Check the automatic spark advance on models equipped with the Magna-Matic ignition system. Place the cam gear in the normal operating position with the movable weight down. Press the weight down and release it. The spring should lift the weight. If not, the spring is stretched or the weight is binding.

INSTALLATION

Aluminum Alloy Engines—Plain Bearing

In aluminum alloy engines, the tappets are inserted first, the crankshaft next, and then the cam gear. When inserting the cam gear, turn the crankshaft and the cam gear so that the timing marks on the gears align.

Aluminum Alloy Engines—Ball Bearing

On crankshafts with ball bearings, the gear teeth are not visible for alignment of the timing marks; therefore, the timing mark is on the counterweight. On ball bearing equipped engines, the tappets are installed first. The crankshaft and the cam gear must be inserted together and their timing marks aligned.

Crankshaft Cover and Crankshaft

INSTALLATION

Models 100900 and 130900

On these models, install the governor slinger onto the cam gear with the spring washer.

To protect the oil seal while assembling the crankcase cover, put oil or grease on the sealing edge of the oil seal. Wrap a piece of thin cardboard around the crankshaft so the seal will slide easily over the shoulder of the crankshaft. If the sharp edge of the oil seal is cut or bent under, the seal may leak.

Cast Iron Engines—Plain Bearings

Assemble the tappets to the cylinder, then insert the cam gear. Push the camshaft into the camshaft hole in the cylinder from the flywheel side through the cam gear. With a blunt punch, press or hammer the camshaft until the end is flush with the outside of the cylinder on the power takeoff side. Place a small amount of sealer on the camshaft plug, then press or hammer it into the camshaft hole in the cylinder at the flywheel side. Install the crankshaft so the timing marks on the teeth and on the cam gear align.

Cast Iron—Ball Bearings

Assemble the tappets, then insert the cam gear into the cylinder, pushing the cam gear forward into the recess in front of the cylinder. Insert the crankshaft into the cylinder. Turn the camshaft and crankshaft until the timing marks align, then push the cam gear back until it engages the gear on the crankshaft with the timing marks together. Insert the camshaft. Place a small amount of sealer on the camshaft plug and press or hammer it into the camshaft hole in the cylinder at the flywheel side.

Crankshaft End-Play Adjustment

The crankshaft end-play on all models, plain and ball bearing, should be 0.002–0.008 in. The method of obtaining the correct end-play varies, however, between cast iron, aluminum, plain, and ball bearing models. New gasket sets include three crankcase cover or bearing support gaskets, 0.005 in., 0.009 in., and 0.015 in. thick.

The end-play of the crankshaft may be checked by assembling a dial indicator on the crankshaft with the pointer against the crankcase. Move the crankshaft in and out. The indicator will show the end-play. Another way to measure the end-play is to assemble a pulley to the crankshaft and measure the end-play with a feeler gauge. Place the feeler gauge between the crankshaft thrust face and the bearing support. The feeler gauge method of measuring crankshaft end-play can only be used on cast iron plain bearing engines with removable bases.

On cast iron engines, the end-play should be 0.002–0.008 in. with one 0.015 in. gasket in place. If the end-play is less than 0.002 in., which would be the case if a new crankcase or sump cover is used, additional gaskets of 0.005 in., 0.009 in., or 0.015 in. may be added in various combinations to obtain the proper end-play.

ALUMINUM ENGINES ONLY

If the end-play is more than 0.008 in. with one 0.015 in. gasket in place, a thrust washer is available to be placed on the crankshaft power take-off end, between the gear and crankcase cover or sump on plain bearing engines. On ball bearing equipped aluminum engines, the thrust washer is added to the magneto end of the crankshaft instead of the power take-off end.

NOTE: Aluminum engines never use less than the 0.015 in. gasket.

Cylinders

INSPECTION

Always inspect the cylinder after the engine has been disassembled. Visual inspection will show if there are any cracks, stripped bolt holes, broken fins, or if the cylinder wall is scored. Use an inside micrometer or telescoping gauge and micrometer to measure the size of the cylinder bore. Measure at right angles.

If the cylinder bore is more than 0.003 in. oversize, or 0.0015 in. out of round on lightweight (aluminum) cylinders, the cylinder must be resized (rebored).

NOTE: Do not deglaze the cylinder walls when installing piston rings in aluminum cylinder engines. Also be aware that there are chrome ring sets available for most engines. These are used to control oil pumping in bores worn to 0.005 in. over standard and do not require honing or glaze breaking to seat.

MEASURE AT
SIX POINTS

CENTER OF
PISTON RING TRAVEL

Checking the cylinder bore

RESIZING

Always resize to exactly 0.010 in., 0.020 in., or 0.030 in. over standard size. If this is done accurately, the stock oversize rings and pistons will fit perfectly and proper clearances will be maintained. Cylinders, either cast iron or lightweight, can be quickly resized with a good hone. Use the stones and lubrication recommended by the hone manufacturer to produce the correct cylinder wall finish for the various engine models.

If a boring bar is used, a hone must be used after the boring operation to produce the proper cylinder wall finish. Honing can be done with a portable electric drill, but it is easaier to use a drill press.

1. Clean the cylinder at top and bottom to remove all burrs and pieces of base and head gaskets.
2. Fasten the cylinder to a heavy iron plate. Some cylinders require shims. Use a level to align the drill press spindle with the bore.
3. Oil the surface of the drill press table liberally. Set the iron plate and the cylinder on the drill press table. Do not anchor the cylinder to the drill press table. If you are using a portable drill, set the plate and the cylinder on the floor.
4. Place the hone driveshaft in the chuck of the drill.
5. Slip the hone into the cylinder. Connect the driveshaft to the hone and set the stop on the drill press so the hone can only extend 3/4–1 in. from the top or bottom of the cylinder. If you are using a portable drill, cut a piece of wood to place in the cylinder as a stop for the hone.
6. Place the hone in the middle of the cylinder bore. Tighten the adjusting knob with your finger or a small screwdriver until the stones fit snugly against the cylinder wall. Do not force the stones against the cylinder wall. The hone should operate at a speed of 300–700 rpm. Lubricate the hone as recommended by the manufacturer.

NOTE: Be sure that the cylinder and the hone are centered and aligned with the driveshaft and the drill spindle.

7. Start the drill and, as the hone spins, move it up and down at the lower end of the cylinder. The cylinder is not worn at the bottom but is round so it will act to guide the hone and straighten the cylinder bore. As the bottom of the cylinder increases in diameter, gradually increase your strokes until the hone travels the full length of the bore.

Cross hatch pattern after honing

NOTE: Do not extend the hone more than 3/4–1 in. past either end of the cylinder bore.

8. As the cutting tension decreases, stop the hone and tighten the adjusting knob. Check the cylinder bore frequently with an accurate micrometer. Hone 0.0005 in. oversize to allow for shrinkage when the cylinder cools.
9. When the cylinder is within 0.0015 in. of the desired size, change from the rough stone to a finishing stone.

The finished resized cylinder should have a cross-hatched appearance. Proper stones, lubrication, and spindle speed along with rapid movement of the hone within the cylinder during the last few strokes, will produce this finish. Cross-hatching provides proper lubrication and ring break-in.

Cylinder Bore Specifications

Basic Engine Model or Series	Std. Bore Size Diameter	
	Max	Min
Aluminum Cylinder		
6B		
60000 before Ser. #5810060	2.3125	2.3115
60000 after Ser. #5810030	2.375	2.374
8B, 80000, 82000	2.375	2.374
92000	2.5625	2.5615
100000	2.500	2.449
110000	2.7812	2.7802
130000	2.5625	2.5615
Cast Iron Cylinder		
5, 6, 5S, N	2.000	1.999
8	2.250	2.249
9	2.250	2.249
14	2.625	2.624

NOTE: It is EXTREMELY important that the cylinder be thoroughly cleaned after honing to eliminate ALL grit. Wash the cylinder carefully in a solvent such as kerosene. The cylinder bore should be cleaned with a brush, soap, and water.

Bearings
INSPECTION

Plain Type

Bearings should be replaced if they are scored or if a plug gauge will enter. Try the gauge at several points in the bearing.

REPLACING PLAIN BEARINGS

Models 9–14

The crankcase cover bearing support should be replaced if the bearing is worn or scored.

REPLACING THE MAGNETO BEARING

Aluminum Cylinder Engines

There are no removable bearings in these engines. The cylinder must be reamed out so a replacement bushing can be installed.

1. Place a pilot guide bushing in the sump bearing, with the flange of the guide bushing toward the inside of the sump.
2. Assemble the sump on the cylinder. Make sure that the pilot guide bushing does not fall out of place.
3. Place the guide bushing into the oil seal recess in the cylinder. This guide bushing will center the counterbore reamer even though the oil bearing surface might be badly worn.
4. Place the counterbore reamer on the pilot and insert them into the cylinder until the tip of the pilot enters the pilot guide bushing in the sump.

Crankshaft Bearing Specifications

Basic Engine Model or Series	PTO Bearing	Bearing Magneto
Aluminum Cylinder		
6B, 8B	.878	.878
60000, 80000	.878	.878
82000, 92000, 110900	.878	.878
100000, 130000	1.003	.878
Cast Iron Cylinder		
5, 6, N	.878	.878
9	.988	.988
14	1.185	1.185

5. Turn the reamer clockwise with a steady, even pressure until it is complletely through the bearing. Lubricate the reamer with kerosene or any other suitable solvent.

NOTE: Counterbore reaming may be performed without any lubrication. However, clean off shavings because aluminum material builds up on the reamer flutes causing eventual damage to the reamer and an oversize counterbore.

6. Remove the sump and pull the reamer out without backing it through the bearing. Clean out the remaining chips. Remove the guide bushing from the oil seal recess.
7. Hold the new bushing against the outer end of the reamed out bearing, with the notch in the bushing aligned with the notch in the cylinder. Note the position of the split in the bushing. At a point in the outer edge of the reamed out bearing opposite to the split in the bushing, make a notch in the cylinder hub at a 45 degree angle to the bearing surface. Use a chisel or a screwdriver and hammer.
8. Press in the new bushing, being careful to align the oil notches with the driver and the support until the outer end of the bushing is flush with the end of the reamed cylinder hub.
9. With a blunt chisel or screwdriver, drive a portion of the bushing into the notch previously made in the cylinder. This is called staking and is done to prevent the bushing from turning.
10. Reassemble the sump to the cylinder with the pilot guide bushing in the sump bearing.
11. Place a finishing reamer on the pilot and insert the pilot into the cylinder bearing until the tip of the pilot enters the pilot guide bushings in the sump bearing.
12. Lubricate the reamer with kerosene, fuel oil, or other suitable solvent, then ream the bushing, turning the reamer clockwise with a steady even pressure until the reamer is completely through the bearing. Improper lubricants will produce a rough bearing surface.
13. Remove the sump, reamer, and the pilot guide bushing. Clean out all reaming chips.

REPLACING THE P.T.O. BEARING

Aluminum Cylinder Engines

The sump or crankcase bearing is repaired the same way as the magneto end bearing. Make sure to complete repair of one bearing before starting to repair the other. Press in new oil seals when bearing repair is completed.

NOTE: On Models 8B–HA, 80590, 81590, 82590, 80790, 81790, 82990, 92590 and 92990, the magneto bearing can be replaced as described above. However, if the sump bearing is worn, the sump must be replaced.

REPLACING OIL SEALS

Note the following points:

1. Assemble the seal with the sharp edge of leather or rubber toward the inside of the engine.

2. Lubricate the inside diameter of the seal with "Lubriplate" or equivalent.

3. Press all seals but those listed below so they are flush with the hub. On models 60000, 80000, 100000 and 13000 with ball bearing which has a mounting flange, the seal must be pressed in until it is $3/16$ in. below the crannkcase mounting flange.

LUBRICATION

The primary purpose of oil is, of course, lubrication. However oil performs three other very important functions as well: cooling, cleaning, and sealing. Oil absorbs and dissipates heat created by combustion and friction. Oil cleans by trapping and holding dirt and by-products of combustion. This dirt is held in suspension by the oil until it is drained. Oil seals the combustion chamber by coating the rings, thus helping increase and maintain compression.

Briggs and Stratton engines are lubricated with a gear driven splash oil slinger or a connecting rod dipper.

Extended Oil Filler Tubes and Dipsticks

When installing the extended oil fill and dipstick assembly, the tube must be installed so the O-ring seal is firmly compressed. To do so, push the tube downward toward the sump, then tighten the blower housing screw, which is used to secure the tube and bracket. When the dipstick assembly is fully depressed, it seals the upper end of the tube.

A leak at the seal between the tube and the sump, or at the seal at the upper end of the dipstick can result in a loss of crankcase vacuum, and a discharge of smoke through the exhaust system.

Breathers

The function of the breather is to maintain a vacuum in the crankcase. The breather has a fiber disc valve which limits the direction of air flow caused by the piston moving back and forth in the cylinder. Air can flow out of the crankcase, but the one-way valve blocks the return flow, thus maintaining a vacuum in the crankcase. A partial vacuum must be maintained in the crankcase to prevent oil from being forced out of the engine at the piston rings, oil seals, breaker plunger, and gaskets.

Installing the connecting rod in a horizontal crankshaft engine

Checking the breather assembly

Installation of an oil slinger

INSPECTION OF THE BREATHER

If the fiber disc valve is stuck or binding, the breather cannot function properly and must be replaced. A 0.045 in. wire gauge should not enter the space between the fiber disc valve and the body. Use a spark plug wire gauge to check the valve. The fiber disc valve is held in place by an internal bracket which will be distorted if pressure is applied to the fiber disc valve. Therefore, do not apply force when checking the valve with the wire gauge.

If the breather is removed for inspection or valve repair, a new gasket should be used when replacing the breather. Tighten the screws securely to prevent oil leakage.

Most breathers are now vented through the air cleaner, to prevent dirt from entering the crankcase. Check to be sure that the venting elbows or the tube are not damaged and that they are properly sealed.

Oil Dippers and Slingers

Oil dippers reach into the oil reservoir in the base of the engine and splash oil onto the internal engine parts. The oil dipper is installed on the connecting rod and has no pump or moving parts.

Oil slingers are driven by the cam gear. Old style slingers using a die cast bracket assembly have a steel bushing between the slinger and the bracket. Replace the bracket on which the oil slinger rides if it is worn to a diameter of 0.490 in. or less. Replace the steel bushing if it is worn. Newer style oil slingers have a stamped steel bracket.

Section 5

Briggs and Stratton

Small Engines — 6-20 hp

ENGINE IDENTIFICATION

The Briggs and Stratton model designation system consists of up to a six digit number. It is possible to determine most of the important mechanical features of the engine by merely knowing the model number. An explanation of what each number means is given below.

1. The first one or two digits indicate the cubic inch displacement (cid).

2. The first digit after the displacement indicates the basic

Briggs and Stratton Model Numbering System

Cubic Inch Displacement	First Digit After Displacement Basic Design Series	Second Digit After Displacement — Crankshaft, Carburetor Governor	Third Digit After Displacement — Bearings, Reduction Gears & Auxiliary Drives	Fourth Digit After Displacement Type of Starter
6	0	0-	0-Plain Bearing	0-Without Starter
8	1	1-Horizontal Vacu-Jet	1-Flange Mounting Plain Bearing	1-Rope Starter
9	2			
10	3	2-Horizontal Pulsa-Jet	2-Ball Bearing	2-Rewind Starter
13	4			
14	5	3-Horizontal Flo-Jet (Pneumatic Governor)	3-Flange Mounting Ball Bearing	3-Electric-110 Volt, Gear Drive
17	6			
19	7	4-Horizontal Flo-Jet (Mechanical Governor)	4-	4-Elec. Starter-Generator-12 Volt, Belt Drive
20	8			
23	9			
24		5-Vertical Vacu-Jet	5-Gear Reduction (6 to 1)	5-Electric Starter Only-12 Volt, Gear Drive
30				
32				
		6-	6-Gear Reduction (6 to 1) Reverse Rotation	6-Wind-up Starter
		7-Vertical Flo-Jet	7-	7-Electric Starter, 12 Volt Gear Drive, with Alternator
		8-	8-Auxiliary Drive Perpendicular to Crankshaft	8-Vertical-pull Starter
		9-Vertical Pulsa-Jet	9-Auxiliary Drive Parallel to Crankshaft	

design series, relating to cylinder construction, ignition and general configuration.

3. The second digit after the displacement indicates the position of the crankshaft and the type of carburetor the engine has.

4. The third digit after the displacement indicates the type of bearings and whether or not the engine is equipped with a reduction gear or auxiliary drive.

5. The last digit indicates the type of starter. The model identification plate is usually located on the air baffle surrounding the cylinder.

General Engine Specifications

Model	Bore Size (In.)	Horsepower
Aluminum Engines		
140000	2.750	6
170000, 171700	3.000	7
190000, 191700	3.000	8
251000	3.4375	10
Cast Iron Engines		
19, 190000, 200000	3.000	8
23, 230000	3.000	9
243000	3.0625	10
300000	3.4375	13
320000	3.5625	16

MAINTENANCE
Air Cleaners

A properly serviced air cleaner protects the engine from dust particles that are in the air. When servicing an air cleaner, check the air cleaner mounting and gaskets for worn or damaged mating surfaces. Replace any worn or damaged parts to prevent dirt and dust from entering the engine through openings caused by improper sealing. Straighten or replace any bent mounting studs.

Oil foam air cleaner

Oil foam air cleaner

SERVICING

Oil Foam Air Cleaners

Clean and re-oil the air cleaner element every 25 hours of operation under normal operating conditions. The capacity of the oil-foam air cleaner is adequate for a full season's use without cleaning. Under very dusty conditions, clean the air cleaner every few hours of operation.

The oil-foam air cleaner is serviced in the following manner:

1. Remove the screw that holds the halves of the air cleaner shell together and retains it to the carburetor.

2. Remove the air cleaner carefully to prevent dirt from entering the carburetor.

3. Take the air cleaner apart (split the two halves).

4. Wash the foam in kerosene or liquid detergent and water to remove the dirt.

5. Wrap the foam in a clean cloth and squeeze it dry.

6. Saturate the foam in clean engine oil and squeeze it to remove the excess oil.

7. Assemble the air cleaner and fasten it to the carburetor with the attaching screw.

Oil Bath Air Cleaner

Pour the old oil out of the bowl. Wash the element thoroughly in solvent and squeeze it dry. Clean the bowl and refill it with the same type of oil used in the crankcase.

Oil bath air cleaner

Dry Element Air Cleaner

Remove the element of the air cleaner and tap (top and bottom) it on a flat surface or wash it in non-sudsing detergent and flush it from the inside until the water coming out is clear. After washing, air dry the element thoroughly before reinstalling it on the engine. NEVER OIL A DRY ELEMENT.

Dry element air cleaner

Heavy Duty Air Cleaner

Clean and re-oil the foam pre-cleaner at three month intervals or every 25 hours, whichever comes first.

Clean the paper element every year or 100 hours, whichever comes first. Use the dry element procedure for cleaning the paper element of the heavy duty air cleaner.

Use the oil foam cleaning procedure to clean the foam sleeve of the heavy duty air cleaner.

If the engine is operated under very dusty conditions, clean the air cleaner more often.

Briggs and Stratton heavy duty air cleaner

LUBRICATION
Oil and Fuel Recommendations

Briggs & Stratton recommends either a leaded regular grade gasoline or low-lead or unleaded fuel. Low-lead or unleaded fuel is preferable because of the reduction in deposits that results from its use, but its use is not required. Nor is it necessary to use leaded fuel occasionally in an engine which runs primarily on unleaded fuel, as is the case with some older automobile engines. Premium fuel is not required, as regular or unleaded will have sufficient knock resistance if the engine is in proper condition. The factory recommends that fuel by purchased in lots small enough to be used up in 30 days or less. When fuel is older than that, it can form gum and varnish, or may be improperly tailored to the prevailing temperature.

You should use a high quality detergent oil designated "For Service SC, SD, SE, or MS." Detergent oil is recommended because of its important ability to keep gum and varnish from clogging the lubrication system. Briggs & Stratton specifically recommends that no special oil additives be used.

Oil Viscosity Recommendations

Winter (under 40°F.)	Summer (above 40°F.)
SAE 5W-20, or SAE 5W-30	SAE 30
If above are not available:	If above is not available:
SAE 10W or SAE 10W-30 (under 0°F.)	SAE 10W-40 or SAE 10W-30
Use SAE 10W or SAE 10W-30 in proportions of 90% motor oil/ 10% kerosene	

Engine Oil Capacity Chart

Basic Model Series	Capacity Pints
Aluminum	
14, 17 Cu. in. Vert. Crankshaft	2¼
14, 17, 19 Cu. in. Horiz. Crankshaft	2¾
25 Cu. in. Vert. Crankshaft	3
25 Cu. in. Horiz. Crankshaft	3
Cast Iron	
19, 20 Cu. in. Horiz. Crank.	3
23, 24, 30, 32 Cu. in. Horiz. Crank.	4

Location of oil fill and level check plugs, 6-1 gear reduction equipped engines

Oil must be changed every 25 hours of operation. If the atmosphere in which the engine is operating is very dirty, oil changes should be made more frequently, as often as every 12 hours, if necessary. Oil should be changed after 5 hours of operation in the case of brand new engines. Drain engine oil when hot.

In hot weather, when under heavy load, or when brand new, engines may consume oil at a rate which will require you to refill the crankcase several times between oil changes. Check the oil level every hour or so until you can accurately estimate how long the engine can go between refills. To check oil level, stop the engine and allow it to sit for a couple of minutes, then remove the dipstick or filler cap. Fill the crankcase to the top of the filler pipe when there is no dipstick, or wipe the dipstick clean, reinsert it, and add oil as necessary until the level reaches the upper mark.

On cast iron engines with a gear reduction unit, crankcase and reduction gears are lubricated by a common oil supply. When draining crankcase, also remove drain plug in reduction unit.

On aluminum engines with reduction gear, a separate oil supply lubricates the gears, although the same type of oil used in the crankcase is used in the reduction gear cover. On these engines, remove the drain plug every fourth oil change (100 hours), then install the plug and refill. The level in the reduction gear cover must be checked during the refill operation by removing the level plug from the side of the gearcase, removing the filler plug, and then filling the case through the filler plug hole until oil runs out the level plug hole. Then, install both plugs.

On 6–1 gear reduction engines (models 6, 8, 8000, 10000, and 13000), no changes are required for the oil in the reduction gear case, but level must be checked and the case refilled, as described in the paragraph above, every 100 hours. Make sure the oil level plug (with screwdriver slot and no vent) is installed in the hole on the side of the case.

TUNE-UP

Spark Plugs

Remove the spark plug with a ¾ in. (1½ in. plug) or a ¹³⁄₁₆ in. (2 in. plug) deep well socket wrench. Clean carbon deposits off the center and side electrodes with a sharp instrument. If possible, you should also attempt to remove deposits from the recess between the insulator and the threaded portion of the plug. If the electrodes are burned away or the insulator is cracked at any point, replace the plug. Using a wire type feeler gauge, adjust the gap by bending the side electrode where it is curved until the gap is 0.030 in.

.030 WIRE GAUGE

Checking spark plug gap with a wire feeler gauge

When installing the plug, make sure the threads of the plug and the threads in the cylinder head are clean. It is best to oil the plug threads very lightly. Be careful not to overtorque the plug, especially if the engine has an aluminum head. If you use a torque wrench, torque to about 15 ft. lbs.

Breaker Points

All Briggs and Stratton engines have magneto ignition systems. Three types are used: Flywheel Type—Internal Breaker Flywheel Type—External Breaker, and Magna-Matic.

FLYWHEEL TYPE—INTERNAL BREAKER

This ignition system has the magneto located on the flywheel and the breaker points located under the flywheel.

The flywheel is located on the crankshaft with a soft metal key. It is held in place by a nut or starter clutch. The flywheel

SPARK PLUG — ARMATURE — COIL — POLE PIECE — MAGNET — CONTACT POINTS — PLUNGER — CONDENSER — FLYWHEEL

Flywheel magneto ignition with internal breaker points and external armature

key must be in good condition to insure proper location of the flywheel for ignition timing. Do not use a steel key under any circumstances. Use only a soft metal key, as originally supplied.

The keyway in both flywheel and crankshaft should not be distorted. Flywheels are made of aluminum, zinc, or cast iron.

Flywheel, Nut, and/or Starter Clutch

REMOVAL AND INSTALLATION

Place a block of wood under the flywheel fins to prevent the flywheel from turning while you are loosening the nut or starter clutch. Be careful not to bend the flywheel. There are special flywheel holders available for this purpose; Briggs & Stratton recommends their use on flywheels of 6¾ in. diameter or less.

On rope starter engines, the ½ in. flywheel nut has a left-hand thread and the ⅝ in. nut has a right-hand thread. The starter clutch used on rewind or wind-up starters has a right-hand thread.

Some flywheels have two holes provided for the use of a flywheel puller. Use a small gear puller or automotive steering wheel puller to remove the flywheel if a flywheel puller is not available. Be careful not to bend the flywheel if a gear puller is used. On rope starter engines leave the nut on for the puller to bear against. Small cast iron flywheels do not require a puller.

Install the flywheel in the reverse order of removal after inspecting the key and keyway for damage or wear.

Breaker Point Removal and Installation

Remove the breaker cover. Care should be taken when removing the cover, to avoid damaging it. If the cover is bent or damaged, it should be replaced to insure a proper seal.

The breaker point gap on all models is 0.020 in. Check the points for contact and for signs of burning or pitting. Points that are set too wide will advance the spark timing and may cause kickback when starting. Points that are set too close will retard the spark timing and decrease engine power.

On models that have a separate condenser, the point set is removed by first removing the condenser and armature wires from the breaker point clip. Loosen the adjusting lock screw and remove the breaker point assembly.

On models where the condenser is incorporated with the breaker points, loosen the screw which holds the post. The

Tune-Up Specifications

Model	Plug Type	Plug Gap (in.)	Point Gap (in.)	Armature Gap 2 leg	Armature Gap 3 leg	Idle Speed
Aluminum Block						
140000, 170000, 190000, 251000	①	.030	.020	.010 – .014	.016 – .019	1750
Cast Iron Block						
19, 190000, 200000	①	.030	.020	.010 – .014	.022 – .026	1200
23, 230000	①	.030	.020	.010 – .014	.022 – .026	1200
243400, 300000, 320000	①	.030	.020	.010 – .014	—	1200

① Manufacturer's Code

1½ in. plug	2 in. plug	Manufacturer
CJ-8	J-8	Champion
RCJ-8	RJ-8	Champion (resistor)
A-7NX	A-71	Autolite
AR-7N	AR-80	Autolite (resistor)
CS-45	GC-46	A.C.
—	R-46	A.C. (resistor)

condenser/point assembly is removed by loosening the screw which holds the condenser clamp.

When installing a point set with the separate condenser, be sure that the small boss on the magneto plate enters the hole in the point bracket. Mount the point set to the magneto plate or the cylinder with a lock screw. Fasten the armature lead wire to the breaker points with the clip and screw. If these lead wires do not have terminals, the bare end of the wires can be inserted into the clip and the screw tightened to make a good connection. Do not let the ends of the wire touch either the point bracket or the magneto plate, or the ignition will be grounded.

To install the integral condenser/point set, place the mounting post of the breaker arm into the recess in the cylinder so that the groove in the post fits the notch in the recess. Tighten the mounting screw securely. Use a ¼ in. wrench. Slip the open loop of the breaker arm spring through the two holes in the arm, then hook the closed loop of the spring over the small post protruding from the cylinder. Push the flat end of the breaker arm into the groove in the mounting post. This places tension on the spring and pulls the arm against the plunger. If the condenser post is threaded, attach the soil primary wire and the ground wire (if furnished) with the lock washer and nut. If the primary wire is fastened to the condenser with a spring fastener, compress the spring and slip the primary wire and ground wire into the hole in the condenser post. Release the spring. Lay the condenser in place and tighten the condenser clamp securely. Install the spring in the breaker arm.

Point Gap Adjustment

Turn the crankshaft until the points are open to the widest gap. When adjusting a breaker point assembly with an integral condenser, move the condenser forward or backward with a screwdriver until the proper gap is obtained (0.020 in.). Point sets with a separate condenser are adjusted by moving the contact point bracket up and down after the lock screw has been loosened. The point gap is set to 0.020 in.

Removing the breaker point assembly

1. FASTEN FLYWHEEL PULLER TO FLYWHEEL WITH TWO SELF TAPPING SCREWS
2. TURN DOWN THESE TWO NUTS TO LOOSEN FLYWHEEL

Removing the flywheel with a puller

TO LOOSEN
WOOD BLOCK
TO TIGHTEN

Removing the flywheel

Removing the integral breaker point and condenser assembly

Breaker Point Plunger

If the breaker point plunger hole becomes excessively worn, oil will leak past the plunger and may get on the points, causing them to burn. To check the hole, loosen the breaker point mounting screw and move the breaker points out of the way. Remove the plunger. If the flat end of the #19055 plug gauge will enter the plunger hole for a distance of ¼ in. or more, the hole should be rebushed.

To install the bushing, it is necessary that the breaker points, armature, and crankshaft be removed. Use a #19056 reamer to ream out the old plunger hole. This should be done by hand. The reamer must be in alignment with the plunger hole. Drive the bushing, #23513, into the hole until the upper end of the bushing is flush with the top of the boss. Remove all metal chips and dirt from the engine.

If the breaker point plunger is worn to a length of 0.870 in. or less, it should be replaced. Plungers must be inserted with the groove at the top or oil will enter the breaker box. Insert the plunger into the hole in the cylinder.

Armature Air Gap Adjustment

Set the air gap between the flywheel and the armature as follows: With the armature up as far as possible and just one screw tightened, slip the proper gauge between the armature and flywheel. Turn the flywheel until the magnets are directly below the armature. Loosen the one mounting screw and the magnets should pull the armature down firmly against the thickness gauge. Tighten the mounting screws.

Adjusting the point gap

Adjusting point gap on the integral point and condenser assembly

Variations in armature positioning

Adjustment of the armature gap

Replacing the breaker plunger bushing

FLYWHEEL TYPE – EXTERNAL BREAKER

Breaker Point Set Removal and Installation

Turn the crankshaft until the points open to their widest gap. This makes it easier to assemble and adjust the points later if the crankshaft is not removed. Remove the condenser and upper and lower mounting screws. Loosen the lock nut and back off the breaker point screw. Install the points in the reverse order of removal.

To avoid the possibility of oil leaking past the breaker point plunger or moisture entering the crankcase between the plunger and the bushing, a plunger seal is installed on the engine models using this type of ignition system. To install a new seal on the plunger, remove the breaker point assembly and condenser. Remove the retainer and eyelet, remove the old seal, and install the new one. Use extreme care when installing the seal on the plunger to avoid damaging the seal. Replace the eyelet and retainer and replace the points and condenser.

NOTE: Apply a small amount of sealer to the threads of both mounting screws and the adjustment screw. The sealer prevents oil from leaking into the breaker point area.

Point Gap Adjustment

Turn the crankshaft until the points open to their widest gap. Turn the breaker point adjusting screw until the points open to 0.020 in. and tighten the lock nut. When the cover is installed, seal the point where the primary wire passes under the cover. This area must be resealed to prevent the entry of dust and moisture.

Armature Timing Adjustment

MODELS 193000, 200000, 230000, 243000, 300400, 320400

Using a puller, remove the flywheel. Set the point gap at 0.020 in. Position the flywheel on the crankshaft taper. Slip the key in place. Install the flywheel and the crankshaft clockwise until the breaker points are just opening. Use a timing light. When the points just start to open, the arrow on the flywheel should line up with the arrow on the armature bracket.

If the arrows do not match, slip off the flywheel without disturbing the position of the crankshaft. Slightly loosen the mounting screw which holds the armature bracket to the cylinder. Slip the flywheel back onto the crankshaft. Insert the flywheel key. Install the flywheel nut finger tight. Move the armature and bracket assembly to align the arrows. Slip off the flywheel and tighten the armature bracket bolts. Install the key and flywheel. Tighten the flywheel nut to 110–118 ft. lbs. on the 193000 and 200000 series. On all the rest, tighten to 138–150 ft. lbs. Set the armature gap at 0.010–0.014 in.

MODELS 19D AND 23D

With the points set at 0.020 in. and the flywheel key screw finger tight together with the flywheel nut, rotate the flywheel clockwise until the breaker points are just opening. The flywheel key drives the crankshaft while doing this. Using a timing light, rotate the flywheel slightly counterclockwise until the edge of the armature lines up with the edge of the flywheel insert. The crankshaft must not turn while doing this. Tighten the key screw and the flywheel nut. Set the armature air gap at 0.022–0.026 in.

Replacing Threaded Breaker Plunger and Bushing

Remove the breaker cover and the condenser and breaker point assembly.

Place a thick ³⁄₈ in. inside diameter washer over the end of the bushing and screw on the ³⁄₈–24 nut. Tighten the nut to pull the bushing out of the hole. After the bushing has been moved

Flywheel magneto ignition with an external breaker assembly

Breaker point gap adjustment

Removing a threaded plunger bushing

about ⅛ in., remove the nut and put on a second thick washer and repeat the procedure. A total stack of ³⁄₈ in. washers will be required to completely remove the bushing. Be sure the plunger does not fall out of the bushing as it is removed.

Place the new plunger in the bushing with the large end of the plunger opposite the threads on the bushing. Screw the ³⁄₈–24 in. nut onto the threads to protect them and insert the bushing into the cylinder. Place a piece of tubing the same diameter as the nut and, using a hammer, drive the bushing into the cylinder until the square shoulder on the bushing is flush with the face of the cylinder. Check to be sure that the plunger operates freely.

Removing an unthreaded plunger bushing

Replacing Unthreaded Breaker Plunger and Bushing

Pull the plunger out as far as possible and use a pair of pliers to break the plunger off as close as possible to the bushing. Use a $\frac{1}{4}$–20 in. tap or a #93029 self threading screw to thread the hole in the bushing to a depth of about $\frac{1}{2}$–$\frac{5}{8}$ in. Use a $\frac{1}{4}$–20 x $\frac{1}{2}$ in. Hex. head screw and two spacer washers to pull the bushing out of the cylinder. The bushing will be free when it has been extracted $\frac{5}{16}$. Carefully remove the bushing and the remainder of the broken plunger. Do not allow the plunger or metal chips to drop into the crankcase.

Correctly insert the new plunger into the new bushing. Insert the plunger and the bushing into the cylinder. Use a hammer and the old bushing to drive the new bushing into the cylinder until the new bushing is flush with the face of the cylinder. Make sure that the plunger operates freely.

Plunger Seal

Later models with Flywheel Type — External Breaker Ignition feature a plunger seal. This seal keeps both oil and moisture from entering the breaker box. If the points have become contaminated on an engine manufactured without this feature, the seal may be installed. Parts, part numbers, and their locations are shown in the illustration. Install the seal onto the plunger very carefully to avoid fracturing it.

Plunger seal used on later model engines

Measuring the armature gap

The Magna-Matic ignition system

MAGNA-MATIC IGNITION SYSTEM

Removing the Flywheel

Flywheels on engines with Magna-Matic ignition are removed with pullers similar to factory designs numbered #19068 and 19203. These pullers employ two bolts, which are screwed into holes tapped into the flywheel. The bolts are turned until the flywheel is forced off the crankshaft. Only this type of device should be used to pull these flywheels.

Armature Air Gap

The armature air gap on engines equipped with Magna-Matic ignition system is fixed and can change only if wear occurs on the crankshaft journal and/or main bearing. Check for wear by inserting a $\frac{1}{2}$ in. wide feeler gauge at several points between the rotor and armature. Minimum feeler gauge thickness is 0.004 in. Keep the feeler gauge away from the magnets on the rotor or you will have a false reading.

Rotor Removal and Installation

The rotor is held in place by a woodruff key and a clamp on later engines, and a woodruff key and set screw on older engines. The rotor clamp must always remain on the rotor, unless the rotor is in place on the crankshaft and within the armature, or a loss of magnetism will occur.

Loosen the socket head screw in the rotor clamp which will allow the clamp to loosen. It may be necessary to use a puller to remove the rotor from the crankshaft. On older models, loosen the small lock screw, then the set screw.

To install the set screw type rotor, place the woodruff key in the keyway on the crankshaft, then slide the rotor onto the crankshaft until the set screw hole in the rotor and the crankshaft are aligned. Be sure the key remains in place. Tighten the set screw securely, then tighten the lock screw to prevent the set screw from loosening. The lock screw is self-threading and the hole does not require tapping.

To install the clamp type rotor, place the woodruff key in place in the crankshaft and align the keyway in the rotor with

Removing the rotor

Installing the rotor

the woodruff key. If necessary, use a short length of pipe and a hammer to drive the rotor onto the shaft until a 0.025 in. feeler gauge can be inserted between the rotor and the bearing support. The split in the clamp must be between the slots in the rotor. Tighten the clamp screws to 60–70 inch lbs.

Rotor Timing Adjustment

The rotor and armature are correctly timed at the factory and require timing only if the armature has been removed from the engine, or if the cam gear or crankshaft has been replaced.

If it is necessary to adjust the rotor, proceed as follows: with the point gap set at 0.020 in., turn the crankshaft in the normal direction of rotation until the breaker points close and just start to open. Use a timing light or insert a piece of tissue paper be-

tween the breaker points to determine when the points begin to open. With the three armature mounting screws slightly loose, rotate the armature until the arrow on the armature lines up with the arrow on the rotor. Align with the corresponding number of engine models, for example, on Model 9, align with #9. Retighten the armature mounting screws.

Coil and/or Armature Replacement

Usually the coil and armature are not separated, but left assembled for convenience. However, if one or both need replacement, proceed as follows: the coil primary wire and the coil ground wire must be unfastened. Pry out the clips that hold the coil and coil core to the armature. The coil core is a slip fit in the coil and can be pushed out of the coil.

To reassemble, push the coil core into the coil with the rounded side toward the ignition cable. Place the coil and core on the armature with the coil retainer between the coil and the armature and with the rounded side toward the coil. Hook the lower end of the clips into the armature, then press the upper end onto the coil core.

Fasten the coil ground wire (bare double wires) to the armature support. Next, place the assembly against the cylinder and around the rotor and bearing support. Insert the three mounting screws together with the washer and lockwasher into the three long oval holes in the armature. Tighten them enough to hold the armature in place but loose enough so the armature can be moved for adjustment of the timing. Attach the primary wires from the coil and the breaker points to the terminal at the upper side of the backing plate. This terminal is insulated from the backing plate. Push the ignition cable through the louvered hole at the left side of the backing plate.

Breaker Point Removal and Installation

Turn the crankshaft until the points open to the widest gap. This makes it easier to assemble and adjust the points later if the crankshaft is not removed. With the terminal screw out, remove the spring screw. Loosen the breaker shaft nut until the nut is flush with the end of the shaft. Tap the nut to free the breaker arm from the tapered end of the breaker shaft. Remove the nut, lockwasher, and breaker arm. Remove the breaker plate screw, breaker plate, pivot, insulating plate, and eccentric. Pry out the breaker shaft seal with a sharp pointed tool.

To install the breaker points, press in the new oil seal with the metal side out. Put the new breaker plate on the top of the insulating plate, making sure that the detent in the breaker plate engages the hole in the insulating plate. Fasten the breaker plate screw enough to put a light tension on the plate. Adjust the eccentric so that the left edge of the insulating plate is parallel to

Adjustment of the timing

The breaker point assembly

Adjusting the breaker point gap

the edge of the box and tighten the screw. This locates the breaker plate so that the proper gap adjustments may be made. Turn the breaker shaft clockwise as far as possible and hold it in this position. Place the new breaker points on the shaft, then the lockwasher, and tighten the nut down on the lockwasher. Replace the spring screw and terminal screw.

Breaker Box Removal and Installation

Remove the two mounting screws, then remove the breaker box, turning it slightly to clear the arm at the inner end of the breaker shaft. The breaker points need not be removed to remove the breaker box.

To install, pull the primary wire through the hole at the lower left corner of the breaker box. See that the primary wire rests in the groove at the top end of the box, then tighten the two mounting screws to hold the box in place.

Breaker Shaft Removal and Installation

The breaker shaft can be removed, after the breaker points are removed, by turning the shaft one half turn to clear the retaining spur at the inside of the breaker box.

Install by inserting the breaker shaft with the arm upward so the arm will clean the retainer boss. Push the shaft all the way in, then turn the arm downward.

Breaker Point Adjustment

To adjust the breaker points, turn the crankshaft until the breaker points open to the widest gap. Loosen the breaker point plate screw slightly. Rotate the eccentric to obtain a point gap of 0.020 in. Tighten the breaker plate screw.

Carburetor adjustment screws

Mixture Adjustment

TWO PIECE FLO-JET

1. Start the engine and run it at 3000 rpm until it warms up.
2. Turn the needle valve (flat handle) to both extremes of operation noting the location of the valve at both points. That is, turn the valve inward until the mixture becomes too lean and the engine starts to slow, then note the position of the valve. Turn it outward slowly until the mixture becomes too rich and the engine begins to slow. Turn the valve back inward to the mid-point between the two extremes.
3. Install a tachometer on the engine. Pull the throttle to the idle position and hold it there through the rest of this step. Adjust the idle speed screw until the engine idles at 1750 rpm if it's an aluminum engine, or 1200 rpm, if it's a cast iron engine. Then, turn the idle valve in and out to adjust mixture, as described in Step 2. If idle valve adjustment changes idle speed, adjust speed to specification.
4. Release the throttle and observe the engine's response. The engine should accelerate without hesitation. If response is poor, one of the mixture adjustments is too lean. Readjust either or both as necessary. If idle speed was changed after idle valve was adjusted, readjust the idle valve first.

ONE PIECE FLO-JET

Follow the instructions for adjusting the Two Piece Flo-Jet carburetor (above). On the large, One Piece Flo-Jet, the needle valve is located under the float bowl, and the idle valve on top of the venturi passage. On the small One Piece Flo-Jet, both valves are adjusted by screws located on top of the venturi passage. The needle valve is located on the air horn side, is centered above the float bowl, and uses a larger screw head.

Idle valve and needle valve locations on the One Piece Flo-Jet

Governor Adjustments

SETTING MAXIMUM GOVERNED SPEED WITH ROTARY LAWNMOWER BLADES

NOTE: Strict limits on engine rpm must be observed when setting top governed speed on rotary lawnmowers. This is done so that blade tip speeds will be kept to less than 19,000 feet per minute. Briggs & Stratton suggests setting the governor 200 rpm low to allow for possible error in the tachometer reading. These figures below, based on blade length, must be strictly adhered to, or a serious accident could result!

MODEL 140000

Loosen the screw which holds the governor lever to the governor shaft. Turn the governor lever counterclockwise until the carburetor throttle is wide open. With a screwdriver, turn the governor shaft counterclockwise as far as it will go. Tighten the screw which holds the governor lever to the governor shaft.

Blade Length (in.)	Max. Governed Speed (R.P.M.)
18	4032
19	3820
20	3629
21	3456
22	3299
23	3155
24	3024
25	2903
26	2791

CAST IRON MODELS 19, 190000, 200000, 23, 230000, 240000, 300000, 320000

Loosen the screw which holds the governor lever to the governor shaft. Push the lever counterclockwise as far as it will go. Hold it in position and turn the governor shaft counterclockwise as far as it will go. This can be done with a screwdriver. Securely tighten the screw that holds the governor lever to the shaft.

ALUMINUM MODELS 140000, 170000, 190000, 251000

Vertical and horizontal shaft engine governors are adjusted by setting the control lever in the high speed position. Loosen the nut on the governor lever. Turn the governor shaft clockwise with a screwdriver to the end of its travel. Tighten the nut. The throttle must be wide open. Check to see if the throttle can be moved from idle to wide open without binding.

Adjusting Top No Load Speed

Set the control lever to the maximum speed position with the engine running. Bend the spring anchor tang to get the desired top speed.

Adjustment For Closer Governing
(Generator Applications Only)

1. Snap knob upward to release adjusting nut.
2. Pull knob out against stop.
3. Then, bend the spring anchor tang to get top no-load speed as described below, depending upon the application.

On models 140400, 146400, 170400, 190400 and 251400 with an 1800 rpm generator: temporarily substituting a #260902 governor spring, set the no load speed at 2600 rpm, and then set throttle stop at 1600 rpm.

On models 140400, 146400, 170400, 190400, 251400 and 251400 with an 3600 rpm generator: set no load speed to 4200 rpm with standard governor spring.

4. Snap knob back into its normal position.
5. Adjust the knob for the desired generator speed.

Bending the spring anchor tang to get desired top speed

You can make a tool like the one shown to adjust spring anchor tang

Obtaining closer governing on generator applications

Choke Adjustment

CHOKE-A-MATIC, PULSA-JET AND VACU-JET CARBURETORS

To check the operation of the choke linkage, move the speed adjustment lever to the choke position. If the choke slide does not fully close, bend the choke link. The speed adjustment lever must make good contact against the top switch.

Install the carburetor and adjust it in the same manner as the Pulsa-Jet carburetor.

TWO-PIECE FLO-JET AUTOMATIC CHOKE

Hold the choke shaft so the thermostat lever is free. At room temperature (68°F), the screw in the thermostat collar should be in the center of the stops. If not, loosen the stop screw and adjust the screw.

Loosen the set screw on the lever of the thermostat assembly. Slide the lever to the right or left on the shaft to ensure free movement of the choke link in any position. Rotate the thermostat shaft clockwise until the stop screw strikes the tube. Hold it in position and set the lever on the thermostat shaft so that the choke valve will be held open about $\frac{1}{8}$ in. from a closed position. Then tighten the set screw in the lever.

Adjusting automatic choke—Two Piece Flo-Jet

Rotate the thermostat shaft counterclockwise until the stop screw strikes the opposite side of the tube. Then open the choke valve manually until it stops against the top of the choke link opening. The choke valve should now be open approximately $\frac{1}{8}$ in. as before.

Check the position of the counterweight lever. With the choke valve in a wide open position (horizontal) the counterweight lever should also be in a horizontal position with the free end toward the right.

Operate the choke manually to be sure that all parts are free to move without binding or rubbing in any position.

Compression Checking

You can check the compression in any Briggs & Stratton engine by performing the following simple procedure: spin the flywheel counterclockwise (flywheel side) against the compression stroke. A sharp rebound indicates that there is satisfactory compression. A slight or no rebound indicates poor compression.

It has been determined that this test is an accurate indication of compression and is recommended by Briggs and Stratton. Briggs & Stratton does not supply compression pressures.

Loss of compression will usually be the result of one or a combination of the following:
1. The cylinder head gasket is blown or leaking.
2. The valves are sticking or not seating properly.
3. The piston rings are not sealing, which would also cause the engine to consume an excessive amount of oil.

Carbon deposits in the combustion chamber should be removed every 100 or 200 hours of use (more often when run at a steady load), or whenever the cylinder head is removed.

FUEL SYSTEM

Carburetors

There are three types of carburetors used on Briggs & Stratton engines. They are the Pulsa-Jet, Vacu-Jet and Flo-Jet. The first two types have three models each and the Flo-Jet has two versions.

Before removing any carburetor for repair, look for signs of air leakage or mounting gaskets that are loose, have deteriorated, or are otherwise damaged.

Note the position of the governor springs, governor link, remote control, or other attachments to facilitate reassembly. Be careful not to bend the links or stretch the springs.

The two types of Flo-Jet carburetors

The three types of Pulsa-Jet carburetors

The three types of Vacu-Jet carburetors

DIAPHRAGM

Automatic choke system

AUTOMATIC CHOKE

The automatic choke operates in conjunction with engine vacuum, similar to the Pulsa-Jet fuel pump.

A diaphragm under the carburetor is connected to the choke shaft by a link. A calibrated spring under the diaphragm holds the choke closed when the engine is not running. Upon starting, vacuum created during the intake stroke is routed to the bottom of the diaphragm through a calibrated passage, thereby opening the choke.

This system also has the ability to respond in the same manner as an accelerator pump. As speed decreases during heavy loads, the choke valve partially closes, enriching the air/fuel mixture, thereby improving low speed performance and lugging power.

To check the automatic choke, remove the air cleaner and replace the stud. Observe the position of the choke valve; it should be fully closed. Move the speed control to the stop position; the governor spring should be holding the throttle in a closed position. Give the starter rope several quick pulls. The choke valve should alternately open and close.

If the choke valve does not react as stated in the previous paragraph, the carburetor will have to be disassembled to determine the problem. Before doing so, however, check the following items so you know what to look for:

ENGINE IS UNDER-CHOKED
1. Carburetor is adjusted too lean.
2. The fuel pipe check valve is inoperative (Vacu-Jet only).
3. The air cleaner stud is bent.
4. The choke shaft is sticking due to dirt.
5. The choke spring is too short or damaged.
6. The diaphragm is not preloaded.

ENGINE IS OVER-CHOKED
1. Carburetor is adjusted too rich.
2. The air cleaner stud is bent.
3. The choke shaft is sticking due to dirt.
4. The diaphragm is ruptured.
5. The vacuum passage is restricted.
6. The choke spring is distorted or stretched.
7. There is gasoline or oil in the vacuum chamber.
8. There is a leak between the link and the diaphragm.
9. The diaphragm was folded during assembly, causing a vacuum leak.
10. The machined surface on the tank top is not flat.

Replacing the Automatic Choke

Inspect the automatic choke for free operation. Any sticking problems should be corrected as proper choke operation depends on freedom of the choke to travel as dictated by engine vacuum.

Remove the carburetor and fuel tank assembly from the engine. The choke link cover may now be removed and the choke link disconnected from the choke shaft. Disassemble the carburetor from the tank top, being careful not to damage the diaphragm.

NOTE

DIAPHRAGM

Assembling the diaphragm spring to the new diaphragm

CHECKING THE DIAPHRAGM AND SPRING

The diaphragm can be reused, provided it has not developed wear spots or punctures. On the Pulsa-Jet models, make sure that the fuel pump valves are not damaged. Also check the choke spring length. The Pulsa-Jet spring minimum length is $1\frac{1}{8}$ in. and the maximum is $1\frac{7}{32}$ in. Vacu-Jet spring length minimum is $\frac{15}{16}$ in., maximum length 1 in. If the spring length is shorter or longer than specified, replace the diaphragm and the spring.

Installing the diaphragm and spring into the spring pocket

CHECKING THE TANK TOP

The machined surface on the top of the tank must be flat in order for the diaphragm to provide an adequate seal between the carburetor and the tank. If the machined surface on the tank is not flat, it is possible for gasoline to enter the vacuum chamber by passing between the machined surface and the diaphragm. Once fuel has entered the vacuum chamber, it can move through the vacuum passage and into the carburetor. The flatness of the machined surface on the tank top can be checked by using a straight-edge and a feeler gauge. The surface should not vary more than 0.002 in. Replace the tank if a 0.002 in. feeler gauge can be passed under the straight-edge.

If a new diaphragm is installed, assemble the spring to the replacement diaphragm, taking care not to bend or distort the spring.

Place the diaphragm on the tank surface, positioning the spring in the spring pocket.

Place the carburetor on the diaphragm ensuring that the choke link and diaphragm are properly aligned between the carburetor and the tank top. On Pulsa-Jet models, place the pump

Checking the tank top for warpage

spring and cap on the diaphragm over the recess or pump chamber in the fuel tank. Thread in the carburetor mounting screws to about two threads. Do not tighten them. Close the choke valve and insert the choke link into the choke shaft.

Remove the air cleaner gasket, if it is in place, before continuing. Insert a 3/8 in. bolt or rod into the carburetor air horn. With the bolt in position, tighten the carburetor mounting screws in a staggered sequence. Please note that the insertion of the 3/8 in. bolt opens the choke to an over-center position, which preloads the diaphragm.

Remove the 3/8 in. bolt. The choke valve should now move to a fully closed position. If the choke valve is not fully closed, make sure that the choke spring is properly assembled to the diaphragm, and also properly inserted in its pocket in the tank top.

All carburetor adjustments should be made with the air cleaner on the engine. Adjustment is best made with the fuel tank half full. See the "Tune-Up" section.

Positioning the diaphragm on top of the fuel tank

Inserting the choke link into the choke shaft

Cutaway view of a two piece Flo-Jet carburetor

Pre-loading the diaphragm to adjust the choke

TWO PIECE FLO-JET CARBURETORS
(LARGE AND SMALL LINE)

Checking the Upper Body for Warpage

With the carburetor assembled and the body gasket in place, try to insert a 0.002 in. feeler gauge between the upper and lower bodies at the air vent boss, just below the idle valve. If the gauge can be inserted, the upper body is warped and should be replaced.

Checking the Throttle Shaft and Bushings

Wear between the throttle shaft and bushings should not exceed 0.010 in. Check the wear by placing a short iron bar on the upper carburetor body so that it just fits under the throttle shaft. Measure the distance with a feeler gauge while holding the shaft down and then holding it up. If the difference is over 0.010 in., either the upper body should be rebushed, the throttle shaft replaced, or both. Wear on the throttle shaft can be checked by comparing the worn and unworn portions of the shaft. To replace the bushings, remove the throttle shaft using a thin punch

Checking throttle shaft wear with a feeler gauge

Removing the throttle shaft bushing

to drive out the pin which holds the throttle stop to the shaft; remove the throttle valve, then pull out the shaft. Place a ¼ in. x 20 tap or an E-Z Out in a vise. Turn the carburetor body so as to thread the tap or E-Z Out into the bushings enough to pull the bushings out of the body. Press the new bushings into the carburetor body with a vise. Insert the throttle shaft to be sure it is free in the bushings. If not, run a size $7/32$ in. drill through both bushings to act as a line reamer. Install the throttle shaft, valve, and stop.

Disassembly of the Carburetor

1. Remove the idle valve.

Two piece Flo-Jet carburetor

2. Loosen the needle valve packing nut.

3. Remove the packing nut and needle valve together. To remove the nozzle, use a narrow, blunt screwdriver so as not to damage the threads in the lower carburetor body. The nozzle projects diagonally into a recess in the upper body and must be removed before the upper body is separated from the lower body, or it may be damaged.

4. Remove the screws which hold the upper and lower bodies together. A pin holds the float in place.

5. Remove the pin to take out the float valve needle. Check the float for leakage. If it contains gasoline or is crushed, it must be replaced. Use a wide, proper fitting screwdriver to remove the float inlet seat.

6. Lift the venturi out of the lower body. Some carburetors have a welch plug. This should be removed only if necessary to remove the choke plate. Some carburetors have nylon choke shaft.

Repair

Use new parts where necessary. Always use new gaskets. Carburetor repair kits are available. Tighten the inlet seat with the gasket securely in place, if used. Some float valves have a spring clip to connect the float valve to the float tang. Others are nylon with a stirrup which fits over the float tang. Older float valves and engines with fuel pumps have neither a spring nor a stirrup.

A viton tip float valve is used in later models of the large, two-piece Flo-Jet carburetor. The seat is pressed into the upper body and does not need replacement unless it is damaged.

Replacing the Pressed-In Float Valve Seat

Clamp the head of a #93029 self threading screw in a vise. Turn the carburetor body to thread the screw into the seat. Continue turning the carburetor body, drawing out the seat. Leave the seat fastened to the screw. Insert the new seat #230996 into the carburetor body. The seat has a starting lead.

Replacing the float valve seat

NOTE: If the engine is equipped with a fuel pump, install a #231019 seat. Press the new seat flush with the body using the screw and old seat as a driver. Make sure that the seat is not pressed below the body surface or improper float-to-float valve contact will occur. Install the float valve.

Checking the Float Level

With the body gasket in place on the upper body and the float valve and float installed, the float should be parallel to the body mounting surface. If not, bend the tang on the float until they are parallel. Do not press on the flat to adjust it.

Assembly of the Carburetor

Assemble the venturi and the venturi gasket to the lower body.

Be sure that the holes in the venturi and the venturi gasket are aligned. Some models do not have a removable venturi. Install the choke parts and welch plug if previously removed. Use a sealer around the welch plug to prevent entry of dirt.

Fasten the upper and lower bodies together with the mounting screws. Screw in the nozzle with a narrow, blunt screwdriver, making sure that the nozzle tip enters the recess in the upper body. Tighten the nozzle securely. Screw in the needle valve and idle valve until they just seat. Back off the needle valve $1\frac{1}{2}$ turns. Do not tighten the packing nut. Back off the idle valve $\frac{3}{4}$ of a turn. These settings are about correct. Final adjustment will be made when the engine is running. See the Tune-Up section for mixture and choke adjustments.

ONE-PIECE FLO-JET CARBURETOR

The large, one-piece Flo-Jet carburetor has its high speed needle valve below the float bowl. All other repair procedures are similar to the small, one-piece Flo-Jet carburetor.

Disassembly

1. Remove the idle and needle valves.

2. Remove the carburetor bowl screw. A pin holds the float in place.

3. Remove the pin to take off the float and float valve needle. Check the float for leakage. If it contains gasoline or is crushed, it must be replaced. Use a screwdriver to remove the carburetor

Small one piece Flo-Jet carburetor

Large one piece Flo-Jet carburetor

nozzle. Use a wide, heavy screwdriver to remove the float valve seat, if used.

If it is necessary to remove the choke valve, venturi throttle shaft, or shaft bushings, proceed as follows.

1. Pry out the welch plug.
2. Remove the choke valve, then the shaft. The venturi will then be free to fall out after the choke valve and shaft have been removed.
3. Check the shaft for wear. (Refer to the "Two-Piece Flo-Jet Carburetor" section for checking wear and replacing bushings).

Repair of the Carburetor

Use new parts where necessary. Always use new gaskets. Carburetor repair kits are available. If the venturi has been removed, install the venturi first, then the carburetor nozzle jets. The nozzle jet holds the venturi in place. Replace the choke shaft and valve. Install a new welch plug in the carburetor body. Use a sealer to prevent dirt from entering.

A viton tip float valve is used in the large, one-piece Flo-Jet carburetor. The seat is pressed in the upper carburetor body and does not need replacement unless it is damaged. Replace the seat in the same manner as for the two-piece Flo-Jet carburetor.

Checking the Float Level

With the body gasket in place on the upper body and float valve and the float installed, the float should be parallel to the body mounting surface. If not, bend the tang on the float until they are parallel. Do not press on the float.

Install the float bowl, idle valve, and needle valve. Turn in the needle valve and the idle valve until they just seat. Open the needle valve $2\frac{1}{2}$ turns and the idle valve $1\frac{1}{2}$ turns. On the large carburetors with the needle valve below the float bowl, open the needle valve and the idle valve $1\frac{1}{8}$ turns.

These settings will allow the engine to start. Final adjustment should be made when the engine is running and has warmed up to operating temperature. See the "Two-Piece Flo-Jet Carburetor" adjustment procedure.

Float Level Chart

Carburetor Number	Float Setting (in.)
2712-S	$^{19}/_{64}$
2713-S	$^{19}/_{64}$
2714-S	$\frac{1}{4}$
*2398-S	$\frac{1}{4}$
2336-S	$\frac{1}{4}$
2336-SA	$\frac{1}{4}$
2337-S	$\frac{1}{4}$
2337-SA	$\frac{1}{4}$
2230-S	$^{17}/_{64}$
2217-S	$^{11}/_{64}$

*When resilient seat is used, set float level at $^9/_{32} \pm ^1/_{64}$.

Governors

The purpose of a governor is to maintain, within certain limits, a desired engine speed even though the load may vary.

AIR VANE GOVERNORS

The governor spring tends to open the throttle. Air pressure against the air vane tends to close the throttle. The engine speed at which these two forces balance is called the governed speed. The governed speed can be varied by changing the governor spring tension.

Worn linkage or damaged governor springs should be replaced to insure proper governor operation. No adjustment is necessary.

Air vane governor installed on horizontal crankshaft engine

Air vane governor installed on vertical crankshaft engine

Mechanical governor exploded view

MECHANICAL GOVERNORS

The governor spring tends to pull the throttle open. The force of the counterweights, which are operated by centrifugal force, tends to close the throttle. The engine speed at which these two forces balance is called the governed speed. The governed speed can be varied by changing the governor spring tension.

GOVERNOR REPAIR

The procedures below describe disassembly and assembly of the various kinds of mechanical governors. Look for gears with worn or broken teeth, worn thrust washers, weight pins, cups, followers, etc. Replace parts that are worn and reassemble.

Model 140000
DISASSEMBLY

1. Loosen the governor lever mounting screw, and pull the lever off the shaft.
2. Remove the two housing mounting screws. Carefully pull the housing off the block, being careful to catch the governor gear, which will slip off the shaft. Pull the steel thrust washer off the shaft.
3. Remove the governor lever roll pin and washer. Unscrew the governor lever shaft by turning it clockwise and remove it.

Assembling the mechanical governor

ASSEMBLY

1. Push the governor lever shaft into the crankcase cover, threaded end first. Assemble the small washer onto the inner end of the shaft, and then screw the shaft into the governor crank follower by turning it counterclockwise. Tighten it securely.
2. Turn the shaft until the follower points down slightly, in a position where it would press against the cup when the housing is installed.
3. Place the washer on the outside end of the shaft. Install the rollpin, so the leading end just reaches the outside diameter of the shaft and the back end protrudes.
4. Install the thrust washer and the governor gear on the shaft in the housing (in that order).
5. Hold the crankcase cover in a vertical (the normal) position and install the housing with the gear in position so the point of the steel cup on the gear contacts the follower. Install and tighten the housing mounting screws.
6. Install the lever on the shaft pointing downward at an angle of about 30 degrees. Adjust as described in the Tune-Up section.

Cast Iron Models 19, 190000, 20000, 23, 230000, 240000, 300000 and 320000
DISASSEMBLY

1. Remove the cotter key and washer from the outer end of the governor shaft. Remove the governor crank from inside the crankcase.
2. Slide the governor gear off the shaft.

ASSEMBLY

1. Install the governor gear onto the shaft inside the crankcase. Then, insert the governor shaft assembly through the bushing from inside the crankcase.

2. Install the governor lever to the shaft loosely, and then adjust it as described in the Tune-Up section.

Aluminum Models 140000, 170000, 190000 and 251000
DISASSEMBLY

On horizontal shaft models: Remove the governor assembly as a unit from the crankcase cover.

On vertical shaft models: Remove the entire assembly as part of the oil slinger (see the Overhaul Section).

ASSEMBLY

1. Assemble governors on horizontal crankshaft models with crankshaft in a horizontal position. The governor rides on a short stationary shaft which is integral with the crankcase cover. The governor shaft keeps the governor from sliding off the shaft after the cover is installed. The governor shaft must hang straight down, or it may jam the governor assembly when the crankcase cover is installed, breaking it when the engine is started. The governor shaft adjustment should be made (see the Tune-Up section) as soon as the crankcase cover is in place so that the governor lever will be clamped in the proper position.

Assembling cover with governor and governor shaft in proper position (horizontal shaft engines)

Adjusting the breaker point gap

2. On both horizontal and vertical crankshaft models, the governor is held together through normal operating forces. For this reason, the governor link and all other external linkages must be in place and properly adjusted whenever the engine is operated.

ENGINE OVERHAUL
Cylinder Head
REMOVAL AND INSTALLATION

Always note the position of the different cylinder head screws so that they can be properly reinstalled. If a screw is used in the wrong position, it may be to short and not engage enough threads. If it is too long, it may bottom on a fin, either breaking the fin, or leaving the cylinder head loose.

MODELS 9-14-19-190000-200000

ALL ALUMINUM CYLINDER ENGINES LONG SCREWS IN THESE 3 HOLES

MODELS 23-230000 240000-300000-320000

Cylinder head bolt tightening sequences

Remove the cylinder screws and then the cylinder head. Be sure to remove the gasket and all remaining gasket material from the cylinder head and the block.

Assemble the cylinder head with a new gasket, cylinder head shield, screws, and washers in their proper places. Graphite grease should be used on aluminum cylinder head screws.

Do not use a sealer of any kind on the head gasket. Tighten the screws down evenly by hand. Use a torque wrench and tighten the head bolts in the correct sequence.

Valves

REMOVAL AND INSTALLATION

Using a valve spring compressor, adjust the jaws so they touch the top and bottom of the valve chamber, and then place one of the jaws over the valve spring and the other underneath, between the spring and the valve chamber. This positioning of the valve spring compressor is for valves that have either pin or collar type retainers. Tighten the jaws to compress the spring. Remove the collars or pin and lift out the valve. Pull out the compressor and the spring.

To remove valves with ring type retainers, position the compressor with the upper jaw over the top of the valve chamber and the lower jaw between the spring and the retainer. Com-

Cylinder Head Bolt Torque Specifications

Basic Model Series	In. lbs. Torque
Aluminum Cylinder	
140000, 170000, 190000, 251000	165
Cast Iron Cylinder	
19, 190000, 200000, 23, 230000, 240000, 300000, 320000	190

The three types of valve spring retainers

Removing the valve springs with the help of a valve spring compressor

press the spring, remove the retainer, and pull out the valve. Remove the compressor and spring.

Before installing the valves, check the thickness of the valve springs. Some engines use the same spring for the intake and exhaust side, while others use a heavier spring on the exhaust side. Compare the springs before installing them.

If the retainers are held by a pin or collars, place the valve spring and retainer and cup (Models 14–19–20–23–24–32) into the valve spring compressor. Compress the spring until it is solid. Insert the compressed spring and retainer into the valve chamber. Then drop the valve into place, pushing the stem through the retainer. Hold the spring up in the chamber, hold the valve down, and insert the retainer pin with needle nose pliers or place the collars in the groove in the valve stem. Loosen the spring until the retainer fits around the pin or collars, then pull out the spring compressor. Be sure the pin or collars are in place.

To install valves with ring type retainers, compress the retainer and spring with the compressor. The large diameter of the retainer should be toward the front of the valve chamber. Insert the compressed spring and retainer into the valve chamber. Drop the valve stem through the larger area of the retainer slot and move the compressor so as to center the small area of the valve retainer slot onto the valve stem shoulder. Release the spring tension and remove the compressor.

Valve Guides

REMOVAL AND INSTALLATION

First check valve guide for wear with a plug gauge, Briggs & Stratton part #19151 or equivalent. If the flat end of the valve guide plug gauge can be inserted into the valve guide for a distance of $5/16$ in., the guide is worn and should be rebushed in the following manner.

See the illustration.

Procure a reamer #19183 and reamer guide bushing #19192, and lubricate the reamer with kerosene. Then, use reamer and reamer guide bushing to ream out the worn guide. Ream to only $1/16$ in. deeper than valve guide bushing #230655 . BE CAREFUL NOT TO REAM THROUGH THE GUIDE.

Press in valve guide bushing #230655 until top end of bushing is flush with top end of valve guide. Use a soft metal driver (brass, copper, etc.) so top end of bushing is not peened over.

The bushing #230655 is finish reamed to size at the factory, so no further reaming is necessary, and a standard valve can be used.

Bushing the valve guide

Bushing the valve guide

— CAUTION —
Valve seating should be checked after bushing the guide, and corrected if necessary by refacing the seat.

REFACING VALVES AND SEATS

Faces on valves and valve seats should be resurfaced with a valve grinder or cutter to an angle of 45 degrees.

NOTE: Some engines have a 30 degree intake valve and seat.

The valve and seat should then be lapped with a fine lapping compound to remove the grinding marks and ensure a good seat. The valve seat width should be $3/64$–$1/16$ in. If the seat is wider, a narrowing stone or cutter should be used. If either the seat or valve is badly burned, it should be replaced. Replace the valve if the edge thickness (margin) is less than $1/64$ in. after it has been resurfaced.

CHECK AND ADJUST TAPPET CLEARANCE

Insert the valves in their respective positions in the cylinder. Turn the crankshaft until one of the valves is at its highest position. Turn the crankshaft one revolution. Check the clearance with a feeler gauge. Repeat for the other valve. Grind off the end of the valve stem if necessary to obtain proper clearance.

Valve Tappet Clearance Chart

Model Series	Intake		Exhaust	
	Max	Min	Max	Min
Aluminum Cylinder				
140000, 170000, 190000, 251000	.007	.005	.011	.009
Cast Iron Cylinder				
19, 190000, 200000	.009	.007	.016	.014
23, 230000, 240000, 300000, 320000	.009	.007	.019	.017

NOTE: Check the valve tappet clearance with the engine cold.

Valve Seat Inserts

Cast iron cylinder engines are equipped with an exhaust valve insert which can be removed and replaced with a new insert. The intake side must be counter-bored to allow the installation of an intake valve seat insert (see below). Aluminum alloy cylinder models are equipped with inserts on both the exhaust and intake valves.

REMOVAL AND INSTALLATION

Valve seat inserts are removed with a special puller.

NOTE: On aluminum alloy cylinder models, it may be necessary to grind the puller nut until the edge is $1/32$ in. thick in order to get the puller nut under the valve insert.

When installing the valve seat insert, make sure that the side with the chamfered outer edge goes down into the cylinder. Install the seat insert and drive it into place with a driver. The seat should then be ground lightly and the valves and seats lapped lightly with grinding compound.

NOTE: Aluminum alloy cylinder models use the old insert as a spacer between the driver and the new insert. Drive in the new insert until it bottoms. The top of the insert will be slightly below the cylinder head gasket surface. Peen around the insert using a punch and hammer.

NOTE: The intake valve seat on cast iron cylinder models has to be counter-bored before installing the new valve seat insert.

COUNTERBORING CYLINDER FOR INTAKE VALVE SEAT ON CAST IRON MODELS

1. Select the proper seat insert, cutter shank, counter bore cutter, pilot and driver from the table. These numbers refer to Briggs & Stratton parts — you may get equivalent parts from other sources if available.

Installing valve seat inserts

2. With cylinder head resting on a flat surface, valve seats up, slide the pilot into the intake valve guide. Then, assemble the correct counterbore cutter to the shank with the cutting blades of the cutter downward.
3. Insert the cutter straight into the valve seat, over the pilot. Cut so as to avoid forcing the cutter to one side, and be sure to stop as soon as the stop on the cutter touches the cylinder head.
4. Blow out all cutting chips thoroughly.

Pistons, Piston Rings, and Connecting Rods

REMOVAL

To remove the piston and connecting rod from the engine, bend down the connecting rod lock. Remove the connecting rod cap. Remove any carbon or ridge at the top of the cylinder bore. This will prevent breaking the rings. Push the piston and rod out of the top of the cylinder.

Valve Seat Inserts Chart

Basic Model Series	Intake Standard	Exhaust Standard	Exhaust Stellite	Insert # Puller Assembly	Puller Nut
Aluminum Cylinder					
140000, 170000, 190000	211661	211661	210940*	19138	19141
250000	211661	211661	210940	19138	19141
Cast Iron Cylinder					
19, 190000	21880	21880	21612	19138	19141
200000, 23, 230000	21880	21880	21612	19138	19141
240000	21880	21612	21612	19138	19141
300000, 320000		21612	21612	19138	19141

*21191 used before serial #5810060 — 210808 used from serial #5810060 — 6012010
Includes puller and #19182, 19141, 19140 and 19139 nuts

Removing the connecting rod cap

Removing the wrist pin and connecting rod from the piston

Pistons used in sleeve bore, aluminum alloy engines are marked with an **L** on top of the piston. These pistons are tin plated and use an expander with the oil ring. This piston assembly is not interchangeable with the piston used in the aluminum bore engines (Kool bore).

Pistons used in aluminum bore (Kool bore) engines are not marked on the top.

To remove the connecting rod from the piston, remove the piston pin lock with thin nose pliers. One end of the pin is drilled to facilitate removal of the lock.

Remove the rings one at a time, slipping them over the ring lands. Use a ring expander to remove the rings.

INSPECTION

Check the piston ring fit. Use a feeler gauge to check the side clearance of the top ring. Make sure that you remove all carbon from the top ring groove. Use a new piston ring to check the side clearance. If the cylinder is to be resized, there is no reason to check the piston, since a new oversized piston assembly will be installed. If the side clearance is more than 0.007 in., the piston is excessively worn and should be replaced.

Check the piston ring end gap by cleaning all carbon from the ends of the rings and inserting them one at a time 1 in. down into the cylinder. Check the end gap with a feeler gauge. If the gap is larger than recommended, the ring should be replaced.

NOTE: When checking the ring gap, do not deglaze the cylinder walls by installing piston rings in aluminum cylinder engines.

Chrome ring sets are available for all current aluminum and cast iron cylinder models. No honing or deglazing is required. The cylinder bore can be a maximum of 0.005 in. oversize when using chrome rings.

Replacing the piston rings

If the crankpin bearing in the rod is scored, the rod must be replaced. 0.005 in. oversize piston pins are available in case the connecting rod and piston are worn at the piston pin bearing. If, however, the crankpin beariang in the connecting rod is worn, the rod should be replaced. Do not attempt to file or fit the rod.

If the piston pin is worn 0.0005 in out of round or below the rejection sizes, it should be replaced.

INSTALLATION

The piston pin is a push fit into both the piston and the connecting rod. On models using a solid piston pin, one end is flat and the other end is recessed. Other models use a hollow piston pin. Place a pin lock in the groove at one side of the piston. From the opposite side of the piston, insert the piston pin, flat end first for solid pins; with hollow pins, insert either end first until it stops against the pin lock. Use thin nose pliers to assemble the pin lock in the recessed end of the piston. Be sure the locks are firmly set in the groove.

Measuring piston ring side gap

Measuring the piston ring gap

Install the rings on the pistons, using a piston ring expander. Make sure that they are installed in the proper position. The scraper groove on the center compression ring should always be down toward the piston skirt. Be sure the oil return holes are clean and all carbon is removed from the grooves.

NOTE: Install the expander under the oil ring in sleeve bore aluminum alloy engines.

Oil the rings and the piston skirt, then compress the rings with a ring compressor. On cast iron engines, install the compressor with the two projections downward; on aluminum engines, install the compressor with the two projections upward. These instructions refer to the piston in normal position—with skirt downward. Turn the piston and compressor upside down on the bench and push downward so the piston head and the edge of the compressor band are even, all the while tightening the compressor. Draw the compressor up tight to fully compress the rings, then loosen the compressor very slightly.

─── **CAUTION** ───

Do not attempt to install the piston and ring assembly without using a ring compressor.

Place the connecting rod and piston assembly, with the rings compressed, into the cylinder bore. Push the piston and rod

Assembling the piston and connecting rod assembly

Cross-sectional views and positioning of the various types of piston rings used in Briggs and Stratton engines

Connecting Rod Bearing Specifications

Basic Model Series	Crank Pin Bearing	Piston Pin Bearing
Aluminum Cylinder		
140000, 170000	1.095	.674
190000	1.127	.674
251000	1.252	.802
Cast Iron Cylinder		
19, 190000	1.001	.674
200000	1.127	.674
23, 230000	1.189	.736
240000	1.314	.674
300000, 320000	1.314	.802

Piston Ring Gap Specifications

Basic Model Series	Comp. Ring	Oil Ring
Aluminum Cylinder		
140000, 170000, 190000, 251000	.035	.045
Cast Iron Cylinder		
19, 190000, 200000, 23, 230000, 240000 300000, 320000	.035	.035

Wrist Pin Specifications

Basic Model Series	Piston Pin	Pin Bore
Aluminum Cylinder		
140000, 170000, 190000	.671	.671
251000	.799	.801
Cast Iron Cylinder		
19, 190000	.671	.673
200000	.671	.673
23, 230000	.734	.736
240000	.671	.673
300000, 320000	.799	.801

down into the cylinder. Oil the crankpin of the crankshaft. Pull the connecting rod against the crankpin and assemble the rod cap so the assembly marks align.

NOTE: **Some rods do not have assembly marks, as the rod and cap will fit together only in one position. Use care to ensure proper installation. On the 251000 engine, the piston has a notch on the top surface. The notch must face the flywheel side of the block when installed. On models 300000 and 320000, the piston has an identification mark "F" located next to the piston pin bore. The mark must appear on the same side as the assembly mark on the rod. The assembly mark on the rod is also used to identify rod and cap alignment. Note, on these pistons, that the top ring has a beveled upper surface on the outside, while the center ring has a flat outer surface. The "F" mark or notch must face the flywheel when the piston is installed.**

Where there are flat washers under the cap screws, remove and discard them prior to installing the rod. Assemble the cap screws and screw locks with the oil dippers (if used), and torque to the figure shown in the chart to avoid breakage or rod scoring later. Turn the crankshaft two revolutions to be sure the rod is correctly installed. If the rod strikes the camshaft, the connecting rod has been installed wrong or the cam gear is out of time. If the crankshaft operates freely, bend the cap screw locks against the screw heads. After tightening the rod screws, the rod should be able to move sideways on the crankpin of the shaft.

Installing the piston and connecting rod assembly into the cylinder block

Connecting rod installation

Connecting Rod Capscrew Torque

Basic Model Series	Inch lbs Avg. Torque
Aluminum Cylinder	
140000, 170000, 190000	165
251000	185
Cast Iron Cylinder	
19, 190000, 200000	190
23, 230000	190
240000, 300000, 320000	190

Crankshaft and Camshaft Gear

REMOVAL

Aluminum Cylinder Engines

To remove the crankshaft from aluminum alloy engines, remove any rust or burrs from the power take-off end of the crankshaft. Remove the crankcase cover or sump. If the sump or cover sticks, tap it lightly with a soft hammer on alternate sides near the dowel. Turn the crankshaft to align the crankshaft and camshaft timing marks, lift out the cam gear, then remove the crankshaft. On models that have ball bearings on the crankshaft, the crankshaft and the camshaft must be removed together with the timing marks properly aligned – see illustration.

Cast Iron Cylinder Models

To remove the crankshaft from cast iron models, remove the crankcase cover. Revolve the crankshaft until the crankpin is pointing upward toward the breather at the rear of the engine (approximately a 45 degree angle). Pull the crankshaft out from the drive side, twisting it slightly if necessary. On models with ball bearings on the crankshaft, both the crankcase cover and bearing support should be removed.

Alignment of the camshaft and crankshaft timing marks

On cast iron models with ball bearings on the drive side, first remove the magneto. Drive into the recess at the front of the engine. Then draw the crankshaft from the magneto side of the engine. Double thrust engines have cap screws inside the crankcase which hold the bearing in place. These must be removed before the crankshaft can be removed.

To remove the camshaft from all cast iron models, except the 300400 and 320400, use a long punch to drive the camshaft out toward the magneto side. Save the plug. Do not burr or peen the end of the shaft while driving it out. Hold the camshaft while driving it out. Hold the camshaft while removing the punch, so it will not drop and become damaged.

CHECKING THE CRANKSHAFT

Discard the crankshaft if it is worn beyond the allowable limit. Check the keyways for wear and make sure they are not spread. Remove all burrs from the keyway to prevent scratching the bearing. Check the three bearing journals, drive end, crankpin, and magneto end, for size and any wear or damage. Check the cam gear teeth for wear. They should not be worn at all. Check the threads at the magneto end for damage. Make sure that the crankshaft is straight.

NOTE: There are 0.020 in. undersize connecting rods available for use on reground crankpin bearings.

Crankshaft Specifications

Basic Model Series	PTO Journal	Mag. Journal	C Crankpin
Aluminum Cylinder			
140000, 170000	1.179	.997 #	1.090
190000	1.179	.997 #	1.122
251000	1.376	1.376	1.247
Cast Iron Cylinder			
19, 190000	1.179	1.179	.996
200000	1.179	1.179	1.122
23, 230000 †	1.376	1.376	1.184
240000	Ball	Ball	1.309
300000, 320000	Ball	Ball	1.309

Synchro balanced magneto bearing reject size— 1.179
†Gear reduction P.T.O.—1.179

Removal and Installation of the Ball Bearings

The ball bearings are pressed onto the crankshaft. If either the bearing or the crankshaft is to be removed, use an arbor press to remove them.

To install, heat the bearing in hot oil (325°F maximum). Don't let the bearing rest on the bottom of the pan in which it is heated. Place the crankshaft in a vise with the bearing side up. When the bearing is quite hot, it will slip fit onto the bearing journal. Grasp the bearing, with the shield down, and thrust it down onto the crankshaft. The bearing will tighten on the shaft while cooling. Do not quench the bearing (throw water on it to cool it).

Alignment of the camshaft and crankshaft timing marks on engines equipped with ball bearings

CHECKING THE CAMSHAFT GEAR

Inspect the teeth for wear and nicks. Check the size of the camshaft and camshaft gear bearing journals. Check the size of the cam lobes. If the cam is worn beyond tolerance, discard it.

Check the automatic spark advance on models equipped with the Magna-Matic ignition system. Place the cam gear in the normal operating position with the movable weight down. Press the weight down and release it. The spring should lift the weight. If not, the spring is stretched or the weight is binding.

Camshaft Specifications

Basic Model Series	Cam Gear or Shaft Journals	Cam Lobe
Aluminum Cylinder		
140000, 170000, 190000	.498	.977
251000	.498	1.184
Cast Iron Cylinder		
19, 190000	.497	1.115
200000	.497	1.115
23, 230000	.497	1.184
240000	.497	1.184
300000	#	1.184
320000	#	1.215

Magneto side— .8105, P.T.O. side— .6145

INSTALLATION

Aluminum Alloy Engines – Plain Bearing

In aluminum alloy engines, the tappets are inserted first, the crankshaft next, and then the cam gear. When inserting the cam gear, turn the crankshaft and the cam gear so that the timing marks on the gears align.

Aluminum Alloy Engines – Ball Bearing

On crankshafts with ball bearings, the gear teeth are not visible for alignment of the timing marks; therefore, the timing mark is on the counterweight. On ball bearing equipped engines, the tappets are installed first. The crankshaft and the cam gear must be inserted together and their timing marks aligned.

Crankshaft Cover and Crankshaft

INSTALLATION

Cast Iron Engines – Plain Bearings

Assemble the tappets to the cylinder, then insert the cam gear. Push the camshaft into the camshaft hole in the cylinder from the flywheel side through the cam gear. With a blunt punch, press or hammer the camshaft until the end is flush with the outside of the cylinder on the power takeoff side. Place a small amount of sealer on the camshaft plug, then press or hammer it into the camshaft hole in the cylinder at the flywheel side. Install the crankshaft so the timing marks on the teeth and on the cam gear align.

Cast Iron – Ball Bearings

Assemble the tappets, then insert the cam gear into the cylinder, pushing the cam gear forward into the recess in front of the cylinder. Insert the crankshaft into the cylinder. Turn the camshaft and crankshaft until the timing marks align, then push the cam gear back until it engages the gear on the crankshaft with the timing marks together. Insert the camshaft. Place a small amount of sealer on the camshaft plug and press or hammer it into the camshaft hole in the cylinder at the flywheel side.

Crankshaft End-Play Adjustment

The crankshaft end-play on all models, plain and ball bearing, should be 0.002–0.008 in. The method of obtaining the correct end-play varies, however, between cast iron, aluminum, plain, and ball bearing models. New gasket sets include three crankcase cover or bearing support gaskets, 0.005 in., 0.009 in. and 0.015 in. thick.

The end-play of the crankshaft may be checked by assembling a dial indicator on the crankshaft with the pointer against the crankcase. Move the crankshaft in and out. The indicator will show the end-play. Another way to measure the end-play is to assemble a pulley to the crankshaft and measure the end-play with a feeler gauge. Place the feeler gauge between the crankshaft thrust face and the bearing support. The feeler gauge

method of measuring crankshaft end-play can only be used on cast iron plain bearing engines with removable bases.

On cast iron engines, the end-play should be 0.002–0.008 in. with one 0.015 in. gasket in place. If the end-play is less than 0.002 in., which would be the case if a new crankcase or sump cover is used, additional gaskets of 0.005 in., 0.009 in. or 0.015 in. may be added in various combinations to obtain the proper end-play.

ALUMINUM ENGINES ONLY

If the end-play is more than 0.008 in. with one 0.015 in. gasket in place, a thrust washer is available to be placed on the crankshaft power take-off end, between the gear and crankcase cover or sump on plain bearing engines. On ball bearing equipped aluminum engines, the thrust washer is added to the magneto end of the crankshaft instead of the power take-off end.

NOTE: Aluminum engines never use less than the 0.015 in. gasket.

Cylinders

INSPECTION

Always inspect the cylinder after the engine has been disassembled. Visual inspection will show if there are any cracks, stripped bolt holes, broken fins, or if the cylinder wall is scored. Use an inside micrometer or telescoping gauge and micrometer to measure the size of the cylinder bore. Measure at right angles.

If the cylinder bore is more than 0.003 in. oversize, or 0.0015 in. out of round on light-weight (aluminum) cylinders, the cylinder must be resized (rebored).

NOTE: Do not deglaze the cylinder walls when installing piston rings in aluminum cylinder engines. Also be aware that there are chrome ring sets available for most engines. These are used to control oil pumping in bores worn to 0.005 in. over standard and do not require honing or glaze breaking to seat.

RESIZING

Always resize to exactly 0.010 in., 0.020 in. or 0.030 in. over standard size. If this is done accurately, the stock oversize rings and pistons will fit perfectly and proper clearances will be maintained. Cylinders, either cast iron or lightweight, can be quickly resized with a good hone. Use the stones and lubrication recommended by the hone manufacturer to produce the correct cylinder wall finish for the various engine models.

If a boring bar is used, a hone must be used after the boring operation to produce the proper cylinder wall finish. Honing can be done with a portable electric drill, but it is easier to use a drill press.

1. Clean the cylinder at top and bottom to remove all burrs and pieces of base and head gaskets.

2. Fasten the cylinder to a heavy iron plate. Some cylinders require shims. Use a level to align the drill press spindle with the bore.

MEASURE AT SIX POINTS

CENTER OF PISTON RING TRAVEL

Checking the cylinder bore

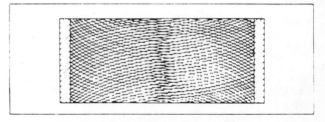

Cross hatch pattern after honing

3. Oil the surface of the drill press table liberally. Set the iron plate and the cylinder on the drill press table. Do not anchor the cylinder to the drill press table. If you are using a portable drill, set the plate and the cylinder on the floor.

4. Place the hone driveshaft in the chuck of the drill.

5. Slip the hone into the cylinder. Connect the driveshaft to the hone and set the stop on the drill press so the hone can only extend $3/4$–1 in. from the top or bottom of the cylinder. If you are using a portable drill, cut a piece of wood to place in the cylinder as a stop for the hone.

6. Place the hone in the middle of the cylinder bore. Tighten the adjusting knob with your finger or a small screwdriver until the stones fit snugly against the cylinder wall. Do not force the stones against the cylinder wall. The hone should operate at a speed of 300–700 rpm. Lubricate the hone as recommended by the manufacturer.

NOTE: Be sure that the cylinder and the hone are centered and aligned with the driveshaft and the drill spindle.

7. Start the drill and, as the hone spins, move it up and down at the lower end of the cylinder. The cylinder is not worn at the bottom but is round so it will act to guide the hone and straighten the cylinder bore. As the bottom of the cylinder increases in diameter, gradually increase your strokes until the hone travels the full length of the bore.

NOTE: Do not extend the hone more than $3/4$–1 in. past either end of the cylinder bore.

8. As the cutting tension decreases, stop the hone and tighten the adjusting knob. Check the cylinder bore frequently with an accurate micrometer. Hone 0.0005 in. oversize to allow for shrinkage when the cylinder cools.

9. When the cylinder is within 0.0015 in. of the desired size, change from the rough stone to a finishing stone.

The finished resized cylinder should have a cross-hatched appearance. Proper stones, lubrication, and spindle speed along with rapid movement of the hone within the cylinder during the last few strokes, will produce this finish. Cross-hatching provides proper lubrication and ring break-in.

NOTE: It is EXTREMELY important that the cylinder be thoroughly cleaned after honing to eliminate ALL grit. Wash the cylinder carefully in a solvent such as kerosene. The cylinder bore should be cleaned with a brush, soap, and water.

Bearings
INSPECTION
Plain Type

Bearings should be replaced if they are scored or if a plug gauge will enter. Try the gauge at several points in the bearings.

REPLACING PLAIN BEARINGS
Models 19–20–23

The crankcase cover bearing support should be replaced if the bearing is worn or scored.

REPLACING THE MAGNETO BEARING
Aluminum Cylinder Engines

There are no removable bearings in these engines. The cylinder must be reamed out so a replacement bushing can be installed.

1. Place a pilot guide bushing in the sump bearing, with the flange of the guide bushing toward the inside of the sump.

2. Assemble the sump on the cylinder. Make sure that the pilot guide bushing does not fall out of place.

3. Place the guide bushing into the oil seal recess in the cylinder. This guide bushing will center the counterbore reamer even though the oil bearing surface might be badly worn.

4. Place the counterbore reamer on the pilot and insert them into the cylinder until the tip of the pilot enters the pilot guide bushing in the sump.

5. Turn the reamer clockwise with a steady, even pressure until it is completely through the bearing. Lubricate the reamer with kerosene or any other suitable solvent.

NOTE: Counterbore reaming may be performed without any lubrication. However, clean off shavings because aluminum material builds up on the reamer flutes causing eventual damage to the reamer and an oversize counterbore.

Cylinder Bore Specifications

Basic Engine Model or Series	Std. Bore Size Diameter	
	Max	Min
Aluminum Cylinder		
140000	2.750	2.749
170000, 190000	3.000	2.999
251000	3.4375	3.4365
Cast Iron Cylinder		
19, 23, 190000, 200000	3.000	2.999
230000	3.000	2.999
243400	3.0625	3.0615
300000	3.4375	3.4365
320000	3.5625	3.5615

Crankshaft Bearing Specifications

Basic Engine Model or Series	PTO Bearing	Bearing Magneto
Aluminum Cylinder		
140000, 170000	1.185	1.004
190000	1.185	1.004
251000	1.383	1.383
Cast Iron Cylinder		
19, 190000, 200000	1.185	1.185
23, 230000	1.382	1.382
240000, 300000	Ball	Ball
320000	Ball	Ball

6. Remove the sump and pull the reamer out without backing it through the bearing. Clean out the remaining chips. Remove the guide bushing from the oil seal recess.

7. Hold the new bushing against the outer end of the reamed out bearing, with the notch in the bushing aligned with the notch in the cylinder. Note the position of the split in the bushing. At a point in the outer edge of the reamed out bearing opposite to the split in the bushing, make a notch in the cylinder hub at a 45 degree angle to the bearing surface. Use a chisel or a screwdriver and hammer.

8. Press in the new bushing, being careful to align the oil notches with the driver and the support until the outer end of the bushing is flush with the end of the reamed cylinder hub.

9. With a blunt chisel or screwdriver, drive a portion of the bushing into the notch previously made in the cylinder. This is called staking and is done to prevent the bushing from turning.

10. Reassemble the sump to the cylinder with the pilot guide bushing in the sump bearing.

11. Place a finishing reamer on the pilot and insert the pilot into the cylinder bearing until the tip of the pilot enters the pilot guide bushings in the sump bearing.

12. Lubricate the reamer with kerosene, fuel oil, or other suitable solvent, then ream the bushing, turning the reamer clockwise with a steady even pressure until the reamer is completely through the bearing. Improper lubricants will produce a rough bearing surface.

13. Remove the sump, reamer, and the pilot guide bushing. Clean out all reaming chips.

REPLACING THE P.T.O. BEARING
Aluminum Cylinder Engines

The sump or crankcase bearing is repaired the same way as the magneto end bearing. Make sure to complete repair of one bearing before starting to repair the other. Press in new oil seals when bearing repair is completed.

REPLACING OIL SEALS

1. Assemble the seal with the sharp edge of leather or rubber toward the inside of the engine.

2. Lubricate the inside diameter of the seal with Lubriplate® or equivalent.

3. Press all seals so they are flush with the hub.

LUBRICATION

The primary purpose of oil is, of course, lubrication. However oil performs three other very important functions as well: cooling, cleaning, and sealing. Oil absorbs and dissipates heat created by combustion and friction. Oil cleans by trapping and holding dirt and by-products of combustion. This dirt is held in suspension by the oil until it is drained. Oil seals the combustion chamber by coating the rings, thus helping increase and maintain compression.

Briggs and Stratton engines are lubricated with a gear driven splash oil slinger or a connecting rod dipper.

Extended Oil Filler Tubes and Dipsticks

When installing the extended oil fill and dipstick assembly, the tube must be installed so the O-ring seal is firmly compressed. To do so, push the tube downward toward the sump, then tighten the blower housing screw, which is used to secure the tube and bracket. When the dipstick assembly is fully depressed, it seals the upper end of the tube.

A leak at the seal between the tube and the sump, or at the seal at the upper end of the dipstick can result in a loss of crankcase vacuum, and a discharge of smoke through the exhaust system.

Breathers

The function of the breather is to maintain a vacuum in the crankcase. The breather has a fiber disc valve which limits the direction of air flow caused by the piston moving back and forth in the cylinder. Air can flow out of the crankcase, but the one-way valve blocks the return flow, thus maintaining a vacuum in the crankcase. A partial vacuum must be maintained in the crankcase to prevent oil from being forced out of the engine at the piston rings, oil seals, breaker plunger, and gaskets.

INSPECTION OF THE BREATHER

If the fiber disc valve is stuck or binding, the breather cannot function properly and must be replaced. A 0.045 in. wire gauge should not enter the space between the fiber disc valve and the body. Use a spark plug wire gauge to check the valve. The fiber disc valve is held in place by an internal bracket which will be distorted if pressure is applied to the fiber disc valve. Therefore, do not apply force when checking the valve with the wire gauge.

If the breather is removed for inspection or valve repair, a new gasket should be used when replacing the breather. Tighten the screws securely to prevent oil leakage.

Most breathers are now vented through the air cleaner, to prevent dirt from entering the crankcase. Check to be sure that the venting elbows or the tube are not damaged and that they are properly sealed.

Oil Dippers and Slingers

Oil dippers reach into the oil reservoir in the base of the engine and splash oil onto the internal engine parts. The oil dipper is installed on the connecting rod and has no pump or moving parts.

Oil slingers are driven by the cam gear. Old style slingers using a die cast bracket assembly have a steel bushing between the slinger and the bracket. Replace the bracket on which the oil slinger rides if it is worn to a diameter of 0.490 in. or less. Replace the steel bushing if it is worn. Newer style oil slingers have a stamped steel brackets.

Checking the breather assembly

Installing the connecting rod in a horizontal crankshaft engine

Section 6

Clinton

Small Engines — 0-6 hp

ENGINE IDENTIFICATION

Clinton engines are identified by a name plate that is installed on the engine at the factory. The name plate contains the serial number and the model number, both of which are necessary when obtaining replacement parts.

Two numbering systems are used to identify Clinton engines. The first one applies to engines made prior to 1961, the second pertains to engines made after that year.

The early numbering system has no practical use to the consumer. It is useful only when ordering parts and only to one who has access to a parts manual. No other information can be gained from the model or serial number than what parts fit a particular engine.

The recent numbering system, however, is quite useful in gaining additional information about a certain engine. The serial number, followed by the type letter, is a numerically sequential number and is used to identify the engine in relation to

Prior to 1961 engine numbering system

Numbering system after 1961

changes that are made in design. The model number consists of a ten digit sequence of numbers that is deciphered in the following manner:

The first digit indicates whether the engine is a two stroke or four stroke design. The number 4 indicates a four stroke engine and the number 5 indicates a two stroke engine.

The second and third digits identify the basic engine series and whether the engine has a vertical or horizontal crankshaft. Odd numbers in the third digit position indicate that the engine has a vertical crankshaft and even numbers indicate a horizontal crankshaft.

The fourth digit identifies the type of starter installed on the engine: 0 — a recoil starter, 1 — a rope starter, 2 — an impulse starter, 3 — a crank starter, 4 — a 12 volt electric starter, 5 — a 12 volt starter generator, 6 — a 110 volt electric starter, 7 — a 12 volt generator, 8 — not assigned any specific meaning, 9 — indicates a short block.

The fifth digit indicates what type of bearing is used on the crankshaft; 1 — aluminum or bronze sleeve bearing with a flange mounting surface and pilot diameter on the engine mounting face for mounting equipment concentric to the crankshaft center line; 2 — ball or roller bearing; 3 — ball or roller bearing with a flange mounting surface and pilot diameter on the engine mounting face for mounting equipment concentric to the crankshaft center line; 4–9 – not assigned any meaning.

The sixth digit identifies whether there are any reduction gears or power-take-off units: 0 – not equipped with any such unit; 1 – auxiliary PTO; 2 – 2:1 reduction gears; 3 – not assigned; 4 – 4:1 reduction gears; 5 – not assigned; 6 – 6:1 reduction gears; 7–9 – not assigned.

The eighth through tenth digits are used to identify model variations.

The type letter identifies parts that are not interchangeable.

Be sure to give the model number and the type letter when obtaining replacement parts.

MAINTENANCE
Air Cleaner Service

METALLIC MESH AIR CLEANER

1. Loosen the air cleaner screw and remove the air cleaner.
2. Place both element and cover parts in a nonvolatile solvent, and agitate the metal mesh vigorously.
3. Dip the mesh into clean engine oil, and then install air cleaner.

DRY PAPER AIR CLEANER

1. Remove the air cleaner can from the engine.
2. Brush lightly with a soft bristle brush (not a wire brush).
3. Blow the dirt out, using compressed air. Blow from inside to outside.
4. Make sure the sealing gasket is in place, and install the air cleaner.

OIL BATH AIR CLEANER

1. Remove the air cleaner from the engine, disassemble, and soak in a nonvolatile solvent.
2. Inspect the plastic blow carefully for cracks, and replace if necessary.
3. Blow the solvent out of the mesh filter with compressed air.
4. Fill to the correct level with SAE 30 oil, and install.

POLYURETHANE AIR CLEANER

1. Remove the element from the air cleaner container, and wash it in soap and water.

Oil bath air cleaner—washing mesh in solvent

2. Dry it thoroughly. Apply enough engine oil to cover the face of the element (about 1 tablespoon), and install the air cleaner.

Oil and Fuel Recommendations
FOUR-CYCLE OIL

Follow the recommendations below. Do not use oil of ML (Mostly Light) rating, as it will void the warranty. MM oil is recommended, but best results are obtained with MS rated oil. Oil rated DG may be used, but is not particularly recommended, and DM oil is specifically not recommended. Its use will void the warranty.
- 32°F and above: SAE 30, MM or MS
- –10°F–32°F: SAE 10W, MM or MS
- Below –10°F: SAE 5W, MM or MS rating

Crankcase Capacity Chart

Model	Capacity
V–100, VS–100, A–300, A–650, VS–2100, VS–3100, VS–4100, and Models 401, 403, 405, 407, 408, 409, 411, 415, 417, 429, 431, 435	1 pint
100, 2100, 3100, H–3100, 4100, 400, 402, 404, 406, 424, 426	1¼ pints
700–A, 700–C, D–700, D–800, A–800, A–900, A–1100, B–1100, C–1100, D–1100, D–1200, A–1200, B–1290, 492, 494, 498	1¼ pints
VS–700, VS–750, VS–800, VS–900, V–1000, V–1100, VS–1100, V–1200, VS–1200, 499	¼ pints
429, 431, 435	1¾ pints

TWO-CYCLE OIL

Use a good quality outboard (two-stroke) motor oil, rated MM or MS. Do not use oils rated DM or DS. Use SEA 30 or SAE 40 viscosity.

On sleeve bearing engines, mix the oil in the proportions: ¾ pint to each gallon of gasoline. On needle bearing engines, use ½ pint to each gallon of gasoline.

NOTE: On outboard motors, during the first five hours of operation (break-in), mix ½ pint of oil to each gallon of gasoline. After that, mix ¼ pint to each gallon of fuel. Mix the oil and fuel thoroughly by first supplying one gallon of fuel, pouring the oil in and shaking the can vigorously, and then completely filling the fuel can.

NOTE: When the engine uses a screw-in type dipstick, check oil level without screwing dipstick into crankcase.

TUNE-UP
Spark Plugs

Remove the spark plug lead and remove the spark plug with a deep well socket. Check for excessive buildup of carbon deposits,

OIL DRAIN

OIL FILLER PLUG

VS-700

OIL FILLER PLUG

KEEP OIL LEVER
TO TOP OF SLOT

DRAIN PLUGS

V-100 crankcase fill and drain points

burned electrodes, and a cracked insulator. Replace the plug if any of these problems exist. If the plug has only light carbon deposits, wire brush the plug clean and set plug gap with a wire feeler gauge to: 0.028–0.033 in. on 2-cycle engines; 0.025–0.028 in. on 4-cycle engines.

Breaker Points

CHECKING

1. To check the breaker points, remove the ball, the magneto box cover, and gasket.
2. Look for evidence of excess oil in the box, which would indicate a leaking oil seal or defective breather assembly.
3. Check the point contacts for excessive wear or pitting. Normally the point assembly will last for many years if it is aligned

properly, the gap is set correctly, and the rubbing block or shuttle is properly aligned on the actuating cam. If it is necessary to replace the points, make sure that the new set is installed properly.

NOTE: Do not file the points and bend only the stationary side of the points to align the contact surfaces. Replace the condenser when the points are replaced.

4. Breaker point gap is critical to engine performance, so make sure it is set to 0.020 in. unless otherwise specified on the bail cover.

REPLACEMENT

1. Remove the flywheel, as described in the "Engine Disassembly" procedure. Remove the bail and remove the breaker box cover and gasket.
2. Disconnect the primary wire, remove the contact set attaching screws, and remove the contacts.
3. Reverse the removal procedure to install the contact set. Leave the contact set mounting screws slightly loose.
4. Rotate the crankshaft until the points are on the highest point of the cam. Set the gap to the specification shown on the bail cover, or to 0.020 in. using a flat feeler gauge, as shown. Tighten the mounting screws.
5. Clean the contact surfaces with carbon tetrachloride and then dry with paper. Align the surfaces, if necessary, as described above under "Checking." Reset the gap, if necessary.
6. Grease the cam, rotate the crankshaft, and then remove the excess grease.
7. Replace the cover gasket if it is damaged or oil soaked. Install the cover and snap the bail into position.
8. Install the flywheel, and torque the nut to specification.

Magneto Air Gap

The magneto air gap is the distance between the stationary laminations of the coil and the rotating magnets of the flywheel. In general, the closer the magnets pass to the laminations, the better the magneto will perform. Some extra clearance must be provided for bearing wear, however. The proper clearance for engines under five horsepower is 0.007–0.017 in. and 0.012–0.020 in. for engines over five horsepower.

The air gap is measured by placing layers of plastic tape over the laminations, replacing the flywheel and turning the flywheel. Remove the flywheel and check to see if the flywheel touched the tape, and adjust the coil assembly accordingly. Use only one layer of tape at a time. The thickness of common plastic electrician's tape is about 0.008–0.009 in.

CARBON
TET

Checking the breaker point gap

General Engine Specifications

Model Number	Block Construction	Piston Displacement	Type	Model Number	Block Construction	Piston Displacement	Type
E65	Alum.	5.76	2 Cycle	A & B 1100	LLCI	8.3	4 Cycle
				C1100	LLCI	8.3	4 Cycle
100	Alum.	7.2	4 Cycle	D1100	LLCI	8.3	4 Cycle
100–1000	Alum.	7.2	4 Cycle	V1100–1000	LLCI	9.5	4 Cycle
100–2000	Alum.	7.2	4 Cycle	VS1100	LLCI	9.5	4 Cycle
V100–1000	LLCI	7.2	4 Cycle	VS1100–1000	LLCI	9.5	4 Cycle
VS100	Alum.	7.2	4 Cycle				
VS100–1000	Alum.	7.2	4 Cycle	1200	LLCI	10.2	4 Cycle
VS100–2000	Alum.	7.2	4 Cycle	1200–1000	LLCI	10.2	4 Cycle
VS100–3000	Alum.	7.2	4 Cycle	1200–2000	LLCI	10.2	4 Cycle
VS100–4000	Alum.	7.2	4 Cycle	A1200	LLCI	10.2	4 Cycle
				B1290–1000	LLCI	10.2	4 Cycle
200	Alum.	4.5	2 Cycle	V1200–1000	LLCI	10.2	4 Cycle
A200	Alum.	4.5	2 Cycle	VS1200	LLCI	10.2	4 Cycle
AVS200	Alum.	4.5	2 Cycle				
AVS200–1000	Alum.	4.5	2 Cycle	2100	Alum.	7.2	4 Cycle
VS200	Alum.	4.5	2 Cycle	A2100	Alum.	7.2	4 Cycle
VS200–1000	Alum.	4.5	2 Cycle	A2100–1000	Alum.	7.2	4 Cycle
VS200–2000	Alum.	4.5	2 Cycle	A2100–2000	Alum.	7.2	4 Cycle
VS200–3000	Alum.	4.5	2 Cycle				
VS200–4000	Alum.	5.76	2 Cycle	VS2100	Alum.	7.2	4 Cycle
				VS2100–1000	Alum.	7.2	4 Cycle
300	LLCI	4.72	4 Cycle	VS2100–2000	Alum.	7.2	4 Cycle
A300	LLCI	4.72	4 Cycle	VS2100–3000	Alum.	7.2	4 Cycle
VS300	LLCI	4.72	4 Cycle				
350	LLCI	4.72	4 Cycle	VS3000	Alum.	7.2	4 Cycle
A400	Alum.	5.76	2 Cycle	3100	Alum.	8.3	4 Cycle
A400–1000	Alum.	5.76	2 Cycle	3100–1000	Alum.	8.3	4 Cycle
AVS400	Alum.	5.76	2 Cycle	3100–2000	Alum.	8.3	4 Cycle
AVS400–1000	Alum.	5.76	2 Cycle	3100–3000	Alum.	8.3	4 Cycle
BVS400	Alum.	5.76	2 Cycle	H3100–1000	LLCI	8.3	4 Cycle
CVS400–1000	Alum.	5.76	2 Cycle	FV3100–1000	LLCI	8.3	4 Cycle
VS400	Alum.	5.76	2 Cycle	AFV3100–1000	LLCI	8.3	4 Cycle
VS400–1000	Alum.	5.76	2 Cycle	AV3100–1000	LLCI	8.3	4 Cycle
VS400–2000	Alum.	5.76	2 Cycle	AV3100–2000	LLCI	8.3	4 Cycle
VS400–3000	Alum.	5.76	2 Cycle	AVS3100	Alum.	8.3	4 Cycle
VS400–4000	Alum.	5.76	2 Cycle	AVS3100–1000	Alum.	8.3	4 Cycle
				AVS3100–2000	Alum.	8.3	4 Cycle
500	LLCI	5.89	4 Cycle	AVS3100–3000	Alum.	8.3	4 Cycle
GK590	Alum.	5.76	2 Cycle	V3100–1000	LLCI	8.3	4 Cycle
				V3100–2000	LLCI	8.3	4 Cycle
650	LLCI	5.89	4 Cycle	VS3100	Alum.	8.3	4 Cycle
				VS3100–1000	Alum.	8.3	4 Cycle
700–A	LLCI	5.89	4 Cycle	VS3100–2000	Alum.	8.3	4 Cycle
B700	LLCI	5.89	4 Cycle	VS3100–3000	Alum.	8.3	4 Cycle
C700	LLCI	5.89	4 Cycle				
D700	LLCI	6.65	4 Cycle	4100	Alum.	8.3	4 Cycle
D700–1000	LLCI	6.65	4 Cycle	4100–1000	Alum.	8.3	4 Cycle
D700–2000	LLCI	6.65	4 Cycle	4100–2000	Alum.	8.3	4 Cycle
D700–3000	LLCI	6.65	4 Cycle	AVS4100–1000	Alum.	8.3	4 Cycle
VS700	LLCI	5.89	4 Cycle	AVS4100–2000	Alum.	8.3	4 Cycle
VS750	LLCI	5.89	4 Cycle	VS4100–1000	Alum.	8.3	4 Cycle
				VS4100–2000	Alum.	8.3	4 Cycle
800	LLCI	8.3	4 Cycle				
A800	LLCI	8.3	4 Cycle	400–0000–000	Alum.	7.2	4 Cycle
VS800	LLCI	8.3	4 Cycle	401–0000–000	Alum.	7.2	4 Cycle
				402–0000–000	Alum.	7.2	4 Cycle
900	LLCI	8.3	4 Cycle	403–0000–000	Alum.	7.2	4 Cycle
900–1000	LLCI	8.3	4 Cycle	404–0000–000	Alum.	8.3	4 Cycle
900–2000	LLCI	8.3	4 Cycle	405–0000–000	Alum.	8.3	4 Cycle
900–3000	LLCI	8.3	4 Cycle	406–0000–000	LLCI	8.3	4 Cycle
900–4000	LLCI	8.3	4 Cycle	407–0000–000	LLCI	8.3	4 Cycle
VS900	LLCI	8.3	4 Cycle	407–0002–000	LLCI	8.3	4 Cycle
				408–0000–000	Alum.	8.3	4 Cycle
V1000–1000	LLCI	8.3	4 Cycle	409–0000–000	Alum.	8.3	4 Cycle
VS1000	LLCI	8.3	4 Cycle				

General Engine Specifications

Model Number	Block Construction	Piston Displacement	Type	Model Number	Block Construction	Piston Displacement	Type
411–0000–000	Alum.	7.2	4 Cycle	492–0300–000	LLCI	8.3	4 Cycle
411–0002–000	Alum.	7.2	4 Cycle	494–0000–000	LLCI	8.3	4 Cycle
415–0000–000	Alum.	8.3	4 Cycle	494–0001–000	LLCI	8.3	4 Cycle
415–0002–000	Alum.	8.3	4 Cycle	497–0000–000	LLCI	10.2	4 Cycle
417–0000–000	LLCI	8.3	4 Cycle	498–0300–000	LLCI	10.2	4 Cycle
424–0000–000	LLCI	8.3	4 Cycle	498–0301–000	LLCI	10.2	4 Cycle
426–0000–000	Alum.	8.3	4 Cycle	499–0000–000	LLCI	10.2	4 Cycle
429–0003–000	LLCI	9.2	4 Cycle	500–0000–000	Alum.	5.76	2 Cycle
431–0003–000	LLCI	9.2	4 Cycle	501–0000–000	Alum.	5.76	2 Cycle
435–0003–000	Alum.	8.3	4 Cycle	501–0001–000	Alum.	5.76	2 Cycle

Alum. = Aluminum
LLCI = Long Life Cast Iron

Breaker Cam

The breaker cam on most Clinton engines is replaceable. Check the fit of the cam over the crankshaft to make sure that it is tight.

Mixture Adjustment

1. If the engine runs very roughly or will not start, first make a preliminary setting as follows:
   a. Very gently turn the main mixture screw in until it seats very lightly.
   b. Turn screw outward (counterclockwise) the specified number of turns:
   - 501 — 1 turn
   - Lift Carburetors — $1\frac{1}{4}$–$1\frac{1}{2}$ turns
   - LMG, LMB, LMV Carburetors — $1\frac{1}{4}$
   - All others — $1\frac{1}{2}$ turns
2. Following this, start and run the engine until hot. Then, turn the mixture screw inward or outward in $\frac{1}{16}$ turn increments, pausing after each adjustment, until best running is obtained. The most accurate adjustment is obtained with a tachometer. Set the mixture so the highest possible rpm is obtained.

Throttle position and dimension

Mechanical Governor Adjustment

1. Stop the engine and set any throttle controls so there is tension on the governor spring.
2. Loosen the adjusting screw. Position the throttle to within exactly $\frac{1}{32}$ in. of the stop on the carburetor casting. Tighten the adjusting screw.

THROTTLE IN OPEN POSITION

GOVERNOR IN CLOSED POSITION

ADJUSTMENT SCREW

Location of governor adjustment screw

Checking compression

quired that the carburetor be disassembled to correct the problem, but at least you will have some direction.

NOTE: Check the throttle shaft/main casting tolerance. This is one place on the carburetor that cannot be repaired.

501 ENGINE CARBURETOR

The 501 engine carburetor is a relatively simple carburetor and can be disassembled after it is removed from the engine.

1. Remove the choke and air filter assembly, remove the mixture adjusting screw and spring, and unscrew the large bolt which holds the float bowl to the rest of the carburetor.

2. Please note that the main fuel nozzle is contained in the top of the large bolt and care should be exercised not to damage the needle seat. Disassemble the float and fuel inlet needle.

NOTE: The needle seat is not replaceable. The bowl cover, needle pin, spring, and seat assembly must be replaced if any part is worn.

3. Clean all parts in solvent, blow dry with compressed air, inspect, and replace any worn or damaged parts.

4. Assemble the carburetor in the reverse order of disassembly. Set the float level with the bowl cover inverted and the float and needle installed so that there is $13/64$ in. plus or minus $1/32$ in. clearance between the outer edge of the bowl cover and the free end of the float. Adjust the level by bending the lip of the float with a screwdriver. Use all new gaskets.

5. Set the mixture adjusting needle one turn open from the seat to start the engine. This carburetor does not have an idle mixture orifice and therefore will not operate at speeds below 3000 engine rpm. The operating range is between 3000 and 3800 rpm. Make the final mixture adjustment with a tachometer if possible. If a tachometer is not available, adjust the mixture screw $1/16$ of a turn at a time until you obtain the best engine performance.

SUCTION LIFT CARBURETOR

Throttle Plate and Shaft Replacement

1. Drill through the plug at the rear of the carburetor body.
2. Force out the plug with a small drift pin.

501 Engine carburetor

Compression Check

1. Remove the spark plug and install a compression gauge in its place.
2. Crank the engine over at normal cranking speed. Gauge readings should be:
   - 2-stroke: above 60 psi
   - 4-stroke up to $4\frac{1}{2}$ hp: 65–70 psi
   - 4-stroke above $4\frac{1}{2}$ hp: 70 psi

FUEL SYSTEM

Carburetor

Clinton engines use a variety of float type carburetors and one type of suction lift carburetor.

A carburetor can only malfunction as a result of these causes: the presence of foreign matter (dirt, water); out of adjustment (too rich or too lean a mixture), and leakage caused by worn parts or cracks in the casting. Look for any of the above causes before disassembling the carburetor. Of course, it may be re-

Suction lift carburetor

3. Remove the plastic plug.

4. Remove the screws which retain the throttle plate to the shaft and remove the plate and shaft.

5. Reinstall in the reverse order of removal.

Make sure that the throttle plate and shaft operate without binding. Use a sealer on the new expansion plug.

NOTE: Horizontal suction lift carburetors do not have a plug.

The choke plate and shaft is replaced in the same manner as the throttle, except that there is no expansion plug to be removed.

Stand Pipe Replacement

The stand pipe is press-fit into the body of the carburetor. Remove the pipe by clamping it in a vise and twisting the carburetor body while at the same time pulling until the pipe is free. To install the pipe, tap it lightly into place until the end of the pipe is 1.94 in. ± 0.45 in. from the body of the carburetor. Use a sealer at the point where the pipe is inserted into the carburetor body.

Idle Needle and Jet Replacement

The idle needle and jet are replaceable and can be removed as follows:

1. Unscrew the needle and remove it.

2. Remove the expansion plug in the bottom of the carburetor body.

3. To remove the idle jet, push a piece of $\frac{1}{16}$ in. diameter rod through the fuel well and up the idle passage until the jet is pushed out of the passage.

To install the idle needle and jet:

4. Make a mark exactly $1\frac{1}{4}$ in. from the end of the $\frac{1}{16}$ in. diameter rod and push the new jet into the idle passage until the mark is exactly in the center of the main fuel well.

5. Install the needle and set it at 4–$4\frac{1}{4}$ turns open from being seated.

The high speed needle and seat can be inspected for wear or damage by removing the needle and checking its taper. If the needle taper is damaged, replace it. Inspect the seat for taper and splits. If the seat is split or tapered from the needle being screwed in too tight, the complete carburetor must be replaced, since the seat cannot be replaced separately. When the needle is replaced, the initial adjustment is $\frac{3}{4}$–$1\frac{1}{2}$ turns open from being lightly seated.

LMG, LMB AND LMV TYPE CARBURETORS

These carburetors are float type carburetors and can be reconditioned in the same manner as the 501 carburetors by taking a note of the following differences in design and specifications.

The main fuel nozzle needle is inserted from the bottom and pushed up through the middle of the float bowl.

The float, fuel inlet needle, and seat are basically the same as that of the 501 carburetor.

Float adjustment for the LMG, LMV, LMB type carburetors

LMG, LMV, LMB type carburetors with an enlarged view of the main jet nozzle

NOTE: Do not remove the main fuel nozzle from the carburetor body unless it is to be replaced. Once it is removed it cannot be reinstalled.

Use all new gaskets.

The throttle valve is installed with the part number or trademark **W** facing toward the mounting flange.

The preliminary setting of the idle adjusting needle is $1\frac{1}{4}$ turns open after being lightly seated.

Adjust the float in the same manner as the 501 carburetor. There should be $\frac{5}{32}$ in. clearance between the float and the casting rim when the carburetor is inverted. When the carburetor is turned over, the float should not drop more than $\frac{3}{16}$ in.

H.E.W. carburetor

The preliminary setting for the main fuel adjusting needle is $1\frac{1}{4}$ turns open after being lightly seated.

H.E.W. CARBURETORS

H.E.W. carburetors are float type carburetors and can be serviced in the same manner as 501 carburetors. After cleaning and inspecting the parts for wear or damage, assemble the carburetor, noting the following differences:

Install the throttle valve with the part number or **W** toward the mounting flange with the throttle in the closed position. Always use new gaskets during assembly.

The initial setting for the idle adjusting screw is $1\frac{1}{2}$ turns open from being lightly seated.

If any part of the float assembly has to be replaced, replace the whole assembly. Do not replace individual parts.

The float setting is made in the same manner as the 501 carburetor, by turning the main body of the carburetor upside down and measuring the clearance between the float and the body casting rim. The distance in this case is $\frac{3}{16}$ in. with $\frac{3}{16}$ in. of travel.

The preliminary setting for the high speed adjusting screw is 1 1/2 turns open from being seated lightly.

U.T. CARBURETORS

These are also float type carburetors and are serviced in the same manner as 501 carburetors, noting the following differences:

The throttle valve is installed with the trademark **C** on the side toward the idle port when viewed from the mounting flange side. Use new screws.

Set the float level with the float installed and the housing inverted. The measurement is to be taken with the needle seated and the gasket removed. Measure between the float seam and the throttle body. Adjust by bending the lip of the float.

The preliminary setting for the high speed adjusting needle is $1\frac{1}{4}$ turns open from being lightly seated.

The preliminary setting for the idle adjusting needle is $1\frac{1}{2}$ turns open from being lightly seated.

The chart below gives the float level setting dimension as related to the carburetor identification number and part number:

Ident No.	Part No.	Float Setting (in.)
2712–S	39–143–500	$\frac{19}{64}$
2713–S	39–144–500	$\frac{19}{64}$
2714–S	39–140–500	$\frac{1}{4}$
*2398–S	39–147–500	$\frac{1}{4}$
2336–S	39–146–500	$\frac{1}{4}$
2336–SA	39–146–500	$\frac{1}{4}$
2337–S	39–145–500	$\frac{1}{4}$
2337–SA	39–145–500	$\frac{1}{4}$
2230–S	39–343–500	$\frac{17}{64}$
2217–S	39–344–500	$\frac{11}{64}$

*With resilient seat—$\frac{9}{32}$ + or − $\frac{1}{64}$

TOUCH 'N' START PRIMER CARBURETOR

Steps in operation of the primer:
1. Seal the bowl vent with a finger.
2. Depress the bulb to pressurize the bowl.
3. Pressure in carburetor bowl forces the fuel into the carburetor throat.
4. When the engine is cranked, the intake valve opens, letting gasoline into the combustion chamber for quick starting.

U.T. carburetor

Operation of Touch 'N' Start primer carburetor

Operation of the Fixed Speed carburetor

5. Gasoline forced from the bowl during priming is replaced by the flow of gasoline from the fuel tank through the gravity fuel inlet.

This primer is applicable to the LMG, LMB, and LMV carburetors. Servicing this carburetor would be the same as listed for the LMG, LMB, and LMV, except a choke lever and choke valve are not used. Note the bowl atmospheric vent is routed back through the primer tube and bulb.

FIXED SPEED CARBURETOR

Steps in operation:
1. Rotate the control knob counterclockwise to open the throttle (4-cycle engines) (2-cycle not equipped with control knob).

2. To stop the engine rotate the control knob clockwise. (4-cycle engines) (2-cycle engines use a shorting device).

NOTE: The governor spring is located on the throttle shaft between the lever and the carburetor casting. The setting on the high speed screw is $1\frac{1}{4}$–$1\frac{1}{2}$ turns. Servicing this carburetor would be the same as listed for the LMG, LMB, or LMV except there is no main nozzle or idle circuit.

CARTER CARBURETORS

Carter float type carburetors are serviced in the same manner as 501 carburetors, taking note of the following differences:

Do not remove the choke valve and shaft unless they are to be replaced. A spring loaded ball holds the choke in the wide open position. Be sure to use a new ball and spring when replacing the choke shaft and plate assembly.

Install the throttle plate with the trademark **C** on the side toward the idle port when viewed from the mounting flange side.

The float setting is made in the same manner as the 501 carburetor and the measurement is $3/16$ in.

The initial setting for the high speed adjustment screw is 2 turns open from being lightly seated.

The initial adjustment for the idle adjusting screw is $1\frac{1}{2}$ turns open from being lightly seated.

Carter carburetor

Fuel Pumps

MECHANICAL FUEL PUMPS

These pumps can be rebuilt. See the exploded view below.

The 220–122–500 and 220–145–500 fuel pumps are simple diaphragm units that screw into the inlet side of the carburetor. They are activated by crankcase pressure pulses. They cannot be rebuilt, but must be replaced if they will not lift the fuel six inches.

The diaphragm type fuel pumps used on VS–1200, V–1200, and 499 series engines can be rebuilt. However, construction is so simple no rebuild specifics are supplied by the factory.

Governors

INSPECTION AND ASSEMBLY OF MECHANICAL GOVERNOR

When assembling an engine equipped with the centrifugal weight governor inspect the governor shaft bearing in the block and the governor arm assembly that goes through the bearing, for wear and replace them if necessary. After inspection, insert the arm through the bearing and fasten the arm and weight assembly into the bearing.

Care should be taken on installation of this arm and weight assembly as they may be locked to the outside linkage 180 degrees from the correct position which would tear out the centrifugal weights and damage the arm and weight assembly upon operation of the engine. The weight and arm or yoke should be as close to the cam axle or governor gear as it can be to be properly installed. In this position, it will operate against the governor collar or thimble assembly and will move in conjunction with the governor spring tension and the centrifugal force of the weights which are attached to the camshaft or governor gear.

The collar should be inspected for wear and possible damage, and the weight assembly itself should be inspected for wear and possible damage or bending of the weights or the weight supports.

When servicing a centrifugal governor, check to be certain that the collar or thimble operates freely on the camshaft or governor gear and that the governor shaft moves freely in the bushing. When the bushing in the block is replaced, check carefully freedom of shaft motion as the bushing may be distorted in installation.

Also the governor shaft can be bent easily on disassembly or reassembly and can be bent in usage. Also check range of movement of collar or thimble after assembly of camshaft or governor gear to determine that these parts do not lock against block.

INSPECTION AND ASSEMBLY OF AIR VANE GOVERNOR

The air vane should be inspected visually when reassembling and replaced if bent or damaged. Use care in the hook-up of the air vane link and throttle plate so that they move freely and do not drag at the connections or on the bearing plate or blower housing. Many of the air vane governors have a spring inside of the vane at the pivot. The spring gives a dampening effect on the vane movement. This spring should be replaced if the engine has been out of service for a period of time so that it retains the pressure on the bushing or vane.

When servicing an air vane governor, the condition of the blower housing should be checked carefully. Dents and bends should be removed from it so that the air stream moves as it should to the air vane. The air vane must be in the same condition as when manufactured and replaced when bent because the governor spring tension and the vane are balanced. If the vane does not sit in the air blast properly, the spring will be too strong for the air vane to stretch.

1. Rocker arm
2. Body
3. Top cover
4. Diaphragm assembly
5. Diaphragm pulsator
6. Mounting gasket
7. Valve gasket
8. Linkage
9. Rocker arm pin
10. Cover plate
11. Pipe plug
12. Valve retainer
13. Valve seat (head)
14. Valve seat (retainer)
15. Rocker arm spring clip
16. Diaphragm spring
17. Rocker arm spring
18. Valve spring
19. Screw & lockwasher assembly (cover plate)
20. Screw & lockwasher assembly (top cover)
21. Screw & lockwasher assembly (valve retainer)
22. Valve

Exploded view of mechanical fuel pump-typical

When assembling the air vane governor, apply tension to the governor spring and close the throttle manually to see that it moves open freely, that it does not bind at the governor linkage, air vane, pivot post, bushings, bearing plate, blower housing, etc. It should move freely from closed to open position by governor spring tension.

ENGINE OVERHAUL

The same basic procedure for disassembling a Clinton engine can be used for all engines. The procedure given below pertains to both 2-stroke and 4-stroke Clinton engines; differences are noted. Procedures for servicing individual components are given at a later point in this section.

Engine Disassembly

1. Remove the engine from the piece of equipment it powers and then remove any brackets, braces, adapters or pulleys.
2. Clean the exterior of the engine.
3. Drain the lubricating oil from the crankcase.
4. Remove the fuel tank and blower housing.
5. Remove the carburetor and governor assembly, marking the spring and link holes for correct reassembly.
6. Remove the muffler assembly.
7. Remove the flywheel nut, using a flywheel holder to hold the flywheel while the nut is removed.
8. While lifting up on the flywheel, gently tap the crankshaft to loosen the flywheel from the crankshaft taper. Remove the flywheel and flywheel key.
9. Remove the complete magneto assembly which includes the coil, breaker points, condenser, and laminations.
10. Remove the cylinder head and gasket on 4-stroke engines.
11. Remove the valve chamber cover and breather assembly from 4-stroke engines.
12. Remove the valve spring keepers after compressing the valve spring with a valve spring compressor.
13. Remove the valves and springs from the block after removing the valve spring compressor.

NOTE: Some valves have a burr on the stem that will prevent the valve from being removed up through the valve guide. If present, it will be necessary to remove

this burr from the stem in order to remove the valve from the engine. To remove the burr, hold a flat file against the burred area and rotate the valve.

14. Remove the base plate or end cover assembly on 4-stroke engines.

NOTE: On some engines the base is an integral part of the block and cannot be removed. If this is the case, remove the side plate on the power take-off side of the engine. Before removing the side plate or crankshaft, make sure all paint, rust, and dirt are cleaned from the area of the crankshaft bearing. This is so the side plate can be easily removed. When removing the side plate from engines that have ball bearings, it will first be necessary to remove the oil seal and the snap-ring from the crankshaft.

15. Remove the mounting plate and reed plate assembly on 2-stroke engines.
16. Remove the connecting rod cap screws and cap. Mark the cap and connecting rod so they can be reassembled in the same position.
17. On 4-stroke engines, check for a carbon or metal ridge at the top of the cylinder. If a ridge is present, remove it with a ridge reamer or hone.
18. On 2-stroke engines, push the piston and connecting rod assembly up into the cylinder as far as it will go so it will not hit the crankshaft when it is removed.
19. Remove the piston and connecting rod assembly from a 4-stroke engine's block.
20. Remove the bearing plate on the flywheel side of the block if the engine being disassembled has one.
21. Remove the crankshaft.

NOTE: On 4-stroke engines with a ball bearing on the PTO side of the engine, it will be necessary to remove the cap screws which hold the bearing in place before the crankshaft can be removed. On some models with ball bearings and tapered roller bearings, it will be necessary to remove the crankshaft oil seal and camshaft axle, and move the camshaft to one side before the crankshaft can be removed. Two stroke engines having a ball bearing on the PTO side of the crankshaft will have to have the retaining ring, which holds the bearing in place, removed before the crankshaft can be removed.

22. Remove the camshaft assembly. Drive the camshaft axle out of the PTO side of the block as the flywheel side is smaller.
23. Remove the piston and connecting rod assembly from 2-stroke engines.
24. After the camshaft is removed from a 4-stroke engine, mark the valve tappets as to whether they are the exhaust tappet or the intake tappet, and then remove them. If no valve work is to be done, be sure to replace the valves in the same position from which they were removed.
25. Remove the piston and rod assembly. After disassembling the engine, clean all parts in a safe solvent, removing all deposits of carbon and oil, etc. Check all operating clearances and replace or rebuild parts as necessary.

Cylinder Bore

After disassembling the engine, inspect the cylinder bore to see if it can be reused. Look for score marks on the cylinder walls. If there are marks and they are too deep to be removed, the block will have to be discarded. If there is a hole in the block due to connecting rod failure, the block will have to be replaced. If there are broken cooling fins on the outside of the block, these can cause overheating and replacing the block should be considered.

Measuring the cylinder bore—a cutaway view

Check the dimensions of the cylinder bore to determine the extent of wear and whether or not it has to be rebored. The cylinder bore can be rebored to 0.010 in. or 0.020 in. oversize since there are oversize pistons available in these sizes. If the cylinder is within serviceable limits and there is no need to rebore it, be sure to deglaze the cylinder before installing the piston assembly with new rings.

Bearings

Clinton engines are equipped with tapered roller bearings, ball bearings, needle bearings, and sleeve bearings. The first thing to determine is whether or not the bearing is worn or damaged and needs replacing. Then clean the bearings in a sofe solvent, inspect them for excessive play due to wear, smoothness of rotation, pitted surfaces, and damage. All of these types of bearings are pressed on the crankshaft and into the bearing plates and are removed by either a bearing splitter and puller or they are driven out of the bearing plates with a punch. In all cases be very careful not to bend, gouge, or otherwise damage the crankshaft or bearing plate when removing and installing the bearings.

Tapered roller, needle, and ball bearing type bearings, all of which are used in Clinton engines

With sleeve bearings, first inspect the bearing surface for scoring and damage to determine whether the bearing has to be replaced. Check the bearing diameter for wear. In some cases you will find that the bearing surface is the same material as the block. These bearings can be reamed out and sleeve bearings installed. If the original bearing is a sleeve type bearing, it can be removed by driving it out with the proper size driving tool. Install the new bearing so that the oil hole in the bearing is aligned with the oil passage in the block or bearing plate. Drive the bearing into the block or bearing plate until it is recessed about $\frac{1}{32}$ in. from the crankshaft thrust face of the block or bearing plate. After installing the new piece, it must be finish reamed. After finish reaming, clean all metal filings and debris from the engine, making sure that all oil passages are free from obstruction.

Valve Seats

Standard valve seats, those without inserts, can be reground to remove all of the oxidized surface metal and gain perfect sealing characteristics. After grinding the valve seats, the valves must be lapped in with lapping compound. Not too much lapping is recommended, just enough to obtain a good seal.

If the engine has had a number of valve jobs and the valve seat is too deep, requiring that too much stock be removed from the valve stem to obtain the proper valve-to-tappet clearance, valve seat inserts may be installed. If over half of the metal between the lock groove and the end of the valve stem has been removed to gain the proper stem-to-tappet clearance, you should consider installing valve seat inserts.

The valve seat width is to be between $\frac{1}{32}$ and $\frac{3}{64}$ in. (0.030–0.045 in.) and the valve seat angle is to be between 43½° and 44½°

On aluminum block engines, iron valve seat inserts are standard equipment. To remove these inserts, it is first necessary to remove the metal that has been rolled over the edge of the insert to hold it in place. This is normally accomplished by using the proper size cutter. If the valve seat insert is loose, a cutter may not be necessary. After the insert has been removed, it is necessary to cut the block to the proper depth of $\frac{3}{16}$–$\frac{7}{32}$ in. This is the depth of the insert plus $\frac{1}{32}$ in. which is used to hold the insert in place. The insert is held to the cylinder block by a definite interference type press fit. The insert should be cooled before attempting to install it in the block. After the insert is fitted in place, with the bevel facing up, the metal around the edge of the insert must be peened over the edge of the insert in order to hold it in place. Do not strike the block too sharply when peening because of the posibility of distorting the cylinder bore. Finish grind the valve seat insert and lap in the valves.

Valve Guides

First, inspect and measure the valve guide diameter to determine whether the guide is worn enough to necessitate rebuilding. The standard guide size for 4-stroke engines under 5 horsepower is 0.2495–0.2510 in. On engines over 5 horsepower, the standard guide size is 0.312–0.313 in.

When the valve stem-to-valve guide clearance is more than the maximum serviceable clearance and cannot be corrected by installing a new valve, you will have to either replace the valve guide (if it is replaceable), oversize it, or knurl it.

Replacing the valve guides

Valves

Inspect the valves for a burned face, warped stem or head, scored or damaged stem, worn keeper groove in the stem, and a head margin of less than $\frac{1}{64}$ in. If any one of these conditions exists, the valve should be replaced. Check the valve stem diameter. On engines with less than 5 horsepower, the stem diameter should be 0.2475–0.2465 in. On engines over 5 horsepower, the stem diameter is to be 0.310–0.309 in. Any time the valve stem-to-guide clearance can be reduced more than 0.001 in. by replacing the valve with a new one, you should do so. Also any time the stem-to-guide clearance is over 0.0045 in. you should consider doing some rework (new valve guides, seats and valves) to bring the clearance below 0.0045 in. but not less than 0.002 in. If it is determined that the old valve can be reused, then it should be refaced, using an automotive type valve grinder to secure a 45 degree face angle on the valve with a $\frac{1}{64}$ in. margin between the head and the face of the valve.

2-Stroke Engine Reed Valves

Inspect the reed valves for the following items: broken reed valves, bent or distorted reed valves, damaged or distorted reed valve seat, or a broken or bent reed valve stop. If any of these conditions exist, the reed valve assembly must be replaced.

Reed valve assembly for two stroke engines

Valve Springs

To check the condition of valve springs, simply remove the spring from the engine and stand it on a flat surface next to a new valve spring. If the old spring is shorter and leans to one side, it should be replaced with a new spring. Some of the cast iron engines have a stronger or stiffer valve spring installed on the exhaust valve. Make sure that a stiffer spring is installed on the exhaust valve or, to be sure, install two stiff springs in the engine. When a valve seat is rebuilt, the valve then seats further down into the block and this results in a loss of spring tension. To restore spring tension, install a thin washer on top of the valve spring.

Valve Tappets

The valve tappets should be inspected for wear on the head of the tappet and score marks or burrs anywhere else. The tappet should be replaced if any defects are found. Measure the dimensions of the tappet, checking for stem diameter and length. Oversize tappets are not available; however, the tappet guide can be knurled and rebored to correct size should the tappet-to-guide clearance become too large.

Measuring the tappet stem diameter

Piston Assembly

Inspect the piston assembly for scored walls, damaged ring lands, worn or damaged wrist pin lock ring grooves, and a cracked or broken piston skirt. If any of these defects are present, the piston must be replaced. Check the dimensions of the piston with a micrometer.

NOTE: The ring land diameter on 4-stroke engines is tapered and the reading at the ring land will be 0.00125 in. smaller per 1 in. of piston length.

Clean all carbon deposits from the ring grooves. An old broken ring will serve as an excellent tool for cleaning ring grooves.

Check the side clearances of the rings with new rings installed. The minimum and maximum clearances for oil, scraper, and compression rings are as follows: two stroke engines —

0.0015–0.004 in.; four stroke engines under five horsepower — 0.002–0.005 in.; four-stroke engines over five horsepower — 0.0025–0.005 in.

If an oversized piston is used, the amount of the oversize is stamped on the top of the piston.

The ring gap on all engines except the $1\frac{7}{8}$ in. bore two stroke engine is 0.007–0.017 in. The ring gap on the $1\frac{7}{8}$ in. bore two stroke engine is 0.005–0.013 in. Oversize rings are available. Install the rings in the following order: oil ring, scraper ring, and compression ring. The oil ring can be installed with either side up, the scraper ring should have the step on the lower side toward the bottom of the engine's crankcase and the compression ring has to be installed with the bevel on the inside circumference facing upward.

NOTE: Two stroke engines have wire retainers or pins located in the ring grooves to keep the ring from moving in the groove. Make sure the ring gap is properly located over these retainers.

Installation sequence and position for the piston rings

Inspect the connecting rod for wear or damage, such as a scored bearing surface, cracks, and damaged threads. Use a micrometer to check all of the connecting rod dimensions. Check the clearance between the wrist pin and the connecting rod at the wrist pin hole. The tolerance for all Clinton engines is 0.0004–0.0011 in. When the clearance reaches 0.002 in., the parts should be replaced with new ones or rebuilt. Clinton, however, does not supply oversize wrist pins.

When installing the piston and connecting rod in a four stroke engine, the piston may go either way, but the rod has an oil hole that must face toward the flywheel side of the engine (with the exception of those engines that use a clearance rod in which the marked side faces toward the camshaft of the engine).

On two stroke engines, the piston is installed with the downward slope of the piston dome facing the exhaust side of the engine. There is no special way for the connecting rod to be installed.

Names of the piston parts

Piston ring retainers in two stroke engines

Install clearance rods with the side marked clearance side facing toward the camshaft

MATCH THESE BOSSES

MATCH THESE BOSSES

Install the connecting rod cap with the match marks opposite each other

Wrist Pins

All Clinton wrist pins are a "hand press fit" into the piston. Care should be taken when removing or installing the wrist pin into rod or piston. It is easy to distort or damage the piston or rod. Never lay a piston on a solid object when removing or installing wrist pin. The piston can be supported in the palm of your hand when servicing to keep from damaging it. There is no special way to install the wrist pin into the piston or rod, except on the 2-cycle engines which in some cases have a hollow wrist pin closed on one end. Make sure the closed end is towards exhaust side.

Crankshaft

Before removing the crankshaft, remove the spark plug and rotate the crankshaft with the starter mechanism, while checking for any wobble of the end of the crankshaft. Any wobbling indicates that the crankshaft is bent and must be replaced. Deviance of 0.001 in. or more is not tolerable. End-play of the crankshaft should be between 0.008–0.018 in. If the endplay exceeds 0.025 in., the condition should be corrected. End-play is adjusted when the engine is assembled by the addition of various size gaskets behind the bearing plate. It is not recommended that the

INTAKE SIDE   EXHAUST SIDE

Installation of a two stroke engine piston

crankshaft be straightened. Check all bearing surfaces for wear with a micrometer. Replace the crankshaft if it is bent or cracked; if the keyway is damaged; if the taper is damaged; if the flywheel end threads are stripped; or if the bearings are scored.

Bearing Plates and Bases

Inspect these visually, to determine whether or not they can be re-used. Reject for the reasons listed:
    a. Broken or cracked, housing mounting flange on the bearing plate or mounting ears or flange base.
    b. Cracked or distorted bearing bases.
    c. Warped or distorted gasket or mounting surface.
    d. Oil seal or bearing pocket oversize.
    e. Stripped threads on the lamination hold-down screw holes in the bearing plates or drain and filler plug holes in the bases.
    f. Worn crankshaft thrust face surfaces on base or bearing plate.

Camshaft

Check the camshaft for extremely worn lobes and broken gear teeth. Oil pump drive camshafts have a pin located below the gear that must have a squared end and must be secure to the crankshaft. Camshafts from engines with vertical crankshafts have a scoop riveted to the bottom of the gear. Make sure that the scoop is secure. Make sure that on those models equipped with centrifugal advance (ignition) that the advance mechanism is free and the springs are not distorted or broken. Check the dimensions of the camshaft axle.

Cylinder

Check the cylinder head for warpage with a straightedge, after removing all dirt and deposits. If the head is warped, place a piece of emery cloth, with the rough side facing up, on a flat surface. Move the cylinder head gasket surface over the emery cloth in a figure eight pattern until the surface of the head is flat. If there are any broken cooling fins or if the spark plug hole threads are stripped, the head must be replaced.

Oil Seals

Oil seals serve two purposes, these being to keep the oil from leaking out of the crankcase on 4-cycle engines, and sealing the crankcase on 2-cycle engines, to keep the vacuum and pressure from being affected by the outside atmospheric pressures.
    Any time an engine is being disassembled and the oil seals are not going to be replaced with new ones, it is a must that oil seal loaders be installed over the crankshaft or cam gear axle to keep from damaging the seal lips when the base, bearing plate, crankshaft or cam gear is removed. Clinton has oil seal loaders that will fit any diameter shaft used in the Clinton line of engines or

Cross section showing use of an oil seal loader

Adjust the valve-to-tappet clearance with the tappets completely off of the cam lobes

you may use an equivalent part designed specifically for this purpose.

Whenever an engine is disassembled or the oil seal removed from it, the oil seal should be given a visual inspection for the following, to determine if it can be reused:

    a. Cut or damaged seal lips.

    b. Distorted or bent seal.

    c. Condition of seal lip to make sure it still is flexible, and has not taken a permanent set.

Any of the above defects, would require replacement of the seal with a new one. To insure that an oil seal will function properly it is recommended that any time an oil seal area of an engine has been worked on new seals be used.

Oil seals should be removed prior to removal and replacement of bearings and reaming of bearings. The oil seals may be removed by prying out or by any means at hand with some care being used not to damage the bearing plate, block, or base to which a new seal must be replaced.

Engine Assembly

FOUR STROKE ENGINES

This is a general procedure and is intended to be only a guide since deviations may be necessary for some models.

1. Insert the tappets into the block.

2. Assemble the oil pump to the cam gear, if so equipped.

3. Install the mechanical governor shaft, if so equipped.

4. Install the crankshaft and cam gear into the engine, making sure that the crankshaft thrust washer is in place if one is used.

5. Align the crankshaft and camshaft timing marks.

6. Install the piston and rod assembly using a ring compressor and great caution not to break rings or damage the piston.

7. Install the rod cap and oil dipper, if so equipped, and the cap screw and lock. Tighten to the correct torque. Crimp the screw locks securely.

8. Install the bearing plate and base plate or end cover assembly to the cylinder block. Check the crankshaft end-play. Engines using sleeve bearings should have 0.005–0.002 in. end-play. Engines using tapered roller bearings should have 0.001–0.006 in. end-play. Engines using roller bearings have no end-play specifications; however, care should be taken not to have the crankshaft too tight after assembly. The end-play is adjusted by the installation of various size gaskets between the plate and block.

9. Install the oil seals in the PTO and flywheel side of the crankshaft.

10. Install the valves into the block and check the valve stem-to-tappet clearance. Clearance is checked with the lobe of the tappet facing away from the valve. Clearance for a four stroke engine is 0.009–0.012 in. Clearance is adjusted by grinding or filing the valve stem.

11. Using a valve spring compressor, assemble the valve springs to the valves.

12. Install the breather assembly into the valve spring chamber and install the cover.

13. Install the cylinder deflector into the engine.

14. Assemble the magneto assembly to the engine block or bearing plate, whichever is applicable, making sure that the points are clean and adjusted to the correct gap.

15. Install the flywheel, flywheel screen, and starter cup to the crankshaft. Tighten the flywheel nut to the proper specification.

16. Install the carburetor assembly and governor, making sure that the governor assembly links and springs are placed in their original holes.

17. Place the cylinder head on the block. Torque the head bolts in three stages, in the proper sequence, and to the proper torque.

Align the crankshaft and camshaft timing marks

Cylinder head bolt tightening sequence

18. Install the blower housing to the engine.
19. Install the spark plug, muffler assembly, and air cleaner.

TWO-STROKE ENGINES

1. Assemble the piston and rod assembly and install it into the block with the help of a piston ring compressor. Be careful not to damage the rings or the piston.
2. Install the crankshaft into the block, installing the crankshaft thrust washer if the engine is so equipped.
3. Assemble the connecting rod and piston assembly to the crankshaft by installing the connecting rod cap and cap screws. Tighten the cap screws to the correct torque.
4. Install the reed valve plate to the engine.
5. Install the bearing plate. Engines that use sleeve bearings should have a crankshaft end-play measuring 0.005–0.020 in. Engines using ball bearings have no specific end-play measurement but make sure that the crankshaft is not tight after assembly. End-play can be adjusted by the addition of various size gaskets behind the bearing plate.
6. Install the bearing oil seals on the PTO side and the flywheel side of the crankshaft.
7. Install the magneto to the bearing plate and adjust the points to the proper gap.
8. Install the flywheel, flywheel screen, and the starter cup to the crankshaft. Tighten the flywheel attaching nut to the correct torque specification.
9. Assemble the carburetor, governor links, springs, and air vane to the engine. Always replace all governor components in the same position from which they were removed.
10. Install the cylinder deflector to the engine.
11. Install the gas tank.
12. Install the air cleaner, spark plug, and muffler.

LUBRICATION SYSTEM

All four stroke engines with vertical crankshafts are lubricated by either a camshaft driven gear type oil pump or an oil scoop attached to the camshaft which rotates in the oil lying in the bottom of the crankcase. The gear type pump forces oil up through a steel tube to the upper main bearing. The oil then falls down onto other parts needing to be lubricated.

The oil scoop attached to the camshaft sprays oil to the top of the engine in a circular path. The upper main bearing in oil scoop engines has an oil access slot in the top of the bearing area to catch the oil as it is splashed up.

On four-stroke engines that have a horizontal crankshaft, there is an oil dipper or distributor located on the bottom of the connecting rod bearing cap. As the crankshaft turns, the dipper churns through the oil in the crankcase and splashes it to all

OIL PUMP

Gear type oil pump

Camshaft mounted oil scoop

ROTATION

ROTATION

Connecting rod cap mounted oil dipper

moving parts that need lubrication. There are oil passages, holes, and slots which the oil passes through on its way to the bearing surfaces.

All connecting rods have oil access holes or chambers to insure adequate lubrication. Make sure that all oil access holes are installed facing toward the top main bearing in engines with vertical crankshafts.

General Engine Specifications

		1	2	3	4	5	6	7	8	9
Cylinder Bore Dia.	Min.	2.3745	2.3745	2.3745	2.3745	2.000	2.000	2.000	2.125	2.375
	Max.	2.3755	2.3755	2.3755	2.3755	2.001	2.001	2.001	2.126	2.376
Piston Skirt Dia.	Min.	2.3690	2.3690	2.3690	2.3690	1.9935	1.9935	1.9935	2.119	2.3690
	Max.	2.3700	2.3700	2.3700	2.3700	1.9945	1.9945	1.9945	2.120	2.3700
Piston Skirt to Cylinder Clearance	Min.	.0045	.0045	.0045	.0045	.0045	.0045	.0045	.005	.0055
	Max.	.0065	.0065	.0065	.0065	.0065	.0065	.0065	.007	.0075
Piston Ring to Groove Clearance	Min.	.002	.002	.002	.002	.002	.002	.002	.002	.002
	Max.	.005	.005	.005	.005	.005	.005	.005	.005	.005
Ring End Gap in Cylinder	Min.	.007	.007	.007	.007	.007	.007	.007	.007	.007
	Max.	.017	.017	.017	.017	.017	.017	.017	.017	.017
Connecting Rod Bore Crankshaft End	Min.	.8140	.8770	.8140	.8770	.7515	.7515	.8770	.8770	.8770
	Max.	.8145	.8775	.8145	.8775	.7520	.7520	.8775	.8775	.8775
Connecting Rod to Crankshaft Clearance	Min.	.0015	.0015	.0015	.0015	.0018	.0018	.0018	.0018	.0018
	Max.	.0030	.0030	.0030	.0030	.0035	.0035	.0035	.0035	.0035

General Engine Specifications

		1	2	3	4	5	6	7	8	9
Connecting Rod to Wrist Pin Clearance	Min.	.0004	.0004	.0004	.0004	.0004	.0004	.0004	.0004	.0004
	Max.	.0011	.0011	.0011	.0011	.0011	.0011	.0011	.0011	.0011
Crankshaft Rod Pin Diameter	Min.	.8119	.8745	.8119	.8745	.7483	.7483	.8745	.8745	.8754
	Max.	.8125	.8752	.8125	.8752	.7490	.7490	.8752	.8752	.8752
Crankpin Out-of-Round	Max.	.001	.001	.001	.001	.001	.001	.001	.001	.001
Crankshaft Main Diameter P.T.O. End	Min.	.8733	.8733	.8733	Note	.7483	.8745	.8745	.8745	.8745
	Max.	.8740	.8740	.8740	#1	.7490	.8752	.8752	.8752	.8752
Crankshaft Main Diameter Flywheel End	Min.	.8120	.8120	.8120	.8120	.7483	.7483	.8745	.8745	.8745
	Max.	.8127	.8127	.8127	.8127	.7490	.7490	.8752	.8752	.8752
Crankshaft to Main Bearing Clearance	Min.	.0018	.0018	.0018	.0018	.0018	.0018	.0018	.0018	.0018
	Max.	.0035	.0035	.0035	.0035	.0035	.0035	.0035	.0035	.0035
Crankshaft End Play	Min.	.008	.008	.008	.008	.004	.004	.008	.008	.008
	Max.	.018	.018	.018	.018	.012	.012	.018	.018	.018
Block or Bearing Plate Main Bearing Bore P.T.O. End	Min.	.8758	.8758	.8758	Note	.7510	.877	.877	.877	.877
	Max.	.8768	.8768	.8768	#5	.7520	.878	.878	.878	.878
Bearing Plate or Block Bearing Bore (Flywheel) End	Min.	.8145	.8145	.8145	.8145	.7510	.7510	.877	.877	.877
	Max.	.8155	.8155	.8155	.8155	.7520	.7520	.878	.878	.878
Valve or Tappet Guide Bore I.D.	Min.	.2495	.2495	.2495	.2495	.2495	.2495	.2495	.2495	.2495
	Max.	.2510	.2510	.2510	.2510	.2510	.2510	.2510	.2510	.2510
Valve Stem to Guide Clearance	Min.	.0015	.0015	.0015	.0015	.0020	.0020	.0020	.0020	.0020
	Max.	.0045	.0045	.0045	.0045	.0045	.0045	.0045	.0045	.0045
Valve Clearance, Intake & Exhaust	Min.	.009	.009	.009	.009	.007	.007	.009	.009	.009
	Max.	.011	.011	.011	.011	.009	.009	.011	.011	.011
Camshaft to Axle Clearance	Min.	.001	.001	.001	.001	.001	.001	.001	.001	.001
	Max.	.003	.003	.003	.003	.003	.003	.003	.003	.003
Camshaft Axle Clearance P.T.O.	Min.	.001	.001	.001	.001					
	Max.	.003	.003	.003	.003					
Camshaft Axle Clearance Flywheel End	Min.	.001	.001	.001	.001					
	Max.	.003	.003	.003	.003					
Point Setting	Min.	.018	.018	.018	.018	.018	.018	.018	.018	.018
	Max.	.021	.021	.021	.021	.021	.021	.021	.021	.021
Spark Plug Gap	Min.	.025	.025	.025	.025	.025	.025	.025	.025	.025
	Max.	.028	.028	.028	.028	.028	.028	.028	.028	.028
Comp. at Cranking Speed, P.S.I.	Min.	65	65	65	65	65	65	65	65	65
Carburetor Float Setting (Clinton)	Min.	$5/32$	$5/32$	$5/32$	$5/32$	$5/32$	$5/32$	$5/32$	$5/32$	$5/32$
	Max.	$11/64$	$11/64$	$11/64$	$11/64$	$11/64$	$11/64$	$11/64$	$11/64$	$11/64$
Carburetor Float Setting (Carter)	Min.	$11/64$	$11/64$	$11/64$	$11/64$	$11/64$	$11/64$	$11/64$	$11/64$	$11/64$
	Max.	$13/64$	$13/64$	$13/64$	$13/64$	$13/64$	$13/64$	$13/64$	$13/64$	$13/64$
Magneto Air Gap	Min.	.007	.007	.007	.007	.007	.007	.007	.007	.007
	Max.	.017	.017	.017	.017	.017	.017	.017	.017	.017
Magneto Edge Gap (Phelon)	Min.	$5/32$	$5/32$	$5/32$	$5/32$	$5/32$	$5/32$	$5/32$	$5/32$	$5/32$
	Max.	$9/32$	$9/32$	$9/32$	$9/32$	$9/32$	$9/32$	$9/32$	$9/32$	$9/32$
Magneto Edge Gap (Clinton)	Min.	$7/64$	$7/64$	$7/64$	$7/64$			$7/64$	$7/64$	$7/64$
	Max.	$1/4$	$1/4$	$1/4$	$1/4$			$1/4$	$1/4$	$1/4$
Oil Recommended A.P.I. Rating	See Bulletin #9 for SAE	MM MS	MM MS	MM MS	MM MS	MM MS	MM MS	MM MS	MM MS	MM MS
Fuel Recommended		Reg. Gas	Reg. Gas	Reg. Gas	Reg. Gas	Reg. Gas	Reg. Gas	Reg. Gas	Reg. Gas	Reg. Gas

Note 1 Crankshaft diameter .8733 to .8740 or .9995 to 1.0002, except 435-0000-000 which is 1.220 to 1.221.
Note 2 Ball, needle or roller bearing used. Check component parts, if worn or defective, replace.
Note 3 Wrist pin clearance, .0004 minimum, .0011 maximum or umits other than listed below. Needle bearing used in D65-1000 Type (C), J9-1000 Type (B), J6-1100 and J6-1101. Check component parts, if worn or defective, replace.
Note 4 412-0000-000, 413-0000-000 with sleeve bearings .002 to .006, 412-0300-000 with roller bearings .001 to .006.
Note 5 Bearing bore diameter .8758 to .8768 or 1.0002 to 1.0003, except 435-0000-000 which is 1.2228 to 1.2235.
Note 6 Valve guides .312 to .313 inside diameter and are replaceble. Tappet guide inside diameter .437 to .438.
Note 7 412-0000-000 .001 minimum, .004 maximum. 413-0000-000 .0005 minimum, .002 maximum.

General Engine Specifications (cont.)

		10	11	12	13	14	15	16	17	18
Cylinder Bore Dia.	Min.	2.375	2.375	2.4685	2.4685	2.4685	2.499	2.8125	2.8125	2.8125
	Max.	2.376	2.376	2.4695	2.4695	2.4695	2.500	2.8135	2.8135	2.8135
Piston Skirt Dia.	Min.	2.3690	2.3690	2.4630	2.4630	2.4630	2.4935	2.8045	2.8045	2.8045
	Max.	2.3700	2.3700	2.4640	2.4640	2.4640	2.4945	2.8055	2.8055	2.8055
Piston Skirt to Cylinder	Min.	.0055	.0055	.0055	.0055	.0055	.0045	.007	.007	.007
Clearance	Max.	.0075	.0075	.0075	.0075	.0075	.0065	.009	.009	.009
Piston Ring to Groove	Min.	.002	.002	.002	.002	.002	.002	.0025	.0025	.0025
Clearance	Max.	.005	.005	.005	.005	.005	.005	.0050	.0050	.0050
Ring End Gap in	Min.	.007	.007	.007	.007	.007	.007	.007	.007	.007
Cylinder	Max.	.017	.017	.017	.017	.017	.017	.017	.017	.017
Connecting Rod Bore	Min.	.8770	.9140	.8770	.9140	.9140	.8770	1.2510	1.1265	1.1265
Crankshaft Ends	Max.	.8775	.9145	.8775	.9145	.9145	.8775	1.2513	1.1268	1.1268
Connecting Rod to Crankshaft	Min.	.0018	.0018	.0018	.0018	.0018	.0015	.001	.0015	.0015
Clearance	Max.	.0035	.0035	.0035	.0035	.0035	.0035	.002	.0025	.0025
Connecting Rod to Wrist	Min.	.0004	.0004	.0004	.0004	.0004	.0004	.0002	.0002	.0002
Pin Clearance	Max.	.0011	.0011	.0011	.0011	.0011	.0011	.0011	.0011	.0011
Crankshaft Rod Pin	Min.	.8745	.9114	.8745	.9114	.9114	.8745	1.2493	1.1243	1.1243
Diameter	Max.	.8752	.9120	.8752	.9120	.9120	.8752	1.2500	1.1250	1.1250
Crankpin Out-of Round	Max.	.001	.001	.001	.001	.001	.001	.001	.001	.001
Crankshaft Main Diameter	Min.	.9995	.9995	Note	Note	.9995	1.220	1.2510	Note	Note
P.T.O. End	Max.	1.0002	1.0002	#2	#2	1.0002	1.221	1.2515	#2	#2
Crankshaft Main Diameter	Min.	.8745	.8745	Note	Note	.8745	.8120	1.2510	Note	Note
Flywheel End	Max.	.8752	.8752	#2	#2	.8752	.8127	1.2515	#2	#2
Crankshaft to Main Bearing	Min.	.0018	.0018	Note	Note	.0018	.0018	.0010	Note	Note
Clearance	Max.	.0035	.0035	#2	#2	.0035	.0035	.0025	#2	#2
Crankshaft End	Min.	.008	.008	.001	.001	.008	.008	Note	.006	.001
Play	Max.	.018	.018	.006	.006	.018	.018	#4	.020	.006
Block or Bearing Plate Main	Min.	1.002	1.002	Note	Note	1.002	1.2228	1.2525	Note	Note
Bearing Bore P.T.O. End	Max.	1.003	1.003	#2	#2	1.003	1.2235	1.2535	#2	#2
Bearing Plate or Block Bearing	Min.	.877	.877	Note	Note	.877	.8145	1.2525	Note	Note
Bore (Flywheel) End	Max.	.878	.878	#2	#2	.878	.8155	1.2535	#2	#2
Valve or Tappet Guide Bore	Min.	.2495	.2495	.2495	.2495	.2495	.2495	.312	Note	Note
I.D.	Max.	.2510	.2510	.2510	.2510	.2510	.2510	.313	#6	#6
Valve Stem to Guide Clearance	Min.	.0020	.0020	.0020	.0020	.0020	.0015	.002	.002	.002
	Max.	.0045	.0045	.0045	.0045	.0045	.0045	.004	.004	.004
Valve Clearance, Intake &	Min.	.009	.009	.009	.009	.009	.009	.011	.010	.010
Exhaust	Max.	.011	.010	.010	.010	.010	.011	.012	.012	.012
Camshaft to Axle	Min.	.001	.001	.001	.001	.001			.0015	.0015
Clearance	Max.	.003	.003	.003	.003	.003			.0035	.0035
Camshaft Axle Clearance	Min.						.001	Note		
P.T.O.	Max.						.003	#7		
Camshaft Axle Clearance	Min.						.001	.001		
Flywheel End	Max.						.003	.003		
Point Setting	Min.	.018	.018	.018	.018	.018	.018	.018	.028	.028
	Max.	.021	.021	.021	.021	.021	.021	.021	.030	.030
Spark Plug Gap	Min.	.025	.025	.025	.025	.025	.025	.025	.025	.025
	Max.	.028	.028	.028	.028	.028	.028	.028	.028	.028
Comp. at Cranking Speed, P.S.I.	Min.	65	65	65	65	65	65	65	70	70
Carburetor Float Setting (Clinton)	Min.	$5/32$	$5/32$	$5/32$	$5/32$	$5/32$	$5/32$	$5/32$		$11/64$
	Max.	$11/64$	$11/64$	$11/64$	$11/64$	$11/64$	$11/64$	$11/64$		$13/64$
Carburetor Float Setting (Carter)	Min.	$11/64$	$11/64$	$11/64$	$11/64$	$11/64$			$15/64$	$15/64$
	Max.	$13/64$	$13/64$	$13/64$	$13/64$	$13/64$			$17/64$	$17/64$
Magneto Air Gap	Min.	.007	.007	.007	.007	.007	.007	.007	.012	.012
	Max.	.017	.017	.017	.017	.017	.017	.017	.020	.020
Magneto Edge Gap (Phelon)	Min.	$5/32$	$5/32$	$5/32$	$5/32$	$5/32$				
	Max.	$9/32$	$9/32$	$9/32$	$9/32$	$9/32$				

General Engine Specifications (cont.)

		10	11	12	13	14	15	16	17	18
Magneto Edge Gap (Clinton)	Min.	7/64	7/64	7/64	7/64	7/64	7/64	7/64		
	Max.	1/4	1/4	1/4	1/4	1/4	1/4	1/4		
Oil Recommended A.P.I. Rating	See Bulletin #9 for SAE	MM MS	MM MS	MM MS	MM MS	MM MS	MM MS	MM MS	MM MS	MM MS
Fuel Recommended		Reg. Gas	Reg. Gas	Reg. Gas	Reg. Gas	Reg. Gas	Reg. Gas	Reg. Gas	Reg. Gas	Reg. Gas

Note 1 Crankshaft diameter .8733 to .8740 or .9995 to 1.0002, except 435-0000-000 which is 1.220 to 1.221.
Note 2 Ball, needle or roller bearing used. Check component parts, if worn or defective, replace.
Note 3 Wrist pin clearance, .0004 minimum, .0011 maximum or umits other than listed below. Needle bearing used in D65-1000 Type (C), J9-1000 Type (B), J6-1100 and J6-1101. Check component parts, if worn or defective, replace.
Note 4 412-0000-000, 413-0000-000 with sleeve bearings .002 to .006, 412-0300-000 with roller bearings .001 to .006.
Note 5 Bearing bore diameter .8758 to .8768 or 1.0002 to 1.0003, except 435-0000-000 which is 1.2228 to 1.2235.
Note 6 Valve guides .312 to .313 inside diameter and are replaceble. Tappet guide inside diameter .437 to .438.
Note 7 412-0000-000 .001 minimum, .004 maximum. 413-0000-000 .0005 minimum, .002 maximum.

General Engine Specification (cont.)

		19	20	21	22	23	24	25	26	27
Cylinder Bore Dia.	Min.	2.9995	3.1245	3.1245	3.1245	3.1245	1.875	1.875	2.125	2.125
	Max.	3.0005	3.1255	3.1255	3.1255	3.1255	1.876	1.876	2.126	2.126
Piston Skirt Dia.	Min.	2.9915	3.117	3.117	3.1185	3.1185	1.8695	1.8695	2.1195	2.1195
	Max.	2.9925	3.118	3.118	3.1195	3.1195	1.8705	1.8705	2.1205	2.1205
Piston Skirt to Cylinder Clearance	Min.	.007	.0065	.0065	.005	.005	.0045	.0045	.005	.005
	Max.	.009	.0085	.0085	.007	.007	.0065	.0065	.007	.007
Piston Ring to Groove Clearance	Min.	.0025	.0025	.0025	.0025	.0025	.0015	.0015	.0015	.0015
	Max.	.0050	.0050	.0050	.0050	.0050	.0040	.0040	.0040	.0040
Ring End Gap in Cylinder	Min.	.007	.010	.010	.010	.010	.005	.005	.007	.007
	Max.	.017	.020	.020	.020	.020	.013	.013	.017	.017
Connecting Rod Bore Crankshaft End	Min.	1.1265	1.2510	1.2510	1.2510	1.2510	.7820	.7820	.7820	.7820
	Max.	1.1268	1.2513	1.2513	1.2513	1.2513	.7827	.7827	.7827	.7827
Connecting Rod to Crankshaft Clearance	Min.	.0015	.0010	.0010	.0010	.0010	.0026	.0026	.0026	.0026
	Max.	.0025	.0018	.0018	.0018	.0018	.0040	.0040	.0040	.0040
Connecting Rod to Wrist Pin Clearance	Min.	.0002	.0002	.0002	.0002	.0002	.0004	.0004	.0004	.0004
	Max.	.0011	.0011	.0011	.0011	.0011	.0011	.0011	.0011	.0011
Crankshaft Rod Pin Diameter	Min.	1.1243	1.2495	1.2495	1.2495	1.2495	.7788	.7788	.7788	.7788
	Max.	1.1250	1.2500	1.2500	1.2500	1.2500	.7795	.7795	.7795	.7795
Crankpin Out-of-Round	Max.	.001	.001	.001	.001	.001	.001	.001	.001	.001
Crankshaft Main Diameter P.T.O. End	Min.	Note #2	Note #2	Note #2	Note #2	Note #2	.8745	.8745	.8745	.9995
	Max.						.8752	.8752	.8752	1.0002
Crankshaft Main Diameter Flywheel End	Min.	Note #2	Note #2	Note #2	Note #2	Note #2	.7495	.7495	.7495	.7495
	Max.						.7502	.7502	.7502	.7502
Crankshaft to Main Bearing Clearance	Min.	Note #2	Note #2	Note #2	Note #2	Note #2	.0015	.0015	.0015	.0015
	Max.						.0035	.0035	.0035	.0035
Crankshaft End Play	Min.	.001	.006	.001	.006	.001	.005	.005	.005	.005
	Max.	.006	.020	.006	.020	.006	.020	.020	.020	.020
Block or Bearing Plate Main Bearing Bore P.T.O. End	Min.	Note #2	Note #2	Note #2	Note #2	Note #2	.8770	.8770	.8770	1.002
	Max.						.8780	.8780	.8780	1.003
Bearing Plate or Block Bearing Bore (Flywheel) End	Min.	Note #2	Note #2	Note #2	Note #2	Note #2	.7517	.7517	.7517	.7517
	Max.						.7525	.7525	.7525	.7525
Valve or Tappet Guide Bore I.D.	Min.	Note #6	Note #6	Note #6	Note #6	Note #6				
	Max.									
Valve Stem to Guide Clearance	Min.	.002	.002	.002	.002	.002				
	Max.	.004	.004	.004	.004	.004				
Valve Clearance, Intake & Exhaust	Min.	.010	.010	.010	.010	.010				
	Max.	.012	.012	.012	.012	.012				
Camshaft to Axle Clearance	Min.	.0015	.0015	.0015	.0015	.0015				
	Max.	.0035	.0035	.0035	.0035	.0035				
Camshaft Axle Clearance P.T.O.	Min.									
	Max.									
Camshaft Axle Clearance Flywheel End	Min.									
	Max.									

General Engine Specification (cont.)

		19	20	21	22	23	24	25	26	27
Point Setting	Min.	.028	.028	.028	.028	.028	.018	.018	.018	.018
	Max.	.030	.030	.030	.030	.030	.021	.021	.021	.021
Spark Plug Gap	Min.	.025	.025	.025	.025	.025	.025	.025	.025	.025
	Max.	.028	.028	.028	.028	.028	.028	.028	.028	.028
Comp. at Cranking Speed. P.S.I.	Min.	70	70	70	70	70	60	60	60	60
Carburetor Float Setting (Clinton)	Min.	11/64		11/64		11/64	5/32	5/32	5/32	5/32
	Max.	13/64		13/64		13/64	11/64	11/64	11/64	11/64
Carburetor Float Setting (Carter)	Min.	15/64	15/64	15/64	15/64	15/64	11/64	11/64	11/64	11/64
	Max.	17/64	17/64	17/64	17/64	17/64	13/64	13/64	13/64	13/64
Magneto Air Gap	Min.	.012	.012	.012	.012	.012	.007	.007	.007	.007
	Max.	.020	.020	.020	.020	.020	.017	.017	.017	.017
Magneto Edge Gap (Phelon)	Min.						5/32	5/32	5/32	5/32
	Max.						9/32	9/32	9/32	9/32
Magneto Edge Gap (Clinton)	Min.						7/64	7/64	7/64	7/64
	Max.						1/4	1/4	1/4	1/4
Oil Recommended A.P.I. Rating	See Bulletin #9 for SAE	MM MS	MM MS	MM MS	MM MS	MM MS	Outboard Motor Oil or SAE 30 Non Detergent			
Fuel Recommended		Reg. Gas	Reg. Gas	Reg. Gas	Reg. Gas	Reg. Gas	Reg. Gas	Reg. Gas	Reg. Gas	Reg. Gas

Note 1 Crankshaft diameter .8733 to .8740 or .9995 to 1.0002, except 435-0000-000 which is 1.220 to 1.221.
Note 2 Ball, needle or roller bearing used. Check component parts, if worn or defective, replace.
Note 3 Wrist pin clearance, .0004 minimum, .0011 maximum or umits other than listed below. Needle bearing used in D65-1000 Type (C), J9-1000 Type (B), J6-1100 and J6-1101. Check component parts, if worn or defective, replace.
Note 4 412-0000-000, 413-0000-000 with sleeve bearings .002 to .006, 412-0300-000 with roller bearings .001 to .006.
Note 5 Bearing bore diameter .8758 to .8768 or 1.0002 to 1.0003, except 435-0000-000 which is 1.2228 to 1.2235.
Note 6 Valve guides .312 to .313 inside diameter and are replaceable. Tappet guide inside diameter .437 to .438.
Note 7 412-0000-000 .001 minimum, .004 maximum. 413-0000-000 .0005 minimum, .002 maximum.

General Engine Specifications (cont.)

		28	29	30	31	32	33	34	35	36
Cylinder Bore Dia.	Min.	2.125	2.125	1.8750	2.1255	1.8750	1.811	1.8750	2.1255	2.1255
	Max.	2.126	2.126	1.8765	2.1270	1.8765	1.814	1.8765	2.1270	2.1270
Piston Skirt Dia.	Min.	2.1195	2.1195	1.8695	2.120	1.871	1.807	1.871	2.120	2.120
	Max.	2.1205	2.1205	1.8705	2.121	1.872	1.809	1.872	2.121	2.121
Piston Skirt to Cylinder Clearance	Min.	.005	.005	.0045	.0045	.0045	.003	.0045	.0045	.0045
	Max.	.007	.007	.0065	.0070	.0065	.007	.0065	.0070	.0070
Piston Ring to Groove Clearance	Min.	.0015	.002	.002	.002	.002	.002	.002	.002	.002
	Max.	.0040	.004	.004	.004	.004	.005	.004	.004	.004
Ring End Gap in Cylinder	Min.	.007	.010	.010	.010	.010	.004	.010	.010	.010
	Max.	.017	.015	.015	.015	.015	.020	.015	.015	.015
Connecting Rod Bore Crankshaft End	Min.	.7820	Note #2	.7816	Note #2	Note #2	Note #2	Note #2	Note #2	Note #2
	Max.	.7827		.7825						
Connecting Rod to Crankshaft Clearance	Min.	.0026	Note #2	.0021	Note #2	Note #2	Note #2	Note #2	Note #2	Note #2
	Max.	.0040		.0037						
Connecting Rod to Wrist Pin Clearance	Min.	.0004	.0004	.0004	Note #3	.0004	Note #2	.0004	Note #3	Note #3
	Max.	.0011	.0011	.0011		.0011		.0011		
Crankshaft Rod Pin Diameter	Min.	.7788	.6594	.7788	.6594	.6594	.4960	.6594	.6594	.6594
	Max.	.7795	.6599	.7795	.6599	.6599	.4965	.6599	.6599	.6599
Crankpin Out-of-Round	Max.	.001	.001	.001	.001	.001	.001	.001	.001	.001
Crankshaft Main Diameter P.T.O. End	Min.	.8745	Note #2	Note #2	Note #2	Note #2	Note #2	Note #2	Note #2	Note #2
	Max.	.8752								
Crankshaft Main Diameter Flywheel End	Min.	.7495	.7495	.7495	.7495	.7495	Note #2	.7495	.7495	.7495
	Max.	.7502	.7502	.7502	.7502	.7502		.7502	.7502	.7502
Crankshaft to Main Bearing Clearance	Min.	.0015	Note #2	.0015	Note #2	Note #2	Note #2	Note #2	Note #2	Note #2
	Max.	.0035		.0035						
Crankshaft End Play	Min.	.005	Note #2	Note #2	Note #2	Note #2	.003	.003	.003	Note #2
	Max.	.020					.020	.020	.020	
Block or Bearing Plate Main Bearing Bore P.T.O. End	Min.	.8770	Note #2	Note #2	Note #2	Note #2	Note #2	Note #2	Note #2	Note #2
	Max.	.8778								

General Engine Specifications (cont.)

		28	29	30	31	32	33	34	35	36
Bearing Plate or Block Bearing Bore (Flywheel) End	Min.	.7517	Note #2	Note #2	Note #2	Note #2	Note #2	Note #2	Note #2	Note #2
	Max.	.7525								
Valve or Tappet Guide Bore I.D.	Min.									
	Max.									
Valve Stem to Guide Clearance	Min.									
	Max.									
Valve Clearance, Intake & Exhaust	Min.									
	Max.									
Camshaft to Axle Clearance	Min.									
	Max.									
Camshaft Axle Clearance P.T.O	Min.									
	Max.									
Camshaft Axle Clearance Flywheel End	Min.									
	Max.									
Point Setting	Min.	.018	.018	.018	.018	.018	.013	.014	.014	.018
	Max.	.021	.021	.021	.021	.021	.017	.016	.016	.021
Spark Plug Gap	Min.	.025	.025	.025	.025	.025	.025	.025	.025	.025
	Max.	.028	.028	.028	.028	.028	.030	.028	.028	.028
Comp. at Cranking Speed. P.S.I.	Min.	60	60	60	60	60	80	60	60	60
Carburetor Float Setting (Clinton)	Min.	$5/32$	$5/32$							$1 25/64$
	Max.	$11/64$	$11/64$							$1 27/64$
Carburetor Float Setting (Carter)	Min.	$11/64$	$11/64$							
	Max.	$13/64$	$13/64$							
Magneto Air Gap	Min.	.007	.007	.007	.007	.007	.007	.007	.007	.007
	Max.	.017	.017	.017	.017	.017	.017	.017	.017	.017
Magneto Edge Gap (Phelon)	Min.	$5/32$								
	Max.	$9/32$								
Magneto Edge Gap (Clinton)	Min.	$7/64$	$7/64$	$7/64$	$7/64$	$7/64$	$3/32$	$3/32$	$3/32$	$7/64$
	Max.	$1/4$	$1/4$	$1/4$	$1/4$	$1/4$	$5/16$	$5/16$	$5/16$	$1/4$
Oil Recommended A.P.I. Rating		Outboard Motor Oil or SAE 30 Non Detergent								
Fuel Recommended		Reg. Gas	Reg. Gas	Reg. Gas	Reg. Gas	Reg. Gas	Reg. Gas	Reg. Gas	Reg Gas	Reg. Gas

Note 1 Crankshaft diameter .8733 to .8740 or .9995 to 1.0002, except 435-0000-000 which is 1.220 to 1.221.
Note 2 Ball, needle or roller bearing used. Check component parts, if worn or defective, replace.
Note 3 Wrist pin clearance, .0004 minimum .0011 maximum or umits other than listed below. Needle bearing used in D65-1000 Type (C), J9-1000 Type (B), J6-1100 and J6-1101. Check component parts, if worn or defective, replace.
Note 4 412-0000-000, 413-0000-000 with sleeve bearings .002 to .006, 412-0300-000 with roller bearings .001 to .006.
Note 5 Bearing bore diameter .8758 to .8768 or 1.0002 to 1.0003, except 435-0000-000 which is 1.2228 to 1.2235.
Note 6 Valve guides .312 to .313 inside diameter and are replaceble. Tappet guide inside diameter .437 to .438.
Note 7 412-0000-000 .001 minimum .004 maximum. 413-0000-000 .0005 minimum, .002 maximum.

Torque Specifications

		All 4 Cycle Aluminum Vertical & Horizontal Shaft	V100, V3100, 406, H3100 407, 417	429 & 431	A300	300, VS300, 350, 500, 650, 700A, C700 D700, VS700, VS750, 800, A800, VS800 900, VS900, V1000, A-B-1100, C1100, D1100, V1100, VS1100, 1200, A1200, B1290, V1200, VS1200, 492, 494, 497, 498, 499	412 & 413	1600, 1800, 2500, 2790 414, 416, 418, 420, 422	200, A200, VS200, VS400, A400, 500, 501	E65 & GK590	E10 — Chainsaw	D15 — Chainsaw D25 — Chainsaw	D35, D55, D65 — Chainsaws E75, E95, — Chainsaws	J5, J6, J7, J8, J9 — Outboards
Connecting Rod Aluminum	Min.	100	100	100	70	70	215	215	35	70	50	55	80	80
	Max.	125	125	125	80	80	235	235	45	80	55	65	90	90

Torque Specifications

Component		All 4 Cycle Aluminum Vertical & Horizontal Shaft	V100, V3100, 406, H3100 407, 417	429 & 431	A300	300, VS300, 350, 500, 650, 700A, C700 D700, VS700, 800, A800, VS800 900, VS900, V1000, A-B-1100, C1100, D1100, V1100, VS1100, 1200, A1200, B1290, V1200, VS1200, 492, 494, 497, 498, 499	412 & 413	1600, 1800, 2500, 2790 414, 416, 418, 420, 422	200, A200, VS200, VS400, A400, 500, 501	E65 & GK590	E10 — Chainsaw	D15 — Chainsaw D25 — Chainsaw	D35, D55, D65 — Chainsaws E75, E95, — Chainsaws	J5, J6, J7, J8, J9 — Outboards
Connecting Rod Forged Steel	Min.									90			90	90
	Max.									100			100	100
Bearing Plate P.T.O. End	Min.	75	75	75			75	160						
	Max.	85	85	85			85	180						
Bearing Plate Flying End	Min.				120	140			75	75		80	80	75
	Max.				150	160			95	95		90	90	95
Back Plate to Block	Min.							70						
	Max.							80						
Head Bolts	Min.	125	225	225	200	200	200	200						
	Max.	150	250	250	220	220	220	220						
Base Bolts	Min.	75	75	75	150	325	75	150	125	125				
	Max.	85	85	85	160	375	85	160	150	150				
End Cover or Gear Box	Min.					120								
	Max.					150								
Speed Reducer Mounting	Min.	110	110		110	110	110	110	110					
	Max.	150	150		150	150	150	150	150					
P.T.O. Housing or Mounting Flange	Min.	75	75	75		120	75	160	125					
	Max.	85	85	85		150	85	180	150					
Carb. Reed Plate or Manifold to Blk.	Min.	60	60	60	60	60	65	60	60	60	45	50	50	50
	Max.	65	65	65	65	65	75	65	65	65	55	60	60	60
Carb. to Reed Plate or Manifold	Min.	35	35	35	60	35	35	60	60	60		50	50	50
	Max.	50	50	50	65	50	50	65	65	65		60	60	60
Blower Housing	Min.	60	60	60	65	65	65	65	65	65	90	80	80	75
	Max.	70	70	70	70	70	75	70	70	70	110	90	90	85
Muffler to Block	Min.	110	110	110	140	140	170	170	40	40	90	60	60	
	Max.	120	120	120	150	150	180	180	60	60	110	70	70	
Flywheel	Min.	375	375	375	375	400	400	*100	375	250	150	375	**250	250
	Max.	400	400	400	400	450	500	*120	400	300	180	425	**300	300
Flywheel Touch & Stop for Brake	Min.	650	650											
	Max.	700	700											
Spark Plug	Min.	275	275	275	275	275	275	275	275	275	250	230	230	230
	Max.	300	300	300	300	300	300	300	300	300	300	270	270	270
Stator Plate	Min.	50	50	50	50	50	50	80	50	50	50	50	50	45
	Max.	60	60	60	60	60	60	100	60	60	60	60	60	65

All torque in inch pounds except those marked with a single*   *Foot Pounds   **D35 Requires Same as D25

CLINTON
SMALL ENGINES – 0–6 HP

Cross Reference Chart

Model No.	Column No.	Model No.	Column No.	Model No.	Column No.	Model No.	Column No.
E–65	29	D1100	9	J–9	36	415–0000–000	4
		V1100–1000	11	E10–1000	33	415–0002–000	4
100	1	VS1100	11	D15–1000	30	416–1300–000	18
100–1000	1	VS1100–1000	11	D25–1000	30	417–0000–000	4
100–2000	1			D35–1000	31	418–1300–000	19
V100–1000	3	1200	12	D55–1000	32	418–1301–000	19
VS100	3	1200–1000	12	D65–1000	31	420–1300–000	21
VS100–1000	3			E75–1000	34	420–1301–000	21
VS100–2000	3	1200–2000	13	E95–1000	35	422–1300–000	23
VS100–3000	3	A1200	13			422–1301–000	23
VS100–4000	3	B1290–1000	13	400–0000–000	1	424–0000–000	2
		V1200–1000	14	401–0000–000	3	426–0000–000	2
200	24	VS1200	14	402–0000–000	1	429–0003–000	15
A200	24			403–0000–000	3	431–0003–000	15
AVS200	24	1600	17	404–0000–000	2	435–0003–000	4
AVS200–1000	25	A1600–1000	18	405–0000–000	4	492–0300–000	9
VS200	24			406–0000–000	2	494–0000–000	9
VS200–1000	24	1800–1000	19	407–0000–000	4	494–0001–000	9
VS200–2000	25			407–0002–000	4	497–0000–000	14
VS200–3000	25	2100	1	408–0000–000	2	498–0300–000	13
VS200–4000	27	A2100	1	409–0000–000	4	498–0301–000	13
		A2100–1000	1	411–0000–000	3	499–0000–000	14
300	5	A2100–2000	1	411–0002–000	3		
A300	5	VS2100	3	412–0000–000	16	500–0000–000	28
VS300	6	VS2100–1000	3	413–0000–000	16	501–0000–000	27
350	5	VS2100–2000	3	414–1300–000	18	501–0001–000	27
		VS2100–3000	3	414–1301–000	18		
A400	28						
A400–1000	28	2500	20				
AVS400	26	A2500	22				
AVS400–1000	27	B2500–1000	21				
BVS400	27						
CVS400–1000	27	2790–1000	23				
VS400	26						
VS400–1000	26	VS3000	3				
VS400–2000	27						
VS400–3000	27	3100	2				
VS400–4000	27	3100–1000	2				
		3100–2000	2				
GK590	29	3100–3000	2				
		H3100–1000	2				
650	7	FV3100–1000	4				
		AFV3100–1000	4				
700–A	7	AV3100–1000	4				
B700	7	AV3100–2000	4				
C700	7	AVS3100	4				
D700	8	AVS3100–1000	4				
D700–1000	8	AVS3100–2000	4				
D700–2000	8	AVS3100–3000	4				
D700–3000	8	V3100–1000	4				
VS700	7	V3100–2000	4				
VS750	7	VS3100	4				
		VS3100–1000	4				
800	9	VS3100–2000	4				
A800	9	VS3100–3000	4				
VS800	9						
		4100	2				
900	9	4100–1000	2				
900–1000	9	4100–2000	2				
900–2000	9	AVS4100–1000	4				
900–3000	9	AVS4100–2000	4				
900–4000	9	VS4100–1000	4				
VS900	10	VS4100–2000	4				
V1000–1000	10	J–5	36				
VS1000	10	J–6	36				
		J–7	36				
A&B1100	9	J–8	36				
C1100	9						

Section 7

Clinton

Small Engines — 6-20 hp

ENGINE IDENTIFICATION

Clinton engines are identified by a name plate that is installed on the engine at the factory. The name plate contains the serial number and the model number, both of which are necessary when obtaining replacement parts.

Two numbering systems are used to identify Clinton engines. The first one applies to engines made prior to 1961, the second pertains to engines made after that year.

The early numbering system has no practical use to the consumer. It is useful only when ordering parts and only to one who has access to a parts manual. No other information can be gained from the model or serial number than what parts fit a particular engine.

The recent numbering system, however, is quite useful in gaining additional information about a certain engine. The serial number, followed by the type letter, is a numerically sequential number and is used to identify the engine in relation to changes that are made in design. The model number consists of a ten digit sequence of numbers that is deciphered in the following manner:

The first digit indicates whether the engine is a two stroke or four stroke design. The number 4 indicates a four stroke engine and the number 5 indicates a two stroke engine.

Numbering system after 1961

The second and third digits identify the basic engine series and whether the engine has a vertical or horizontal crankshaft. Odd numbers in the third digit position indicate that the engine has a vertical crankshaft and even numbers indicate a horizontal crankshaft.

The fourth digit identifies the type of starter installed on the engine: 0 — a recoil starter, 1 — a rope starter, 2 — an impulse starter, 3 — a crank starter, 4 — a 12 volt electric starter, 5 — a 12 volt starter generator, 6 — a 110 volt electric starter, 7 — a 12 volt generator, 8 — not assigned any specific meaning, 9 — indicates a short block.

The fifth digit indicates what type of bearing is used on the crankshaft; 1 — aluminum or bronze sleeve bearing with a flange mounting surface and pilot diameter on the engine mounting face for mounting equipment concentric to the crankshaft center line; 2 — ball or roller bearing; 3 — ball or roller bearing with a flange mounting surface and pilot diameter on the engine

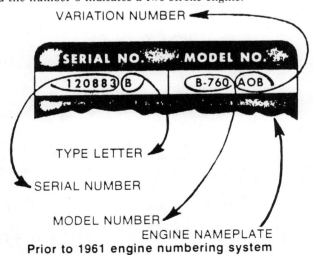

Prior to 1961 engine numbering system

7–1

mounting face for mounting equipment concentric to the crankshaft center line; 4–9 – not assigned any meaning.

The sixth digit identifies whether there are any reduction gears or power-take off units: 0 – not equipped with any such unit; 1 – auxiliary PTO; 2 – 2:1 reduction gears; 3 – not assigned; 4 – 4:1 reduction gears; 5 – not assigned; 6 – 6:1 reduction gears; 7 – 9 – not assigned.

The eighth through tenth digits are used to identify model variations.

The type letter identifies parts that are not interchangeable. Be sure to give the model number and the type letter when obtaining replacement parts.

MAINTENANCE

Air Cleaner Service

METALLIC MESH AIR CLEANER

1. Loosen the air cleaner screw and remove the air cleaner.
2. Place both element and cover parts in a nonvolatile solvent, and agitate the metal mesh vigorously.
3. Dip the mesh into clean engine oil, and then install air cleaner.

DRY PAPER AIR CLEANER

1. Remove the air cleaner can from the engine.
2. Brush lightly with a soft bristle brush (not a wire brush).
3. Blow the dirt out, using compressed air. Blow from inside to outside.
4. Make sure the sealing gasket is in place, and install the air cleaner.

OIL BATH AIR CLEANER

1. Remove the air cleaner from the engine, disassemble, and soak in a nonvolatile solvent.
2. Inspect the plastic blow carefully for cracks, and replace if necessary.
3. Blow the solvent out of the mesh filter with compressed air.
4. Fill to the correct level with SAE 30 oil, and install.

Oil bath air cleaner—washing mesh in solvent

POLYURETHANE AIR CLEANER

1. Remove the element from the air cleaner container, and wash it in soap and water.
2. Dry it thoroughly. Apply enough engine oil to cover the face of the element (about 1 tablespoon), and install the air cleaner.

Oil and Fuel Recommendations

FOUR-CYCLE OIL

Follow the recommendations below. Do not use oil of ML (Mostly Light) rating, as it will void the warranty. MM oil is recommended, but best results are obtained with MS rated oil. Oil rated DG may be used, but is not particularly recommended, and DM oil is specifically not recommended. Its use will void the warranty.

- 32°F and above: SAE 30, MM or MS
- –10°F–32°F: SAE 10W, MM or MS
- Below –10°F: SAE 5W, MM or MS rating

Crankcase Capacity Chart

Model	Capacity
V – 100, VS – 100, A – 300, A – 650, VS – 2100, VS – 3100, VS – 4100, and Models 401, 403, 405, 407, 408, 409, 411, 415, 417, 429, 431, 435	1 pint
100, 2100, 3100, H – 3100, 4100, 400, 402, 404, 406, 424, 426	1¼ pints
700 – A, 700 – C, D – 700, D – 800, A – 800, A – 900, A – 1100, B – 1100, C – 1100, D – 1100, D – 1200, A – 1200, B – 1290, 492, 494, 498	1¼ pints
VS – 700, VS – 750, VS – 800, VS – 900, V – 1000, V – 1100, VS – 1100, V – 1200, VS – 1200, 499	¼ pints
429, 431, 435	1¾ pints
413	2 pints
412	2½ pints
1600, 1800, 414, 418	3 pints
2500, 2790, 420, 422	4½ pints

TWO-CYCLE OIL

Use a good quality outboard (two-stroke) motor oil, rated MM or MS. Do not use oils rated DM or DS. Use SEA 30 or SAE 40 viscosity.

On sleeve bearing engines, mix the oil in the proportions: ¾ pint to each gallon of gasoline. On needle bearing engines, use ½ pint to each gallon of gasoline.

NOTE: On outboard motors, during the first five hours of operation (break-in), mix ½ pint of oil to each gallon of gasoline. After that, mix ¼ pint to each gallon of fuel. Mix the oil and fuel thoroughly by first supplying one gallon of fuel, pouring the oil in and shaking the can vigorously, and then completely filling the fuel can.

NOTE: When the engine uses a screw-in type dipstick, check oil level without screwing dipstick into crankcase.

V-100 crankcase fill and drain points

VS-700

TUNE-UP

Spark Plugs

Remove the spark plug lead and remove the spark plug with a deep well socket. Check for excessive buildup of carbon deposits, burned electrodes, and a cracked insulator. Replace the plug if any of these problems exist. If the plug has only light carbon deposits, wire brush the plug clean and set plug gap with a wire feeler gauge to : 0.028–0.033 in. on 2-cycle engines; 0.025–0.028 in. on 4-cycle engines.

Breaker Points

CHECKING

1. To check the breaker points, remove the ball, the magneto box cover, and gasket.
2. Look for evidence of excess oil in the box, which would indicate a leaking oil seal or defective breather assembly.
3. Check the point contacts for excessive wear or pitting. Normally the point assembly will last for many years if it is aligned

Checking the breaker point gap

properly, the gap is set correctly, and the rubbing block or shuttle is properly aligned on the actuating cam. If it is necessary to replace the points, make sure that the new set is installed properly.

NOTE: Do not file the points and bend only the stationary side of the points to align the contact surfaces. Replace the condenser when the points are replaced.

4. Breaker point gap is critical to engine performance, so make sure it is set to 0.020 in. unless otherwise specified on the bail cover.

REPLACEMENT

1. Remove the flywheel, as described in the "Engine Disassembly" procedure. Remove the bail and remove the breaker box cover and gasket.
2. Disconnect the primary wire, remove the contact set attaching screws, and remove the contacts.
3. Reverse the removal procedure to install the contact set. Leave the contact set mounting screws slightly loose.
4. Rotate the crankshaft until the points are on the highest point of the cam. Set the gap to the specification shown on the bail cover, or to 0.020 in. using a flat feeler gauge, as shown. Tighten the mounting screws.
5. Clean the contact surfaces with carbon tetrachloride and then dry with paper. Align the surfaces, if necessary, as described above under "Checking." Reset the gap, if necessary.
6. Grease the cam, rotate the crankshaft, and then remove the excess grease.
7. Replace the cover gasket if it is damaged or oil soaked. Install the cover and snap the bail into position.
8. Install the flywheel, and torque the nut to specification.

Magneto Air Gap

The magneto air gap is the distance between the stationary laminations of the coil and the rotating magnets of the flywheel. In general, the closer the magnets pass to the laminations, the better the magneto will perform. Some extra clearance must be provided for bearing wear, however. The proper clearance for

General Engine Specifications

Model Number	Block Construction	Piston Displacement	Type	Model Number	Block Construction	Piston Displacement	Type
E65	Alum.	5.76	2 Cycle	A & B 1100	LLCI	8.3	4 Cycle
				C1100	LLCI	8.3	4 Cycle
100	Alum.	7.2	4 Cycle	D1100	LLCI	8.3	4 Cycle
100 – 1000	Alum.	7.2	4 Cycle	V1100 – 1000	LLCI	9.5	4 Cycle
100 – 2000	Alum.	7.2	4 Cycle	VS1100	LLCI	9.5	4 Cycle
V100 – 1000	LLCI	7.2	4 Cycle	VS1100 – 1000	LLCI	9.5	4 Cycle
VS100	Alum.	7.2	4 Cycle				
VS100 – 1000	Alum.	7.2	4 Cycle	1200	LLCI	10.2	4 Cycle
VS100 – 2000	Alum.	7.2	4 Cycle	1200 – 1000	LLCI	10.2	4 Cycle
VS100 – 3000	Alum.	7.2	4 Cycle	1200 – 2000	LLCI	10.2	4 Cycle
VS100 – 4000	Alum.	7.2	4 Cycle	A1200	LLCI	10.2	4 Cycle
				B1290 – 1000	LLCI	10.2	4 Cycle
200	Alum.	4.5	2 Cycle	V1200 – 1000	LLCI	10.2	4 Cycle
A200	Alum.	4.5	2 Cycle	VS1200	LLCI	10.2	4 Cycle
AVS200	Alum.	4.5	2 Cycle				
AVS200 – 1000	Alum.	4.5	2 Cycle	1600	LLCI	16.3	4 Cycle
VS200	Alum.	4.5	2 Cycle	A1600 – 1000	LLCI	16.3	4 Cycle
VS200 – 1000	Alum.	4.5	2 Cycle				
VS200 – 2000	Alum.	4.5	2 Cycle	1800 – 1000	LLCI	18.6	4 Cycle
VS200 – 3000	Alum.	4.5	2 Cycle				
VS200 – 4000	Alum.	5.76	2 Cycle	2100	Alum.	7.2	4 Cycle
				A2100	Alum.	7.2	4 Cycle
300	LLCI	4.72	4 Cycle	A2100 – 1000	Alum.	7.2	4 Cycle
A300	LLCI	4.72	4 Cycle	A2100 – 2000	Alum.	7.2	4 Cycle
VS300	LLCI	4.72	4 Cycle				
350	LLCI	4.72	4 Cycle	VS2100	Alum.	7.2	4 Cycle
				VS2100 – 1000	Alum.	7.2	4 Cycle
A400	Alum.	5.76	2 Cycle	VS2100 – 2000	Alum.	7.2	4 Cycle
A400 – 1000	Alum.	5.76	2 Cycle	VS2100 – 3000	Alum.	7.2	4 Cycle
AVS400	Alum.	5.76	2 Cycle				
AVS400 – 1000	Alum.	5.76	2 Cycle	2500	LLCI	25	4 Cycle
BVS400	Alum.	5.76	2 Cycle	A2500	LLCI	25	4 Cycle
CVS400 – 1000	Alum.	5.76	2 Cycle	B2500 – 1000	LLCI	25	4 Cycle
VS400	Alum.	5.76	2 Cycle				
VS400 – 1000	Alum.	5.76	2 Cycle	2790 – 1000	LLCI	25	4 Cycle
VS400 – 2000	Alum.	5.76	2 Cycle				
VS400 – 3000	Alum.	5.76	2 Cycle	VS3000	Alum.	7.2	4 Cycle
VS400 – 4000	Alum.	5.76	2 Cycle				
				3100	Alum.	8.3	4 Cycle
500	LLCI	5.89	4 Cycle	3100 – 1000	Alum.	8.3	4 Cycle
GK590	Alum.	5.76	2 Cycle	3100 – 2000	Alum.	8.3	4 Cycle
				3100 – 3000	Alum.	8.3	4 Cycle
650	LLCI	5.89	4 Cycle	H3100 – 1000	LLCI	8.3	4 Cycle
				FV3100 – 1000	LLCI	8.3	4 Cycle
700 – A	LLCI	5.89	4 Cycle	AFV3100 – 1000	LLCI	8.3	4 Cycle
B700	LLCI	5.89	4 Cycle	AV3100 – 1000	LLCI	8.3	4 Cycle
C700	LLCI	5.89	4 Cycle	AV3100 – 2000	LLCI	8.3	4 Cycle
D700	LLCI	6.65	4 Cycle	AVS3100	Alum.	8.3	4 Cycle
D700 – 1000	LLCI	6.65	4 Cycle	AVS3100 – 1000	Alum.	8.3	4 Cycle
D700 – 2000	LLCI	6.65	4 Cycle	AVS3100 – 2000	Alum.	8.3	4 Cycle
D700 – 3000	LLCI	6.65	4 Cycle	AVS3100 – 3000	Alum.	8.3	4 Cycle
VS700	LLCI	5.89	4 Cycle	V3100 – 1000	LLCI	8.3	4 Cycle
VS750	LLCI	5.89	4 Cycle	V3100 – 2000	LLCI	8.3	4 Cycle
				VS3100	Alum.	8.3	4 Cycle
800	LLCI	8.3	4 Cycle	VS3100 – 1000	Alum.	8.3	4 Cycle
A800	LLCI	8.3	4 Cycle	VS3100 – 2000	Alum.	8.3	4 Cycle
VS800	LLCI	8.3	4 Cycle	VS3100 – 3000	Alum.	8.3	4 Cycle

General Engine Specifications (cont.)

Model Number	Block Construction	Piston Displacement	Type	Model Number	Block Construction	Piston Displacement	Type
900	LLCI	8.3	4 Cycle				
900–1000	LLCI	8.3	4 Cycle	4100	Alum.	8.3	4 Cycle
900–2000	LLCI	8.3	4 Cycle	4100–1000	Alum.	8.3	4 Cycle
900–3000	LLCI	8.3	4 Cycle	4100–2000	Alum.	8.3	4 Cycle
900–4000	LLCI	8.3	4 Cycle	AVS4100–1000	Alum.	8.3	4 Cycle
VS900	LLCI	8.3	4 Cycle	AVS4100–2000	Alum.	8.3	4 Cycle
V1000–1000	LLCI	8.3	4 Cycle	VS4100–1000	Alum.	8.3	4 Cycle
VS1000	LLCI	8.3	4 Cycle	VS4100–2000	Alum.	8.3	4 Cycle
400–0000–000	Alum.	7.2	4 Cycle	420–1300–000	LLCI	25	4 Cycle
401–0000–000	Alum.	7.2	4 Cycle	420–1301–000	LLCI	25	4 Cycle
402–0000–000	Alum.	7.2	4 Cycle	422–1300–000	LLCI	25	4 Cycle
403–0000–000	Alum.	7.2	4 Cycle	422–1301–000	LLCI	25	4 Cycle
404–0000–000	Alum.	8.3	4 Cycle	424–0000–0000	LLCI	8.3	4 Cycle
405–0000–000	Alum.	8.3	4 Cycle	426–0000–000	Alum.	8.3	4 Cycle
406–0000–000	LLCI	8.3	4 Cycle	429–0003–000	LLCI	9.2	4 Cycle
407–0000–000	LLCI	8.3	4 Cycle	431–0003–000	LLCI	9.2	4 Cycle
407–0002–000	LLCI	8.3	4 Cycle	435–0003–000	Alum.	8.3	4 Cycle
408–0000–000	Alum.	8.3	4 Cycle	492–0300–000	LLCI	8.3	4 Cycle
409–0000–000	Alum.	8.3	4 Cycle	494–0000–000	LLCI	8.3	4 Cycle
411–0000–000	Alum.	7.2	4 Cycle	494–0001–000	LLCI	8.3	4 Cycle
411–0002–000	Alum.	7.2	4 Cycle	497–0000–000	LLCI	10.2	4 Cycle
412–0000–000	LLCI	15.5	4 Cycle	498–0300–000	LLCI	10.2	4 Cycle
413–0000–000	LLCI	15.5	4 Cycle	498–0301–000	LLCI	10.2	4 Cycle
414–1300–000	LLCI	16.3	4 Cycle	499–0000–000	LLCI	10.2	4 Cycle
414–1301–000	LLCI	16.3	4 Cycle	500–0000–000	Alum.	5.76	2 Cycle
415–0000–000	Alum.	8.3	4 Cycle	501–0000–000	Alum.	5.76	2 Cycle
415–0002–000	Alum.	8.3	4 Cycle	501–0001–000	Alum.	5.76	2 Cycle
416–1300–000	LLCI	16.3	4 Cycle				
417–0000–000	LLCI	8.3	4 Cycle				
418–1300–000	LLCI	18.6	4 Cycle				
418–1301–000	LLCI	18.6	4 Cycle				

Alum. = Aluminum
LLCI = Long Life Cast Iron

engines under five horsepower is 0.007–0.017 in. and 0.012–0.020 in. for engines over five horsepower.

The air gap is measured by placing layers of plastic tape over the laminations, replacing the flywheel and turning the flywheel. Remove the flywheel and check to see if the flywheel touched the tape, and adjust the coil assembly accordingly. Use only one layer of tape at a time. The thickness of common plastic electrician's tape is about 0.008–0.009 in.

Breaker Cam

The breaker cam on most Clinton engines is replaceable. Check the fit of the cam over the crankshaft to make sure that it is tight.

NOTE: On Model 412 and 413 engines, the breaker cam is machined onto the crankshaft.

Mixture Adjustment

1. If the engine runs very roughly or will not start, first make a preliminary setting as follows:
   a. Very gently turn the main mixture screw in until it seats very lightly.

b. Turn screw outward (counterclockwise) the specified number of turns:
- 501 – 1 turn
- Lift Carburetors – $1\frac{1}{4}$–$1\frac{1}{2}$ turns
- LMG, LMB, LMV Carburetors – $1\frac{1}{4}$
- All others – $1\frac{1}{2}$ turns

2. Following this, start and run the engine until hot. Then, turn the mixture screw inward or outward in $\frac{1}{16}$ turn increments, pausing after each adjustment, until best running is obtained. The most accurate adjustment is obtained with a tachometer. Set the mixture so the highest possible rpm is obtained.

Mechanical Governor Adjustment

1. Stop the engine and set any throttle controls so there is tension on the governor spring.
2. Loosen the adjusting screw. Position the throttle to within exactly $\frac{1}{32}$ in. of the stop on the carburetor casting. Tighten the adjusting screw.

Compression Check

1. Remove the spark plug and install a compression gauge in its place.

THROTTLE
IN
OPEN POSITION

GOVERNOR
IN
CLOSED POSITION

ADJUSTMENT
SCREW

Location of governor adjustment screw

$1/32''$

Throttle position and dimension

2. Crank the engine over at normal cranking speed. Gauge readings should be:
- 2-stroke: above 60 psi
- 4-stroke up to $4\frac{1}{2}$ hp: 65–70 psi
- 4-stroke above $4\frac{1}{2}$ hp: 70 psi

FUEL SYSTEM

Carburetor

Clinton engines use a variety of float type carburetors and one type of suction lift carburetor.

A carburetor can only malfunction as a result of these causes: the presence of foreign matter (dirt, water); out of adjustment (too rich or too lean a mixture), and leakage caused by worn parts or cracks in the casting. Look for any of the above causes before disassembling the carburetor. Of course, it may be required that the carburetor be disassembled to correct the problem, but at least you will have some direction.

NOTE: Check the throttle shaft/main casting tolerance. This is one place on the carburetor that cannot be repaired.

501 ENGINE CARBURETOR

The 501 engine carburetor is a relatively simple carburetor and can be disassembled after it is removed from the engine.

1. Remove the choke and air filter assembly, remove the mixture adjusting screw and spring, and unscrew the large bolt which holds the float bowl to the rest of the carburetor.

2. Please note that the main fuel nozzle is contained in the top of the large bolt and care should be exercised not to damage the needle seat. Disassemble the float and fuel inlet needle.

Checking compression

501 Engine carburetor

NOTE: The needle seat is not replaceable, The bowl cover, needle pin, spring, and seat assembly must be replaced if any part is worn.

3. Clean all parts in solvent, blow dry with compressed air, inspect, and replace any worn or damaged parts.

4. Assemble the carburetor in the reverse order of disassembly. Set the float level with the bowl cover inverted and the float and needle installed so that there is $\frac{13}{64}$ in. plus or minus $\frac{1}{32}$ in. clearance between the outer edge of the bowl cover and the free end of the float. Adjust the level by bending the lip of the float with a screwdriver. Use all new gaskets.

5. Set the mixture adjusting needle one turn open from the seat to start the engine. This carburetor does not have an idle mixture orifice and therefore will not operate at speeds below 3000 engine rpm. The operating range is between 3000 and 3800 rpm. Make the final mixture adjustment with a tachometer if possible. If a tachometer is not available, adjust the mixture screw $\frac{1}{16}$ of a turn at a time until you obtain the best engine performance.

SUCTION LIFT CARBURETOR

Throttle Plate and Shaft Replacement

1. Drill through the plug at the rear of the carburetor body.
2. Force out the plug with a small drift pin.
3. Remove the plastic plug.
4. Remove the screws which retain the throttle plate to the shaft and remove the plate and shaft.
5. Reinstall in the reverse order of removal.

Make sure that the throttle plate and shaft operate without binding. Use a sealer on the new expansion plug.

NOTE: Horizontal suction lift carburetors do not have a plug.

The choke plate and shaft is replaced in the same manner as the throttle, except that there is no expansion plug to be removed.

Stand Pipe Replacement

The stand pipe is press-fit into the body of the carburetor. Remove the pipe by clamping it in a vise and twisting the carburetor body while at the same time pulling until the pipe is free. To install the pipe, tap it lightly into place until the end of the pipe is 1.94 in. ± 0.45 in. from the body of the carburetor. Use a sealer at the point where the pipe is inserted into the carburetor body.

Idle Needle and Jet Replacement

The idle needle and jet are replaceable and can be removed as follows:
1. Unscrew the needle and remove it.
2. Remove the expansion plug in the bottom of the carburetor body.
3. To remove the idle jet, push a piece of $\frac{1}{16}$ in. diameter rod through the fuel well and up the idle passage until the jet is pushed out of the passage.
To install the idle needle and jet:
4. Make a mark exactly $1\frac{1}{4}$ in. from the end of the $\frac{1}{16}$ in. diameter rod and push the new jet into the idle passage until the mark is exactly in the center of the main fuel well.
5. Install the needle and set it at 4–$4\frac{1}{4}$ turns open from being seated.

The high speed needle and seat can be inspected for wear or damage by removing the needle and checking its taper. If the needle taper is damaged, replace it. Inspect the seat for taper and splits. If the seat is split or tapered from the needle being screwed in too tight, the complete carburetor must be replaced,

Suction lift carburetor

since the seat cannot be replaced separately. When the needle is replaced, the initial adjustment is $\frac{3}{4}$–$1\frac{1}{2}$ turns open from being lightly seated.

LMG, LMB AND LMV TYPE CARBURETORS

These carburetors are float type carburetors and can be reconditioned in the same manner as the 501 carburetors by taking a note of the following differences in design and specifications.

The main fuel nozzle needle is inserted from the bottom and pushed up through the middle of the float bowl.

The float, fuel inlet needle, and seat are basically the same as that of the 501 carburetor.

NOTE: Do not remove the main fuel nozzle from the carburetor body unless it is to be replaced. Once it is removed it cannot be reinstalled.

Use all new gaskets.

The throttle valve is installed with the part number or trademark **W** facing toward the mounting flange.

The preliminary setting of the idle adjusting needle is $1\frac{1}{4}$ turns open after being lightly seated.

Adjust the float in the same manner as the 501 carburetor. There should be $\frac{5}{32}$ in. clearance between the float and the casting rim when the carburetor is inverted. When the carburetor is turned over, the float should not drop more than $\frac{3}{16}$ in.

The preliminary setting for the main fuel adjusting needle is $1\frac{1}{4}$ turns open after being lightly seated.

Float adjustment for the LMG, LMV, LMB type carburetors

H.E.W. CARBURETORS

H.E.W. carburetors are float type carburetors and can be serviced in the same manner as 501 carburetors. After cleaning and inspecting the parts for wear or damage, assemble the carburetor, noting the following differences:

Install the throttle valve with the part number or **W** toward the mounting flange with the throttle in the closed position. Always use new gaskets during assembly.

The initial setting for the idle adjusting screw is $1\frac{1}{2}$ turns open from being lightly seated.

If any part of the float assembly has to be replaced, replace the whole assembly. Do not replace individual parts.

LMG, LMV, LMB type carburetors with an enlarged view of the main jet nozzle

H.E.W. carburetor

The float setting is made in the same manner as the 501 carburetor, by turning the main body of the carburetor upside down and measuring the clearance between the float and the body casting rim. The distance in this case is $\frac{3}{16}$ in. with $\frac{3}{16}$ in. of travel.

The preliminary setting for the high speed adjusting screw is $1\frac{1}{2}$ turns open from being seated lightly.

U.T. CARBURETORS

These are also float type carburetors and are serviced in the same manner as 501 carburetors, noting the following differences:

The throttle valve is installed with the trademark **C** on the side toward the idle port when viewed from the mounting flange side. Use new screws.

Set the float level with the float installed and the housing inverted. The measurement is to be taken with the needle seated and the gasket removed. Measure between the float seam and the throttle body. Adjust by bending the lip of the float.

The preliminary setting for the high speed adjusting needle is $1\frac{1}{4}$ turns open from being lightly seated.

The preliminary setting for the idle adjusting needle is $1\frac{1}{2}$ turns open from being lightly seated.

The chart below gives the float level setting dimension as related to the carburetor identification number and part number:

Ident No.	Part No.	Float Setting (in.)
2712–S	39–143–500	$\frac{19}{64}$
2713–S	39–144–500	$\frac{19}{64}$
2714–S	39–140–500	$\frac{1}{4}$
*2398-S	39–147–500	$\frac{1}{4}$
2336–S	39–146–500	$\frac{1}{4}$
2336–SA	39–146–500	$\frac{1}{4}$
2337-S	39–145–500	$\frac{1}{4}$
2337–SA	39–145–500	$\frac{1}{4}$
2230-S	39–343–500	$\frac{17}{64}$
2217-S	39–344–500	$\frac{11}{64}$

* With resilient seat—$\frac{9}{32}$ + or − $\frac{1}{64}$

U.T. carburetor

Operation of Touch 'N' Start primer carburetor

TOUCH 'N' START PRIMER CARBURETOR

Steps in operation of the primer:
1. Seal the bowl vent with a finger.
2. Depress the bulb to pressurize the bowl.
3. Pressure in carburetor bowl forces the fuel into the carburetor throat.
4. When the engine is cranked, the intake valve opens, letting gasoline into the combustion chamber for quick starting.
5. Gasoline forced from the bowl during priming is replaced by the flow of gasoline from the fuel tank through the gravity fuel inlet.

This primer is applicable to the LMG, LMB, and LMV carburetors. Servicing this carburetor would be the same as listed for the LMG, LMB, and LMV, except a choke lever and choke valve are not used. Note the bowl atmospheric vent is routed back through the primer tube and bulb.

FIXED SPEED CARBURETOR

Steps in operation:

1. Rotate the control knob counterclockwise to open the throttle (4-cycle engines) (2-cycle not equipped with control knob).

2. To stop the engine rotate the control knob clockwise. (4-cycle engines) (2-cycle engines use a shorting device).

NOTE: The governor spring is located on the throttle shaft between the lever and the carburetor casting. The setting on the high speed screw is $1\frac{1}{4}$–$1\frac{1}{2}$ turns. Servicing this carburetor would be the same as listed for the LMG, LMB, or LMV except there is no main nozzle or idle circuit.

Carter carburetor

CONTROL KNOB

GOVERNOR SPRING MAINTAINS SPEED UNDER CHANGING LOAD

TO INTAKE VALVE

Operation of the Fixed Speed carburetor

CARTER CARBURETORS

Carter float type carburetors are serviced in the same manner as 501 carburetors, taking note of the following differences:

Do not remove the choke valve and shaft unless they are to be replaced. A spring loaded ball holds the choke in the wide open position. Be sure to use a new ball and spring when replacing the choke shaft and plate assembly.

Install the throttle plate with the trademark **C** on the side toward the idle port when viewed from the mounting flange side.

The float setting is made in the same manner as the 501 carburetor and the measurement is $\frac{3}{16}$ in.

The initial setting for the high speed adjusting screw is 2 turns open from being lightly seated.

The initial adjustment for the idle adjusting screw is $1\frac{1}{2}$ turns open from being lightly seated.

Fuel Pumps

MECHANICAL FUEL PUMPS

These pumps can be rebuilt. See the exploded view.

The 220–122–500 and 220–145–500 fuel pumps are simple diaphragm units that screw into the inlet side of the carburetor. They are activated by crankcase pressure pulses. They cannot be rebuilt, but must be replaced if they will not lift the fuel six inches.

The diaphragm type fuel pumps used on VS–1200, V–1200, and 499 series engines can be rebuilt. However, construction is so simple no rebuild specifics are supplied by the factory.

Governors

INSPECTION AND ASSEMBLY OF MECHANICAL GOVERNOR

When assembling an engine equipped with the centrifugal weight governor inspect and governor shaft bearing in the block and the governor arm assembly that goes through the bearing, for wear and replace them if necessary. After inspection, insert

1. Rocker arm
2. Body
3. Top cover
4. Diaphragm assembly
5. Diaphragm pulsator
6. Mounting gasket
7. Valve gasket
8. Linkage
9. Rocker arm pin
10. Cover plate
11. Pipe plug
12. Valve retainer
13. Valve seat (head)
14. Valve seat (retainer)
15. Rocker arm spring clip
16. Diaphragm spring
17. Rocker arm spring
18. Valve spring
19. Screw & lockwasher assembly (cover plate)
20. Screw & lockwasher assembly (top cover)
21. Screw & lockwasher assembly (valve retainer)
22. Valve

Exploded view of mechanical fuel pump-typical

the arm through the bearing and fasten the arm and weight assembly into the bearing.

Care should be taken on installation of this arm and weight assembly as they may be locked to the outside linkage 180 degrees from the correct position which would tear out the centrifugal weights and damage the arm and weight assembly upon operation of the engine. The weight and arm or yoke should be as close to the cam axle or governor gear as it can be to be properly installed. In this position, it will operate against the governor collar or thimble assembly and will move in conjunction with the governor spring tension and the centrifugal force of the weights which are attached to the camshaft or governor gear.

The collar should be inspected for wear and possible damage, and the weight assembly itself should be inspected for wear and possible damage or bending of the weights or the weight supports.

When servicing a centrifugal governor, check to be certain that the collar or thimble operates freely on the camshaft or governor gear and that the governor shaft moves freely in the bushing. When the bushing in the block is replaced, check carefully freedom of shaft motion as the bushing may be distorted in installation.

Also the governor shaft can be bent easily on disassembly or reassembly and can be bent in usage. Also check range of movement of collar or thimble after assembly of camshaft or governor gear to determine that these parts do not lock against block.

INSPECTION AND ASSEMBLY OF AIR VANE GOVERNOR

The air vane should be inspected visually when reassembling and replaced if bent or damaged. Use care in the hook-up of the air vane link and throttle plate so that they move freely and do not drag at the connections or on the bearing plate or blower housing. Many of the air vane governors have a spring inside of the vane at the pivot. The spring gives a dampening effect on the vane movement. This spring should be replaced if the engine has been out of service for a period of time so that it retains the pressure on the bushing or vane.

When servicing an air vane governor, the condition of the blower housing should be checked carefully. Dents and bends should be removed from it is that the air stream moves as it should to the air vane. The air vane must be in the same condition as when manufactured and replaced when bent because the governor spring tension and the vane are balanced. If the vane does not sit in the air blast properly, the spring will be too strong for the air vane to stretch.

When assembling the air vane governor, apply tension to the governor spring and close the throttle manually to see that it moves open freely, that it does not bind at the governor linkage, air vane, pivot post, bushings, bearing plate, blower housing, etc. It should move freely from closed to open position by governor spring tension.

ENGINE OVERHAUL

The same basic procedure for disassembling a Clinton engine can be used for all engines. The procedure given below pertains to both 2-stroke and 4-stroke Clinton engines; differences are noted. Procedures for servicing individual components are given at a later point in this section.

Engine Disassembly

1. Remove the engine from the piece of equipment it powers and then remove any brackets, braces, adapters or pulleys.
2. Clean the exterior of the engine.
3. Drain the lubricating oil from the crankcase.
4. Remove the fuel tank and blower housing.
5. Remove the carburetor and governor assembly, marking the spring and link holes for correct reassembly.
6. Remove the muffler assembly.
7. Remove the flywheel nut, using a flywheel holder to hold the flywheel while the nut is removed.
8. While lifting up on the flywheel, gently tap the crankshaft to loosen the flywheel from the crankshaft taper. Remove the flywheel and flywheel key.
9. Remove the complete magneto assembly which includes the coil, breaker points, condenser, and laminations.
10. Remove the cylinder head and gasket on 4-stroke engines.
11. Remove the valve chamber cover and breather assembly from 4-stroke engines.
12. Remove the valve spring keepers after compressing the valve spring with a valve spring compressor.
13. Remove the valves and springs from the block after removing the valve spring compressor.

NOTE: Some valves have a burr on the stem that will prevent the valve from being removed up through the valve guide. If present, it will be necessary to remove this burr from the stem in order to remove the valve from the engine. To remove the burr, hold a flat file against the burred area and rotate the valve.

14. Remove the base plate or end cover assembly on 4-stroke engines.

NOTE: On some engines the base is an integral part of the block and cannot be removed. If this is the case, remove the side plate on the power take-off side of the engine. Before removing the side plate or crankshaft, make sure all paint, rust, and dirt are cleaned from the area of the crankshaft bearing. This is so the side plate can be easily removed. When removing the side plate from engines that have ball bearings, it will first be necessary to remove the oil seal and the snap-ring from the crankshaft.

15. Remove the mounting plate and reed plate assembly on 2-stroke engines.
16. Remove the connecting rod cap screws and cap. Mark the cap and connecting rod so they can be reassembled in the same position.
17. On 4-stroke engines, check for a carbon or metal ridge at the top of the cylinder. If a ridge is present, remove it with a ridge reamer or hone.
18. On 2-stroke engines, push the piston and connecting rod assembly up into the cylinder as far as it will go so it will not hit the crankshaft when it is removed.
19. Remove the piston and connecting rod assembly from a 4-stroke engine's block.
20. Remove the bearing plate on the flywheel side of the block if the engine being disassembled has one.
21. Remove the crankshaft.

NOTE: On 4-stroke engines with a ball bearing on the PTO side of the engine, it will be necessary to remove the cap screws which hold the bearing in place before the crankshaft can be removed. On some models with ball bearings and tapered roller bearings, it will be necessary to remove the crankshaft oil seal and camshaft axle, and move the camshaft to one side before the crankshaft can be removed. Two stroke engines have a ball bearing on the PTO side of the crankshaft will have to have the retaining ring, which holds the bearing in place, removed before the crankshaft can be removed.

22. Remove the camshaft assembly. Drive the camshaft axle out of the PTO side of the block as the flywheel side is smaller.
23. Remove the piston and connecting rod assembly from 2-stroke engines.
24. After the camshaft is removed from a 4-stroke engine, mark the valve tappets as to whether they are the exhaust tappet or the intake tappet, and then remove them. If no valve work is to be done, be sure to replace the valves in the same position from which they were removed.
25. Remove the piston and rod assembly. After disassembling the engine, clean all parts in a safe solvent, removing all deposits of carbon and oil, etc. Check all operating clearances and replace or rebuild parts as necessary.

Cylinder Bore

After disassembling the engine, inspect the cylinder bore to see if it can be reused. Look for score marks on the cylinder walls. If there are marks and they are too deep to be removed, the block will have to be discarded. If there is a hole in the block due to connecting rod failure, the block will have to be replaced. If there are broken cooling fins on the outside of the block, these can cause overheating and replacing the block should be considered.

Check the dimensions of the cylinder bore to determine the extent of wear and whether or not it has to be rebored. The cylinder bore can be rebored to 0.010 in. or 0.020 in. oversize since

Measuring the cylinder bore—a cutaway view

there are oversize pistons available in these sizes. If the cylinder is within serviceable limits and there is no need to rebore it, be sure to deglaze the cylinder before installing the piston assembly with new rings.

Bearings

Clinton engines are equipped with tapered roller bearings, ball bearings, needle bearings, and sleeve bearings. The first thing to determine is whether or not the bearing is worn or damaged and needs replacing. Then clean the bearings in a safe solvent, inspect them for excessive play due to wear, smoothness of rotation, pitted surfaces, and damage. All of these types of bearings are pressed on the crankshaft and into the bearing plates and are removed by either a bearing splitter and puller or they are driven out of the bearing plates with a punch. In all cases by very careful not to bend, gouge, or otherwise damage the crankshaft or bearing plate when removing and installing the bearings.

With sleeve bearings, first inspect the bearing surface for scoring and damage to determine whether the bearing has to be replaced. Check the bearing diameter for wear. In some cases you will find that the bearing surface is the same material as the block. These bearings can be reamed out and sleeve bearings installed. If the original bearing is a sleeve type bearing, it can be removed by driving it out with the proper size driving tool. Install the new bearing so that the oil hole in the bearing is aligned with the oil passage in the block or bearing plate. Drive the bearing into the block or bearing plate until it is recessed about $\frac{1}{32}$ in. from the crankshaft thrust face of the block or bearing plate. After installing the new piece, it must be finish reamed. After finish reaming, clean all metal filings and debris from the engine, making sure that all oil passages are free from obstruction.

Valve Seats

Standard valve seats, those without inserts, can be reground to remove all of the oxidized surface metal and gain perfect sealing characteristics. After grinding the valve seats, the valves must be lapped in with lapping compound. Not too much lapping is recommended, just enough to obtain a good seal.

If the engine has had a number of valve jobs and the valve seat is too deep, requiring that too much stock be removed from the valve stem to obtain the proper valve-to-tappet clearance, valve seat inserts may be installed. If over half of the metal between the lock groove and the end of the valve stem has been removed to gain the proper stem-to-tappet clearance, you should consider installing valve seat inserts.

On aluminum block engines, iron valve seat inserts are standard equipment. To remove these inserts, it is first necessary to remove the metal that has been rolled over the edge of the insert to hold it in place. This is normally accomplished by using the proper size cutter. If the valve seat insert is loose, a cutter may not be necessary. After the insert has been removed, it is necessary to cut the block to the proper depth of $\frac{3}{16}$–$\frac{7}{32}$ in. This is the depth of the insert plus $\frac{1}{32}$ in. which is used to hold the insert in place. The insert is held to the cylinder block by a definite interference type press fit. The insert should be cooled before at-

Tapered roller, needle, and ball bearing type bearings, all of which are used in Clinton engines

tempting to install it in the block. After the insert is fitted in place, with the bevel facing up, the metal around the edge of the insert must be peened over the edge of the insert in order to hold it in place. Do not stroke the block too sharply when peening because of the possibility of distorting the cylinder bore. Finish grind the valve seat insert and lap in the valves.

Valve Guides

First, inspect and measure the valve guide diameter to determine whether the guide is worn enough to necessitate rebuilding. The standard guide size for 4-stroke engines under 5 horsepower is 0.2495–0.2510 in. On engines over 5 horsepower, the standard guide size is 0.312–0.313 in.

When the valve stem-to-valve guide clearance is more than the maximum serviceable clearance and cannot be corrected by installing a new valve, you will have to either replace the valve guide (if it is replaceable), oversize it, or knurl it.

Valve guides are replaceable in the 1600, 1800, 2500, 2790, 414–0000–000, 418–0000–000, 420–0000–000, and the 422–0000–000 series engines. The valve guides are pressed out of the block from the base of the cylinder head side. Note the position of the guide before removing it. Install the new guide by reversing the removal procedure. The new guides should be installed at least $1\frac{1}{4}$ in. below the top of the cylinder block. Remove any burrs that might have been created by the installation procedure with a $\frac{5}{16}$ in. (0.312) reamer.

The valve seat width is to be between $\frac{1}{32}$ and $\frac{3}{64}$ in. (0.030–0.045 in.) and the valve seat angle is to be between $43\frac{1}{2}$ and $44\frac{1}{2}$

Replacing the valve guides

Valves

Inspect the valves for a burned face, warped stem or head, scored or damaged stem, worn keeper groove in the stem, and a head margin of less than $\frac{1}{64}$ in. If any one of these conditions exists, the valve should be replaced. Check the valve stem diameter. On engines with less than 5 horsepower, the stem diameter should be 0.2475–0.2465 in. On engines over 5 horsepower, the stem diameter is to be 0.310–0.309 in. Any time the valve stem-to-guide clearance can be reduced more than 0.001 in. by replacing the valve with a new one, you should do so. Also any time the stem-to-guide clearance is over 0.0045 in. you should consider doing some rework (new valve guides, seats and valves) to bring the clearance below 0.0045 in. but not less than 0.002 in. If it is determined that the old valve can be reused, then it should be refaced, using an automotive type valve grinder to secure a 45 degree face angle on the valve with a $\frac{1}{64}$ in. margin between the head and the face of the valve.

Reed valve assembly for two stroke engines

2-Stroke Engine Reed Valves

Inspect the reed valves for the following items: broken reed valves, bent or distorted reed valves, damaged or distorted reed valve seat, or a broken or bent reed valve stop. If any of these conditions exist, the reed valve assembly must be replaced.

Valve Springs

To check the condition of valve springs, simply remove the spring from the engine and stand it on a flat surface next to a new valve spring. If the old spring is shorter and leans to one side, it should be replaced with a new spring. Some of the cast iron engines have a stronger or stiffer valve spring installed on the exhaust valve. Make sure that a stiffer spring is installed on the exhaust valve or, to be sure, install two stiff springs in the

engine. When a valve seat is rebuilt, the valve then seats further down into the block and this results in a loss of spring tension. To restore spring tension, install a thin washer on top of the valve spring.

Valve Tappets

The valve tappets should be inspected for wear on the head of the tappet and score marks or burrs anywhere else. The tappet should be replaced if any defects are found. Measure the dimensions of the tappet, checking for stem diameter and length. Oversize tappets are not available; however, the tappet guide can be knurled and rebored to correct size should the tappet-to-guide clearance become too large.

Piston Assembly

Inspect the piston assembly for scored walls, damaged ring lands, worn or damaged wrist pin lock ring grooves, and a cracked or broken piston skirt. If any of these defects are present, the piston must be replaced. Check the dimensions of the piston with a micrometer.

NOTE: The ring land diameter on 4-stroke engines is tapered and the reading at the ring land will be 0.00125 in. smaller per 1 in. of piston length.

Clean all carbon deposits from the ring grooves. An old broken ring will serve as an excellent tool for cleaning ring grooves.

Check the side clearances of the rings with new rings installed. The minimum and maximum clearances for oil, scraper, and compression rings are as follows: two stroke engines—0.0015–0.004 in.; four stroke engines under five horsepower—0.002–0.005 in.; four stroke engines over five horsepower—0.0025–0.005 in.

If an oversized piston is used, the amount of the oversize is stamped on the top of the piston.

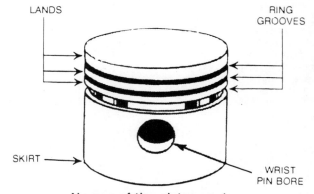

Names of the piston parts

Installation sequence and position for the piston rings

Measuring the tappet stem diameter

The ring gap on all engines except the $1\frac{7}{8}$ in. bore two stroke engine is 0.007–0.017 in. The ring gap on the $1\frac{7}{8}$ in. bore two stroke engine is 0.005–0.013 in. Oversize rings are available. Install the rings in the following order: oil ring, scraper ring, and compression ring. The oil ring can be installed with either side up, the scraper ring should have the step on the lower side toward the bottom of the engine's crankcase and the compression ring has to be installed with the bevel on the inside circumference facing upward.

NOTE: Two stroke engines have wire retainers or pins located in the ring grooves to keep the ring from moving in the groove. Make sure the ring gap is properly located over these retainers.

Inspect the connecting rod for wear or damage, such as a scored bearing surface, cracks, and damaged threads. Use a micrometer to check all of the connecting rod dimensions. Check the clearance between the wrist pin and the connecting rod at the wrist pin hole. The tolerance for all Clinton engines is 0.0004–0.0011 in. When the clearance reaches 0.002 in., the parts should be replaced with new ones or rebuilt. Clinton, however, does not supply oversize wrist pins.

When installing the piston and connecting rod in a four stroke engine, the piston may go either way, but the rod has an oil hole that must face toward the flywheel side of the engine (with the exception of those engines that use a clearance rod in which the marked side faces toward the camshaft of the engine).

On two stroke engines, the piston is installed with the downward slope of the piston dome facing the exhaust side of the engine. There is no special way for the connecting rod to be installed.

Installation of a two stroke engine piston

Wrist Pins

All Clinton wrist pins are a "hand press fit" into the piston. Care should be taken when removing or installing the wrist pin into rod or piston. It is easy to distort or damage the piston or rod. Never lay a piston on a solid object when removing or installing wrist pin. The piston can be supported in the palm of your hand when servicing to keep from damaging it. There is no special way to install the wrist pin into the piston or rod, except on the 2-cycle engines which in some cases have a hollow wrist pin closed on one end. Make sure the closed end is towards exhaust side.

Crankshaft

Before removing the crankshaft, remove the spark plug and rotate the crankshaft with the starter mechanism, while checking for any wobble of the end of the crankshaft. Any wobbling indicates that the crankshaft is bent and must be replaced. Deviance

Piston ring retainers in two stroke engines

Install clearance rods with the side marked clearance side facing toward the camshaft

Install the connecting rod cap with the match marks opposite each other

of 0.001 in. or more is not tolerable. End-play of the crankshaft should be between 0.008 in. and 0.0018 in. If the endplay exceeds 0.025 in., the condition should be corrected. End-play is adjusted when the engine is assembled by the addition of various size gaskets behind the bearing plate. It is not recommended that the crankshaft be straightened. Check all bearing surfaces for wear with a micrometer. Replace the crankshaft if it is bent or cracked; if the keyway is damaged; if the taper is damaged; if the flywheel end threads are stripped; or if the bearings are scored.

Bearing Plates and Bases

Inspect these visually, to determine whether or not they can be re-used. Reject for the reasons listed:

a. Broken or cracked, housing mounting flange on the bearing plate or mounting ears or flange base.

b. Cracked or distorted bearing bases.

c. Warped or distorted gasket or mounting surface.

d. Oil seal or bearing pocket oversize.

e. Stripped threads on the lamination hold-down screw holes in the bearing plates or drain and filler plug holes in the bases.

f. Worn crankshaft thrust face surfaces on base or bearing plate.

Camshaft

Check the camshaft for extremely worn lobes and broken gear teeth. Oil pump drive camshafts have a pin located below the gear that must have a squared end and must be secure to the crankshaft. Camshafts from engines with vertical crankshafts have a scoop riveted to the bottom of the gear. Make sure that the scoop is secure. Make sure that on those models equipped with centrifugal advance (ignition) that the advance mechanism is free and the springs are not distorted or broken. Check the dimensions of the camshaft axle.

Cylinder

Check the cylinder head for warpage with a straight-edge, after removing all dirt and deposits. If the head is warped, place a piece of emery cloth, with the rough side facing up, on a flat surface. Move the cylinder head gasket surface over the emery cloth in a figure eight pattern until the surface of the head is flat. If there are any broken cooling fins or if the spark plug hole threads are stripped, the head must be replaced.

Oil Seals

Oil seals serve two purposes, these being to keep the oil from leaking out of the crankcase on 4-cycle engines, and sealing the crankcase on 2-cycle engines, to keep the vacuum and pressure from being affected by the outside atmospheric pressures.

Any time an engine is being disassembled and the oil seals are not going to be replaced with new ones, it is a must that oil seal loaders be installed over the crankshaft or cam gear axle to keep from damaging the seal lips when the base, bearing plate, crankshaft or cam gear is removed. Clinton has oil seal loaders that will fit any diameter shaft used in the Clinton line of engines or you may use an equivalent part designed specifically for this purpose.

Cross section showing use of an oil seal loader

Whenever an engine is disassembled or the oil seal removed from it, the oil seal should be given a visual inspection for the following, to determine if it can be reused:

a. Cut or damaged seal lips.

b. Distorted or bent seal.

c. Condition of seal lip to make sure it still is flexible, and has not taken a permanent set.

Any of the above defects, would require replacement of the seal with a new one. To insure that an oil seal will function properly it is recommended that any time an oil seal area of an engine has been worked on new seals be used.

Oil seals should be removed prior to removal and replacement of bearings and reaming of bearings. The oil seals may be removed by prying out or by any means at hand with some care being used not to damage the bearing plate, block, or base to which a new seal must be replaced.

Engine Assembly
FOUR STROKE ENGINES

This is a general procedure and is intended to be only a guide since deviations may be necessary for some models.

1. Insert the tappets into the block.
2. Assemble the oil pump to the cam gear, if so equipped.

TIMING MARKS

Align the crankshaft and camshaft timing marks

3. Install the mechanical governor shaft, if so equipped.

4. Install the crankshaft and cam gear into the engine, making sure that the crankshaft thrust washer is in place if one is used.

5. Align the crankshaft and camshaft timing marks.

6. Install the piston and rod assembly using a ring compressor and great caution not to break rings or damage the piston.

7. Install the rod cap and oil dipper, if so equipped, and the cap screw and lock. Tighten to the correct torque. Crimp the screen locks securely.

8. Install the bearing plate and base plate or end cover assembly to the cylinder block. Check the crankshaft end-play. Engines using sleeve bearings should have 0.005–0.002 in. end-play. Engines using tapered roller bearings should have 0.001–0.006 in. end-play. Engines using roller bearings have no end-play specifications; however, care should be taken not to have the crankshaft too tight after assembly. The end-play is adjusted by the installation of various size gaskets between the plate and block.

9. Install the oil seals in the PTO and flywheel side of the crankshaft.

10. Install the valves into the block and check the valve stem-to-tappet clearance. Clearance is checked with the lobe of the tappet facing away from the valve. Clearance for a four stroke engine is 0.009–0.012 in. Clearance is adjusted by grinding or filing the valve stem.

11. Using a valve spring compressor, assemble the valve springs to the valves.

12. Install the breather assembly into the valve spring chamber and install the cover.

Adjust the valve-to-tappet clearance with the tappets completely off of the cam lobes

13. Install the cylinder deflector into the engine.

14. Assemble the magneto assembly to the engine block or bearing plate, whichever is applicable, making sure that the points are clean and adjusted to the correct gap.

15. Install the flywheel, flywheel screen, and starter cup to the crankshaft. Tighten the flywheel nut to the proper specification.

16. Install the carburetor assembly and governor, making sure that the governor assembly links and springs are placed in their original holes.

17. Place the cylinder head on the block. Torque the head bolts in three stages, in the proper sequence, and to the proper torque.

18. Install the blower housing to the engine.

19. Install the spark plug, muffler assembly, and air cleaner.

TWO-STROKE ENGINES

1. Assemble the piston and rod assembly and install it into the block with the help of a piston ring compressor. Be careful not to damage the rings or the piston.

2. Install the crankshaft into the block, installing the crankshaft thrust washer if the engine is so equipped.

3. Assemble the connecting rod and piston assembly to the crankshaft by installing the connecting rod cap and cap screws. Tighten the cap screws to the correct torque.

4. Install the reed valve plate to the engine.

5. Install the bearing plate. Engines that use sleeve bearings should have a crankshaft end-play measuring 0.005–0.020 in. Engines using ball bearings have no specific end-play measurement but make sure that the crankshaft is not tight after assembly. End-play can be adjusted by the addition of various size gaskets behind the bearing plate.

6. Install the bearing oil seals on the PTO side and the flywheel side of the crankshaft.

7. Install the magneto to the bearing plate and adjust the points to the proper gap.

8. Install the flywheel, flywheel screen, and the starter cup to the crankshaft. Tighten the flywheel attaching nut to the correct torque specification.

9. Assemble the carburetor, governor links, springs, and air vane to the engine. Always replace all governor components in the same position from which they were removed.

10. Install the cylinder deflector to the engine.

11. Install the gas tank.

12. Install the air cleaner, spark plug, and muffler.

LUBRICATION SYSTEM

All four stroke engines with vertical crankshafts are lubricated by either a camshaft driven gear type oil pump or an oil scoop attached to the camshaft which rotates in the oil lying in the

Cylinder head bolt tightening sequence

Connecting rod cap mounted oil dipper

Gear type oil pump

Camshaft mounted oil scoop

bottom of the crankcase. The gear type pump forces oil up through a steel tube to the upper main bearing. The oil then falls down onto other parts needing to be lubricated.

The oil scoop attached to the camshaft sprays oil to the top of the engine in a circular path. The upper main bearing in oil scoop engines has an oil access slot in the top of the bearing area to catch the oil as it is splashed up.

On four-stroke engines that have a horizontal crankshaft, there is an oil dipper or distributor located on the bottom of the connecting rod bearing cap. As the crankshaft turns, the dipper churns through the oil in the crankcase and splashes it to all moving parts that need lubrication. There are oil passages, holes, and slots which the oil passes through on its way to the bearing surfaces.

General Engine Specifications

		1	2	3	4	5	6	7	8	9
Cylinder Bore Dia.	Min.	2.3745	2.3745	2.3745	2.3745	2.000	2.000	2.000	2.125	2.375
	Max.	2.3755	2.3755	2.3755	2.3755	2.001	2.001	2.001	2.126	2.376
Piston Skirt Dia.	Min.	2.3690	2.3690	2.3690	2.3690	1.9935	1.9935	1.9935	2.119	2.3690
	Max.	2.3700	2.3700	2.3700	2.3700	1.9945	1.9945	1.9945	2.120	2.3700
Piston Skirt to Cylinder Clearance	Min.	.0045	.0045	.0045	.0045	.0045	.0045	.0045	.005	.0055
	Max.	.0065	.0065	.0065	.0065	.0065	.0065	.0065	.007	.0075
Piston Ring to Groove Clearance	Min.	.002	.002	.002	.002	.002	.002	.002	.002	.002
	Max.	.005	.005	.005	.005	.005	.005	.005	.005	.005
Ring End Gap in Cylinder	Min.	.007	.007	.007	.007	.007	.007	.007	.007	.007
	Max.	.017	.017	.017	.017	.017	.017	.017	.017	.017
Connecting Rod Bore Crankshaft End	Min.	.8140	.8770	.8140	.8770	.7515	.7515	.8770	.8770	.8770
	Max.	.8145	.8775	.8145	.8775	.7520	.7520	.8775	.8775	.8775
Connecting Rod to Crankshaft Clearance	Min.	.0015	.0015	.0015	.0015	.0018	.0018	.0018	.0018	.0018
	Max.	.0030	.0030	.0030	.0030	.0035	.0035	.0035	.0035	.0035
Connecting Rod to Wrist Pin Clearance	Min.	.0004	.0004	.0004	.0004	.0004	.0004	.0004	.0004	.0004
	Max.	.0011	.0011	.0011	.0011	.0011	.0011	.0011	.0011	.0011
Crankshaft Rod Pin Diameter	Min.	.8119	.8745	.8119	.8745	.7483	.7483	.8745	.8745	.8754
	Max.	.8125	.8752	.8125	.8752	.7490	.7490	.8752	.8752	.8752
Crankpin Out-of-Round	Max.	.001	.001	.001	.001	.001	.001	.001	.001	.001
Crankshaft Main Diameter P.T.O. End	Min.	.8733	.8733	.8733	Note #1	.7483	.8745	.8745	.8745	.8745
	Max.	.8740	.8740	.8740		.7490	.8752	.8752	.8752	.8752
Crankshaft Main Diameter Flywheel End	Min.	.8120	.8120	.8120	.8120	.7483	.7483	.8745	.8745	.8745
	Max.	.8127	.8127	.8127	.8127	.7490	.7490	.8752	.8752	.8752
Crankshaft to Main Bearing Clearance	Min.	.0018	.0018	.0018	.0018	.0018	.0018	.0018	.0018	.0018
	Max.	.0035	.0035	.0035	.0035	.0035	.0035	.0035	.0035	.0035
Crankshaft End Play	Min.	.008	.008	.008	.008	.004	.004	.008	.008	.008
	Max.	.018	.018	.018	.018	.012	.012	.018	.018	.018
Block or Bearing Plate Main Bearing Bore P.T.O. End	Min.	.8758	.8758	.8758	Note #5	.7510	.877	.877	.877	.877
	Max.	.8768	.8768	.8768		.7520	.878	.878	.878	.878
Bearing Plate or Block Bearing Bore (Flywheel) End	Min.	.8145	.8145	.8145	.8145	.7510	.7510	.877	.877	.877
	Max.	.8155	.8155	.8155	.8155	.7520	.7520	.878	.878	.878
Valve or Tappet Guide Bore I.D.	Min.	.2495	.2495	.2495	.2495	.2495	.2495	.2495	.2495	.2495
	Max.	.2510	.2510	.2510	.2510	.2510	.2510	.2510	.2510	.2510
Valve Stem to Guide Clearance	Min.	.0015	.0015	.0015	.0015	.0020	.0020	.0020	.0020	.0020
	Max.	.0045	.0045	.0045	.0045	.0045	.0045	.0045	.0045	.0045
Valve Clearance. Intake & Exhaust	Min.	.009	.009	.009	.009	.007	.007	.009	.009	.009
	Max.	.011	.011	.011	.011	.009	.009	.011	.011	.011
Camshaft to Axle Clearance	Min.	.001	.001	.001	.001	.001	.001	.001	.001	.001
	Max.	.003	.003	.003	.003	.003	.003	.003	.003	.003
Camshaft Axle Clearance P.T.O.	Min.	.001	.001	.001	.001					
	Max.	.003	.003	.003	.003					
Camshaft Axle Clearance Flywheel End	Min.	.001	.001	.001	.001					
	Max.	.003	.003	.003	.003					
Point Setting	Min.	.018	.018	.018	.018	.018	.018	.018	.018	.018
	Max.	.021	.021	.021	.021	.021	.021	.021	.021	.021
Spark Plug Gap	Min.	.025	.025	.025	.025	.025	.025	.025	.025	.025
	Max.	.028	.028	.028	.028	.028	.028	.028	.028	.028
Comp. at Cranking Speed. P.S.I.	Min.	65	65	65	65	65	65	65	65	65
Carburetor Float Setting (Clinton)	Min.	$\frac{5}{32}$	$\frac{5}{32}$	$\frac{5}{32}$	$\frac{5}{32}$	$\frac{5}{32}$	$\frac{5}{32}$	$\frac{5}{32}$	$\frac{5}{32}$	$\frac{5}{32}$
	Max.	$\frac{11}{64}$	$\frac{11}{64}$	$\frac{11}{64}$	$\frac{11}{64}$	$\frac{11}{64}$	$\frac{11}{64}$	$\frac{11}{64}$	$\frac{11}{64}$	$\frac{11}{64}$
Carburetor Float Setting (Carter)	Min.	$\frac{11}{64}$	$\frac{11}{64}$	$\frac{11}{64}$	$\frac{11}{64}$	$\frac{11}{64}$	$\frac{11}{64}$	$\frac{11}{64}$	$\frac{11}{64}$	$\frac{11}{64}$
	Max.	$\frac{13}{64}$	$\frac{13}{64}$	$\frac{13}{64}$	$\frac{13}{64}$	$\frac{13}{64}$	$\frac{13}{64}$	$\frac{13}{64}$	$\frac{13}{64}$	$\frac{13}{64}$
Magneto Air Gap	Min.	.007	.007	.007	.007	.007	.007	.007	.007	.007
	Max.	.017	.017	.017	.017	.017	.017	.017	.017	.017
Magneto Edge Gap (Phelon)	Min.	$\frac{5}{32}$	$\frac{5}{32}$	$\frac{5}{32}$	$\frac{5}{32}$	$\frac{5}{32}$	$\frac{5}{32}$	$\frac{5}{32}$	$\frac{5}{32}$	$\frac{5}{32}$
	Max.	$\frac{9}{32}$	$\frac{9}{32}$	$\frac{9}{32}$	$\frac{9}{32}$	$\frac{9}{32}$	$\frac{9}{32}$	$\frac{9}{32}$	$\frac{9}{32}$	$\frac{9}{32}$
Magneto Edge Gap (Clinton)	Min.	$\frac{7}{64}$	$\frac{7}{64}$	$\frac{7}{64}$	$\frac{7}{64}$			$\frac{7}{64}$	$\frac{7}{64}$	$\frac{7}{64}$
	Max.	$\frac{1}{4}$	$\frac{1}{4}$	$\frac{1}{4}$	$\frac{1}{4}$			$\frac{1}{4}$	$\frac{1}{4}$	$\frac{1}{4}$
Oil Recommended A.P.I. Rating	See Bulletin #9 for SAE	MM MS	MM MS	MM MS	MM MS	MM MS	MM MS	MM MS	MM MS	MM MS

General Engine Specifications

Fuel Recommended	1	2	3	4	5	6	7	8	9
	Reg. Gas	Reg. Gas	Reg. Gas	Reg. Gas	Reg. Gas	Reg. Gas	Reg. Gas	Reg. Gas	Reg. Gas

Note 1 Crankshaft diameter .8733 to .8740 or .9995 to 1.0002. except 435-0000-000 which is 1.220 to 1.221.
Note 2 Ball, needle or roller bearing used. Check component parts. if worn or defective. replace.
Note 3 Wrist pin clearance. .0004 minimum. .0011 maximum or units other than listed below. Needle bearing used in D65-1000 Type (C), J9-1000 Type (B). J6-1100 and J6-1101. Check component parts. if worn or defective. replace.
Note 4 412-0000-000. 413-0000-000 with sleeve bearings .002 to .006. 412-0300-000 with roller bearings .001 to .006.
Note 5 Bearing bore diameter .8758 to .8768 or 1.0002 to 1.0003. except 435-0000-000 which is 1.2228 to 1.2235.
Note 6 Valve guides .312 to .313 inside diameter and are replaceable. Tappet guide inside diameter .437 to .438.
Note 7 412-0000-000 .001 minimum. .004 maximum. 413-0000-000 .0005 minimum. .002 maximum.

General Engine Specifications

		10	11	12	13	14	15	16	17	18
Cylinder Bore Dia.	Min.	2.375	2.375	2.4685	2.4685	2.4685	2.499	2.8125	2.8125	2.8125
	Max.	2.376	2.376	2.4695	2.4695	2.4695	2.500	2.8135	2.8135	2.8135
Piston Skirt Dia.	Min.	2.3690	2.3690	2.4630	2.4630	2.4630	2.4935	2.8045	2.8045	2.8045
	Max.	2.3700	2.3700	2.4640	2.4640	2.4640	2.4945	2.8055	2.8055	2.8055
Piston Skirt to Cylinder Clearance	Min.	.0055	.0055	.0055	.0055	.0055	.0045	.007	.007	.007
	Max.	.0075	.0075	.0075	.0075	.0075	.0065	.009	.009	.009
Piston Ring to Groove Clearance	Min.	.002	.002	.002	.002	.002	.002	.0025	.0025	.0025
	Max.	.005	.005	.005	.005	.005	.005	.0050	.0050	.0050
Ring End Gap in Cylinder	Min.	.007	.007	.007	.007	.007	.007	.007	.007	.007
	Max.	.017	.017	.017	.017	.017	.017	.017	.017	.017
Connecting Rod Bore Crankshaft Ends	Min.	.8770	.9140	.8770	.9140	.9140	.8770	1.2510	1.1265	1.1265
	Max.	.8775	.9145	.8775	.9145	.9145	.8775	1.2513	1.1268	1.1268
Connecting Rod to Crankshaft Clearance	Min.	.0018	.0018	.0018	.0018	.0018	.0015	.001	.0015	.0015
	Max.	.0035	.0035	.0035	.0035	.0035	.0035	.002	.0025	.0025
Connecting Rod to Wrist Pin Clearance	Min.	.0004	.0004	.0004	.0004	.0004	.0004	.0002	.0002	.0002
	Max.	.0011	.0011	.0011	.0011	.0011	.0011	.0011	.0011	.0011
Crankshaft Rod Pin Diameter	Min.	.8745	.9114	.8745	.9114	.9114	.8745	1.2493	1.1243	1.1243
	Max.	.8752	.9120	.8752	.9120	.9120	.8752	1.2500	1.1250	1.1250
Crankpin Out-of Round	Max.	.001	.001	.001	.001	.001	.001	.001	.001	.001
Crankshaft Main Diameter P.T.O. End	Min.	.9995	.9995	Note #2	Note #2	.9995	1.220	1.2510	Note #2	Note #2
	Max.	1.0002	1.0002			1.0002	1.221	1.2515		
Crankshaft Main Diameter Flywheel End	Min.	.8745	.8745	Note #2	Note #2	.8745	.8120	1.2510	Note #2	Note #2
	Max.	.8752	.8752			.8752	.8127	1.2515		
Crankshaft to Main Bearing Clearance	Min.	.0018	.0018	Note #2	Note #2	.0018	.0018	.0010	Note #2	Note #2
	Max.	.0035	.0035			.0035	.0035	.0025		
Crankshaft End Play	Min.	.008	.008	.001	.001	.008	.008	Note #4	.006	.001
	Max.	.018	.018	.006	.006	.018	.018		.020	.006
Block or Bearing Plate Main Bearing Bore P.T.O. End	Min.	1.002	1.002	Note #2	Note #2	1.002	1.2228	1.2525	Note #2	Note #2
	Max.	1.003	1.003			1.003	1.2235	1.2535		
Bearing Plate or Block Bearing Bore (Flywheel) End	Min.	.877	.877	Note #2	Note #2	.877	.8145	1.2525	Note #2	Note #2
	Max.	.878	.878			.878	.8155	1.2535		
Valve or Tappet Guide Bore I.D.	Min.	.2495	.2495	.2495	.2495	.2495	.2495	.312	Note #6	Note #6
	Max.	.2510	.2510	.2510	.2510	.2510	.2510	.313		
Valve Stem to Guide Clearance	Min.	.0020	.0020	.0020	.0020	.0020	.0015	.002	.002	.002
	Max.	.0045	.0045	.0045	.0045	.0045	.0045	.004	.004	.004
Valve Clearance. Intake & Exhaust	Min.	.009	.009	.009	.009	.009	.009	.011	.010	.010
	Max.	.011	.010	.010	.010	.010	.011	.012	.012	.012
Camshaft to Axle Clearance	Min.	.001	.001	.001	.001	.001			.0015	.0015
	Max.	.003	.003	.003	.003	.003			.0035	.0035
Camshaft Axle Clearance P.T.O.	Min.						.001	Note #7		
	Max.						.003			
Camshaft Axle Clearance Flywheel End	Min.						.001	.001		
	Max.						.003	.003		
Point Setting	Min.	.018	.018	.018	.018	.018	.018	.018	.028	.028
	Max.	.021	.021	.021	.021	.021	.021	.021	.030	.030
Spark Plug Gap	Min.	.025	.025	.025	.025	.025	.025	.025	.025	.025
	Max.	.028	.028	.028	.028	.028	.028	.028	.028	.028

General Engine Specifications (cont.)

		10	11	12	13	14	15	16	17	18
Comp. at Cranking Speed. P.S.I.	Min.	65	65	65	65	65	65	65	70	70
Carburetor Float Setting (Clinton)	Min.	$5/32$	$5/32$	$5/32$	$5/32$	$5/32$	$5/32$	$5/32$		$11/64$
	Max.	$11/64$	$11/64$	$11/64$	$11/64$	$11/64$	$11/64$	$11/64$		$13/64$
Carburetor Float Setting (Carter)	Min.	$11/64$	$11/64$	$11/64$	$11/64$	$11/64$			$15/64$	$15/64$
	Max.	$13/64$	$13/64$	$13/64$	$13/64$	$13/64$			$17/64$	$17/64$
Magneto Air Gap	Min.	.007	.007	.007	.007	.007	.007	.007	.012	.012
	Max.	.017	.017	.017	.017	.017	.017	.017	.020	.020
Magneto Edge Gap (Phelon)	Min.	$5/32$	$5/32$	$5/32$	$5/32$	$5/32$				
	Max.	$9/32$	$9/32$	$9/32$	$9/32$	$9/32$				
Magneto Edge Gap (Clinton)	Min.	$7/64$	$7/64$	$7/64$	$7/64$	$7/64$	$7/64$	$7/64$		
	Max.	$1/4$	$1/4$	$1/4$	$1/4$	$1/4$	$1/4$	$1/4$		
Oil Recommended A.P.I. Rating	See Bulletin #9 for SAE	MM MS	MM MS	MM MS	MM MS	MM MS	MM MS	MM MS	MM MS	MM MS
Fuel Recommended		Reg. Gas	Reg. Gas	Reg. Gas	Reg. Gas	Reg. Gas	Reg. Gas	Reg. Gas	Reg. Gas	Reg. Gas

Note 1 Crankshaft diameter .8733 to .8740 or .9995 to 1.0002, except 435-0000-000 which is 1.220 to 1.221.
Note 2 Ball, needle or roller bearing used. Check component parts, if worn or defective, replace.
Note 3 Wrist pin clearance, .0004 minimum, .0011 maximum or umits other than listed below. Needle bearing used in D65-1000 Type (C), J9-1000 Type (B), J6-1100 and J6-1101. Check component parts, if worn or defective, replace.
Note 4 412-0000-000, 413-0000-000 with sleeve bearings .002 to .006, 412-0300-000 with roller bearings .001 to .006.
Note 5 Bearing bore diameter .8758 to .8768 or 1.0002 to 1.0003, except 435-0000-000 which is 1.2228 to 1.2235.
Note 6 Valve guides .312 to .313 inside diameter and are replaceble. Tappet guide inside diameter .437 to .438.
Note 7 412-0000-000 .001 minimum, .004 maximum. 413-0000-000 .0005 minimum, .002 maximum.

General Engine Specification

		19	20	21	22	23	24	25	26	27
Cylinder Bore Dia.	Min.	2.9995	3.1245	3.1245	3.1245	3.1245	1.875	1.875	2.125	2.125
	Max.	3.0005	3.1255	3.1255	3.1255	3.1255	1.876	1.876	2.126	2.126
Piston Skirt Dia.	Min.	2.9915	3.117	3.117	3.1185	3.1185	1.8695	1.8695	2.1195	2.1195
	Max.	2.9925	3.118	3.118	3.1195	3.1195	1.8705	1.8705	2.1205	2.1205
Piston Skirt to Cylinder Clearance	Min.	.007	.0065	.0065	.005	.005	.0045	.0045	.005	.005
	Max.	.009	.0085	.0085	.007	.007	.0065	.0065	.007	.007
Piston Ring to Groove Clearance	Min.	.0025	.0025	.0025	.0025	.0025	.0015	.0015	.0015	.0015
	Max.	.0050	.0050	.0050	.0050	.0050	.0040	.0040	.0040	.0040
Ring End Gap in Cylinder	Min.	.007	.010	.010	.010	.010	.005	.005	.007	.007
	Max.	.017	.020	.020	.020	.020	.013	.013	.017	.017
Connecting Rod Bore Crankshaft End	Min.	1.1265	1.2510	1.2510	1.2510	1.2510	.7820	.7820	.7820	.7820
	Max.	1.1268	1.2513	1.2513	1.2513	1.2513	.7827	.7827	.7827	.7827
Connecting Rod to Crankshaft Clearance	Min.	.0015	.0010	.0010	.0010	.0010	.0026	.0026	.0026	.0026
	Max.	.0025	.0018	.0018	.0018	.0018	.0040	.0040	.0040	.0040
Connecting Rod to Wrist Pin Clearance	Min.	.0002	.0002	.0002	.0002	.0002	.0004	.0004	.0004	.0004
	Max.	.0011	.0011	.0011	.0011	.0011	.0011	.0011	.0011	.0011
Crankshaft Rod Pin Diameter	Min.	1.1243	1.2495	1.2495	1.2495	1.2495	.7788	.7788	.7788	.7788
	Max.	1.1250	1.2500	1.2500	1.2500	1.2500	.7795	.7795	.7795	.7795
Crankpin Out-of-Round	Max.	.001	.001	.001	.001	.001	.001	.001	.001	.001
Crankshaft Main Diameter P.T.O. End	Min.	Note #2	Note #2	Note #2	Note #2	Note #2	.8745	.8745	.8745	.9995
	Max.						.8752	.8752	.8752	1.0002
Crankshaft Main Diameter Flywheel End	Min.	Note #2	Note #2	Note #2	Note #2	Note #2	.7495	.7495	.7495	.7495
	Max.						.7502	.7502	.7502	.7502
Crankshaft to Main Bearing Clearance	Min.	Note #2	Note #2	Note #2	Note #2	Note #2	.0015	.0015	.0015	.0015
	Max.						.0035	.0035	.0035	.0035
Crankshaft End Play	Min.	.001	.006	.001	.006	.001	.005	.005	.005	.005
	Max.	.006	.020	.006	.020	.006	.020	.020	.020	.020
Block or Bearing Plate Main Bearing Bore P.T.O. End	Min.	Note #2	Note #2	Note #2	Note #2	Note #2	.8770	.8770	.8770	1.002
	Max.						.8780	.8780	.8780	1.003
Bearing Plate or Block Bearing Bore (Flywheel) End	Min.	Note #2	Note #2	Note #2	Note #2	Note #2	.7517	.7517	.7517	.7517
	Max.						.7525	.7525	.7525	.7525

General Engine Specification (cont.)

		19	20	21	22	23	24	25	26	27
Valve or Tappet Guide Bore I.D.	Min.	Note #6	Note #6	Note #6	Note #6	Note #6				
	Max.									
Valve Stem to Guide Clearance	Min.	.002	.002	.002	.002	.002				
	Max.	.004	.004	.004	.004	.004				
Valve Clearance. Intake & Exhaust	Min.	.010	.010	.010	.010	.010				
	Max.	.012	.012	.012	.012	.012				
Camshaft to Axle Clearance	Min.	.0015	.0015	.0015	.0015	.0015				
	Max.	.0035	.0035	.0035	.0035	.0035				
Camshaft Axle Clearance P.T.O.	Min.									
	Max.									
Camshaft Axle Clearance Flywheel End	Min.									
	Max.									
Point Setting	Min.	.028	.028	.028	.028	.028	.018	.018	.018	.018
	Max.	.030	.030	.030	.030	.030	.021	.021	.021	.021
Spark Plug Gap	Min.	.025	.025	.025	.025	.025	.025	.025	.025	.025
	Max.	.028	.028	.028	.028	.028	.028	.028	.028	.028
Comp. at Cranking Speed. P.S.I.	Min.	70	70	70	70	70	60	60	60	60
Carburetor Float Setting (Clinton)	Min.	$\frac{11}{64}$		$\frac{11}{64}$		$\frac{11}{64}$	$\frac{5}{32}$	$\frac{5}{32}$	$\frac{5}{32}$	$\frac{5}{32}$
	Max.	$\frac{13}{64}$		$\frac{13}{64}$		$\frac{13}{64}$	$\frac{11}{64}$	$\frac{11}{64}$	$\frac{11}{64}$	$\frac{11}{64}$
Carburetor Float Setting (Carter)	Min.	$\frac{15}{64}$	$\frac{15}{64}$	$\frac{15}{64}$	$\frac{15}{64}$	$\frac{15}{64}$	$\frac{11}{64}$	$\frac{11}{64}$	$\frac{11}{64}$	$\frac{11}{64}$
	Max.	$\frac{17}{64}$	$\frac{17}{64}$	$\frac{17}{64}$	$\frac{17}{64}$	$\frac{17}{64}$	$\frac{13}{64}$	$\frac{13}{64}$	$\frac{13}{64}$	$\frac{13}{64}$
Magneto Air Gap	Min.	.012	.012	.012	.012	.012	.007	.007	.007	.007
	Max.	.020	.020	.020	.020	.020	.017	.017	.017	.017
Magneto Edge Gap (Phelon)	Min.						$\frac{5}{32}$	$\frac{5}{32}$	$\frac{5}{32}$	$\frac{5}{32}$
	Max.						$\frac{9}{32}$	$\frac{9}{32}$	$\frac{9}{32}$	$\frac{9}{32}$
Magneto Edge Gap (Clinton)	Min.						$\frac{7}{64}$	$\frac{7}{64}$	$\frac{7}{64}$	$\frac{7}{64}$
	Max.						$\frac{1}{4}$	$\frac{1}{4}$	$\frac{1}{4}$	$\frac{1}{4}$
Oil Recommended A.P.I. Rating	See Bulletin #9 for SAE	MM MS	MM MS	MM MS	MM MS	MM MS	Outboard Motor Oil or SAE 30 Non Detergent			
Fuel Recommended		Reg. Gas	Reg. Gas	Reg. Gas	Reg. Gas	Reg. Gas	Reg. Gas	Reg. Gas	Reg. Gas	Reg. Gas

Note 1 Crankshaft diameter .8733 to .8740 or .9995 to 1.0002. except 435-0000-000 which is 1.220 to 1.221.
Note 2 Ball. needle or roller bearing used. Check component parts. if worn or defective. replace
Note 3 Wrist pin clearance. .0004 minimum. .0011 maximum or umits other than listed below. Needle bearing used in D65-1000 Type (C), J9-1000 Type (B). J6-1100 and J6-1101. Check component parts. if worn or defective. replace.
Note 4 412-0000-000. 413-0000-000 with sleeve bearings .002 to .006. 412-0300-000 with roller bearings .001 to .006.
Note 5 Bearing bore diameter .8758 to .8768 or 1.0002 to 1.0003. except 435-0000-000 which is 1.2228 to 1.2235.
Note 6 Valve guides .312 to .313 inside diameter and are replaceble. Tappet guide inside diameter .437 to .438
Note 7 412-0000-000. .001 minimum. .004 maximum. 413-0000-000. .0005 minimum. .002 maximum.

General Engine Specifications

		28	29	30	31	32	33	34	35	36
Cylinder Bore Dia.	Min.	2.125	2.125	1.8750	2.1255	1.8750	1.811	1.8750	2.1255	2.1255
	Max.	2.126	2.126	1.8765	2.1270	1.8765	1.814	1.8765	2.1270	2.1270
Piston Skirt Dia	Min.	2.1195	2.1195	1.8695	2.120	1.871	1.807	1.871	2.120	2.120
	Max.	2.1205	2.1205	1.8705	2.121	1.872	1.809	1.872	2.121	2.121
Piston Skirt to Cylinder Clearance	Min.	.005	.005	.0045	.0045	.0045	.003	.0045	.0045	.0045
	Max.	.007	.007	.0065	.0070	.0065	.007	.0065	.0070	.0070
Piston Ring to Groove Clearance	Min.	.0015	.002	.002	.002	.002	.002	.002	.002	.002
	Max.	.0040	.004	.004	.004	.004	.005	.004	.004	.004
Ring End Gap in Cylinder	Min.	.007	.010	.010	.010	.010	.004	.010	.010	.010
	Max.	.017	.015	.015	.015	.015	.020	.015	.015	.015
Connecting Rod Bore Crankshaft End	Min.	.7820	Note #2	.7816	Note #2	Note #2	Note #2	Note #2	Note #2	Note #2
	Max.	.7827		.7825						
Connecting Rod to Crankshaft Clearance	Min.	.0026	Note #2	.0021	Note #2	Note #2	Note #2	Note #2	Note #2	Note #2
	Max.	.0040		.0037						
Connecting Rod to Wrist Pin Clearance	Min.	.0004	.0004	.0004	Note #3	.0004	Note #2	.0004	Note #3	Note #3
	Max.	.0011	.0011	.0011		.0011		.0011		

General Engine Specifications (cont.)

		28	29	30	31	32	33	34	35	36
Crankshaft Rod Pin Diameter	Min.	.7788	.6594	.7788	.6594	.6594	.4960	.6594	.6594	.6594
	Max.	.7795	.6599	.7795	.6599	.6599	.4965	.6599	.6599	.6599
Crankpin Out-of-Round	Max.	.001	.001	.001	.001	.001	.001	.001	.001	.001
Crankshaft Main Diameter P.T.O. End	Min.	.8745	Note #2	Note #2	Note #2	Note #2	Note #2	Note #2	Note #2	Note #2
	Max.	.8752	Note #2	Note #2	Note #2	Note #2	Note #2	Note #2	Note #2	Note #2
Crankshaft Main Diameter Flywheel End	Min.	.7495	.7495	.7495	.7495	.7495	Note #2	.7495	.7495	.7495
	Max.	.7502	.7502	.7502	.7502	.7502	Note #2	.7502	.7502	.7502
Crankshaft to Main Bearing Clearance	Min.	.0015	Note #2	.0015	Note #2	Note #2	Note #2	Note #2	Note #2	Note #2
	Max.	.0035	Note #2	.0035	Note #2	Note #2	Note #2	Note #2	Note #2	Note #2
Crankshaft End Play	Min.	.005	Note #2	Note #2	Note #2	Note #2	.003	.003	.003	Note #2
	Max.	.020	Note #2	Note #2	Note #2	Note #2	.020	.020	.020	Note #2
Block or Bearing Plate Main Bearing Bore P.T.O. End	Min.	.8770	Note #2	Note #2	Note #2	Note #2	Note #2	Note #2	Note #2	Note #2
	Max.	.8778	Note #2	Note #2	Note #2	Note #2	Note #2	Note #2	Note #2	Note #2
Bearing Plate or Block Bearing Bore (Flywheel) End	Min.	.7517	Note #2	Note #2	Note #2	Note #2	Note #2	Note #2	Note #2	Note #2
	Max.	.7525	Note #2	Note #2	Note #2	Note #2	Note #2	Note #2	Note #2	Note #2
Valve or Tappet Guide Bore I.D.	Min.									
	Max.									
Valve Stem to Guide Clearance	Min.									
	Max.									
Valve Clearance, Intake & Exhaust	Min.									
	Max.									
Camshaft to Axle Clearance	Min.									
	Max.									
Camshaft Axle Clearance P.T.O.	Min.									
	Max.									
Camshaft Axle Clearance Flywheel End	Min.									
	Max.									
Point Setting	Min.	.018	.018	.018	.018	.018	.013	.014	.014	.018
	Max.	.021	.021	.021	.021	.021	.017	.016	.016	.021
Spark Plug Gap	Min.	.025	.025	.025	.025	.025	.025	.025	.025	.025
	Max.	.028	.028	.028	.028	.028	.030	.028	.028	.028
Comp. at Cranking Speed, P.S.I.	Min.	60	60	60	60	60	80	60	60	60
Carburetor Float Setting (Clinton)	Min.	$5/32$	$5/32$							$1\ 25/64$
	Max.	$11/64$	$11/64$							$1\ 27/64$
Carburetor Float Setting (Carter)	Min.	$11/64$	$11/64$							
	Max.	$13/64$	$13/64$							
Magneto Air Gap	Min.	.007	.007	.007	.007	.007	.007	.007	.007	.007
	Max.	.017	.017	.017	.017	.017	.017	.017	.017	.017
Magneto Edge Gap (Phelon)	Min.	$5/32$								
	Max.	$9/32$								
Magneto Edge Gap (Clinton)	Min.	$7/64$	$7/64$	$7/64$	$7/64$	$7/64$	$3/32$	$3/32$	$3/32$	$7/64$
	Max.	$1/4$	$1/4$	$1/4$	$1/4$	$1/4$	$5/16$	$5/16$	$5/16$	$1/4$
Oil Recommended A.P.I. Rating		Outboard Motor Oil or SAE 30 Non Detergent								
Fuel Recommended		Reg. Gas	Reg. Gas	Reg. Gas	Reg. Gas	Reg. Gas	Reg. Gas	Reg. Gas	Reg. Gas	Reg. Gas

Note 1 Crankshaft diameter .8733 to .8740 or .9995 to 1.0002, except 435-0000-000 which is 1.220 to 1.221.
Note 2 Ball, needle or roller bearing used. Check component parts, if worn or defective, replace.
Note 3 Wrist pin clearance .0004 minimum, .0011 maximum or umits other than listed below. Needle bearing used in D65-1000 Type (C), J9-1000 Type (B), J6-1100 and J6-1101. Check component parts, if worn or defective, replace.
Note 4 412-0000-000, 413-0000-000 with sleeve bearings .002 to .006, 412-0300-000 with roller bearings .001 to .006.
Note 5 Bearing bore diameter .8758 to .8768 or 1.0002 to 1.0003, except 435-0000-000 which is 1.2228 to 1.2235.
Note 6 Valve guides .312 to .313 inside diameter and are replaceble. Tappet guide inside diameter .437 to .438.
Note 7 412-0000-000 .001 minimum, .004 maximum, 413-0000-000 .0005 minimum, .002 maximum.

Torque Specifications

Part		All 4 Cycle Aluminum Vertical & Horizontal Shaft	V100, V3100, 406, H3100 407, 417	429 & 431	A300	300, VS300, 350, 500, 650, 700A, C700 D700, VS700, VS750, 800, A800, VS800 900, VS900, V1000, A-B-1100, C1100, D1100, V1100, VS1100, 1200, A1200, B1290, V1200, VS1200, 492, 494, 497, 498, 499	412 & 413	1600, 1800, 2500, 2790 414, 416, 418, 420, 422	200, A200, VS200, VS400, A400, 500, 501	E65 & GK590	E10 — Chainsaw	D15 — Chainsaw D25 — Chainsaw	D35, D55, D65 — Chainsaws E75, E95, — Chainsaws	J5, J6, J7, J8, J9 — Outboards
Connecting Rod Aluminum	Min.	100	100	100	70	70	215	215	35	70	50	55	80	80
	Max.	125	125	125	80	80	235	235	45	80	55	65	90	90
Connecting Rod Forged Steel	Min.									90			90	90
	Max.									100			100	100
Bearing Plate P.T.O. End	Min.	75	75	75			75	160						
	Max.	85	85	85			85	180						
Bearing Plate Flying End	Min.				120	140			75	75		80	80	75
	Max.				150	160			95	95		90	90	95
Back Plate to Block	Min.							70						
	Max.							80						
Head Bolts	Min.	125	225	225	200	200	200	200						
	Max.	150	250	250	220	220	220	220						
Base Bolts	Min.	75	75	75	150	325	75	150	125	125				
	Max.	85	85	85	160	375	85	160	150	150				
End Cover or Gear Box	Min.					120								
	Max.					150								
Speed Reducer Mounting	Min.	110	110		110	110	110	110	110					
	Max.	150	150		150	150	150	150	150					
P.T.O. Housing or Mounting Flange	Min.	75	75	75		120	75	160	125					
	Max.	85	85	85		150	85	180	150					
Carb. Reed Plate or Manifold to Blk.	Min.	60	60	60	60	60	65	60	60	60	45	50	50	50
	Max.	65	65	65	65	65	75	65	65	65	55	60	60	60
Carb. to Reed Plate or Manifold	Min.	35	35	35	60	35	35	60	60	60		50	50	50
	Max.	50	50	50	65	50	50	65	65	65		60	60	60
Blower Housing	Min.	60	60	60	65	65	65	65	65	65	90	80	80	75
	Max.	70	70	70	70	70	75	70	70	70	110	90	90	85
Muffler to Block	Min.	110	110	110	140	140	170	170	40	40	90	60	60	
	Max.	120	120	120	150	150	180	180	60	60	110	70	70	
Flywheel	Min.	375	375	375	375	400	400	*100	375	250	150	375	**250	250
	Max.	400	400	400	400	450	500	*120	400	300	180	425	**300	300
Flywheel Touch & Stop for Brake	Min.	650	650											
	Max.	700	700											
Spark Plug	Min.	275	275	275	275	275	275	275	275	275	250	230	230	230
	Max.	300	300	300	300	300	300	300	300	300	300	270	270	270

Torque Specifications

		All 4 Cycle Aluminum Vertical & Horizontal Shaft	V100, V3100, 406, H3100 407, 417	429 & 431	A300	300, VS300, 350, 500, 650, 700A, C700 D700, VS700, VS750, 800, A800, VS800 900, VS900, V1000, A-B-1100, C1100, D1100, V1100, VS1100, 1200, A1200, B1290, V1200, VS1200, 492, 494, 497, 498, 499	412 & 413	1600, 1800, 2500, 2790 414, 416, 418, 420, 422	200, A200, VS200, VS400, A400, 500, 501	E65 & GK590	E10 — Chainsaw	D15 — Chainsaw D25 — Chainsaw	D35, D55, D65 — Chainsaws E75, E95, — Chainsaws	J5, J6, J7, J8, J9 — Outboards
Stator Plate	Min.	50	50	50	50	50	50	80	50	50	50	50	50	45
	Max.	60	60	60	60	60	60	100	60	60	60	60	60	65

All torque in inch pounds except those marked with a single *
* Foot Pounds
** D35 Requires Same as D25

Cross Reference Chart

Model No.	Column No.
E–65	29
100	1
100–1000	1
100–2000	1
V100–1000	3
VS100	3
VS100–1000	3
VS100–2000	3
VS100–3000	3
VS100–4000	3
200	24
A200	24
AVS200	24
AVS200–1000	25
VS200	24
VS200–1000	24
VS200–2000	25
VS200–3000	25
VS200–4000	27
300	5
A300	5
VS300	6
350	5

Model No.	Column No.
D1100	9
V1100–1000	11
VS1100	11
VS1100–1000	11
1200	12
1200–1000	12
1200–2000	13
A1200	13
B1290–1000	13
V1200–1000	14
VS1200	14
1600	17
A1600–1000	18
1800–1000	19
2100	1
A2100	1
A2100–1000	1
A2100–2000	1
VS2100	3
VS2100–1000	3
VS2100–2000	3
VS2100–3000	3

Cross Reference Chart

Model No.	Column No.	Model No.	Column No.
A400	28	2500	20
A400–1000	28	A2500	22
AVS400	26	B2500–1000	21
AVS400–1000	27		
BVS400	27	2790–1000	23
CVS400–1000	27		
VS400	26	VS3000	3
VS400–1000	26		
VS400–2000	27	3100	2
VS400–3000	27	3100–1000	2
VS400–4000	27	3100–2000	2
		3100–3000	2
GK590	29	H3100–1000	2
		FV3100–1000	4
650	7	AFV3100–1000	4
		AV3100–1000	4
700–A	7	AV3100–2000	4
B700	7	AVS3100	4
C700	7	AVS3100–1000	4
D700	8	AVS3100–2000	4
D700–1000	8	AVS3100–3000	4
D700–2000	8	V3100–1000	4
D700–3000	8	V3100–2000	4
VS700	7	VS3100	4
VS750	7	VS3100–1000	4
		VS3100–2000	4
800	9	VS3100–3000	4
A800	9		
VS800	9	4100	2
		4100–1000	2
900	9	4100–2000	2
900–1000	9	AVS4100–1000	4
900–2000	9	AVS4100–2000	4
900–3000	9	VS4100–1000	4
900–4000	9	VS4100–2000	4
VS900	10		
		J–5	36
V1000–1000	10	J–6	36
VS1000	10	J–7	36
		J–8	36
A&B1100	9		
C1100	9	415–0000–000	4
		415–0002–000	4
J–9	36	416–1300–000	18
E10–1000	33	417–0000–000	4
D15–1000	30	418–1300–000	19
D25–1000	30	418–1301–000	19
D35–1000	31	420–1300–000	21
D55–1000	32	420–1301–000	21
D65–1000	31	422–1300–000	23
E75–1000	34	422–1301–000	23
E95–1000	35	424–0000–000	2
		426–0000–000	2
400–0000–000	1	429–0003–000	15
401–0000–000	3	431–0003–000	15
402–0000–000	1	435–0003–000	4
403–0000–000	3	492–0300–000	9
404–0000–000	2	494–0000–000	9
405–0000–000	4		
406–0000–000	2		

Cross Reference Chart (cont.)

Model No.	Column No.	Model No.	Column No.
407 – 0000 – 000	4		
407 – 0002 – 000	4	494 – 0001 – 000	9
408 – 0000 – 000	2	497 – 0000 – 000	14
409 – 0000 – 000	4	498 – 0300 – 000	13
411 – 0000 – 000	3	498 – 0301 – 000	13
411 – 0002 – 000	3	499 – 0000 – 000	14
412 – 0000 – 000	16		
413 – 0000 – 000	16	500 – 0000 – 000	28
414 – 1300 – 000	18	501 – 0000 – 000	27
414 – 1301 – 000	18	501 – 0001 – 000	27

Section 8

Jacobson

Small Engines — 0-6 hp

ENGINE IDENTIFICATION

The Jacobsen 321 and 501 engines are almost identical, with only minor differences. These engines are 2-cycle design and can be used in both horizontal and vertical applications. Both engines are single cylinder models. The 501 is a heavy duty version of the 321. A stamped plate showing the engine type is affixed to the crankcase.

MAINTENANCE
Air Cleaners

Four types of air cleaners are used on these engines. The type of air cleaner depends on the engine application. An aluminum foil, paper, oil bath and polyurethane filter can be found on the engine. Proper cleaning and prescribed replacement are necessary to maintain good engine performance.

CLEANING AND REPLACEMENT

Aluminum Foil Type

Periodically wash the filter in a safe solvent. Shake off the excess solvent and air-dry the filter. Dip the filter in 30W engine oil. Allow the excess oil to drain off, and install the filter.

Paper Filter

If the filter is only lightly soiled, tap it gently to remove any loose dirt, then blow compressed air through it from the inside out. If the filter is dirty, replace it.

Oil Bath Type

Remove the cover from the reservoir and dump out the old oil. Wash the air cleaner with a safe solvent. Dry it thoroughly. Refill the air cleaner to the indicated level with the same grade and weight oil used in the engine.

Typical air cleaner

Polyurethane Type

The foam type cleaner should be removed and washed in kerosene. Ring out all the excess kerosene and add about 2 tablespoons of clean 30W engine oil directly into the foam. Squeeze the foam gently to evenly distribute the oil.

Oil and Fuel Recommendations

The proper mixture of gasoline and oil is critical to the performance of a 2-cycle engine. Jacobsen recommends the use of regular, leaded gasoline, mixed with 30W, non-detergent oil meeting specification MM or SB (stamped on top of the can). For the 501, the ratio is $\frac{1}{2}$ pint of oil to 1 gallon of gasoline; for the 321, the ratio is $\frac{1}{4}$ pint of oil to 1 gallon of gasoline. Always mix the oil and fuel in a safe container intended for that purpose. Never mix the oil and fuel in the engine's fuel tank.

General Engine Specifications

Model	Displacement Cu. In. (cc.)	Hp.	Bore (In.)	Stroke (In.)	Compression Ratio	Compression (psi)
321 & 501	6.2 (101)	3.0	2.125	1.75	5.0:1	75 – 85

TUNE-UP

These engines use three different types of magnetos: the standard breaker type, the capacitive discharge breakerless (CD) type, and the solid state breakerless type called the Blaser. The Blaser unit is not serviceable, nor is any of its components, therefore it is not covered in this section. If any part fails, the entire solid state system must be replaced as a unit. To determine which system you have, check the accompanying illustrations.

Breakerless CD magneto

A. Breaker points
B. Condenser
C. Coil (primary and secondary)
D. Coil secondary wire (ground)
E. Coil primary wire (live)
F. Ignition short-out wire (live)
G. Condenser lead (live)
H. High tension lead
J. Stator assembly (includes points, coil condenser)

Breaker type magneto

NOTE: To convert a Blaser magneto to a point-type system, the flywheel must be replaced.

Stator Assembly

REMOVAL

1. Remove the flywheel.
2. Remove the metal cover and dust shield from the stator.

NOTE: Before removing the stator, check the position of the spark plug wire and the ignition short-out wire. Take note of their position for replacement. If there is a film of fuel around the crankcase seal, it means that the seal must be replaced.

3. Using a $\frac{3}{8}$ in. socket, remove the 2 screws, lockwashers and flat washers from the stator.
4. Disconnect the wire from the spark plug and the coil.
5. Lift out the stator, while guiding the plug wire through the backplate.

INSPECTION

NOTE: Points should be filed only as an emergency procedure. Filing removes a protective coating which prevents corrosion. Bad points should be replaced.

Blaser solid state breakerless magneto

Stator assembly

1. Check for grounding of the silver leaf tension band on the point arm.
2. Check for pitting corrosion or dirt on the points.
3. Check for an improperly grounded secondary wire.
4. Check for a loose or grounded primary wire.
5. Check to see if the ignition short-out wire is grounded to the stator plate.
6. Check to see if the condenser wire is loose or grounded to the stator plate.
7. Check for weak or non-existing spark caused by coil damage and/or loose or broken primary wires and secondary wires.
8. Check for timing delay or a weak spark caused by insulation missing from the coil lead.
9. Check for magneto shorting caused by pin-hole leaks in the wiring insulation.
10. Check for improper centering of the stator on top of the seal.
11. Check for a warped stator plate caused by improper centering of the stator around the crankshaft seal.

INSTALLATION

1. Carefully position the stator assembly while guiding the spark plug wire through the hole in the backplate.
2. Install the retaining screws and washers.
3. Make sure that the spark plug wire and ignition short-out wire are properly positioned.
4. Insert the high tension lead wire through the hole in the backplate and install the metal spark plug clip to the end of the lead wire.

Ignition Timing

NOTE: The engine will run with the stator assembly set anywhere within the slotted holes, however, good engine performance can be achieved only with the timing set as follows:

1. Position the high tension lead wire and the ignition short-out wire under the stator plate.

Setting the timing

NOTE: The only exception to this timing process is the 20 inch Snow Jet. On this unit the stator must be rotated fully counterclockwise, which will set the ignition timing at full advance 28 degrees or $1/8$ in. before top dead center.

2. From the ignition end of the crankshaft, turn the stator clockwise, as far as it will go. kThe stator must be seated perfectly flatly around and not on top of the crankshaft seal.
3. Install the two mounting screws. The timing should now be set at 22 degrees, which is $1/16$ in. before top dead center.

Adjusting the point gap; A is the adjusting screw

Ignition Points

NOTE: Be very careful in setting the points. Any slight misalignment will adversely affect performance.

1. Examine the cam for wear. If wear is obvious, the cam should be replaced. Install the cam so that the shoulder on the cam is facing away from the engine.

NOTE: Most engines have a shoulder or an arrow on one side of the cam. It is standard practice to install the cam with the arrow or shoulder facing away from the engine.

2. Install the key that holds the cam and flywheel in place. Rotate the crankshaft clockwise until the point breaker arm is on the highest spot of the cam.
3. Lightly loosen the screw which retains the breaker arm base plate. Insert a 0.020 feeler gauge between the contact points and use the gauge to separate the points, moving tyhe base plate until the points are properly gapped at 0.020 in. apart.
4. Tighten the point retaining screw.

NOTE: Pulling a clean piece of paper through the point opening will wipe off any residue left by your fingers or any oily preservative put on the points to retard corrosion.

5. Replace the fiber gasket and metal cover.
6. Make sure that there is good spring tension on the cover hold down spring.

Idle Speed and Mixture Adjustments

NEEDLE ADJUSTMENT

1. Check and make sure that the high speed adjusting needle screw is positioned so that the flat side of the screw head is next to the notch in the housing.
2. Turn the high speed needle screw one full turn counterclockwise.
3. Turn the idle mixture screw one full turn counterclockwise.
4. Start the engine and set the throttle to full speed and allow the engine to warm up for a few minutes.
5. Turn the high speed needle clockwise $1/4$ turn. The engine should slow down.
6. Turn the needle back $1/4$ turn to the 1 turn open position, then turn it counterclockwise $1/4$ turn. The engine should slow slightly. If not, the carburetor needs repair.
7. Turn the high speed needle back $1/4$ turn. Then, turn it

COUNTERCLOCKWISE

NOTCH

FLAT SIDE
OF SCREW

CLOCKWISE
High speed needle adjustment

60 MIN.
POSITION (1
TURN COUNTER-
CLOCKWISE
FROM CLOSED)

Adjustment increments

ALL NUMBERS ILLUSTRATE THE MINUTES OF TRAVEL
ON A CLOCK FACE WITH "0" REPRESENTING THE
STARTING NOTCH FOR ADJUSTMENT

Final adjustment

slowly, clockwise until the engine begins to slow down. This should not take more than $\frac{1}{4}$ turn. From this point, turn the needle counterclockwise about 1 full turn, or until the engine is running at maximum speed. Then, continue turning the needle counterclockwise, slowly, until the engine gives a slight stuttering sound in the exhaust. No more than $\frac{1}{16}$–$\frac{1}{8}$ turn should be needed.

NOTE: Final high speed adjustment, or fine tuning, can be made only after about 20 minutes at full work load. Fine adjustment should give smooth acceleration and just barely give a light exhaust smoke.

8. Idle adjustment is set after high speed adjustment. Seat the idle adjusting screw gently. Open the screw, counterclockwise, one full turn.

9. Start the engine and move the throttle control to the slow position.

10. Turn the idle screw clockwise in $\frac{1}{16}$ turn increments until a satisfactory idle is obtained, then turn it back $\frac{1}{16}$ turn.

FUEL SYSTEM

Carburetors

CHOKE ADJUSTMENT

This adjustment is possible only on those engines with handle-mounted throttle control levers. Any time that the handle or control wires are removed and/or reassembled the choke must be adjusted.

1. On units with a single, four position control on the handle, set the lever to the **FAST** position.

THROTTLE
HOUSING
SCREW

1/16"

Choke adjustment

2. Check the hook in the control link to be sure that it is in contact with, or no more than $\frac{1}{16}$ in. away from the first loop in the choke spring.

3. If the choke is partially on the loop, or if there is more than $\frac{1}{16}$ in. clearance between the hook and the loop, loosen the screw holding the throttle wire housing.

4. At this point, the housing may be moved forward or rearward to make the correct setting. After the housing is positioned properly, tighten the throttle housing screw securely.

ENGINE CONTROL CABLE ADJUSTMENT

1. Loosen the cable clamp screw so that the engine control cable will slide back and forth.
2. Move the engine control lever on the handle to the **FAST** position.
3. Move the control cable forward or backward until the choke control lever just touches the carburetor choke, with the choke valve fully open.
4. Tighten the clamp screw.
5. Move the engine control lever to the **STOP** position and make sure that the choke control lever makes contact with the spring **STOP** switch. Move the engine control lever to the **STOP (CHOKE)** position and see if the choke valve is fully closed. If not, redo the adjustments.

DISASSEMBLY

NOTE: The following overhaul procedure applies to all carburetors used with these engines. Differences among carburetors will be noted where necessary.

1. Remove the idle fuel regulating screw and the high speed adjustment needle, where applicable.
2. Using a box wrench, remove the brass retaining screw at the bottom of the float bowl housing.
3. Hold the carburetor upsidedown and remove the float bowl housing from the carburetor body.
4. Inspect the housing for dirt or other contaminants. Discard all gaskets.
5. Remove the filter screen in the bowl housing by tapping the bowl lightly on a flat surface.

NOTE: Fixed jet carburetors do not have a screen.

6. Remove the hinge pin and lift the float from the carburetor.
7. Using a screwdriver of exactly the right blade size, remove the main metering nozzle from the stem of the carburetor body. Make sure that the four idle bleed holes and the four air bleed holes are not restricted. On fixed jet carburetors, there are six air bleed holes.
8. Remove the brass needle and seat, along with the gasket.
9. Using a safe solvent, clean the carburetor body. Dry the body with a lint-free rag or compressed air.

ASSEMBLY

1. Inspect the needle and seat. If any wear is suspected, replace it.
2. Inspect the float. If it appears damaged in any way, or if, when shaking it, any rattling is heard within, replace it.
3. Install the needle and seat with a new gasket.
4. Install the float.
5. If the spring between the needle and float has a fastener, make sure it is connected to the tab on the float.
6. Adjust the clearance between the float and the rim of the body casting, to $5/32$ in., by bending the tab on the float arm. The adjustment is made with the carburetor in an upsidedown position.

NOTE: On fixed jet carburetors, the float gap is $1/16$–$3/32$ in.

7. Install the metering nozzle.
8. Carefully, install the filter screen in the housing. Make sure that the screen bottoms in the housing.
9. Install a new gasket on the threaded stem of the carburetor body.
10. Place the float bowl housing on the carburetor body with the high speed adjustment needle hole on the same side as the idle adjusting needle hole.
11. Center the inside gasket on the threaded hole by revolving a screwdriver in the retainer screw hole on the float bowl housing. Place a new gasket on the retaining screw and turn the screw in finger tight.

NOTE: It is very important to further adjustments that the retaining screw be installed securely. Do not, however, tighten it too tightly, as it can strip the threads.

CHOKE CONTROL LEVER

MAIN ADJUSTMENT SCREW (HIGH SPEED)

CARBURETOR CONTROL LEVER

'STOP' POSITION

'JUST TOUCHES'

'STOP' SWITCH (CONTACT POINT)

CARBURETOR CHOKE (FULLY OPEN)

SWITCH WIRE

CABLE CLAMP

ENGINE CONTROL CABLE

CABLE SCREW

IDLE SPEED REGULATING SCREW

FAST (HIGH SPEED) POSITION

Engine control cable adjustment

A. Throttle plate and throttle shaft assembly
B. Idle speed regulating screw
C. Idle fuel regulating screw
D. Choke plate and choke shaft assembly
E. Atmospheric vent hole
F. Float bowl housing drain valve
G. Float bowl housing retainer screw
H. High-speed adjusting needle

J. Float bowl housing
K. Idle chamber
L. Choke end view of Carburetors 1 and 3 (see Figure 3-23)
M. Throttle plate end view, (all three carburetors)
N. Primary and secondary idle passages, (all three carburetors)
P. Main nozzle air vent hole, (all three carburetors)

The three types of carburetor used

A. Inlet fitting
B. Inlet valve
C. Fuel bowl
D. Float
E. Float hinge pin
F. Filter screen
G. Main adjustment screw
H. Fuel bowl retainer

J. Main nozzle
K. Nozzle air bleed holes
L. Venturi
M. Idle channel
N. Idle fuel adjustment screw
P. Emulsion channel
Q. Idle air pickup hole
R. Idle chamber

S. Secondary idle hole
T. Throttle bore
U. Primary idle hole
V. Choke plate
W. Throttle plate
X. Throttle shaft seal
Y. Drain valve
Z. Idle "RPM" adjusting screw

Cross section of the Walbro carburetor

A. Main Orifice (Current Models)
B. Main Orifice (Early Models)

Main orifice

A. Needle   C. Tab
B. Spring   D. Float travel setting

Float, and needle

A. Threaded stem
B. Metering nozzle   D. Air bleed holes
C. Idle fuel holes   E. Brass seat

Meter nozzle installation

Float adjustment

A. Retaining Screw

Retainer screw installation

12. Hold the carburetor in the palm of your hand and tighten the retaining screw with a box wrench until you can feel the carburetor begin digging into the palm of your hand.
13. Install the choke plate and throttle plate.
14. While holding the throttle plate closed, install the idle speed screw until it just contacts the stop on the throttle disc.
15. Install the idle fuel regulating screw. Check the orifice to be sure that the loose wire in the housing is not visible, then turn the screw in until it gently bottoms.
16. Install the high speed adjustment needle until it gently bottoms. Make sure that the screwdriver slot in the needle is aligned with the notch in the housing.

NOTE: If the alignment cannot be made easily, the screw is overtightened. In this event, you've ruined the float bowl housing. The housing must be replaced.

ENGINE

NOTE: The following procedures apply to both the 321 and 501 engines. Where differences in procedure between the two exist, they will be noted.

Disassembly

GOVERNOR, CARBURETOR AND REED ADAPTER

1. Disconnect the governor and throttle linkage.
2. On the 321, pull out the brass retaining button.
3. Lift the notched, nylon throttle link and disconnect it.
4. Unbolt and remove the carburetor.
5. Remove the four hex-head mounting screws and lift off the reed assembly.
6. Inspect the reeds for proper seating. If there is any noticeable clearance or the reeds appear bent or damaged in any way, replace them.
7. Remove the throttle linkage from the engine backplate.
8. Remove the governor vane.

STARTER

9. Unbolt and remove the starter assembly.

------------------------ CAUTION ------------------------

When working on any recoil starter, safety glasses must be worn. Also, extreme caution must be used, since any misstep can cause the recoil spring to fly out with extreme force, resulting in personal injury.

Side Pull Types

10. Relax the tension on the starter cord by holding the starter assembly with the pulley up.
11. Pull the starter rope until about 12 in. of rope extends.
12. Hold the pulley steady with thumb pressure and hook the rope with a screwdriver at the point at which it exits the housing.
13. Hold the rope away from the pulley and turn the pulley clockwise with the rope. Allow it to uncoil until the spring is no longer applying tension to the rope.

------------------------ CAUTION ------------------------

When removing the pulley use extreme caution to keep the starter spring confined in the housing.

NOTE: Early models use a split pulley sheave, so the tension spring is not enclosed in a case.

14. Insert a new rope through the pulley and then through the exit hole in the housing.
15. Install the handle grip and tie a knot at both ends of the rope.
16. Pull all the excess rope through the exit hole and grasp the rope where it enters the pulley.
17. Hold the rope securely and wrap the pulley, using spring tension, by pulling the rope counterclockwise 4 revolutions.
18. Hold the pulley stationary with your thumb and remove all slack from the rope. Release the thumb pressure on the pulley and slowly allow the rope to rewind into the pulley.

NOTE: To avoid spring damage by overwinding, pull the starter rope to its full extent and rotate the starter pulley counterclockwise by hand. If it will not rotate $1/4$ turn counterclockwise before bottoming out, the spring is overwound and must be relaxed 1 turn.

19. Install the spring by relaxing tension on the pulley as previously described.

NOTE: On certain types of starters, the holder slips off the spring as it is installed in the housing. Always install the new spring with the windings in the same direction as the old one. If any damage is suspected to any components, replace it.

20. The remainder of starter assembly is the reverse of disassembly.

Up-and-Away Starters

21. Relax the tension on the rope by pulling it out about 12 in. Hold the pulley to prevent it from retracting.
22. Raise the rope with your finger at the point where it exits the housing. Release the pulley and the spring will automatically unwind. The rope can now be removed.
23. Using snap-ring pliers, remove the snap-ring then the pawl lever and pawl.

NOTE: Since one end of the coil spring is attached to the hub of the pulley, a slight turn, clockwise, may aid in removal.

24. Lift the cup and recoil spring from the pulley hub.
25. Remove the pulley from the fan housing.
26. Inspect all parts if wear is evident or damage is suspected, replace the parts. Clean all parts in a safe solvent.
27. Assemble the starter is reverse order of disassembly.

NOTE: When installing the spring cup, make sure that the point at which the spring fastens to the cup is 180 degrees from the hole at which the rope passes through the housing.

FLYWHEEL

28. On the 321 engine, remove the 4 hex head capscrews securing the starter housing, using a $3/8$ in. socket. On horizontal engines, remove the exhaust manifold before removing the housing.
29. Hold the flywheel with a strap wrench.
30. Using an $11/16$ in. socket, remove the flywheel nut, then remove the cupped washer and starter cup.
31. Using another, less valuable nut, place the nut on the end of the crakshaft to protect the threads. Strike this nut sharply with a hammer while applying firm, upward pressure on the flywheel by hand. The flywheel should pop off.

NOTE: Mishandling of the flywheel could cause damage or loss of magnetism.

32. Inspect the flywheel for cracks, warping or other damage. Test its magnetic strength by placing a $1/2$ in. socket on the inner rim and gently shaking the flywheel. The socket should not fall off.

STATOR

33. Remove the stator as described in the Tune-Up Section.

BAFFLE AND CYLINDER HEAD

34. Remove the spark plug. Discard it if necessary.
35. Remove the four head bolts.
36. Lift off the baffle and head. Discard the gasket.
37. Check the head for damage, cracks, warping and other problems. Clean the head in a safe solvent.

PISTON AND CONNECTING ROD

NOTE: Before removing the rod cap, note the metal staking on the cap. When the cap is installed, the screw slots must be staked to lock them in place. If the cap and rod have been disassembled 3 times, they must be replaced.

38. Place a clean, lint-free cloth under the rear of the engine. This is to catch any of the 28 needle bearings which may fall from the rod cap.

A. Retaining ring
B. Piston pin (installed)

Removing the piston pin retainer

A. Rod cap
B. Metal staking
C. Rod cap screws

Rod cap removal

Checking ring gap

A. Needle bearings
B. Half circle bearing guides
C. Bearing liners
D. Reference marks

Needle bearings

39. Remove the 2 rod cap screws and push the piston and rod out through the bore. COUNT THE NEEDLE BEARINGS!
40. If wear is suspected, replace the needle bearings.
41. Remove the piston pin retaining rings.
42. Using a soft drift (brass or wood) tap the pin out of the piston.
43. Inspect the pin and rod for scoring, wear or cracks. If necessary, replace them.

NOTE: A clogged or faulty air cleaner can cause the pin to develop a concave surface. This can be detected by running your finger the length of the pin. Replace the pin if this condition exists.

44. If the pin revolves or even turns slightly in the piston, using finger pressure, replace it.
45. Using a ring expander, carefully remove the rings.
46. Check the ring gaps by inserting the rings about 1/2 in. down the bore. A gap exceeding 0.070 in. means ring replacement.
47. Check the bore for damage, scoring, wear, scuffing, heat discoloration or any other sign of wear. Remove all carbon from the port areas.
48. Measure the cylinder bore with an inside micrometer. The bore should be 2.1265 in. If the bore is worn more than .003 in. over this, rebore the cylinder or get a new short block. If reboring is done, the bore may be honed up to 0.020 in. over size. An oversized piston should be used.

Cylinder bore sizing

49. When reringing the piston, always use a glaze breaker on the bore to establish a new bore finish.
50. If any damage or excessive wear is noted on the piston, replace it.
51. Remove all carbon from the ring grooves.

Cylinder hone

A. Seal spreader
B. Backplate screws
Seal spreader installed

Backplate seal removal

BACKPLATE

52. Insert a seal spreader (supplied as a part of the Jacobsen overhaul kit) in the backplate seal before removing the backplate.

53. Using a $\frac{3}{8}$ in. socket, remove the 6 screws attaching the backplate.

54. Place the backplate in an arbor press with the seal facing downward.

55. Press the bearing out with gradual pressure.

56. Discard the seal.

CRANKCASE HEAD

57. Using a $\frac{1}{2}$ in. socket, remove the 4 hex-head capscrews that secure the crankcase head assembly.

58. With a soft mallet, tap the crankshaft and crankcase head assembly from the engine ignition side.

Crankcase head removal

A. Cap screws and locking tab
B. Flat spot
Crankshaft removal

CRANKSHAFT

59. On the 321 engine, use a $\frac{5}{16}$ in. socket to remove the 2 hex-head capscrews and locking tabs. On the 501 engine, remove the snapring.

60. Place the crankcase head in an arbor press with the crank-

shaft downward. Apply light, steady pressure until the bearing and crankshaft are pressed out.

61. If the bearings show any sign of wear, replace them. To remove the bearings, remove the retainer and press the bearing off the crankshaft with the arbor press.

Assembly

CRANKSHAFT AND CRANKCASE HEAD

1. Using the arbor press, install the crankshaft bearing and retaining ring.
2. Coat the outside of a new seal with sealer and gently position it until it bottoms. For engines with reduction gear boxes, install the gearbox housing seal in the same manner.
3. Install the crankcase head, making sure that the flat spot on the head faces the cylinder bore. Torque the bolts to 80–100 inch lbs.

BACKPLATE

4. Press the bearing into the backplate to a depth of $\frac{3}{16}$ in. from the surface.
5. Coat the outside of a new seal with sealer and gently seat the seal until it bottoms. Take great care to avoid cocking the seal. Install the backplate and torque the capscrews to 60–80 inch lbs.

PISTON AND CONNECTING ROD

6. Coat all parts with clean engine oil.
7. Place the connecting rod in the piston. Fit the pin through the piston and rod, so that the hole in the pin is on the same side as the flat face of the piston crown. This corresponds to the intake side. The pin may be hammered into place using a brass drift.

Connecting rod end cap and liners

8. Check the pin with a micrometer to make sure that it was not deformed during installation.
9. Install the retaining rings in the piston to secure the pin.
10. Using a ring expander, carefully install the rings so that the chamfered edge of the top ring faces downward, and the chamfered edge of the bottom ring faces upward.
11. The rings must be positioned with the gaps on opposite sides of the piston, prior to installation.
12. Compress the rings with a ring compressor and carefully install the piston and rod assembly into the cylinder bore. Take care to avoid scratching the new bore finish with the rod end.

Needle bearings and bearing guides

Take care to avoid striking the rod journal with the rod end.

13. The piston assembly should slide in easily. Make sure that the face of the piston crown is on the intake side of the engine. Do not yet seat the rod on the journal.
14. Secure a new needle bearing set. The set is usually sold in a handy pack wherein the bearings are held together in a wax strip. With the strip at 70–80°F, wrap the strip around the crankpin after removing the paper backing.
15. Apply petroleum jelly around the ends of the bearings and slide the semi-circular bearing guides over the ends of the bearings. Four guides are required.
16. Carefully seat the rod over the journal. Install the rod cap. The mating marks (dots) on the rod and cap, must be on the same side. Torque the attaching bolts to 40–50 inch lbs. Stake the bolts to prevent loosening. THIS STAKING MUST BE DONE!

CYLINDER HEAD

17. Install the head, using a new gasket. Torque the head bolts, in a cross-wise fashion, to 150–180 inch lbs. (12–15 ft. lbs.).
18. Install the air baffle.
19. Install the carburetor and stator as described previously.
20. Set the ignition timing.

Piston ring chamfer

EXHAUST →

← ROTATION

Correct piston and rod installation

FLYWHEEL

21. On horizontal engines, install the starter cup, cupped washer and flywheel retaining nut along with the flywheel.

NOTE: Install the cupped washer with the high portion at the center, facing outward.

22. Use a strap wrench to hold the flywheel while tightening the nut. Torque the nut to 300–360 inch lbs. (25–30 ft. lbs.).
23. Install the flywheel starter cowling.

FINAL ASSEMBLY

24. Install the spark plug and torque it to 180–200 inch lbs. (15–18 ft. lbs.).
25. Install the spark plug wire, making sure it isn't cracked.
26. Install all other parts in reverse order of removal.
27. Check all adjustments.

Section 9
John Deere
Small Engines—6-20 hp

MODELS 8, 9, 61, 81 AND 91

The engine covered in this section is used in models 8, 9, 61, 81, and 91 chain saws. The engine is a 2-cycle, high speed model with a displacement of 2.1 cu. in. (34.42cc). The bore of the single cylinder is 1.5 in. (3.81 cm).

Engine Disassembly

NOTE: The following procedure assumes prior removal of the chain and chain bar.

1. Remove the clutch assembly and sprocket. A special spanner made for this job, makes it a lot easier.

NOTE: The clutch has left-handed threads. Remove it by turning clockwise.

2. Unbolt and remove the starter housing.
3. Unbolt and remove the handle.
4. On models 61 through serial number TL2058 81 through serial number TL1599 and 91 through serial number TL509, remove the starter cup by removing the hex nut and lock washer.
5. Remove the fan using a gear puller.
6. Pull out the choke and remove the carburetor cover and air filter.
7. On models 61, 81 and 91, remove the carburetor cover. The air filter may then be pried out.
8. Unbolt and remove the carburetor bracket.
9. Remove the choke linkage by lifting it out, away from the carburetor and pulling it out through the grommet.
10. On models 9 and 91, remove the compression release retaining ring rod, washer and linkage.
11. Disconnect the throttle linkage and fuel line, then remove the carburetor.
12. Remove the grommet from the carburetor adjusting screws and the spacer from the intake ports.
13. On models 61 from ser. no. TL2058, 81 from ser. no. TL1600 and 91 from ser. no. TL510, unbolt and remove the carburetor bracket.
14. Snap off the choke button and slide the choke shaft out through the rear handle and carburetor chamber.
15. Remove the carburetor and plastic chamber.
16. Remove the rear handle.
17. Remove the spark plug wire and plug.
18. Remove the woodruff key from the crankshaft and remove the points cover.
19. Remove the ignition coil.
20. Remove the condenser and condenser lead.

NOTE: Remove the coil and condenser as a unit on those engines with integral units.

21. On models 61 through TL2058, 81 through TL1599 and 91 through TL509, remove the wire clip, washer and points from the breaker pivot shaft. Pry loose the breaker block and compress the breaker spring to remove it. Unbolt and remove the breaker pivot shaft, plate and lubricating wick.
22. On models with point modules, unbolt and remove the module.
23. Unbolt and remove the crankcase cover.
24. Remove the fuel line and filter.
25. Remove the cylinder and piston. The cylinder is retained by four socket-head bolts and one hex bolt at the coil mount.
26. Remove the muffler.
27. Remove the piston from the cylinder.
28. Remove the piston pin retaining clips and drive out the pin.
29. Unbolt the connecting rod cap and lift out the rod and cap. Note that there are 24 roller bearings in the cap and rod.
30. Remove the crankshaft, thrust bearings and on models 61, 81 and 91, the thrust bushings.
31. Remove the oil pump housing.

Assembling the Engine

1. Before beginning, make sure all parts are clean and any reused parts are free of wear or damage.

MAIN BEARINGS

2. Press main bearing into inside of crankcase cover. Numbered side of bearing must face inside and be 0.010 in. (0.254mm) from inside surface of cover.
3. Press other main bearing into crankcase. Numbered side of bearing must face in and be 0.010 in. (0.254mm) from inside surface of crankcase.

63. O-ring (2 used)
64. Oil pick up tube
65. Oil pick up filter
66. Retaining ring
67. Oil pump O-ring
68. Oil pump plunger
69. Oil pump spring
70. Oip pump set screw
71. Check valve spring
72. Oil pump ball
73. O-ring
74. Oil adjusting needle
75. Oil tank cover assembly
76. No. 6 x ³/₈" flat head screw
77. Guide bar stud
78. Seal
79. Roller bearing
80. Crankcase assembly
81. No. 8 lock nut
82. Idle screw
83. Oil pump plunger rod
84. Oil tank bellows
85. Oil pump link
86. No. 10 square nut (2 used)
87. Threottle link
88. Grommet
89. Bellows
90. Thrust washer
91. Sprocket bearing
92. Sprocket
93. Guide bar plate, inside
94. Guide bar plate, outside
95. Clutch cover
96. ¼" UNF lock nut
97. Guide bar washer
98. Sprocket cover
99. Clutch spring (2 used)
100. Drive plate
101. Friction shoe
102. Clutch
103. No. 6 x ⁷/₁₆" socket hd. screw
104. Roller bearing
105. Rod with bearing and screws
106. Roller strip (24 needle bearings)
107. Retaining ring
108. Piston
109. Ring
110. Cylinder gasket
111. Cylinder
112. Muffler body
113. Spacer
114. Baffle
115. Spark arrestor (attachment)
116. Muffler cover
117. No. 10 x 1⅛" scre with lock washer
118. Socket hd. screw with lock washer
119. Spark plug
120. Rear handle cover
121. No. 6 x ⁵/₈" pan head screw (2 used)
122. Rear handle insert
123. Throttle
124. No. 6 x ³/₈" flat hd. screw (2 used)
125. Oil pump shaft
126. Throttle pin
127. Handle with throttle pin
128. No. 10 x ⁵/₈" flat head screw
129. Oil lever
130. Choke

1. No. 8 x ³/₈" machine screw (2 used)
2. Motor cover
3. Starting motor
4. No. 10 x ³/₈" machine screw
5. Steel washer
6. Start-stop switch
7. Ground wire
8. Switch wire
9. No. 10 x 1⅛" flat head screw
10. Starter housing
11. Driven gear
12. Drive gear washer (2 used)
13. Brake spring
14. Fiber washer
15. Friction shoe
16. Friction shoe washer
17. No. 5 x ⁵/₁₆" machine screw (2 used)
18. Intermediate gear assembly
19. No. 8 x ³/₈" machine screw
20. Connector clip
21. Connector
22. No. 10 x ³/₄" flat head screw
23. No. 10 x ³/₁₆" set screw
24. Drive gear
25. ⁵/₁₆" hex. jam nut
26. ⁵/₁₆" int. tooth lock washer
27. Fan and starter cup assembly
28. Breaker point module screws
29. No. 10 x ⁵/₈" screws with lock washer
30. Breaker point module
31. No. 10 x ⁵/₈" hex. hd. screw with washer

32. Coil-condensor
33. Oil seal
34. Roller bearing
35. Crankcase cover
36. ³/₃₂" x ³/₈" woodruff key
37. Crankshaft
38. Thrust bushings (2 used)
39. Plastic gas tank
40. Check valve
41. Carburetor cover
42. Air filter
43. Carburetor bracket
44. No. 10 x 2⅛" hex. hd. mach. screws
45. Carburetor
46. Oil cap
47. Gas cap
48. Fuel pick up tube
49. Steel washer
50. Fuel filter
51. Fuel pick up
52. No. 8 x ³/₈" pan hd. mach. screw
53. Sticker
54. Rubber handle grip
55. No. 12 x ³/₄" binding hd. screw
56. Front handle
57. No. 12 x ⁹/₁₆" binding hd. machine screw
58. No. 8 x 1³/₄" UNF fillister hd. screw
59. Adjusting block
60. Check valve
61. Oil tank cover gasket
62. Oil tank gasket

Exploded view of the manually started Series 8, 9, 61, 81, and 91 engines

1. No. 10 x ³⁄₄″ flat head screw (2 used)
2. No. 10 x 1¹⁄₈″ flat head screw
3. Starter bushing
4. Rewind spring
5. Rotor
6. Brake spring
7. Fiber washer
8. Friction shoe
9. Steel washer
10. No. 5 x ¹⁄₄″ screw with lock washer
11. Cord
12. Starter handle
13. Handle insert
14. Housing
15. Starter assembly
16. ⁵⁄₁₆″ UNF, jam nut
17. ⁵⁄₁₆″ int. tooth lock washer
18. Fan and starter cup assembly
19. Breaker point module screws (2 used)
20. Breaker point module
21. No. 10 x ⁵⁄₈″ screws with lock washers (2 used)
22. Coil-condenser
23. No. 10 x ⁵⁄₈″ hex. hd. screw with washer
24. Oil seal
25. Roller bearing
26. Crankcase cover
27. ³⁄₃₂″ x ³⁄₈″ woodruff key
28. Crankshaft
29. Thrust bushings (2 used)
30. Plastic gas tank
31. Check valve
32. Carburetor cover
33. Air filter
34. Carburetor bracket
35. No. 10 x 2¹⁄₈″ hex. hd. mach. screws (2 used)
36. Carburetor
37. Gas tank
38. Oil tank
39. Fuel pick up tube
40. Steel washer
41. Fuel filter
42. Fuel pick up
43. No. 8 x ³⁄₈″ pan head machine screws (2 used)
44. Sticker
45. Rubber handle grip
46. No. 12 x ³⁄₄″ binding head screw
47. Front handle
48. No. 12 x ⁹⁄₁₆″ binding head machine screw
49. No. 8 x 1³⁄₄″ UNF fillister hd. screw
50. Adjusting block
51. Check valve
52. Oil tank cover gasket
53. Oil tank gasket
54. O-ring (2 used)
55. Oil pick up tube
56. Oil pick up filter
57. Retaining ring
58. Oil pump O-ring
59. Oil pump plunger
60. Oil pump spring
61. Oil pump set screw
62. Check valve spring
63. Oil pump ball
64. O-ring
65. Oil adjusting needle
66. Oil tank cover assembly
67. No. 6 x ³⁄₈″ flat head screw
68. Guide bar stud
69. Seal
70. Roller bearing
71. Crankcase assembly
72. No. 8 lock nut
73. Idle screw

74. Oil pump plunger rod
75. Oil tank bellows
76. Oil pump link
77. Bellows
78. Grommet
79. Throttle link
80. No. 10 square nut (2 used)
81. Thrust washer
82. Sprocket bearing
83. Sprocket
84. Guide bar plate, inside
85. Guide bar plate, outside
86. Clutch cover
87. ¹⁄₄″ UNF lock nut
88. Guide bar washer
89. Sprocket cover
90. Clutch spring (2 used)
91. Drive plate
92. Friction shoe
93. Clutch
94. No. 6 x ⁷⁄₁₆″ socket head screw
95. Roller bearing
96. Rod with bearing and screws
97. Roller strip (24 needle bearings)

98. Retaining ring
99. Piston
100. Ring
101. Cylinder gasket
102. Cylinder
103. Muffler body
104. Spacer
105. Baffle
106. Spark arrestor (attachment)
107. Muffler cover
108. No. 10 x 1¹⁄₈″ screw with lock washer (2 used)
109. Spark plug
110. Socket head screw with lock washer (4 used)
111. Rear handle cover
112. No. 6 x ⁵⁄₈″ pan head screw (2 used)
113. Rear handle insert
114. Throttle
115. No. 6 x ³⁄₈″ flat head screw (2 used)
116. Oil pump shaft
117. Throttle pin
118. Handle with throttle pin
119. No. 10 x ⁵⁄₈″ flat head screw
120. Oil lever
121. Choke

Exploded view of the model 81 electric start engine

MAIN SEALS

4. When installing main seals press the seals on the metal side, from the outside of each casting, pressing inward.

5. Position each main seal flush with the outside of its respective casting boss.

CRANKSHAFT

6. Position the thrust bearings next to the crankshaft with the thrust race between the bearing and casting on 8 and 9 saws. On 61, 81, and 92 Saws, position thrust bushing next to crankcase.

7. Insert crankshaft into crankcase, being careful not to damage the oil seal with the threaded end of the crankshaft. Check to be sure keyed end of crankshaft is on fan side.

CONNECTING ROD

Make sure that connecting rod and cap are free of oil and dirt, and that the identification marks are aligned.

8. Position the crankshaft with the throw toward the cylinder opening.

9. Press a waxed strip of twelve ($\frac{1}{2}$ set) roller bearings into the cap race and the remaining strip into the connecting rod race.

NOTE: Whenever the connecting rod is removed, use new needle bearings when assembling.

10. Using connecting rod cap holder, install connecting rod and cap on crankshaft with two socket head screws, being certain that the fractured mating surface of the rod and cap are properly aligned.

11. Tighten screws with $\frac{7}{64}$ in. hex wrench to 35 inch lbs. (0.454 Nm) torque.

RINGS

The No. 8 and 9 Chain Saws have two piston rings, and the 61, 81, 81 Electric Start, and 91 have one ring. The rings are very different in design and cannot be substituted for each other. However, the piston with proper ring(s) can be substituted.

12. Install the ring(s) on the piston with the beveled edge on each ring positioned to the top of the piston.

13. (8 and 9 Saws) Position the ring openings next to the locating pin.

PISTON

14. Examine piston, cylinder and rings for score marks, chipping, or excessive wear. Replace if necessary.

15. (8 and 9 Saws). When assembling the piston to the engine, make sure that the side marked **EXH** on the piston is to the exhaust side of the cylinder.

16. Coat cylinder wall with a light film of oil before installing piston.

17. Using fingers or an adjustable pipe strap for ring compression, guide piston head into cylinder.

18. Compress the rings and press the cylinder down on the piston until the wrist pin just contacts the flange base of the cylinder. Do not rotate the piston after it has been inserted into the cylinder.

WRIST PIN

19. Install one piston retaining ring in piston boss, making certain that the sharp edge and identifying marks and the ends of the retaining ring are facing out. Gap must face skirt (bottom) of the piston.

IMPORTANT: Always install a new gasket between cylinder and crankcase.

20. Position the piston over the connecting rod. Support the piston boss on a piece of hardwood, and drive wrist pin through connecting rod bearing into piston boss.

21. Install the remaining wrist pin retaining ring in the piston boss with sharp edge and identifying marks facing out.

CYLINDER

22. Install two socket head screws into cylinder recess on sprocket side of saw and two socket head screws on the starter side before installing cylinder. Tighten screws with $\frac{5}{32}$ in. long hex wrench to 60 inch lbs. (0.779 Nm) torque.

CRANKCASE COVER

23. First install thrust bearing and washer (8 and 9 Saw) or thrust bushing (61, 81, and 91 Saws) against crankshaft shoulder. Check the bearing in the crankcase cover and replace if necessary.

24. Check all parting surfaces for nicks and dirt deposits and examine crankcase cover gasket. Replace gasket if necessary.

25. Lower cover into position over crankshaft, making sure that threads of crankshaft do not contact the oil seal.

26. Install crankcase cover on saws (serial numbers through TL2058) (61 Saw), (serial numbers through TL1599) (81 Saw), (serial numbers through TL509) (81 Saw), (serial numbers through TL509) (91 Saw), and 8 and 9 Saws with four No. 10 x $\frac{5}{8}$ in. screws, into coil support "A" and seven No. 6 x $\frac{5}{8}$ in. screws "B".

NOTE: Tighten three No. 10 x $\frac{5}{8}$ in. screws around breaker points and one No. 10 x $\frac{5}{8}$ in. screw into cylinder to 45 inch lbs. (0.584 Nm) torque, then tighten self-tapping screws alternately on opposite sides to assure uniform compression of the gasket and to prevent air or fuel leakage.

27. Before installing crankcase cover on saws (TL2059 and later) (61 Saw) (TL1600 and later) (81 Saw), 81 Electric Start, and (TL510 and later) (91 Saw) install plastic gas tank. Insert rubber fuel line through port in carburetor chamber, then slide gas tank into crankcase recess and pull fuel line end through port.

28. To install crankcase cover on saws (TL2059 and later) (61 Saw), (TL1600 and later) (81 Saw), 81 Electric Start and (TL510 and later) (91 Saw), install three No. 10 x $\frac{5}{8}$ in. screws around breaker point module at "A" and one No. 10 x $\frac{3}{8}$ in. screw into cylinder at "B". Tighten screws to 45 inch lbs. (0.584 Nm) Torque.

BREAKER POINTS

61 Chain Saw (serial numbers through TL2058), 81 Chain Saw (serial numbers through TL 1599), 91 Chain Saw (serial numbers through TL509)

29. To install breaker points on saws (through serial numbers TL2058) (61 Saw), (through serial numbers TL1599) (81 Saw), and (through serial numbers TL509) (91 Saw), position new felt washer in breaker compartment.

30. Install original retaining plate, then press pivot bracket shaft into position through retainer plate.

31. Install terminal screw through insulator block and slide block into breaker case recess. Compress breaker spring assembly and press into position and align points.

NOTE: Insulator block is properly positioned when the large shoulder of the block is up and toward the inside of the breaker compartment.

1. Ground wire
2. Ignition switch
3. Slotted hex. head screws
4. Starter
5. Hex. nut
6. Lock washer
7. Starter cup
8. Fan
9. Slotted hex. head screws
10. Stator
11. Felt washer
12. O-ring
13. Cylinder gasket
14. Fuel cap
15. Fuel intake line with filter
16. Crankcase
17. Thrust washer
18. Thrust bearing
19. Woodruff key
20. Crankshaft
21. Oil exhaust line
22. O-ring (14, 15 saws)
23. Needle adjusting spring (14, 15 saws)
24. Metering needle (14, 15 saws)
25. Oil exhause fitting (14, 15 saws)
26. Oil exhaust fitting (12 saw)
27. Crankcase cover
28. Guide bar adjusting block
29. Adjusting screw
30. Clutch cover
31. Clutch
32. Thrust washer
33. Sprocket race
34. Sprocket
35. Sprocket bearing
36. Inside guide bar plate
37. Outside guide bar plate
38. Sprocket cover
39. Hex. nut
40. Washer
41. Lock nut
42. Washer
43. Slotted hex. head screws
44. Hex. lock nut
45. O-ring
46. Thrust washer
47. Thrust bearing
48. Machine screw
49. Square nut
50. Machine screw
51. Front handle
52. Front handle grip (14, 15 saws)
53. Oil tank cap
54. Needle bearings (26)
55. Connecting rod
56. Roller bearing
57. Snap ring for wrist pin
58. Piston
59. Cap screw
60. Muffler cover (14, 15 saws)
61. Muffler baffle (14, 15 saws)
62. Muffler body (14, 15 saws)
63. Muffler spacer (14, 15 saws)

64. Muffler stud (14, 15 saws)
65. Muffler stack (12 saw)
66. Spark plug connector spring
67. Spark plug cover
68. Spark plug
69. Cylinder
70. Hex. nut
71. Piston rings
72. Oil pickup filter (12 saw—screen, 14, 15 saws—felt)
73. Oil pickup tube
74. Oil intake line (12 saw)
75. Oil intake fitting
76. Hex. socket screw
77. Rear handle grip (14, 15 saws)
78. Self-tapping screw
79. Trigger latch button (12, 14 saws)

80. Trigger latch button spring (12, 14 saws)
81. Trigger latch retaining ring (12, 14 saws)
82. Rear handle cover
83. Plunger button
84. Oil pump plunger
85. O-ring
86. Oil pump spring
87. Grommet
88. Idle—speed screw
89. Idle—speed screw spring
90. Rear handle
91. Compression release rod (15 saw)
92. Cap screw
93. Choke button
94. Grommet
95. Throttle trigger
96. Shut-off valve link (14, 15 saws)

97. Shut-off valve (14, 15 saws)
98. Machine screw
99. Reed limiter
100. Reed
101. Carburetor gasket
102. Choke link
103. Cap screw
104. Split-back screen (12, 15 saws)
105. Cap screw (12, 15 saws)
106. Cap screw
107. Trigger link bellows
108. Trigger link
109. Carburetor
110. Air intake filter
111. Carburetor cover snap ring
112. Carburetor cover
113. Carburetor cover knob

Exploded view of the Series 12, 14 and 15 engines

32. Secure breaker spring assembly to pivot shaft with washer and wire clip.

33. Install two retaining screws, tighten, then back off one-half turn.

34. Replace lubricating wick.

35. Replace coil and condenser leads on terminal screw. Secure with lockwasher and nut.

61 Chain Saw (TL2059 and later), 81 Chain Saw (TL1600 and later), 81 Electric Start Chain Saw, and 91 Chain Saw (TL510 and later)

36. To install breaker point module on saws (TL2059 and later) (61 Saw), (TL1600 and later) (81 Saw), 81 Electric Start, and (TL510 and later) (91 Saw), install blue breaker lead wire from coil to terminal on side of module.

37. Secure breaker point module with two retainer screws.

COIL AND CONDENSER

61 Chain Saw (through serial numbers TL2058), 81 Chain Saw (through serial numbers TL 1599), 91 Chain Saw (through serial numbers TL509) and 8 and 9 Chain Saws

38. On saws (through serial numbers TL2058) (61 Saw), (through serial numbers TL1599) (81 Saw), (through serial numbers TL509) (91 Saw), and all 8 and 9 Saws, press the coil lead wire into crankcase recess under coil and stator assembly and secure stator with two No. 10 x $\frac{5}{8}$ in. screws with washers. Do not tighten.

39. Install condenser in recess beside breaker points with retainer clip and screw. Connect coil and condenser lead to breaker block terminal.

40. To replace coil, first remove lead wire from contact block and move this wire out of the way. Remove the screw holding ground wire and retaining strip. Then slide old coil off stator legs and install new coil, pushing as far back as possible. Replace ground wire above retaining strip and secure. Replace lead wire on the contact block post along with the condenser lead.

61 Chain Saw (TL2059 and later), 81 Chain Saw (TL1600 and later), 81 Electric Start Chain Saw, and 91 Chain Saw (TL510 and later)

41. On saws (TL2059 and later) (61 Saw), (TL1600 and later) (81 Saw), 81 Electric Start, and (TL510 and later) (91 Saw), the coil has a built-in condenser. The condenser lead is soldered to the coil.

42. Install blue breaker lead wire under retainer tab on crankcase cover and secure coil-condenser with two No. 10 x 5-inch screws with washers. Do not tighten.

ADJUSTING BREAKER POINTS

NOTE: Set breaker point gap at 0.015 in. (0.381mm).

43. To set gap, first turn crankshaft in the direction of rotation until keyway aligns with the center of the coil. Insert a 0.015 in. (0.381mm) feeler gauge between points and adjust for light contact with gauge.

NOTE: Double-check gap by inserting the next smaller gauge to make certain that the adjustment is not too close.

44. When properly adjusted, tighten the adjusting screw. Recheck gap after tightening since rotation of screw can change gap.

NOTE: Be sure to wipe feeler gauge blade free of oil and to clean points with a strip of soft paper.

45. Replace gasket, dust cover, and two cover retaining screws on saws (through serial numbers TL2058) (61 Saw), (through serial numbers TL1599) (81 Saw), (through serial numbers TL509) (91 Saw), and all 8 and 9 Saws, or replace plastic dust cover on breaker point module on saws (TL2059 and later) (61 Saw), TL1600 and later) (81 Saw), 81 Electric Start, and (TL510 and later) (91 Saw).

REAR HANDLE

46. Position rear handle into place and insert oil pump link through bellows grommet and into oil pump plunger rod. Attach the throttle link to the throttle and position throttle over spring pin.

NOTE: Secure rear handle only after oil pump link and throttle are connected and install choke link and compression release link (9,91 Saws) only after rear handle cover and rubber insert are in place.

47. Position two square nuts in carburetor chamber slots and secure rear handle with two No. 10 x $\frac{5}{8}$ in. flat head screws at "A" and two No. 6 x $\frac{3}{8}$ in. flat head screws at "B".

NOTE: Leave the rear handle cover and rubber insert off until rear handle and carburetor assembly is complete.

CARBURETOR

48. Install plastic spacer into recess in carburetor chamber and fasten carburetor to spacer with two No. 10 x $2\frac{1}{8}$ in. cap screws through carburetor bracket.

49. Install fuel line and attach throttle link and oil pump link. Replace rear handle cover and rubber insert.

50. Rotate choke link (8, 9, 61, 81, and 91 Saws) through grommet in rear wall of carburetor chamber and push into position. Connect link.

51. Replace air filter and screw down carburetor cover.

52. On saws (TL2059 and later) (61 Saw), (TL1600 and later) (81 Saw), 81 Electric Start, and (TL510 and later) (91 Saw), install choke into slot in carburetor bracket and slide shaft through rear wall of carburetor chamber and rear handle grommet. Snap choke button over end of shaft.

53. Fasten carburetor bracket with two No. 10 x $2\frac{1}{8}$ in. cap screws and press filter screen into place.

54. Snap carburetor cover into position.

55. Maneuver compression release link (9, 91 Saws) through grommet in rear wall of carburetor chamber and install pivot rod and washer over link, then press into place.

56. Using magnetic screwdriver or screwdriver and tape, press retaining ring into recess on compression release pivot rod.

FAN

———————— CAUTION ————————

Whenever installing or removing fan, use shop rags, or gloves to protect hands from possible injury from sharp fan edges.

NOTE: Before installing fan, inspect the key, keyway, and crankshaft taper for damage. Inspect the fins for cracking or breakage. Do not use a damaged fan.

57. Install key in crankshaft slot and align fan keyway with crankshaft keyway.

NOTE: On saws (through serial numbers TL2058) (61 Saw) (through serial numbers TL1599) (81 Saw), (through serial numbers TL509) (91 Saw), and all 8 and 9 Saws, install fan, then position starter cup lugs to straddle the screw bosses in the fan. On saws (TL2059 and lat-

1. Ground wire
2. Ignition switch
3. Slotted hex. head screws
4. Starter
5. Hex. nut
6. Lock washer
7. Starter cup
8. Fan
9. Slotted hex. head screws
10. Stator
11. Felt washer
12. Cap screw
13. Fuel cap
14. Fuel intake line
15. Fuel line fitting
16. Fuel intake line with filter
17. Fuel tank
18. O-ring
19. Crankcase
20. Thrust washer
21. Thrust bearing
22. Woodruff key
23. Crankshaft
24. Machine screw
25. Front handle
26. Front handle grip (19 saw)
27. Machine screw
28. Square nut
29. O-ring (18, 19 saws)
30. Needle adjusting
    spring (18, 19 saws)
31. Metering needle
    (18, 19 saws)
32. Oil exhaust fitting
    (18, 19 saws)
33. Oil exhaust fitting (17 saw)
34. Machine screw
35. Guide bar adjusting block
36. Hex. nut
37. Inside guide bar plate
38. Guide bar plate spring clip
39. Outside guide bar plate
40. Washer
41. Lock nut
42. Sprocket cover
43. Hex. nut
44. Washer
45. Outer sprocket washer
46. Sprocket bearing
47. Sprocket
48. Sprocket race
49. Thrust washer
50. Clutch
51. Clutch cover
52. Slotted hex. head screws
53. Crankcase cover
54. O-ring
55. Thrust washer
56. Thrust bearing
57. Needle bearings (28)
58. Roller bearing
59. Connecting rod
60. Hex. socket screw
61. Snap ring for wrist pin
62. Piston
63. Piston rings
64. Cylinder gasket
65. Muffler body
66. Muffler stud (18, 19 saws)
67. Muffler baffle (18, 19 saws)
68. Muffler spacer (18, 19 saws)
69. Muffler cover (18, 19 saws)
70. Cap screw
    (18, 19 and 23 saws)

71. Cap screw (17 saw)
72. Muffler cover (17 saw)
73. Muffler stud (17 saw)
74. Muffler baffle (17, 23 saws)
75. Muffler plate (17, 23 saws)
76. Hex. nut
77. Cylinder
78. Spark plug cover
79. Spark plug connector spring
80. Spark plug
81. Cylinder elbow fitting (19 saw)
82. Machine screw
83. Oil exhaust line
84. Oil line retaining bracket
85. Oil tank cap
86. Oil pickup filter
87. Oil pickup tube
88. Check-valve cap
    (18, 19 saws)
89. Check-valve gasket
    (18, 19 saws)
90. Check-valve screen
    (18, 19 saws)
91. Check-valve (18, 19 saws)
92. Check-valve
    fitting (18, 19 saws)
93. Oil pressure line (18, 19 saws)
94. Oil intake line (17 saw)
95. Oil intake fitting
96. Carburetor front gasket
97. Carburetor spacer
98. Carburetor rear gasket
99. Carburetor
100. Grommet
101. Choke link
102. Cotter pin

103. Vent connector (18, 19.
    23 saws)
104. Reed (19, 23 saws)
105. Reed plate gasket (19, 23 saws)
106. Reed (17, 18 saws)
107. Reed spacer (17, 18 saws)
108. Reed limiter (17, 18 saws)
109. Machine screw (17, 18 saws)
110. Shut-off valve (18, 19 saws)
111. Throttle trigger
112. Throttle latch retaining
    ring (17, 18 saws)
113. Rear handle cover
114. Throttle latch
    spring (17, 18 saws)
115. Throttle latch
    button (17, 18 saws)
116. Cap screw
117. Rear handle support
118. Cap screw
119. Self-tapping screw
120. Shut-off valve
    link (18, 19 saws)
121. Choke button
122. Rear handle grip
123. Compression release
    rod (19, 23 saws)
124. Square nut

125. Rear handle
126. Grommet
127. Oil pump spring
128. Idle-speed screw spring
129. Idle-speed screw
130. O-ring
131. Oil pump plunger
132. Plunger button
133. Throttle link bellows
134. Cap screw
135. Throttle link
136. Cap screw
137. Cap screw
138. Spit-back screen
139. Air intake filter
140. Carburetor cover snap ring
141. Carburetor cover
142. Carburetor cover knob
143. Metering needle (23 saw)
144. Needle adjusting spring
    (23 saw)
145. O-ring (23 saw)

Exploded view of the Series 17, 18, 19 and 23 engines

er) (61 Saw), (TL1600 and later) (81 Saw), 81 Electric Start, and (TL510 and later) (91 Saw), the fan and starter cup are one-piece.

NOTE: For easier installation of fan, rotate crankshaft so that flywheel magnets are away from stator legs.

58. Press fan into position, install internal tooth lock washer and $5/16$ in. jam nut, and seat fan by tightening nut to 150 inch lbs. (1.948 Nm) torque.

NOTE: If nut is not sufficiently tightened, it may back off, allowing fan to loosen and cause possible damage to fan, crankshaft, and keyway.

MAGNETO AIR GAP

59. To set magneto air gap, first align stator legs and flywheel magnet.
60. Insert a 0.010 in. (0.254mm) plastic or brass feeler gauge between stator legs and flywheel magnet. Loosen stator retaining screws and allow magnets to draw stator legs into position. Then tighten retaining screws and remove feeler gauge.

OILER ASSEMBLY

NOTE: Make sure that the pick-up line is not kinked and that the oil pump plunger engages the recess in the pump piston.

NOTE: Push plunger back as far as it will go before installing cover.

OIL PUMP COVER

61. Install oil pump cover, pump, and pick-up line with four No. 6 x $3/8$ in. screws.

FUEL INTAKE LINE

8, 9, 61 (through serial numbers TL2058), 81 (through serial numbers TL1599), and 91 (through serial numbers TL509) Saws

62. Assemble parts removed during fuel intake line repair. 61 (TL2059 and later), 81 (TL1600 and later), and 91 (TL510 and later) Saws.
63. Make a loop in the end of a copper wire. Put wire through opening in carburetor chamber and out through the fuel tank opening. Push looped end of wire about 3 in. into fuel line end that connects to carburetor. Place drop of oil on fuel line holder. Push fuel line to carburetor chamber using wire as a guide. Remove wire from end of fuel line and pull fuel line until fuel line holder snaps into place. Put fuel filter on other end of fuel line and place down into fuel tank.
64. Make sure that the filter is at bottom of tank.
65. Install fuel intake line in saws (TL2059 and later) (61 Saw), (TL1600 and later) (81 Saw), 81 Electric Start, and (TL510 and later) (91 Saw). Installation is the same for all saws.

CLUTCH

66. When installing the clutch springs, install one spring at a time. Place the shoe over the drive plate; then set the base of the spring to the base of the shoe. Compress the spring and force the free end of the spring against the spring holder on the drive plate. Use the same procedure to assemble other spring to clutch.

NOTE: Check clutch shoe spring tension. If any clutch shoe appears to be loose, replace all springs.

67. Install thrust washer, sprocket bearing, sprocket, clutch, and sprocket cover.
68. Install clutch on crankshaft by turning on counterclockwise. Tighten to 150 inch lbs. (1.948 Nm) torque.

GUIDE BAR AND CHAIN

69. Install inside guide bar plate (8, 9, 81, 81 Electric Start, and 91 Saws), being sure that tang faces to outside.
70. Install chain on guide bar, fit chain into sprocket and install guide bar over stud. Rotate chain and guide bar into position.

NOTE: Be sure to install chain with chipper link cutting edges forward.

SPROCKET COVER

71. Replace sprocket cover, guide bar washer, and $1/4$ in. UNF lock nut. Tighten lock nut to 150 inch lbs. (1.95 Nm) torque.

STARTER

(8, 9, 61, 81, and 91 Saws)

NOTE: Do not lubricate starter assembly in any way except as indicated. Excessive oil and grease will defeat the purpose of the friction shoes and cause the starter to slip.

72. Install cord in rotor and wind cord around rotor in direction of rotation until you cannot make another complete wrap around. Hold remaining cord in notch of rotor.
73. Move rewind spring from starter housing center post with pliers. Grease post lightly.
74. With letter side of rotor up, place rotor on rewind spring, making sure that spring is in spring holder of rotor. Prewind the rotor three turns clockwise; then insert cord through starter bushing.
75. Tie slip knot in cord and install handle.
76. Install brake spring, fiber washer, friction shoe, fiber washer, friction shoe washer, and No. 5 x $1/4$ in. screw with lock washer.
77. When installing friction shoes, make sure "head" of shoe is facing the direction of rotation (clockwise) when pulling the starter handle.

STARTER COVER

78. Install starter cover with two No. 10 x $3/4$ in. screws at "A" and one No. 10 x $1 1/8$ in. screw in front of housing at "B".

81 ELECTRIC STARTER

NOTE: Do not lubricate starter assembly in any way except as indicated. Excessive oil and grease will defeat the purpose of the friction shoes and cause starter to slip.

79. Install starter drive gear on shaft, two drive gear washers, brake spring, fiber washer, friction shoe assembly, fiber washer, friction shoe washer, and No. 5 x $1/4$ in. screw with lock washer.
80. Slide start-stop switch into starter housing grooves.
81. Install battery wire (white) exposed (-) lead to stop terminal and (white) exposed (+) lead through port in starter housing. Press connector into recess in housing and fasten with clip.
82. Connect black (-) wire to switch ground terminal and green wire to ground on starter housing with one No. 5 x $5/16$ in. machine screw.
83. Connect blue wire at start terminal of switch.
84. Slide driven gear shaft through bushing in starter housing and secure with one No. 10 x $3/8$ in. matching screw and washer.

85. Install starting motor drive gear on motor shaft with one No. 10 x $\frac{3}{16}$ in. set screw.

86. Install starting motor with two No. 8 x $\frac{3}{8}$ in. machine screws, then connect battery $\frac{3}{8}$ in. machine screws, then connect battery white lead (+) to (+) motor terminal and black lead (–) to motor (–) terminal.

87. Install motor cover with two No. 8 x $\frac{3}{8}$ in. machine screws.

88. Connect blue wire from switch start terminal to blue wire from coil before installing starter housing.

81 ELECTRIC STARTER COVER

89. Install starter cover with two No. 10 x $\frac{3}{4}$ in. flat head screws at "A" and one No. 10 x $1\frac{1}{8}$ in. flat head screw at "B".

81 ELECTRIC START FRONT HANDLE

90. Install front handle with one No. 12 x $\frac{3}{8}$ in. screw at "A" and one No. 12 x $\frac{9}{16}$ in. at "B".

81 ELECTRIC START MUFFLER

91. The muffler can be assembled to the cylinder before or after the rear handle is in place.

NOTE: To inspect piston for scoring or other damage without disassembling the saw, remove muffler and inspect piston through the cylinder exhaust port.

92. Assemble covers, baffle, and spacers; then carefully slide the assembly into position. Install two No. 10 x $1\frac{1}{8}$ in. being sure that spacers are aligned with cover holes. Tighten to 60 inch lbs. (0.779 Nm) torque.

SPARK PLUG

93. Check spark plug to be certain that it is clean and properly gapped. Install spark plug into cylinder, finger tight.

NOTE: If there is any doubt that the plug is not good use a new plug. Two-cycle engines are hard on spark plugs and they are frequently the cause of service problems.

MODELS 12, 14, 15, 17, 18, 19 AND 23

The engines covered in this section are virtually identical. Any differences are noted. Repair procedures listed below apply to all engines except as noted.

Engine Disassembly

1. Remove two nuts then remove washers, sprocket cover, outside guide bar plate, guide bar, chain, and inside guide bar plate.

2. Remove lock nut and one washer (12, 14, and 15 Saws) or two washers (17, 18, 19, and 23 Saws), sprocket, sprocket race, and thrust washer.

3. Remove the clutch and clutch cover by using a spanner. (The clutch has left-hand threads—remove clockwise.)

NOTE: When using a hand spanner, remove the starter (Step 6) and hold fan tight.

———————————— CAUTION ————————————
Always place a cloth over the fan to prevent cutting your hand.

4. Loosen the crankcase cover by removing the top handle bar screw and the five slotted hex. head screws.

5. Unscrew the oil exhaust line nut and remove the crankcase cover, thrust washer, and thrust bearing.

6. Remove the starter by removing four slotted hex. head screws and pulling switch terminal off the stator terminal.

7. Remove starter cup by removing hex. nut and lock washer.

8. Remove the fan by using a puller.

NOTE: Use two of the hex. head screws removed in Step 6 in the puller.

9. Pull spark plug wire out of stator and off spark plug. Remove stator by removing three slotted hex. head screws. Remove the felt washer on shaft behind stator.

10. Remove muffler by removing cap screws.

NOTE: Even though muffler designs vary, removal is very similar.

11. Remove the front handle.

12. Remove cover and air filter.

13. (12, 15, 17, 18, 19 and 23 Saws only) Remove the spitback screen by removing one screw.

14. Remove carburetor by removing two cap screws; tip rear of carburetor up and out to the right.

NOTE: On 17, 18 and 19 Saws, remove the choke link cotter pin and disconnect fuel line as you remove the carburetor. The 17, 18, 19 and 23 Saws also have a spacer block.

15. (14, 15, 18, 19, and 23 Saws) Remove the rubber rear handle grip.

NOTE: On 17, 18, 19, and 23 Saws, remove three slotted hex. head screws to remove bottom part of rear handle first.

16. Remove oil intake line from the oil intake fitting.

17. (14, 15, 18, 19, and 23 Saws) Remove handle cover by removing two self-tapping screws.

NOTE: On 23 Saw remove compression release rod first.

18. (15 and 19 Saws) Remove the compression release rod.

19. Remove the spark plug.

20. Remove the rear handle by removing four cap screws in the carburetor chamber and one cap screw (12, 14, and 15 Saws) in the bottom handle brace.

NOTE: On 17, 18 and 19 Saws, remove the oil exhaust line retainer bracket.

21. (17, 18, and 19 Saws) Separate the fuel tank from the cylinder and crankcase by removing three screws.

22. Remove cylinder and gasket by removing four nuts with lock washers.

23. Remove piston by removing two snap rings using snap ring pliers; then push out wrist pin.

24. Remove the connecting rod by removing two hex. socket screws.

NOTE: The 12, 14 and 15 Saws have 26 needle bearings and the 17, 18 and 19 Saws have 28 needle bearings. The 23 Saw has 24 needle bearings. Count the needles to make sure they are removed from the engine.

25. Remove crankshaft thrust bearings, and thrust washers.

Overhaul

IGNITION SYSTEM

Disassembly

1. To set the points or to replace the points, coil, or condenser, refer to the "DISASSEMBLY" section and perform Steps 6–8.

To replace the complete stator, also perform Step 9.

NOTE: The condenser must be removed before removing the coil.

2. Remove the coil by disconnecting the two terminal wires. Slide coil off the core.

Cleaning and Adjusting Points

1. Place a clean piece of paper between the points, close the gap and pull the paper away. Repeat the procedure until no oil is noted on the paper.
2. To adjust points, turn crankshaft to TDC (follower arm is on the high point of cam, keyway on crankshaft will face center of condenser). Loosen two mounting screws (12, 14 and 15 Saws) or one mounting screw (17, 18, 19 and 23 Saws) and use a feeler gauge to set points to specifications. Tighten screws and recheck point gap.

NOTE: Do not set point gap over .015 in.. Do not lubricate.

Inspection

1. Inspect coil assembly for cracks on insulation, evidence of overheating, or other damage. Make certain that all coil electrical leads are intact.
2. Inspect the condenser for visible damage, such as damaged terminal lead, dents or gouges in can, or broken mounting clip.
3. Check all electrical leads for wear, insulation breaks, and insecure terminals. Check the switch wire terminal on the coil to make certain that it is not bent and touching the stator frame causing a ground.

STARTER

Disassembly

1. To remove the starter, refer to the "DISASSEMBLY" section and perform Step 6.
2. Remove the retaining ring using a screwdriver. Remove fiber washers, friction shoe, and spring.
3. To remove starter rotor with cord, pull out on cord, tie a slip knot, remove washer, insert, and handle; then untie knot. Untie slip knot and let cord wrap around rotor.
4. Carefully work starter rotor back to find end of spring tension, then slowly remove starter rotor. BE CAREFUL WITH SPRING!

―――― **CAUTION** ――――
If the rewind spring is removed, use special precaution to prevent the spring from flying out.

Inspection

1. Check friction shoe edges for wear. Edges must be sharp.
2. Check fiber washers; if glazed or oil soaked, replace.
3. Check the rewind, shoe compression, and brake springs for breaks and dirt.
4. Check dust shield and rewind spring shield for tears, wear, or dirt.
5. Check cord for fraying.

Assembly

For special starter assembly instructions, refer to the "ASSEMBLY" section.

CLUTCH AND SPROCKET

Disassembly

To remove the sprocket and clutch, refer to the "DISASSEMBLY" section and perform Steps 1–3.

Inspection

1. Check sprocket teeth for wear.
2. Check sprocket drum for wear.

NOTE: If the chain slips at operating speed, a worn drum or worn clutch shoes are indicated.

3. Check sprocket bearing for wear.

NOTE: Check sprocket race for discoloration which indicates a worn or defective roller bearing.

4. Check sprocket thrust washer for cracks and damage.
5. If experiencing clutch slippage, check clutch shoes and also sprocket drum; replace the necessary components.
6. Check for broken clutch springs.
7. Check for wear on the drive pin.
8. Check for broken, or burred clutch shoes.

Assembly

For special clutch assembly instructions, refer to the "ASSEMBLY" section.

CARBURETOR

NOTE: When adjusting the carburetor, be sure the air filter is clean and properly installed over the carburetor. The proper setting for the carburetor at free speed is obtained with the guide bar and chain assembled.

Initial Carburetor Adjustment

NOTE: The 14 saw does not have a hi-speed mixture screw.

1. Carefully turn the hi-speed mixture screw and lo-speed mixture screw clockwise until they just touch the seat. DO NOT OVERTIGHTEN, or the valve seat will be damaged.
2. Turn the hi-speed mixture screw one full turn counterclockwise.
3. Turn the lo-speed mixture screw one full turn counterclockwise.
4. Start the engine.
5. Adjust the idle-speed screw for maximum idle speed with chain not moving.
6. Allow the engine to warm up for two or three minutes.

Final Carburetor Adjustment

1. With throttle released, turn lo-speed mixture screw to obtain maximum consistent idling speed. When squeezing the throttle to full power, the saw engine should accelerate without hesitation. If the engine hesitates, turn lo-speed mixture screw counterclockwise (usually no more than $\frac{1}{8}$ of a turn) to obtain proper acceleration. Make final idle-speed screw setting so chain stops moving.
2. Squeeze the throttle to full power, then try a cut. If necessary, adjust the hi-speed mixture screw to obtain the best cutting performance.
3. Idle the engine for approximately ten seconds. Check to be sure the chain is not moving; then squeeze the throttle to full power. The engine should accelerate to full power without hesitation. The carburetor is now properly adjusted.

NOTE: Never operate the engine with the hi-speed mixture screw less than $\frac{7}{8}$ turn open. Too lean a mixture may damage the engine due to a lack of lubrication. A white deposit in the muffler indicates the engine is set too lean.

Disassembly

To remove the carburetor, refer to the "DISASSEMBLY" section and perform Steps 12–14.

Inspection

1. Check all mixture screws for wear and replace screws that are worn.
2. Check carburetor inlet needle seat for wear, scoring or other damage.
3. Check inlet needle for dirt or imbedded chips.
4. Check inlet screen for dirt.
5. When assembling the inlet needle to the inlet control lever and control spring, care should be taken to see that the spring is set in the well of the carburetor body and located on the dimple of the inlet control lever.

NOTE: Do not stretch control spring.

6. The inlet control lever is properly set when flush with the floor of the diaphragm chamber.
7. If diaphragm end of control lever is low, pry up. If lever is high, depress diaphragm end and push on needle for proper adjustment.
8. Check carburetor body for cracks, clogged passages, and defective gaskets.
9. Check diaphragms for cracks, punctures, distortion, or deterioration.
10. Check the reed for nicks, chips, burrs, rust, pits, or cracks. Also, the reed should not be bent.

ENGINE
Disassembly

1. To replace the main bearings and the oil seals, refer to the "DISASSEMBLY" section and perform Steps 1–5 for the right-hand bearing and seal or Steps 1–25 for the left-hand bearing and seal.

NOTE: The seals can be replaced independently of the bearing by using a screwdriver to pry out on the lip. Bearings are a press fit-remove with a press.

2. To replace the cylinder, refer to the "DISASSEMBLY" section and perform Steps 1–22.
3. To replace the pistons and rings, refer to the "DISASSEMBLY" section and perform Steps 1–23.
4. To replace the rod, refer to the "DISASSEMBLY" section and perform Steps 1–24.
5. To replace the crankshaft, refer to the "DISASSEMBLY" section and perform Steps 1–25.

Inspection

1. Check crankshaft threads for damage.
2. Check crankshaft for worn, scratched, or damaged bearing surfaces.
3. Check crankshaft for flats.
4. Check crankshaft for discoloration.
5. Check seals for glazing and scoring.
6. Check bearings for damage.
7. Check cylinder for scoring.
8. Check cylinder for chrome peeling.
9. Check for broken rings.
10. Check for worn ring ends; they should be sharp.
11. Check for elongation of wrist pin holes in piston.

Assembly

For special engine assembly instructions, refer to the "ASSEMBLY" section.

Assembly

To assemble the saw, perform the disassembly steps in reverse, making note of the special assembly instructions listed with each component below.

MAIN BEARINGS

1. When installing main bearings, press fit the bearings on the numbered side from inside the casting-pressing outward.
2. Position the main bearings in the casting to 0.010 in. below the surface.

CRANKSHAFT

1. Position the thrust bearings next to the crankshaft with the thrust washers between the bearing and the casting.
2. Put the tapered end of the crankshaft into the crankcase.

CONNECTING ROD

1. Whenever the connecting rod is removed, use new needle bearings when assembling.
2. The connecting rod has match marks on one side.
3. When tightening the hex, socket screws to specifications, make certain that the fractured mating surface of the rod and cap properly align.

RINGS

1. When installing the rings on the piston, the beveled edge on each ring must be positioned to the top of the piston.
2. Position the ring openings next to the locating pin.

MAIN SEALS

1. When installing main seals, press the seals on the metal side, from the outside of the casting-pressing inward.
2. Position the main seals flush with the outside of the casting boss.

WRIST PIN

When installing the snap rings to the wrist pin, position the ring gap to the skirt (bottom) of the piston. The flat side of the snap rings must face out.

HANDLE

When assembling the handle, make certain that the spark plug wire is positioned under the metal tab. If not, the tab may cut into the wire and ground the electrical system.

CRANKCASE COVER

Always use a new O-ring when replacing the crankcase cover.

CLUTCH

When installing the clutch springs install one spring at a time.

Place the shoe over the drive plate, and set the base of the spring at the base of the shoe. Compress the spring and force the free end of the spring against the spring holder on the drive plate. Use the same procedure to assemble other springs to clutch.

STARTER

1. When replacing rotor and cord, install the cord in the rotor. Wind the cord around the rotor until you cannot make a complete wrap around. Unwrap cord to notch of rotor. Hold remaining cord in notch. Move rewind spring $\frac{1}{16}$ in. away from center post. Grease center post lightly. Place the rotor on the rewind spring and prewind the rotor two turns on 12, 14 and 15 Models and three turns on 17, 18, 19 and 23 Models. Hold rotor and insert cord through metal bushing. Pull out cord; tie a slip knot and install the handle, insert, and washer. Untie slip knot.

2. When installing the friction shoes, the notched side of the drive plate must face the direction of rotation (clockwise) when pulling the starter handle.

NOTE: Starter will not operate properly if friction shoes are installed backwards.

Section 10

Honda

Small Engines — 6-20 hp

G300 AND G400

The G300 is a single cylinder, 4-stroke engine of 16.6 cu. in., developing 7hp at 3600 rpm and available in four torque ranges from 10.8 to 65.1 ft. lbs.

The G400 is of similar configuration, developing 10 horsepower at 3600 rpm with torque ranges of 16.6 ft. lbs. and 32.5 ft. lbs. Displacement is 28.1 cu. in.

Changing the Oil

Oil change is accomplished by removing the filler and drain plugs and allowing all the oil to drain into a container for disposal. Replace the drain plug and fill the sump with 2.5 pts. of SAE 10W-40 SE grade engine oil.

Air Cleaner

Three types of air cleaners are used: the dry, semi-dry and oil bath types.

DRY TYPE

The dry type has two segments, a paper and a foam element. The paper element may be blown clean with compressed air or washed with soap and water. The foam element may also be washed in soap and water, but is then soaked in engine oil and wrung out. Make sure that the paper element is completely dry before installation.

Dry type air cleaner

SEMI-DRY TYPE

Wash the element in a safe, non-flammable, solvent. Allow it to dry, soak it in clean engine oil and wring it out.

OIL BATH TYPE

Dump the old oil into a container for disposal. Wash the element and oil cup in a safe, non-flammable solvent. Add fresh 10W–40 engine oil to the oil cup, filling it to the indicator line (0.1 pt.).

Fuel Filter

1. Turn off the fuel valve.
2. Loosen the ring nut and remove the filter cup.
3. Wash the cup in kerosene (not gasoline or other low flash point solvent!).
4. Install the cup and gasket, making sure the gasket is in good condition.

5. Tighten the ring nut securely and turn on the fuel valve. Check for leaks.

Ignition Timing

1. Remove the fan cover. Make a connection between the timing tester terminal and the engine switch and cylinder.
2. Rotate the flywheel to align the **F** mark on the flywheel with the index mark on the cylinder barrel. The timing is correct if the tester operates when these marks align.

Valve Clearance Adjustment

1. With the engine cold, bring the piston to TDC on the compression stroke, and rotate the flywheel until the **T** mark aligns with the index mark on the cylinder.
2. Check the clearance on both valves by inserting a feeler gauge between the lifter and valve stem. Clearance should be 0.0031–0.0063 in. for intake and 0.0043–0.0075 in. for exhaust.
3. If the clearance is too great, replace the lifter. If the clearance is too small, grind the end of the valve stem.

Semi-dry type air cleaner

Oil bath type air cleaner

Idle Speed Adjustment

1. Turn the idle speed screw in until it bottoms lightly.
2. Back it out 1.5 turns for the G300 and 2 turns for the G400.
3. Start the engine and allow it to warm up completely. About 30 minutes should be enough. Turn the idle screw in or out to obtain 1350–1550 rpm.

Governor Adjustment

1. Loosen the governor arm adjusting bolt and nut.
2. Turn the governor arm shaft counterclockwise to the fully closed position. Tighten the bolt and nut with the throttle valve fully opened.
3. Start the engine and run it for 30 minutes.
4. Turn the stop screw in or out to obtain a speed of 3850–4000 rpm.

(A) Fuel valve
(B) Ring nut
(C) To loosen
(D) Strainer cup

Fuel filter

Engine Disassembly

1. The governor and carburetor linkage are removed together. Simply remove the wingnut from the linkage and disassemble the parts.
2. Drain the carburetor before unbolting and removing it.

(A) Feeler gauge
(B) "T" mark

Valve adjustment

(A) Throttle stop screw
(B) Pilot screw

Idle speed adjustment

3. Unbolt and remove the recoil starter, fan cover, starter pulley and fan.
4. Using a flywheel puller, remove the flywheel.
5. Remove the key from the crankshaft and keep it with the flywheel to avoid loss.
6. Remove the ignition point cover and unbolt and remove the coil. Disconnect the spark plug lead.
7. Disconnect the condenser lead and unbolt and remove the ignition points.
8. Remove the spark plug.
9. Remove the head bolts in the sequence shown and lift off the head. Discard the gasket.
10. Remove the tappet cover and gasket. Lift out the breather case.
11. Rotate the crankshaft to fully lift the valves one at a time. With the valve at its maximum lift place a valve holder tool or equivalent under the valve head to hold it in position.
12. Remove the adjuster from the end of the valve stem. Keep the adjusters in order for reuse.
13. Compress the valve spring and remove the spring retainer.
14. Lift out the valve and mark it int. or exh.

(A) Throttle valve   (D) Attaching bolt and nut
(B) Throttle lever   (E) Governor arm shaft
(C) Stop screw

Governor adjustment

Fuel tank removal and installation

10–4

15. Bolt a valve guide remover plate, Honda tool #07969–8960000, or its equivalent, to the head. Screw the stud into the plate. Install the washer and left-hand threaded nut on the end of the handle which extends to the tappet chamber. Oil the handle threads. Back out the threaded stud to pull the guides.

NOTE: Do not remove the remover plate, until the new guide is in place. This will assure correct alignment of the new guide.

16. Remove the crankcase cover. Avoid damaging the seal when sliding it over the shaft.
17. Mark the relationship of the camshaft and crankshaft gears.
18. Raise and hold the lifters, remove the retaining clip and slide out the camshaft.
19. Remove the lifters and mark them for reuse.
20. Remove the lockpin and lift out the governor arm.
21. Remove the connecting rod cap nuts, washers and plate. Discard the washers. Use new ones for installation.

22. Rotate the crankshaft journal away from the connecting rod and slide the crankshaft out.
23. Remove the piston and rod.
24. Remove the retaining clips and slide the piston pin out of the rod and piston.
25. Using a ring expander, remove the rings.
26. Remove the cotter pin and remove the governor weights from the weight holder.
27. Remove the E-clip and remove the weight holder from the shaft.
28. Mark the relationship of the timing gear to the crankshaft, and using a press, remove the gear.

INSPECTION

1. Clean all parts in a safe (non-flammable) solvent.
2. Spin the crankshaft bearing by hand. If it is noisy or has excessive play, replace it.

Carburetor and governor linkage removal and installation

SET BOLT

DRAIN SCREW

FLOAT CHAMBER

MAIN NOZZLE

MAIN JET

FLOAT VALVE

FLOAT

FLOAT PIN

CARBURETOR BODY

Carburetor removal and installation

Breaker point set, exploded view

S TOOL

VALVE SPRING COMPRESSOR

VALVE

VALVE SPRING

SPRING RETAINER

Valve removal

VALVE GUIDE

Valve guide installation depth

3. Using a micrometer, check the OD of the piston skirt. The diameter should be 2.991–2.992 in. for the 300 and 3.385–3.386 in. for the 400. If the skirt is less than 2.986 in. (300) or 3.380 in. (400) replace it.

4. Using a micrometer, check the cylinder bore. The diameter should be 2.992 (300) or 3.386 in. (400). If the piston to skirt clearance exceeds 0.005 in. use oversize piston and rings.

5. Place the rings, one at a time in the cylinder bore and check the end gap. The gap should be 0.0079–0.0157 in. If the gap exceeds 0.0236 in., replace the rings.

6. Check the piston-to-rings side clearance. Clearance should be 0.0008–0.0024 for each of the compression rings and 0.0004–0.0020 for the oil ring. If the clearance exceeds 0.059 for any ring, replace the rings.

7. Check the piston pin bore in the rod. The bore should be 0.6300–0.6302 (300) or 0.7481–0.7483 (400) in. If the bore exceeds 0.6317 (300) or 0.7498 (400), replace the rod.

8. Check the piston pin OD. Diameter should be 0.6297–0.6299 (300) or 0.7478–0.7480 (400) in. If the diameter is less than 0.6287 or 0.7469 replace the pin.

9. Fasten the connecting rod cap in place on the rod and

check the big end bore. Bore should be 0.001 in. larger than the journal diameter. If the bore exceeds this, replace the rod.

10. Check the crankpin journal OD. The diameter should be 1.1789–1.1795 (300) or 1.4544–1.4551 (400) in. If the journal diameter is less than 1.1750 (300) or 1.4508 (400) in., use oversize bearings.

11. Check the crankpin to bearing oil clearance. Clearance should be 0.0019–0.0026 (300/400) or 0.0016–0.0024 (400 KI) in.

12. Install the rod and cap on the journal and check the side clearance. Side clearance should be 0.0236–0.0394 (300/400) or 0.0059–0.0157 (400 KI) in. If the clearance exceeds 0.0394 in., replace the rod.

13. Check the camshaft lobe height. The height should be 1.5142–1.5244 in. for intake and 1.154–1.5256 in. for exhaust.

14. Check the valve stem OD. The diameter should be

FAN COVER

6 × 12 (4)

6 × 10 (3)

FLYWHEEL

RECOIL STARTER

ENGINE SWITCH   STARTER PULLEY   COOLING FAN

Pull start, fan cover and fan exploded view

0.2738–0.2744 for the intake and 0.2720–0.2726 in. for the exhaust. If the diameter is less than 0.2720 for the intake or 0.2713 in. for the exhaust, replace the valve.

15. Valve guide ID should be 0.2756–0.2762 in. If ID exceeds 0.2783, replace the guide.

16. Valve seat width should be 0.0417 in. (0.0557 in. for 400 exh.). If the width exceeds 0.0787 in., replace the seat.

17. The valve spring free length should be 1.6811 in. If it is less than 1.6142 in. replace it.

18. If you're going to cut the valve seat, use a 45 degree, then 30 degree, then 45 degree cutter.

Assembly

1. Using a press, install the timing gear on the crankshaft lining up the marks previously made.

2. Install the governor weight holder on the shaft and secure it with the E-clip.

3. Place the governor weights over the slider and install the slider onto the shaft. Attach the weights to the weight holder with a cotter pin.

4. Place the rod in the piston and coat the pin with clean engine oil. Insert the pin into the piston and rod. Install the pin clips.

5. Install the rings, as illustrated, using a ring expander. Space the ring gaps at 120 degree intervals.

6. Compress the rings and insert the piston into the cylinder with the **896** mark on the top of the piston facing the valve side of the engine.

7. Install the crankshaft.

8. Turn the crankshaft to mate with the connecting rod. Install the rod cap and torque the bolts to 3–4 ft. lbs.

9. Install the governor arm and lockpin.

10. Install the camshaft and lifters, mating the timing marks with the crankshaft timing gear.

11. Install the crankcase cover and torque the bolts to 7 ft. lbs.

12. Install a thrust washer and new guide onto the stud of the valve guide tool. Coat the inner and outer faces of the guise with clean engine oil. Turn the stud to force the guide into the proper depth, which should be 0.7 in. from the end of the guide to the face of the crankcase.

13. Slide the valves into the guides and install the springs, retainers and adjusters.

14. Install the breather case, tappet cover and gasket.

15. Install the head and torque the bolts in the sequence shown, to 7 ft. lbs.

16. Install all other parts in reverse of removal. Torque the flywheel nut to 36 ft. lbs. Check the timing as described below.

BREAKER POINTS SHROUD

WOODRUFF KEY

6 × 12 (3)

5 × 10 (2)

SPARK PLUG CAP

6 × 32 (2)

HIGH TENSION CABLE

POINT COVER

16 mm NUT

IGNITION COIL   8 × 20 (2)   SIDE COVER

Flywheel and ignition removal and installation

S TOOL

VALVE GUIDE REMOVER

Valve guide replacement

SPARK PLUG

CYLINDER HEAD

CYLINDER HEAD GASKET

INTAKE VALVE

EXHAUST VALVE

VALVE GUIDE

VALVE SPRING

SPRING RETAINER

BREATHER CASE

TAPPET COVER

Cylinder head removal and installation

Measuring piston skirt

Checking ring end gap

PISTON

"MARKING"

CYLINDER BARREL

CRANKSHAFT

CAMSHAFT

CONNECTING ROD DIPPER

CONNECTING ROD WASHER
(Except for G400KI)

CONNECTING ROD CAP BOLTS

CRANKCASE COVER

< Camshaft P.T.O.>

GOVERNOR ARM SHAFT

LOCK PIN

Piston and crankcase exploded view

Checking connecting rod side clearance

Checking pin OD

Checking pin bore

Checking crankpin OD

Checking ring side clearance

Checking crankpin oil clearance

Checking valve spring free length

Checking valve guide ID

Checking valve seat width

Checking valve stem OD

OIL RING

MARK

PISTON RINGS

20mm 20mm

PISTON PIN CLIPS

PISTON PIN

CONNECTING ROD

Piston, pin and rings

Section 11

Kohler 4-Stroke

Small Engines—0-6 hp

ENGINE IDENTIFICATION

An engine identification plate is mounted on the carburetor side of the engine blower housing. The numbers that are important, as far as ordering replacement parts is concerned, are the model, serial, and specification numbers.

The model number indicates the engine model series. It also is a code indicating the cubic inch displacement and the number of cylinders. The model number K181, for instance, indicates the engine is 18 cu. in. in displacement and that it has 1 cylinder. The letters following the model number indicate that a variety of other equipment is installed on the engine. The letters and what they mean are as follows:

- C Clutch model
- G Housed with fuel tank
- H Housed less fuel tank
- P Pump model
- R Reduction gear
- S Electric start
- T Retractable start

NOTE: A model number without a suffix letter indicates a basic rope start version.

General Engine Specifications

Model	Bore & Stroke (In.)	Displacement	Horsepower
K91	2⅜ × 2	8.86	4.0

The specification number indicates a model variation. It indicates a combination of various groups used to build the engine. It may have a letter preceding it which is sometimes important in determining superseding parts. The first two numbers of the specifications number is the code designation the engine model: the remaining numbers are issued in numerical sequence as each new specification is released, for example, 2899, 28100, 28101, etc. The current specification number model code is K91.

The serial number lists the order in which the engine was built. If a change takes place to a model or a specification, the serial number is used to indicate the points at which the change takes place. The first letter or number in the serial number indicates what year the engine was built. The letter prefix to the engine serial number was dropped in 1969 and thereafter the prefix is a number. Engines made in 1969 have either the letter **E** or the number **1**. The code is as follows:

- A — 1965
- B — 1966
- C — 1967
- D — 1968
- E — 1969

First Digit Numbers:

- 1 — 1969
- 2 — 1970
- 3 — 1971
- 4 — 1972
- 5 — 1973
- 6 — 1974

MAINTENANCE

Air Cleaners

A dirty air cleaner can cause rich fuel/air mixture and consequent poor engine operation and sludge deposits. If the filter becomes dirty enough, dirt that otherwise would be trapped can pass through and may wear the engine's moving parts prematurely. It is therefore necessary that all maintenance work be performed precisely as specified.

DRY AIR CLEANERS

Clean dry element air cleaners every 50 hours of operation, or every 6 months (whichever comes first) under good operating conditions. Service more frequently if the operating area is dusty. Remove the element and tap it lightly against a hard surface to remove the bulk of the dirt. If dirt will not drop off easily, replace the element. Do not use compressed air or solvents. Replace the air cleaner every 100–200 hours, under good conditions, and more frequently if the air is dusty.

Observe the following precautions:

1. Handle the element carefully-do not allow the gasket surfaces to become bent or twisted.

2. Make sure gasket surfaces seal against back plate and cover.

DRY ELEMENT

PRECLEANER (OPTION)

Exploded view of dry type air cleaner with pre-cleaner

3. Tighten wing nut only finger tight—if it is too tight, cleaner may not seal properly.

If the dry type air cleaner is equipped with a precleaner, service this unit when cleaning the paper element. Servicing consists of cleaning the precleaner in soap and water, squeezing the excess out, and then allowing it to air dry before installation. Do not oil!

OIL BATH AIR CLEANERS

This type of unit may be used to replace the dry type in applications where very frequent replacement of the element is required. The conversion is simple and requires the use of an elbow to fit the oil bath unit onto the engine in a vertical position.

Service the unit every 25 hours of operation under good conditions and, under dusty conditions, as often as every 8 hours of operation.

Service as follows:
1. Remove cover and lift element out of bowl.
2. Drain dirty oil from bowl, and then wash thoroughly in clean solvent.
3. Swish the element in the solvent and then allow it to drip dry. Do not dry with compressed air. Lightly oil the element with engine oil.
4. Inspect air horn, filter bowl, and cover gaskets, and replace as necessary (if grooved or cracked).
5. Install filter bowl gasket on air horn, then put the bowl into position. Fill bowl to indicated level with engine oil.
6. Install element, put the cover in position, and then install copper gasket (if used) and wingnut. Tighten wingnut with fingers only to avoid distorting housing. Make sure all joints in the unit seal tightly.

Lubrication

CRANKCASE

Oil level must be maintained between **F** and **L** marks—do not overfill. Check every day and add as necessary. On new engines, be especially careful to stop engine and check level frequently. When checking, make sure regular type dipstick is inserted fully. On screw type dipstick, check level with dipstick inserted fully but not screwed in. On this type, however, make sure to screw dipstick back in tightly when oil level check is completed.

Use SC type oils meeting viscosity specifications according to the prevailing temperature as shown in the chart below.

Change initial fill of oil on new engines after five hours of operation. Then, change oil every 25 hours of operation. Change oil when engine is hot. Change more frequently in dusty areas. If the engine has just been overhauled, it is best to fill it initially with a non-detergent oil. Then, after 5 hours, refill with SC type oil.

Oil capacities are:
K91 — ½ qt.

Oil Viscosity Chart

Air Temperature	Oil Viscosity	Oil Type
Above 30° F	SAE 30	API Service SC*
30° to 0° F	SAE 10W-30	API Service SC*
Below 0° F	SAE 5W-20	API Service SC*

*SC standard recommendation—CC (MIL-2104B) and SD class oils may also be used.

REDUCTION GEAR UNITS

Every 50 hours, remove the oil plug on the lower part of the reduction unit cover to check level. If oil does not reach the level of the oil plug, remove the vented fill plug from the top of the cover and refill with engine oil until level is correct. This oil need not be changed unless unit has been out of service for several months. In this situation, remove the drain plug, drain oil, then replace plug and fill to proper level as described above.

FUEL RECOMMENDATIONS

Use either leaded or unleaded regular grade fuel of at least 90 octane. Unleaded fuel produces fewer combustion chamber deposits, so its use is preferred.

Purchase fuel from a reputable dealer, and make sure to use only fresh fuel (fuel less than 30 days old). If the engine is stored, drain the fuel system or use a fuel stabilizer that is compatible with the type of fuel tank the engine is equipped with.

TUNE-UP

Spark Plugs

SERVICE

The spark plug should be removed and serviced every 100 hours of engine operation. The plug should have a light coating of light gray colored deposits. If deposits are black, fuel/air mixture could be too rich due to improper carburetor adjustment or a dirty air cleaner. If deposits are white, the engine may be overheating or a spark plug of too high a heat range could be in use. Kohler recommends that the plug be replaced rather than sandblasted or scraped if there are excessive deposits. Torque plugs to 18–22 ft. lbs.

TESTING

To test a plug for adequate performance, remove it from the engine, attach the ignition wire, and then rest the side electrode against the cylinder head. Crank the engine vigorously. If there is a sharp spark, the plug and ignition system are allright, although ignition timing should be checked if the engine fires irregularly.

Adjusting breaker gap

Breaker Points

INSPECTION

Remove the breaker cover and inspect the points for pitting or buildup of metal on either the movable or stationary contact every 100 hours of operation. Replace the points if they are badly burned. If there is a great deal of metal buildup on either contact, the condenser may be faulty and should be replaced.

To replace points, remove the primary wiring connector screw and pull off the primary wire. Then, remove the contact set mounting screws and remove the contact set. Install the new set of points in reverse order, leaving upper mounting screw slightly loose. Then set point gap and timing as described below.

SETTING BREAKER GAP AND TIMING

1. Remove the breaker cover and disconnect the spark plug lead. Rotate the engine in direction of normal rotation until the points reach the maximum opening.
2. Using a clean, flat feeler gauge of 0.020 in. size, check the gap between the points. Gauge should just slide between the contacts without opening them when flat between them. If the gap is incorrect, loosen the upper mounting screw (if necessary), and shift the breaker base with the blade of the screwdriver until gap is correct.
3. There is a timing sight hole in either the bearing plate or the blower housing. If there is a snap button in the hole, pry it out with a screwdriver.
4. While observing the sight hole, turn the engine slowly in normal direction of rotation. When the T mark appears in the hole, the points should just be beginning to open. If timing is incorrect, breaker gap will have to be reset slightly (0.018–0.022 in.). If the points are not yet opening when the timing mark is centered in the hole, make the point gap wider. If points open too early, narrow it. Recheck the setting after tightening the upper breaker mounting screw by turning the engine in normal direction of rotation past the firing point and checking that the points open at just the right time.

NOTE: This procedure may be performed with the engine running at 1200–1800 rpm if a timing light is available. Connect the timing light according to manufacturer's instructions. You may have to chalk the timing mark to see it adequately.

Adjusting timing

TRIGGER AIR GAP

Trigger air gap is set within the range 0.005–0.010 in. As long as the gap falls within this range, the ignition system should perform adequately. Optimum ignition performance during cold weather starting is provided if the gap is adjusted to 0.005 in. If you wish to adjust this or to ensure that the gap falls within the proper range, rotate the flywheel until the flywheel projection is lined up with the trigger assembly. Then, loosen the trigger bracket capscrews and slide the trigger back and forth to get the proper gap, as measured with a flat feeler gauge. Then, retighten capscrews.

IGNITION COILS

Coils do not require regular service, except to make sure they are kept clean, that the connections are tight, and that rubber insulators are in good condition (replace if cracked). If you suspect poor performance of a breakerless type ignition system and trigger air gap is correct, check resistance with an ohmmeter. To do this, disconnect the high tension lead at the coil and connect the meter between coil terminal and coil mounting bracket. If resistance is not about 11,500 ohms, replace the coil. Also, check the reading with the meter lead going to the coil terminal pulled off and connected to the spark plug connector of the high tension lead. If there is continuity here, replace the coil.

PERMANENT MAGNETS

These may be checked for magnet strength by holding a screwdriver (non-magnetic) blade within one inch of the magnet. If the magnetic field is good, the blade will be attracted to the magnet. Otherwise, replace it.

Mixture Adjustments

NOTE: Before making any adjustments, be sure that the carburetor air cleaner is not clogged. A clogged air cleaner will cause an over-rich mixture, black exhaust smoke, and may lead you to believe that the carburetor is out of adjustment when, in reality, it is not. The carburetor is set at the factory and rarely needs adjustment unless, of course, it has been disassembled or rebuilt.

1. With the engine stopped, turn the main and idle fuel adjusting screws all the way in until they bottom lightly. Do not force the screws or you will damage the needles.

Adjustment screws on the side draft carburetor

Adjustment screws on the updraft carburetor

2. For a preliminary setting, turn the main fuel screw out 2 full turns and the idle screw out 1¼ turns.

3. Start the engine and allow it to reach operating temperatures; then operate the engine at full throttle and under a load, if possible.

Tune-Up Specifications

Model	Plug Gap (in.)	Breaker Point Gap (in.)	Trigger Air Gap (in.)	Normal Timing (deg)	Retard Timing (deg)
K91	.025①	.020	—	20B	—

B—Before
① Shielded plug gap—.020 in.

Spark Plug Specifications

| Engine Model | Plug Size | Hex Size | Plug Reach | Standard Plugs | | Resistor Plugs | |
				Solid Post	Knurled Nut	Non-Shielded	Shielded
K91	14 mm	13/16″	3/8″	J-8 270321-S	J-8 220040-S	XJ-8 232604-S	XEJ-8 220258-S

4. For final adjustment, turn the main fuel adjustment screw in until the engine slows down (lean mixture), then out until it slows down again (rich mixture). Note the positions of the screw at both settings, then set it about halfway between the two positions.

5. Set the idle mixture adjustment screw in the same manner. The idle speed (no-load) on most engines is 1200 rpm; however, on engines with a parasitic load (hydrastatic drives) the engine idle speed may have to be increased to as much as 1700 rpm for best no-load idle.

Governor Adjustment

All Kohler engines use mechanical, camshaft driven governors.

INITIAL ADJUSTMENT

1. Loosen, but do not remove, the nut that holds the governor arm to the governor cross shaft.

2. Grasp the end of the cross shaft with a pair of pliers and turn it in counterclockwise as far as it will go. The tab on the cross shaft will stop against the rod on the governor gear assembly.

3. Pull the governor arm away from the carburetor, then re-tighten the nut which holds the governor arm to the shaft. With updraft carburetors, lift the arm as far as possible, then retighten the arm nut.

FINAL ADJUSTMENT

After making the initial adjustment and connecting the throttle wire on the variable speed applications, start the engine and check the maximum operating speed with a tachometer.

If adjustment is necessary:

1. Loosen the bushing nut slightly.

2. Move the throttle bracket in a counterclockwise direction to increase speed, or in a clockwise direction to decrease engine speed. Maximum speed is 4000 rpm.

3. With the speed set to the proper range, tighten the bushing nut to lock the throttle bracket in position.

Choke Adjustment

THERMOSTATIC TYPE

If the engine does not start when cranked, continue cranking and move the choke lever first to one side and then to the other to determine whether the setting is too lean or too rich. Once the direction in which lever must be moved has been determined, loosen the adjusting screw on the choke body. Then, move the bracket downward to increase choking or upward to decrease it. Then, tighten the lockscrew. Try starting it again and readjust as necessary.

ELECTRIC-THERMOSTATIC TYPE

Remove the air cleaner from the carburetor and check the position of the choke plate. The choke should be fully closed when engine is at outside temperature and the temperature is very low. In milder temperatures, slightly less closure is required.

If adjustment is required, move the choke arm until the hole in the brass shaft lines up with the slot in the bearings. Insert a #43 (0.089 in.) drill through the shaft and push it downward so it engages the notch in the base of the choke unit. Then, loosen the clamp bolt on the choke lever and push the arm upward to move the choke plate toward the closed position. When the desired position is obtained, tighten the clamp bolt. Then, remove the drill.

GOVERNOR ARM

CROSS-SHAFT

NUT

Initial adjustment of the governor installed on the K91, K141, K161, and the K181 engines

CHOKE LEVER ARM

DRILL (.089")

CLAMP BOLT

MOVE ARM UP TO INCREASE CHOKE— MOVE DOWNWARD TO DECREASE CHOKE

HOLE IN SHAFT

CROSS SHAFT

NOTCH IN BASE OF CHOKE

Adjusting electric-thermostatic choke

Remount the air cleaner, and then check for any binding in the choke linkage. Correct as necessary. Finally, run the engine until hot, and make sure the choke opens fully. If not, readjust it toward the open position as necessary.

Compression Check

Compression is checked by removing the spark plug lead and spinning the flywheel forward against compression. If the piston does not bounce baceward with considerable force, checking with a gauge may be necessary. On Automatic Compression Release engines, rotate the flywheel backward against power stroke-if little resistance is felt, check compression with a gauge.

The compression gauge check requires rapid motoring (spinning) of the crankshaft, at about 1000 rpm. Install the gauge in the spark plug hole and motor the engine. Gauge should read 110–120 psi. If reading is less than 100 psi, the engine requires major repair to piston rings or valves.

FUEL SYSTEM
Carburetor

If a carburetor will not respond to mixture screw adjustments, then you can assume that there are dirt, gum, or varnish deposits in the carburetor or worn/damaged parts. To remedy these problems, the carburetor will have to be completely disassembled, cleaned, and worn parts replaced and reassembled.

Parts should be cleaned with solvent to remove all deposits. Replace worn parts and use all new gaskets. Carburetor rebuilding kits are available.

DISASSEMBLY

Side Draft Carburetors

1. Remove the carburetor from the engine.
2. Remove the bowl nut, gasket, and bowl. If the carburetor has a bowl drain, remove the drain spring, spacer and plug, and gasket from inside the bowl.

Exploded view of the updraft carburetor

Exploded view of the sidedraft carburetor. Inset shows float adjustment procedure

3. Remove the float pin, float, needle, and needle seat. Check the float for dents, leaks, and wear on the float lip or in the float pin holes.

4. Remove the bowl ring gasket.

5. Remove the idle fuel adjusting needle, main fuel adjusting needle, and springs.

6. Do not remove the choke and throttle plates or shafts. If these parts are worn, replace the entire carburetor assembly.

Updraft Carburetors

1. Remove the carburetor from the engine.

2. Remove the bowl cover and the gasket.

3. Remove the float pin, float, needle and needle seat. Check the float pin for wear.

4. Remove the idle fuel adjustment needle, main fuel adjustment needle, and the springs. Do not remove the choke plate or the shaft unless the replacement of these parts is necessary.

ASSEMBLY

Side Draft Carburetor

1. Install the needle seat, needle, float, and float pin.

2. Set the float level. With the carburetor casting inverted and the float resting against the needle in its seat, there should be $^{11}/_{64}$ in. plus or minus $^{1}/_{32}$ in. clearance between the machined surface of the casting and the free end of the float.

3. Adjust the float level by bending the lip of the float with a small screwdriver.

4. Install the new bowl ring gasket, new bowl nut gasket, and bowl nut. Tighten the nut securely.

5. Install the main fuel adjustment needle. Turn it in until the needle seats in the nozzle and then back out two turns.

6. Install the idle fuel adjustment needle. Back it out about $1^{1}/_{4}$ turns after seating it lightly against the jet.

7. Install the carburetor on the engine.

Updraft Carburetor

1. Install the throttle shaft and plate. The elongated side of the valve must be toward the top.

2. Install the needle seat. A $^{5}/_{16}$ in. socket should be used. Do not overtighten.

3. Install the needle, float, and float pins.

4. Set the float level. With the bowl cover casting inverted and the float resting lightly against the needle in its seat, there should be $^{7}/_{16}$ in. plus or minus $^{1}/_{32}$ in. clearance between the machined surface casting and the free end of the float.

5. Adjust the float level by bending the lip of the float with a small screwdriver.

6. Install the new carburetor bowl gasket, bowl cover, and bowl cover screws. Tighten the screws securely.

7. Install the main fuel adjustment needle. Turn it in until the screw seats in the nozzle and then back it out 2 turns.

8. Install the idle fuel adjustment needle. Back it out about $1^{1}/_{2}$ turns after seating the screw lightly against the jet. Install the idle speed screw and spring. Adjust the idle to the desired speed with the engine running.

9. Install the carburetor on the engine.

Fuel Pump

Fuel pumps used on single cylinder Kohler engines are either the mechanical or vacuum actuated type. The mechanical type is operated by an eccentric on the camshaft and the vacuum type is operated by the pulsating negative pressures in the crankcase. The K91 vacuum type pump is not serviceable and must be replaced when faulty. The mechanical pump is serviceable and rebuilding kits are available.

1. Disconnect fuel lines, remove mounting screws, and pull the pump off the engine.

2. File a mark across some point at the union of pump body and cover. Remove the screws and remove the cover.

ASSEMBLY SCREW
PUMP COVER
VALVE GASKET
MATCH MARK
PUMP VALVE
VALVE SPRING
VALVE SPRING
PUMP VALVE
VALVE RETAINER
VALVE PLATE SCREW
DIAPHRAGM
MOUNTING GASKET
DIAPHRAGM SPRING
PUMP BODY
MATCH MARK

Exploded view of the fuel pump

3. Turn the cover upside down and remove the valve plate screw and washer. Remove the valve retainer, valves, valve springs, and valve gasket, after noting the position of each part. Discard the valve springs, valves and valve retainer gasket.

4. Clean the fuel head with solvent and a soft wire brush. Hold the pump cover with the diaphragm surface upward; position a new gasket into the cavity. Put the valve spring and valves into position in the cavity and reassemble the valve retainer. Lock the retainer into position by installing the fuel pump valve retainer screw.

5. Rebuild the lower diaphragm section.

6. Hold the mounting bracket and press down on the diaphragm to compress the spring underneath. Turn the bracket 90 degrees to unhook the diaphragm and remove it.

7. Clean the mounting bracket with solvent and a wire brush.

8. Stand a new diaphragm spring in the casting, put the diaphragm into position, and push downward to compress the spring. Turn the diaphragm 90 degrees to reconnect it.

9. Position the pump cover on top of the mounting bracket with the indicating marks lined up. Install the screws loosely on mechanical pumps; on vacuum pumps, tighten the screws.

10. Holding only the mounting bracket, push the pump lever to the limit of its travel, hold it there, and then tighten the four screws.

11. Remount the fuel pump on the engine with a new gasket, tighten the mounting bolts, and reconnect the fuel lines.

ENGINE OVERHAUL

Disassembly

The following procedure is designed to be a general guide rather than a specific and all imclusive disassembly procedure. The sequence may have to be varied slightly to allow for the removal of

Left side—measuring the cylinder bore. Right side—honing must produce a cross-hatch pattern as shown

special equipment or accessory itmes such as motor/generators, starters, instrument panels, etc.

1. Disconnect the high tension spark plug lead and remove the spark plug.

2. Close the valve on the fuel sediment bowl and remove the fuel line at the carburetor.

3. Remove the air cleaner from the carburetor intake.

4. Remove the carburetor.

5. Remove the fuel tank. The sediment bowl and brackets remain attached to the fuel tank.

6. Remove the blower housing, cylinder baffle, and head baffle.

7. Remove the rotating screen and the starter pulley.

8. The flywheel is mounted on the tapered portion of the crankcase and is removed with the help of a puller. Do not strike the flywheel with any type of hammer.

9. Remove the breaker point cover, breaker point lead, breaker assembly, and the push-rod that operates the points.

10. Remove the magneto assembly.

11. Remove the valve cover and breather assembly.

12. Remove the cylinder head.

13. Raise the valve springs with a valve spring compressor and remove the valve spring keepers from the valve stems. Remove the valve spring retainers, springs, and valves.

14. Remove the oil pan base and unscrew the connecting rod can screws. Remove the connecting rod cap and piston assembly from the cylinder block.

NOTE: It will probably be necessary to use a ridge reamer on the cylinder walls before removing the piston assembly, to avoid breaking the piston rings.

15. Remove the crankshaft, oil seals and, if necessary, the anti-friction bearings.

NOTE: It may be necessary to press the crankshaft out of the cylinder block. The bearing plate should be removed first, if this is the case.

16. Turn the cylinder block upside down and drive the camshaft pin out from the power take-off side of the engine with a small punch. The pin will slide out easily once it is driven free of the cylinder block.

17. Remove the camshaft and the valve tappets.

18. Loosen and remove the governor arm from the governor shaft.

19. Unscrew the governor bushing nut and remove the governor shaft from the inside of the cylinder block.

20. Loosen, but do not remove, the screw located at the lower right of the governor bushing nut until the governor gear is free to slide off of the stub shaft.

Engine Rebuilding

CYLINDER BLOCK SERVICE

Make sure that all surfaces are free of gasket fragments and sealer materials. The crankshaft bearings are not to be removed unless replacement is necessary. One bearing is pressed into the cylinder block and the other is located in the bearing plate. If there is no evidence of scoring or grooving and the bearings turn easily and quietly it is not necessary to replace them.

The cylinder bore must not be worn, tapered, or out-of-round more than 0.005 in. Check at two locations 90 degrees apart and compare with specifications. If it is, the cylinder must be rebored. If the cylinder is very badly scored or damaged it may have to be replaced, since the cylinder can only be rebored to either 0.010 in. or 0.020 in. and 0.030 in. maximum. Select the nearest suitable oversize and bore it to that dimension. On the other hand, if the cylinder bore is only slightly damaged, only a light deglazing may be necessary.

HONING THE CYLINDER BORE

1. The hone must be centered in relation to the crankshaft crossbore. It is best to use a low speed drill press. Lubricate the hone with kerosene and lower it into the bore. Adjust the stones so they contact the cylinder walls.

2. Position the lower edge of the stones even with the lower edge of the bore, hone at about 600 rpm. Move the hone up and down continuously. Check bore size frequently.

3. When the bore reaches a dimension 0.0025 in. smaller than desired size, replace the coarse stones with burnishing stones. Use burnishing stones until the dimension is within 0.0005 in. of desired size.

4. Use finishing stones and polish the bore to final size, moving the stones up and down to get a 60 degree cross-hatch pattern. Wash the cylinder wall thoroughly with soap and water, dry, and apply a light coating of oil.

CRANKSHAFT SERVICE

Inspect the keyway and the gears that drive the camshaft. If the keyways are badly worn or chipped, the crankshaft should be replaced. If the cam gear teeth are excessively worn or if any are broken, the crankshaft must be replaced.

Check the crankpin for score marks or metal pickup. Slight score marks can be removed with a crocus cloth soaked in oil. If the crankpin is worn more than 0.002 in., the crankshaft is to be either replaced or the crankpin reground to 0.010 in. undersize. If the crankpin is reground to 0.010 in. undersize, a 0.010 in. undersize connecting rod must be used to achieve proper running clearance.

CONNECTING ROD SERVICE

Check the bearing area for wear, score marks, and excessive running and side clearance. Replace the rod and cap if they are worn beyond the limits allowed.

PISTON AND RINGS SERVICE

Production and Service Type

Rings are available in the standard size as well as 0.010 in., 0.020 in., and 0.030 in. oversize sets.

NOTE: Never reuse old rings.

The standard size rings are to be used when the cylinder is not worn or out-of-round. Oversize rings are only to be used when the cylinder has been rebored to the corresponding oversize. Service type rings are used only when the cylinder is worn but within the wear and out-of-round limitations; wear limit is 0.005 in. oversize and out-of-round limit is 0.004 in.

The old piston may be reused if the block does not need reboring and the piston is within wear limits. Never reuse old rings. After removing old rings, thoroughly remove deposits from ring grooves. New rings must each be positioned in its running area of the cylinder bore for an end clearance check, and each must meet specifications.

The cylinder must be deglazed before replacing the rings. If chrome plated rings are used, the chrome plated ring must be installed in the top groove. Make sure that the ring grooves are free from all carbon deposits. Use a ring expander to install the rings. Then check side clearance.

PISTON AND ROD SERVICE

Normally very little wear will take place at the piston boss and piston pin. If the original piston and connecting rod can be used after rebuilding, the piston pin may also be used. However if a new piston or connecting rod or both have to be used, a new piston pin must also be installed. Lubricate the pin before installing it with a loose to light interference fit. Use new piston pin retainers whether or not the pin is new. Make sure they're properly engaged.

VALVES AND VALVE MECHANISM SERVICE

Inspect the valve mechanism, valves, and valve seats or inserts for evidence of wear, deep pitting, cracks or distortion. Check the clearance between the valve stems and the valve guides.

Valve guides must be replaced if they are worn beyond the limit allowed. K91 model engines do not use valve guides. To remove valve guides, press the guide down into the valve chamber and carefully break off the protruding end until the guide is completely removed. Be careful not to damage the block when removing the old guides. Use an arbor press to install the new guides. Press the new guides to the depth specified, then use a valve guide reamer to gain the proper inside diameter.

Make sure that replacement valves are the correct type (special hard faced valves are needed in some cases). Exhaust valves are always hard faced.

Intake valve seats are usually machined into the block, although inserts are used in some engines. Exhaust valve seats are made of special hardened material. The seating surfaces should be held as close to $\frac{1}{32}$ in. in width as possible. Seats more than $\frac{1}{16}$ in. wide must be reground with 45 degree and 15 degree cutters to obtain the proper width. Reground or new valves and seats must be lapped in for a proper fit.

Positioning of production and service type piston rings on the piston

After resurfacing valves and seats and lapping them in, check the valve clearance. Hold the valve down on its seat and rotate the camshaft until it has no effect on the tappet, then check the clearance between the end of the valve stem and the tappet. If the clearance is not sufficient (it will always be less after grinding), it will be necessary to grind off the end of the valve stem until the correct clearance is obtained.

CYLINDER HEAD SERVICE

Remove all carbon deposits and check for pitting from hot spots. Replace the head if metal has been burned away because of head gasket leakage. Check the cylinder head for flatness. If the head is slightly warped, it can be resurfaced by rubbing it on a piece of sandpaper placed on a flat surface. Be careful not to nick or scratch the head when removing carbon deposits.

Engine Assembly

REAR MAIN BEARING

Install the rear main bearing by pressing it into the cylinder block with the shielded side toward the inside of the block. If it does not have a shielded side, then either side may face inside.

GOVERNOR SHAFT

1. Place the cylinder block on its side and slide the governor shaft into place from the inside of the block. Place the speed control disc on the governor bushing nut and thread the nut into the block, clamping the throttle bracket into place.
2. There should be a slight end-play in the governor shaft and that can be adjusted by moving the needle bearing in the block.
3. Place a space washer on the stub shaft and slide the governor gear assembly into place.
4. Tighten the holding screw from outside the cylinder block.
5. Rotate the governor gear assembly to be sure that the holding screw does not contact the weight section of the gear.

CAMSHAFT

1. Turn the cylinder block upside down.
2. The tappets must be installed before the camshaft is installed. Lubricate and install the tappets into the valve guides.
3. Position the camshaft inside the block.

NOTE: Align the marks on the camshaft and the automatic spark advance, if so equipped.

Installing the camshaft

4. Lubricate the rod and insert it into the bearing plate side of the block. Install one 0.005 in. washer between the end of the camshaft and the block. Push the rod through the camshaft and tap it lightly until the rod just starts to enter the bore at the PTO end of the block. Check the endplay and adjust it with additional washers if necessary. Press the rod into its final position.
5. The fit at the bearing plate for the camshaft rod is a light to loose fit to allow oil that might leak past to drain back into the block.

CRANKSHAFT

1. Place the block on the base of an arbor press and carefully insert the tapered end of the crankshaft through the inner race of the anti-friction bearing.

Timing marks for the dynamic balance system

Timing marks for the automatic spark advance

2. Turn the crankshaft and camshaft until the timing mark in the shoulder of the crankshaft lines up with the mark on the cam gear.

3. When the marks are aligned, press the crankshaft into the bearing, making sure that the gears mesh as it is being pressed in. Recheck the alignment of the timing marks on the crankshaft and the camshaft.

4. The end-play of the crankshaft is controlled by the application of various thickness gaskets between the bearing plate and the block. Normal end-play is achieved by installing 0.020 in. and 0.010 in. gaskets, with the thicker gaskets on the inside.

BEARING PLATE

1. Press the front main bearing into the bearing plate. Make sure that the bearing is straight.

2. Press the bearing plate onto the crankshaft and into position on the block. Install the cap screws and secure the plate to the block. Draw up evenly on the screws.

3. Measure the crankshaft end-play, which is very critical on gear reduction engines.

PISTON AND ROD ASSEMBLY

1. Lubricate the pin and assemble it to the connecting rod and piston. Install the wrist pin retaining ring. Use new retaining rings.

Alignment of the timing marks for the crankshaft and the camshaft

INSTALL WITH
MARKS TOWARD
BEARING PLATE

Connecting rod and cap alignment marks

2. Lubricate the entire assembly, stagger the ring gaps and, using a ring compressor, slide the piston and rod assembly into the cylinder bore with the connecting rod marks on the flywheel side of the engine.

3. Place the block on its end and oil the connecting rod end and the crankpin.

4. Attach the rod cap, lock or lock washers, and the cap screws. Tighten the screws to the correct torque.

NOTE: Align the marks on the cap and the connecting rod.

5. Bend the lock tabs to lock the screws.

CRANKSHAFT OIL SEALS

Apply a coat of grease to the lip and guide the oil seals onto the crankshaft. Make sure no foreign material gets onto the knife edges of seal, and make sure the seal does not bend. Place the block on its side and drive the seals squarely into the bearing plate and block.

OIL PAN BASE

Using a new gasket on the base, install pilot studs to align the cylinder block, gasket, and base. Tighten the four attaching screws to the correct torque.

VALVES

1. See the engine rebuilding section of the "Clinton" chapter for details concerning installation of the seats and guides. Clean the valves, seats, and parts thoroughly. Grind and lap-in the valves and seats for proper seating. Valve seat width must be $\frac{1}{32}$–$\frac{1}{16}$ in. After grinding and lapping, slide the valves into position and check the clearance between stem and tappet. If the clearance is too small, grind the stem ends square and remove all burrs. On engines with adjustable valves, make the adjustment at this time.

2. Place the valve springs, retainers, and rotators under the valve guides. Lubricate the valve stems, and then install the

Exploded views of two common types of breathers showing assembly sequence

valves down through the guides, compress the springs, and place the locking keys or pins in the grooves of the valve stems.

CYLINDER HEAD

1. Use a new cylinder head gasket.
2. Lubricate and tighten the head bolts evenly, and in sequence, to the proper torque.
3. Install the spark plug.

BREATHER ASSEMBLY

Assemble the breather assembly, making sure that all parts are clean and the cover is securely tightened to prevent oil leakage.

MAGNETO

On flywheel magneto systems, the coil-core assembly is secured onto the bearing plate. On magneto-alternator systems, the coil is part of the stator assembly, which is secured to the bearing plate. On rotor type magneto systems, the rotor has a keyway and is press fitted onto the crankshaft. The magnet rotor is marked "engine-side" for proper assembly. Run all leads through the hole provided at the 11 o'clock position on the bearing plate.

FLYWHEEL

1. Place the washer in place on the crankshaft and place the flywheel in position. Install the key.

Cylinder head torquing sequences

2. Install the starter pulley, lock washer, and retaining nut. Tighten the retaining nut to the specified torque.

BREAKER POINTS

1. Install the pushrod.
2. Position the breaker points and fasten them with the two screws.
3. Place the cover gasket into position and attach the magneto lead.
4. Set the gap and install the cover.

CARBURETOR

Insert a new gasket and assemble the carburetor to the intake port with the two attaching screws.

GOVERNOR ARM AND LINKAGE

1. Insert the carburetor linkage in the throttle arm.
2. Connect the governor arm to the carburetor linkage and slide the governor arm into the governor shaft.
3. Position the governor spring in the speed control disc.
4. Before tightening the clamp bolt, turn the shaft counter-clockwise with pliers as far as it will go; pull the arm as far as it will go to the left (away from the carburetor), tighten the nut, and check for freedom of movement. Adjust the governor.

BLOWER HOUSING AND FUEL TANK

Install the head baffle, cylinder baffle, and the blower housing, in that order. The smaller cap screws are used on the bottom of the crankcase. Install the fuel tank and connect the fuel line.

RUN-IN PROCEDURE

1. Fill the crankcase with a non-detergent oil and run it under load for 5 hours to break it in.
2. Drain oil and refill crankcase with the recommended detergent type oil. Non-detergent oil must not be used except for break-in.

Engine Rebuilding Specifications

Specification	K91
Displacement	
Cubic Inches	8.86
Cubic Centimeters	145.19
Horsepower (Max RPM)	4.0
Cylinder Bore	
New Diameter	2.375
Maximum Wear Diameter	2.378
Maximum Taper	.0025
Maximum out of Round	.005
Crankshaft	
End Play (Free)	.0228/.0038
Crankpin	
New Diameter	.936
Maximum Out of Round	.0005
Maximum Taper	.001

Engine Rebuilding Specifications (cont.)

Specification	K91
Camshaft	
Run Clearance on Pin	.001/.0025
End Play	.005/.020
Connecting Rod	
Big End Maximum Diameter	.9385
Rod-Crankpin Max Clear	.0035
Small (Pin) End-New Dia	.56315
Rod to Pin Clearance	.0007/.0008
Piston	
Thrust Face-Max Wear Dia*	2.359
Thrust Face*-Bore Clear	.0035/.006
Ring-Max Slide Clearance	.006
Ring-End Gap in New Bore	.007/.017
Ring-End Gap in Used Bore	.027
Valve-Intake	
Valve-Tappet Cold Clear	.005/.009
Valve Lift (Zero Lash)	.2095
Stem to Guide Max Wear Clear	.004
Valve-Exhaust	
Valve-Tappet Cold Clear	.011/.015
Valve Lift (Zero Lash)	.1828
Stem to Guide Max Wear Clear	.006
Tappet	
Clearance in Guide	.0005/.002
Ignition	
Spark Plug Gap-Gasoline	.025
Spark Plug Gap-Gas	.018
Spark Plug Gap (Shielded)	.020
Breaker Point Gap	.020
Trigger Air Gap (Breakerless)	NOT USED
SparkRun ° BTDC	20°
Spark Retard	NO RETARD
Torque Valves (Also See Pages 15.3)	
Spark Plug (foot lbs)	18–22
Cylinder Head	200 in. lbs
Connecting Rod	140 in. lbs
Flywheel Nut	40–50 ft lbs

*Measured just below oil ring and at right angles to piston pin

Valve Specifications

Dimension		Model K91	
		Intake	Exhaust
A	Seat Angle	89°	89°
B	Seat Width	.037/.045	.037/.045
C	Insert OD	—	.972/.973
D	Guide Depth	None	None
E	Guide ID	None	None
F	Valve Head Diameter	.979/.989	.807/.817
G	Valve Face Angle	45°	45°
H	Valve Stem Diameter	.2480/.2485	.2460/.2465

Key to valve specifications chart

Section 12

Kohler 4-Stroke

Small Engines — 6-20 hp

ENGINE IDENTIFICATION

An engine identification plate is mounted on the carburetor side of the engine blower housing. The numbers that are important, as far as ordering replacement parts is concerned, are the model, serial, and specification numbers.

The model number indicates the engine model series. It also is a code indicating the cubic inch displacement and the number of cylinders. The model number K181, for instance, indicates the engine is 18 cu in. in displacement and that it has 1 cylinder. The letters following the model number indicate that a variety of other equipment is installed on the engine. The letters and what they mean are as follows:

- C Clutch model
- G Housed with fuel tank
- H Housed less fuel tank
- P Pump model
- R Reduction gear
- S Electric Start
- T Retractable start

NOTE: A model number without a suffix letter indicates a basic rope start version.

The specification number indicates a model variation. It indicates a combination of various groups used to build the engine. It may have a letter preceding it which is sometimes important in determining superseding parts. The first two numbers of the specifications number is the code designating the engine model; the remaining numbers are issued in numerical sequence as each new specification is released, for example, 2899, 28100, 28101, etc. The current specification number model code is as follows:

    K141–29
    K161–28
    K181–30
    K241–46
    K301–47
    K321–60
    K341–71

The serial number lists the order in which the engine was built. If a change takes place to a model or a specification, the serial number is used to indicate the points at which the change takes place. The first letter or number in the serial number indicates what year the engine was built. The letter prefix to the en-

General Engine Specifications

Model	Bore & Stroke (in.)	Displacement	Horsepower
K141 (—29355)	$2^7/_8 \times 2^1/_2$	16.22	6.25
K141 (29356—)	$2^{15}/_{16} \times 2^1/_2$	16.9	6.25
K161 (—281161)	$2^7/_8 \times 2^1/_2$	16.22	6.25
K161 (281162—)	$2^{15}/_{16} \times 2^1/_2$	16.9	6.25
K181	$2^{15}/_{16} \times 2^3/_4$	18.6	8.0
K241	$3^1/_4 \times 2^7/_8$	23.9	10.0
K241A	$3^1/_4 \times 2^7/_8$	23.9	8.0
K301	$3^3/_8 \times 3^1/_4$	29.07	12.0
K301A	$3^3/_8 \times 3^1/_4$	29.07	12.0
K321	$3^1/_2 \times 3^1/_4$	31.27	14.0
K321A	$3^1/_2 \times 3^1/_4$	31.27	14.0
K341	$3^3/_4 \times 3^1/_4$	35.89	16.0
K341A	$3^3/_4 \times 3^1/_4$	35.89	16.0

gine serial number was dropped in 1969 and thereafter the prefix is a number. Engines made in 1969 have either the letter **E** or the number **1**. The code is as follows:

	First Digit Numbers:
• A – 1965	• 1 – 1969
• B – 1966–	• 2 – 1970
• C – 1967	• 3 – 1971
• D – 1968	• 4 – 1972
• E – 1969	• 5 – 1973
	• 6 – 1974

MAINTENANCE

Air Cleaners

A dirty air cleaner can cause rich fuel/air mixture and consequent poor engine operation and sludge deposits. If the filter becomes dirty enough, dirt that otherwise would be trapped can pass through and may wear the engine's moving parts prematurely. It is therefore necessary that all maintenance work be performed precisely as specified.

DRY AIR CLEANERS

Clean dry element air cleaners every 50 hours of operation, or every 6 months (whichever comes first) under good operating conditions. Service more frequently if the operating area is dusty. Remove the element and tap it lightly against a hard surface to remove the bulk of the dirt. If dirt will not drop off easily, replace the element. Do not use compressed air or solvents. Replace the air cleaner every 100–200 hours, under good conditions, and more frequently if the air is dusty.

Observe the following precautions:
1. Handle the element carefully – do not allow the gasket surfaces to become bent or twisted.
2. Make sure gasket surfaces seal against back plate and cover.
3. Tighten wing nut only finger tight – if it is too tight, cleaner may not seal properly.

If the dry type air cleaner is equipped with a precleaner, service this unit when cleaning the paper element. Servicing consists of cleaning the precleaner in soap and water, squeezing the excess out, and then allowing it to air dry before installation. Do not oil!

DRY ELEMENT

PRECLEANER (OPTION)

Exploded view of dry type air cleaner with precleaner

OIL BATH AIR CLEANERS

This type of unit may be used to replace the dry type in applications where very frequent replacement of the element is required. The conversion is simple and requires the use of an elbow to fit the oil bath unit onto the engine in a vertical position.

Service the unit every 25 hours of operation under good conditions and, under dusty conditions, as often as every 8 hours of operation.
Service as follows:
1. Remove cover and lift element out of bowl.
2. Drain dirty oil from bowl, and then wash thoroughly in clean solvent.
3. Swish the element in the solvent and then allow it to drip dry. Do not dry with compressed air. Lightly oil the element with engine oil.

Oil Viscosity Chart

Air Temperature	Oil Viscosity	Oil Type
Above 30° F	SAE 30	API Service SC*
30° to 0° F	SAE 10W-30	API Service SC*
Below 0° F	SAE 5W-20	API Service SC*

* SC standard recommendation—CC (MIL-2104B) and SD class oils may also be used.

4. Inspect air horn, filter bowl, and cover gaskets, and replace as necessary (if grooved or cracked).
5. Install filter bowl gasket on air horn, then put the bowl into position. Fill bowl to indicated level with engine oil.
6. Install element, put the cover in position, and then install copper gasket (if used) and wingnut. Tighten wingnut with fingers only to avoid distorting housing. Make sure all joints in the unit seal tightly.

Lubrication

CRANKCASE

Oil level must be maintained between **F** and **L** marks—do not overfill. Check every day and add as necessary. On new engines, be especially careful to stop engine and check level frequently. When checking, make sure regular type dipstick is inserted fully. On screw type dipstick, check level with dipstick inserted fully but not screwed in. On this type, however, make sure to screw dipstick back in tightly when oil level check is completed. Use SC type oils meeting viscosity specifications according to the prevailing temperature as shown in the chart below.

Change initial fill of oil on new engines after five hours of operation. Then, change oil every 25 hours of operation. Change oil when engine is hot. Change more frequently in dusty areas. If the engine has just been overhauled, it is best to fill it initially with a non-detergent oil. Then, after 5 hours, refill with SC type oil.

Oil capacities are:
- K141, K161, K181 – 1qt.
- K241, K301, K321 – 2 qts.
- On K241A, K301A, K321A and K341A, install 1 qt., then fill to **F** mark on dipstick.
- K331 – 3 qts.

REDUCTION GEAR UNITS

Every 50 hours, remove the oil plug on the lower part of the reduction unit cover to check level. If oil does not reach the level of the oil plug, remove the vented fill plug from the top of the cover and refill with engine oil until level is correct. This oil need not be changed unless unit has been out of service for several months. In this situation, remove the drain plug, drain oil, then replace plug and fill to proper level as described above.

FUEL RECOMMENDATIONS

Use either leaded or unleaded regular grade fuel of at least 90 octane. Unleaded fuel produces fewer combustion chamber deposits, so its use is preferred.

Purchase fuel from a reputable dealer, and make sure to use only fresh fuel (fuel less than 30 days old). If the engine is stored, drain the fuel system or use a fuel stabilizer that is compatible with the type of fuel tank the engine is equipped with.

Tune-Up Specifications

Model	Plug Gap (in.)	Breaker Point Gap (in.)	Trigger Air Gap (in.)	Normal Timing (deg)	Retard Timing (deg)
K141 (small bore)	.025 ①	.020	.005 – .010	20	3B
K141 (large bore)	.025 ①	.020	.005 – .010	20	—
K161 (small bore)	.025 ①	.020	.005 – .010	20	3B
K161 (large bore)	.025 ①	.020	.005 – .010	20	—
K181	.025 ①	.020	.005 – .010	20	3B
K241	.025 ①	.020	.005 – .010	20	3A
K301	.025 ①	.020	.005 – .010	20	3A
K321	.025 ①	.020	.005 – .010	20	—
K341	.025 ①	.020	.005 – .010	20	—

B—Before
A—After
① Shielded plug gap—.020 in.

Spark Plug Specifications

Engine Model	Plug Size	Hex Size	Plug Reach	Standard Plugs		Resistor Plugs	
				Solid Post	Knurled Nut	Non-Shielded	Shielded
K141	14 mm	13/16″	3/8″	J-8 270321-S	J-8 220040-S	XJ-8 232604-S	XEJ-8 220258-S
K161	14 mm	13/16″	3/8″	J-8 270321-S	J-8 220040-S	XJ-8 232604-S	XEJ-8 220258-S
K181	14 mm	13/16″	3/8″	J-8 270321-S	J-8 220040-S	XJ-8 232604-S	XEJ-8 220258-S
K241	14 mm	13/16″	7/16″	H-10 235040-S	Not Available	XH-10 235041-S	XEH-10 235259-S
K301	14 mm	13/16″	7/16″	H-10 235040-S	Not Available	XH-10 235041-S	XEH-10 235259-S
K321	14 mm	13/16″	7/16″	H-10 235040-S	Not Available	XH-10 235041-S	XEH-10 235259-S
K341	14 mm	13/16″	7/16″	H-10 235040-S	Not Available	XH-10 235041-S	XEH-10 235259-S

Gap Setting—Gasoline .025″ (Shielded .020″) Tightening Torque—All plugs 18 to 22 foot lbs.
(Champion plugs listed—use Champion or equivalent plugs.)

TUNE-UP
Spark Plugs
SERVICE

The spark plug should be removed and serviced every 100 hours of engine operation. The plug should have a light coating of light gray colored deposits. If deposits are black, fuel/air mixture could be too rich due to improper carburetor adjustment or a dirty air cleaner. If deposits are white, the engine may be over-heating or a spark plug of too high a heat range could be in use.

Kohler recommends that the plug be replaced rather than sandblasted or scraped if there are excessive deposits. Torque plugs to 18–22 ft. lbs.

TESTING

To test a plug for adequate performance, remove it from the engine, attach the ignition wire, and then rest the side electrode against the cylinder head. Crank the engine vigorously. If there is a sharp spark, the plug and ignition system are allright, although ignition timing should be checked if the engine fires irregularly.

Breaker Points
INSPECTION

Remove the breaker cover and inspect the points for pitting or buildup of metal on either the movable or stationary contact ev-

Adjusting breaker gap

ery 100 hours of operation. Replace the points if they are badly burned. If there is a great deal of metal buildup on either contact, the condenser may be faulty and should be replaced.

To replace points, remove the primary wiring connector screw and pull off the primary wire. Then, remove the contact set mounting screws and remove the contact set. Install the new set of points in reverse order, leaving upper mounting screw slightly loose. Then set point gap and timing as described below.

SETTING BREAKER GAP AND TIMING

1. Remove the breaker cover and disconnect the spark plug lead. Rotate the engine in direction of normal rotation until the points reach the maximum opening.
2. Using a clean, flat feeler gauge of 0.020 in. size, check the gap between the points. Gauge should just slide between the contacts without opening them when flat between them. If the gap is incorrect, loosen the upper mounting screw (if necessary), and shift the breaker base with the blade of the screwdriver until gap is correct.
3. There is a timing sight hole in either the bearing plate or the blower housing. If there is a snap button in the hole, pry it out with a screwdriver.

4. While observing the sight hole, turn the engine slowly in normal direction of rotation. When the **S** or **SP** mark (engines with Automatic Compression Release) or the **T** mark on engines without ACR, appears in the hole, the points should just be beginning to open. If timing is incorrect, breaker gap will have to be reset slightly (0.018–0.022 in.). If the points are not yet opening when the timing mark is centered in the hole, make the point gap wider. If points open too early, narrow it. Recheck the setting after tightening the upper breaker mounting screw by turning the engine in normal direction of rotation past the firing point and checking that the points open at just the right time.

NOTE: This procedure may be performed with the engine running at 1200–1800 rpm if a timing light is available. Connect the timing light according to manufacturer's instructions. You may have to chalk the timing mark to see it adequately.

TRIGGER AIR GAP

Trigger air gap is set within the range 0.005–0.010 in. As long as the gap falls within this range, the ignition system should per-

Adjusting timing

form adequately. Optimum ignition performance during cold weather starting is provided if the gap is adjusted to 0.005 in. If you wish to adjust this or to ensure that the gap falls within the proper range, rotate the flywheel until the flywheel projection is lined up with the trigger assembly. Then, loosen the trigger bracket capscrews and slide the trigger back and forth to get the proper gap, as measured with a flat feeler gauge. Then, retighten capscrews.

IGNITION COILS

Coils do not require regular service, except to make sure they are kept clean, that the connections are tight, and that rubber insulators are in good condition (replace if cracked). If you suspect poor performance of a breakerless type ignition system and trigger air gap is correct, check resistance with an ohmmeter. To do this, disconnect the high tension lead at the coil and connect the meter between coil terminal and coil mounting bracket. If resistance is not about 11,500 ohms, replace the coil. Also, check the reading with the meter lead going to the coil terminal pulled off and connected to the spark plug connector of the high tension lead. If there is continuity here, replace the coil.

PERMANENT MAGNETS

These may be checked for magnet strength by holding a screwdriver (non-magnetic) blade within one inch of the magnet. If the magnetic field is good, the blade will be attracted to the magnet. Otherwise, replace it.

Mixture Adjustments

NOTE: Before making any adjustments, be sure that the carburetor air cleaner is not clogged. A clogged air cleaner will cause an over-rich mixture, black exhaust smoke, and may lead you to believe that the carburetor is out of adjustment when, in reality, it is not. The carburetor is set at the factory and rarely needs adjustment unless, of course, it has been disassembled or rebuilt.

1. With the engine stopped, turn the main and idle fuel adjusting screws all the way in until they bottom lightly. Do not force the screws or you will damage the needles.
2. For a preliminary setting, turn the main fuel screw out 2 full turns and the idle screw out 1¼ turns.
3. Start the engine and allow it to reach operating temperatures; then operate the engine at full throttle and under a load, if possible.
4. For final adjustment, turn the main fuel adjustment screw in until the engine slows down (lean mixture), then out until it

Adjustment screws on the side draft carburetor

Adjustment screws on the updraft carburetor

slows down again (rich mixture). Note the positions of the screw at both settings, then set it about halfway between the two positions.

5. Set the idle mixture adjustment screw in the same manner. The idle speed (no-load) on most engines is 1200 rpm; however, on engines with a parasitic load (hydrastatic drives) the engine idle speed may have to be increased to as much as 1700 rpm for best no-load idle.

Governor Adjustment

All Kohler engines use mechanical, camshaft driven governors.

INITIAL ADJUSTMENT

1. Loosen, but do not remove, the nut that holds the governor arm to the governor cross shaft.
2. Grasp the end of the cross shaft with a pair of pliers and turn it in counterclockwise as far as it will go. The tab on the cross shaft will stop against the rod on the governor gear assembly.
3. Pull the governor arm away from the carburetor, then retighten the nut which holds the governor arm to the shaft. With updraft carburetors, lift the arm as far as possible, then retighten the arm nut.

FINAL ADJUSTMENT

K141–K181

After making the initial adjustment and connecting the throttle wire on the variable speed applications, start the engine and check the maximum operating speed with a tachometer.
If adjustment is necessary:
1. Loosen the bushing nut slightly.
2. Move the throttle bracket in a counterclockwise direction to increase speed, or in a clockwise direction to decrease engine speed. Maximum speed for the K141 and K181 is 3600 rpm.
3. With the speed set to the proper range, tighten the bushing nut to lock the throttle bracket in position.

K241

Engine must be adjusted to 3600 rpm.
1. Start the engine and measure the speed with a tachometer.
2. If the speed is incorrect, adjust as follows:
   a. Constant Speed Governor—Tighten the governor adjusting screw to increase speed, or loosen to decrease speed until the correct speed is attained.

Initial adjustment of the governor installed on the K91, K141, K161, and the K181 engines

b. Variable Speed Governor—Loosen the capscrew, move the high speed stop bracket until the correct speed is attained, and then retighten the capscrew.

If the governor is too sensitive (causing hunting or surging), or not sensitive enough (causing too great a drop in speed when load is applied), the governor sensitivity should be adjusted. Make the governor more sensitive by moving the spring to holes further apart. Make it less sensitive by moving it to holes that are closer together. Standard setting is the third hole from the bottom on the governor arm and second hole from the top on the speed control bracket.

Choke Adjustment

THERMOSTATIC TYPE

If the engine does not start when cranked, continue cranking and move the choke lever first to one side and then to the other to determine whether the setting is too lean or too rich. Once the direction in which lever must be moved has been determined, loosen the adjusting screw on the choke body. Then, move the bracket downward to increase choking or upward to decrease it. Then, tighten the lockscrew. Try starting it again and readjust as necessary.

ELECTRIC-THERMOSTATIC TYPE

Remove the air cleaner from the carburetor and check the position of the choke plate. The choke should be fully closed when engine is at outside temperature and the temperature is very low. In milder temperatures, slightly less closure is required.

If adjustment is required, move the choke arm until the hole in the brass shaft lines up with the slot in the bearings. Insert a #43 (0.089 in.) drill through the shaft and push it downward so it engages the notch in the base of the choke unit. Then, loosen the clamp bolt on the choke lever and push the arm upward to move the choke plate toward the closed position. When the desired position is obtained, tighten the clamp bolt. Then, remove the drill.

Remount the air cleaner, and then check for any binding in the choke linkage. Correct as necessary. Finally, run the engine until hot, and make sure the choke opens fully. If not, readjust it toward the open position as necessary.

Adjusting electric-thermostatic choke

Valve Adjustment

On K241, K301, K321 and K341 engines, adjustable valve tappets are provided. With the engine cold, turn crankshaft until it reaches Top Dead Center timing mark. If valves are slightly open, turn the crankshaft another turn until valves are closed and engine is again at Top Center. Check valve clearances with a flat feeler gauge. Note that exhaust and intake clearances are different, and make sure you're using the right gauge for each valve. If the valve clearance is correct, a gauge can just be inserted between tappet and valve stem. A slight pull is required to bring it back out. If clearance is incorrect, loosen the locking nut and turn the adjusting nut in or out to get the proper clearance. Hold the adjusting nut while tightening the locknut and recheck clearance.

Compression Check

Compression is checked by removing the spark plug lead and spinning the flywheel forward against compression. If the piston does not bounce backward with considerable force, checking with a gauge may be necessary. On Automatic Compression Release engines, rotate the flywheel backward against power stroke-if little resistance is felt, check compression with a gauge.

The compression gauge check requires rapid motoring (spinning) of the crankshaft, at about 1000 rpm. Install the gauge in the spark plug hole and motor the engine. Gauge should read 110–120 psi. If reading is less than 100 psi, the engine requires major repair to piston rings or valves.

FUEL SYSTEM

Carburetor

If a carburetor will not respond to mixture screw adjustments, then you can assume that there are dirt, gum, or varnish deposits in the carburetor or worn/damaged parts. To remedy these problems, the carburetor will have to be completely disassembled, cleaned, and worn parts replaced and reassembled.

Exploded view of the updraft carburetor

Exploded view of the sidedraft carburetor. Inset shows float adjustment procedure

Parts should be cleaned with solvent to remove all deposits. Replace worn parts and use all new gaskets. Carburetor rebuilding kits are available.

DISASSEMBLY

Side Draft Carburetors

1. Remove the carburetor from the engine.
2. Remove the bowl nut, gasket, and bowl. If the carburetor has a bowl drain, remove the drain spring, spacer and plug, and gasket from inside the bowl.
3. Remove the float pin, float, needle, and needle seat. Check the float for dents, leaks, and wear on the float lip or in the float pin holes.
4. Remove the bowl ring gasket.
5. Remove the idle fuel adjusting needle, main fuel adjusting needle, and springs.
6. Do not remove the choke and throttle plates or shafts. If these parts are worn, replace the entire carburetor assembly.

Updraft Carburetors

1. Remove the carburetor from the engine.
2. Remove the bowl cover and the gasket.
3. Remove the float pin, float, needle and needle seat. Check the float pin for wear.
4. Remove the idle fuel adjustment needle, main fuel adjustment needle, and the springs. Do not remove the choke plate or the shaft unless the replacement of these parts is necessary.

ASSEMBLY

Side Draft Carburetor

1. Install the needle seat, needle, float, and float pin.
2. Set the float level. With the carburetor casting inverted and the float resting against the needle in its seat, there should be $^{11}/_{64} \pm ^1/_{32}$ in. clearance between the machined surface of the casting and the free end of the float.
3. Adjust the float level by bending the lip of the float with a small screwdriver.
4. Install the new bowl ring gasket, new bowl nut gasket, and bowl nut. Tighten the nut securely.
5. Install the main fuel adjustment needle. Turn it in until the needle seats in the nozzle and then back out two turns.
6. Install the idle fuel adjustment needle. Back it out about $1^1/_4$ turns after seating it lightly against the jet.
7. Install the carburetor on the engine.

Updraft Carburetor

1. Install the throttle shaft and plate. The elongated side of the valve must be toward the top.
2. Install the needle seat. A $^5/_{16}$ in. socket should be used. Do not over-tighten.
3. Install the needle, float, and float pins.
4. Set the float level. With the bowl cover casting inverted and the float resting lightly against the needle in its seat, there should be $^7/_{16}$ in. plus or minus $^1/_{32}$ in. clearance between the machined surface casting and the free end of the float.
5. Adjust the float level by bending the lip of the float with a small screwdriver.
6. Install the new carburetor bowl gasket, bowl cover, and bowl cover screws. Tighten the screws securely.
7. Install the main fuel adjustment needle. Turn it in until the screw seats in the nozzle and then back it out 2 turns.
8. Install the idle fuel adjustment needle. Back it out about $1^1/_2$ turns after seating the screw lightly against the jet. Install the idle speed screw and spring. Adjust the idle to the desired speed with the engine running.
9. Install the carburetor on the engine.

Fuel Pump

Fuel pumps used on single cylinder Kohler engines are either the mechanical or vacuum actuated type. The mechanical type is operated by an eccentric on the camshaft and the vacuum type is operated by the pulsating negative pressures in the crankcase. The K91 vacuum type pump is not serviceable and must be replaced when faulty. The mechanical pump is serviceable and rebuilding kits are available.

1. Disconnect fuel lines, remove mounting screws, and pull the pump off the engine.
2. File a mark across some point at the union of pump body and cover. Remove the screws and remove the cover.
3. Turn the cover upside down and remove the valve plate screw and washer. Remove the valve retainer, valves, valve springs, and valve gasket, after noting the position of each part. Discard the valve springs, valves and valve retainer gasket.
4. Clean the fuel head with solvent and a soft wire brush. Hold the pump cover with the diaphragm surface upward; position a new gasket into the cavity. Put the valve spring and valves into position in the cavity and reassemble the valve retainer. Lock the retainer into position by installing the fuel pump valve retainer screw.
5. Rebuild the lower diaphragm section.
6. Hold the mounting bracket and press down on the diaphragm to compress the spring underneath. Turn the bracket 90 degrees to unhook the diaphragm and remove it.
7. Clean the mounting bracket with solvent and a wire brush.
8. Stand a new diaphragm spring in the casting, put the diaphragm into position, and push downward to compress the spring. Turn the diaphragm 90 degrees to reconnect it.
9. Position the pump cover on top of the mounting bracket with the indicating marks lined up. Install the screws loosely on mechanical pumps; on vacuum pumps, tighten the screws.

Exploded view of the fuel pump

10. Holding only the mounting bracket, push the pump lever to the limit of its travel, hold it there, and then tighten the four screws.

11. Remount the fuel pump on the engine with a new gasket, tighten the mounting bolts, and reconnect the fuel lines.

ENGINE OVERHAUL
Disassembly

The following procedure is designed to be a general guide rather than a specific and all inclusive disassembly procedure. The sequence may have to be varied slightly to allow for the removal of special equipment or accessory items such as motor/generators, starters, instrument panels, etc.

1. Disconnect the high tension spark plug lead and remove the spark plug.
2. Close the valve on the fuel sediment bowl and remove the fuel line at the carburetor.
3. Remove the air cleaner from the carburetor intake.
4. Remove the carburetor.
5. Remove the fuel tank. The sediment bowl and brackets remain attached to the fuel tank.
6. Remove the blower housing, cylinder baffle, and head baffle.
7. Remove the rotating screen and the starter pulley.
8. The flywheel is mounted on the tapered portion of the crankcase and is removed with the help of a puller. Do not strike the flywheel with any type of hammer.
9. Remove the breaker point cover, breaker point lead, breaker assembly, and the push-rod that operates the points.
10. Remove the magneto assembly.
11. Remove the valve cover and breather assembly.
12. Remove the cylinder head.
13. Raise the valve springs with a valve spring compressor and remove the valve spring keepers from the valve stems. Remove the valve spring retainers, springs, and valves.
14. Remove the oil pan base and unscrew the connecting rod can screws. Remove the connecting rod cap and piston assembly from the cylinder block.

NOTE: It will probably be necessary to use a ridge reamer on the cylinder walls before removing the piston assembly, to avoid breaking the piston rings.

15. Remove the crankshaft, oil seals and, if necessary, the anti-friction bearings.

NOTE: It may be necessary to press the crankshaft out of the cylinder block. The bearing plate should be removed first, if this is the case.

16. Turn the cylinder block upside down and drive the camshaft pin out from the power take-off side of the engine with a small punch. The pin will slide out easily once it is driven free of the cylinder block.
17. Remove the camshaft and the valve tappets.
18. Loosen and remove the governor arm from the governor shaft.
19. Unscrew the governor bushing nut and remove the governor shaft from the inside of the cylinder block.
20. Loosen, but do not remove, the screw located at the lower right of the governor bushing nut until the governor gear is free to slide off of the stub shaft.

Engine Rebuilding
CYLINDER BLOCK SERVICE

Make sure that all surfaces are free of gasket fragments and sealer materials. The crankshaft bearings are not to be removed unless replacement is necessary. One bearing is pressed into the cylinder block and the other is located in the bearing plate. If there is no evidence of scoring or grooving and the bearings turn easily and quietly it is not necessary to replace them.

The cylinder bore must not be worn, tapered, or out-of-round more than 0.005 in. Check at two locations 90 degrees apart and compare with specifications. If it is, the cylinder must be rebored. If the cylinder is very badly scored or damaged it may have to be replaced, since the cylinder can only be rebored to either 0.010 in. or 0.020 in. and 0.030 in. maximum. Select the nearest suitable oversize and bore it to that dimension. On the other hand, if the cylinder bore is only slightly damaged, only a light deglazing may be necessary.

HONING THE CYLINDER BORE

1. The hone must be centered in relation to the crankshaft crossbore. It is best to use a low speed drill press. Lubricate the hone with kerosene and lower it into the bore. Adjust the stones so they contact the cylinder walls.
2. Position the lower edge of the stones even with the lower edge of the bore, hone at about 600 rpm. Move the hone up and down continuously. Check bore size frequently.

Left side—measuring the cylinder bore. Right side—honing must produce a cross-hatch pattern as shown

3. When the bore reaches a dimension 0.0025 in. smaller than desired size, replace the coarse stones with burnishing stones. Use burnishing stones until the dimension is within 0.0005 in. of desired size.

4. Use finishing stones and polish the bore to final size, moving the stones up and down to get a 60 degree cross-hatch pattern. Wash the cylinder wall thoroughly with soap and water, dry, and apply a light coating of oil.

CRANKSHAFT SERVICE

Inspect the keyway and the gears that drive the camshaft. If the keyways are badly worn or chipped, the crankshaft should be replaced. If the cam gear teeth are excessively worn or if any are broken, the crankshaft must be replaced.

Check the crankpin for score marks or metal pickup. Slight score marks can be removed with a crocus cloth soaked in oil. If the crankpin is worn more than 0.002 in., the crankshaft is to be either replaced or the crankpin reground to 0.010 in. undersize. If the crankpin is reground to 0.010 in. undersize, a 0.010 in. undersize connecting rod must be used to achieve proper running clearance.

CONNECTING ROD SERVICE

Check the bearing area for wear, score marks, and excessive running and side clearance. Replace the rod and cap if they are worn beyond the limits allowed.

PISTON AND RINGS SERVICE

Production and Service Type

Rings are available in the standard size as well as 0.010 in., 0.020 in., and 0.030 in. oversize sets.

NOTE: Never reuse old rings.

The standard size rings are to be used when the cylinder is not worn or out-of-round. Oversize rings are only to be used when the cylinder has been rebored to the corresponding oversize. Service type rings are used only when the cylinder is worn but within the wear and out-of-round limitations; wear limit is 0.005 in. oversize and out-of-round limit is 0.004 in.

NOTE: Never reuse old rings.

The standard size rings are to be used when the cylinder is not worn or out-of-round. Oversize rings are only to be used when the cylinder has been rebored to the corresponding oversize. Service type rings are used only when the cylinder is worn but within the wear and out-of-round limitations; wear limit is 0.005 in. oversize and out-of-round limit is 0.004 in.

The old piston may be reused if the block does not need reboring and the piston is within wear limits. Never reuse old rings. After removing old rings, thoroughly remove deposits from ring grooves. New rings must each be positioned in its running area of the cylinder bore for an end clearance check, and each must meet specifications.

The cylinder must be deglazed before replacing the rings. If chrome plated rings are used, the chrome plated ring must be installed in the top groove. Make sure that the ring grooves are free from all carbon deposits. Use a ring expander to install the rings. Then check side clearance.

PISTON AND ROD SERVICE

Normally very little wear will take place at the piston boss and piston pin. If the original piston and connecting rod can be used after rebuilding, the piston pin may also be used. However if a new piston or connecting rod or both have to be used, a new piston pin must also be installed. Lubricate the pin before installing it with a loose to light interference fit. Use new piston pin retainers whether or not the pin is new. Make sure they're properly engaged.

VALVES AND VALVE MECHANISM SERVICE

Inspect the valve mechanism, valves, and valve seats or inserts for evidence of wear, deep pitting, cracks or distortion. Check the clearance between the valve stems and the valve guides.

Valve guides must be replaced if they are worn beyond the limit allowed. K91 model engines do not use valve guides. To remove valve guides, press the guide down into the valve chamber

Positioning of production and service type piston rings on the piston

and carefully break off the protruding end until the guide is completely removed. Be careful not to damage the block when removing the old guides. Use an arbor press to install the new guides. Press the new guides to the depth specified, then use a valve guide reamer to gain the proper inside diameter.

Make sure that replacement valves are the correct type (special hard faced valves are needed in some cases). Exhaust valves are always hard faced.

Intake valve seats are usually machined into the block, although inserts are used in some engines. Exhaust valve seats are made of special hardened material. The seating surfaces should be held as close to $\frac{1}{32}$ in. in width as possible. Seats more than $\frac{1}{16}$ in. wide must be reground with 45 degree and 15 degree cutters to obtain the proper width. Reground or new valves and seats must be lapped in for a proper fit.

After resurfacing valves and seats and lapping them in, check the valve clearance. Hold the valve down on its seat and rotate the camshaft until it has no effect on the tappet, then check the clearance between the end of the valve stem and the tappet. If the clearance is not sufficient (it will always be less after grinding), it will be necessary to grind off the end of the valve stem until the correct clearance is obtained. This is necessary on all engines except the K141, K301, K321 and K341 engines which all have adjustable tappets.

CYLINDER HEAD SERVICE

Remove all carbon deposits and check for pitting from hot spots. Replace the head if metal has been burned away because of head gasket leakage. Check the cylinder head for flatness. If the head is slightly warped, it can be resurfaced by rubbing it on a piece of sandpaper placed on a flat surface. Be careful not to nick or scratch the head when removing carbon deposits.

DYNAMIC BALANCE SYSTEM SERVICE

The dynamic balance system consists of two balance gears which run on needle bearings. The gears are assembled on two stub shafts that are pressed into special bosses in the crankcase. Snap-rings hold the gears and spacer washers are used to control end-play. The gears are driven off of the crankgear. The dynamic balance system is found on special versions of K241 and K301 models and is standard equipment on K321 engines.

If the stub shaft is worn or damaged, press the old shaft out. The new shafts must be pressed in a specified distance which depends upon the distance between the stub shaft boss and main bearing boss. Measure the distance the stub shaft boss protrudes above the main bearing boss and then press the shaft in for a protrusion of the shaft end beyond stub shaft boss as specified. If stub shaft boss protrudes about $\frac{7}{16}$ in. beyond main bearing boss, press the shaft in until it is 0.735 in. above stub shaft boss. If protrusion is about $\frac{1}{16}$ in., press the stub shaft in until it is 1.110 in. above the stub shaft boss, and then use a $\frac{3}{8}$ in. spacer.

When installing the balance gears, slip one 0.010 in. spacer onto the stub shaft, then install the gear/bearing assembly onto the stub shaft with the timing marks facing out. Proper end-play of 0.002–0.010 in. is attained with one 0.005 in. spacer, one 0.010 in. spacer, and one 0.020 in. spacer which are all installed on the snap-ring retainer end of the shaft. Install the thickest spacer next to the retainer. Check the end-play and adjust it by adding or subtracting 0.005 in. spacers.

To time the balance gears, first press the crankshaft into the block and align the primary timing mark on the top of the balance gear with the standard timing mark next to the crankgear. Press the shaft in until the crankgear is engaged $\frac{1}{16}$ in. into the top gear (narrow side). Rotate the crankshaft to align the timing marks on the crankgear and camgear. Press the crankshaft the remainder of the way into the block.

Timing marks for the dynamic balance system

Rotate the crankshaft until it is about 15 degrees past BDC and slip one 0.010 in. spacer over the stub shaft before installing the bottom gear/bearing assembly.

Align the secondary timing mark on this gear with the secondary timing mark on the counterweight of the crankshaft and then install the gear on the shaft. The secondary timing mark will also be aligned with the standard timing mark on the crankshaft after installation. Use one 0.005 in. spacer and one 0.020 in. spacer (with larger spacer next to retainer) to get proper end play of 0.002–0.010 in. Install the snap-ring retainer, then check and adjust the endplay.

Engine Assembly

REAR MAIN BEARING

Install the rear main bearing by pressing it into the cylinder block with the shielded side toward the inside of the block. If it does not have a shielded side, then either side may face inside.

GOVERNOR SHAFT

1. Place the cylinder block on its side and slide the governor shaft into place from the inside of the block. Place the speed control disc on the governor bushing nut and thread the nut into the block, clamping the throttle bracket into place.
2. There should be a slight end-play in the governor shaft and that can be adjusted by moving the needle bearing in the block.
3. Place a space washer on the stub shaft and slide the governor gear assembly into place.
4. Tighten the holding screw from outside the cylinder block.
5. Rotate the governor gear assembly to be sure that the holding screw does not contact the weight section of the gear.

CAMSHAFT

1. Turn the cylinder block upside down.
2. The tappets must be installed before the camshaft is installed. Lubricate and install the tappets into the valve guides.
3. Position the camshaft inside the block.

NOTE: Align the marks on the camshaft and the automatic spark advance, if so equipped.

Installing the camshaft

4. Lubricate the rod and insert it into the bearing plate side of the block. Install one 0.005 in. washer between the end of the camshaft and the block. Push the rod through the camshaft and tap it lightly until the rod just starts to enter the bore at the PTO end of the block. Check the endplay and adjust it with additional washers if necessary. Press the rod into its final position.
5. The fit at the bearing plate for the camshaft rod is a light to loose fit to allow oil that might leak past to drain back into the block.

CRANKSHAFT

1. Place the block on the base of an arbor press and carefully insert the tapered end of the crankshaft through the inner race of the anti-friction bearing, or sleeve bearing on the K141.
2. Turn the crankshaft and camshaft until the timing mark in the shoulder of the crankshaft lines up with the mark on the cam gear.

Timing marks for the automatic spark advance

3. When the marks are aligned, press the crankshaft into the bearing, making sure that the gears mesh as it is being pressed in. Recheck the alignment of the timing marks on the crankshaft and the camshaft.
4. The end-play of the crankshaft is controlled by the application of various thickness gaskets between the bearing plate and the block. Normal end-play is achieved by installing 0.020 in. and 0.010 in. gaskets, with the thicker gaskets on the inside.

BEARING PLATE

1. Press the front main bearing into the bearing plate. Make sure that the bearing is straight.
2. Press the bearing plate onto the crankshaft and into position on the block. Install the cap screws and secure the plate to the block. Draw up evenly on the screws.
3. Measure the crankshaft end-play, which is very critical on gear reduction engines.

PISTON AND ROD ASSEMBLY

1. Lubricate the pin and assemble it to the connecting rod and piston. Install the wrist pin retaining ring. Use new retaining rings.
2. Lubricate the entire assembly, stagger the ring gaps and, using a ring compressor, slide the piston and rod assembly into

Alignment of the timing marks for the crankshaft and the camshaft

the cylinder bore with the connecting rod marks on the flywheel side of the engine.

3. Place the block on its end and oil the connecting rod end and the crankpin.

4. Attach the rod cap, lock or lock washers, and the cap screws. Tighten the screws to the correct torque.

NOTE: Align the marks on the cap and the connecting rod.

5. Bend the lock tabs to lock the screws.

CRANKSHAFT OIL SEALS

Apply a coat of grease to the lip and guide the oil seals onto the crankshaft. Make sure no foreign material gets onto the knife edges of seal, and make sure the seal does not bend. Place the block on its side and drive the seals squarely into the bearing plate and block.

OIL PAN BASE

Using a new gasket on the base, install pilot studs to align the cylinder block, gasket, and base. Tighten the four attaching screws to the correct torque.

VALVES

1. See the engine rebuilding section of the "Clinton" chapter for details concerning installation of the seats and guides. Clean the valves, seats, and parts thoroughly. Grind and lap-in the valves and seats for proper seating. Valve seat width must be $\frac{1}{32}$–$\frac{1}{16}$ in. After grinding and lapping, slide the valves into position and check the clearance between stem and tappet. If the clearance is too small, grind the stem ends square and remove all burrs. On engines with adjustable valves, make the adjustment at this time.

2. Place the valve springs, retainers, and rotators under the valve guides. Lubricate the valve stems, and then install the valves down through the guides, compress the springs, and place the locking keys or pins in the grooves of the valve stems.

CYLINDER HEAD

1. Use a new cylinder head gasket.
2. Lubricate and tighten the head bolts evenly, and in sequence, to the proper torque.
3. Install the spark plug.

INSTALL WITH MARKS TOWARD BEARING PLATE

Connecting rod and cap alignment marks

BREATHER ASSEMBLY

Assemble the breather assembly, making sure that all parts are clean and the cover is securely tightened to prevent oil leakage.

MAGNETO

On flywheel magneto systems, the coil-core assembly is secured onto the bearing plate. On magneto-alternator systems, the coil is part of the stator assembly, which is secured to the bearing plate. On rotor type magneto systems, the rotor has a keyway and is press fitted onto the crankshaft. The magnet rotor is marked "engine-side" for proper assembly. Run all leads through the hole provided at the 11 o'clock position on the bearing plate.

FLYWHEEL

1. Place the washer in place on the crankshaft and place the flywheel in position. Install the key.

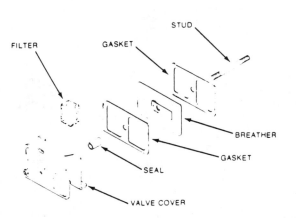

FILTER
STUD
GASKET
BREATHER
GASKET
SEAL
VALVE COVER

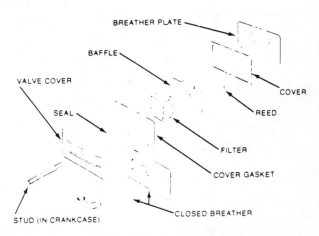

BREATHER PLATE
BAFFLE
VALVE COVER
SEAL
COVER
REED
FILTER
COVER GASKET
STUD (IN CRANKCASE)
CLOSED BREATHER

Exploded views of two common types of breathers showing assembly sequence

Cylinder head torquing sequences

Key to valve specifications chart

Valve Specifications

Dimension	Model K141, K161, K181		K241, K301, K321, K341	
	Intake	Exhaust	Intake	Exhaust
A Seat Angle	89°	89°	89°	89°
B Seat Width	.037/.045	.037/.045	.037/.045	.037/.045
C Insert OD	—	1.2535/1.2545	—	1.2535/1.2545
D Guide Depth	1.312	1.312	1.586	1.497
E Guide ID	.312/.313	.312/.313	.312/.313	.312/.313
F Valve Head Diameter	$1\frac{3}{8}$	$1\frac{1}{8}$	1.370/1.380	1.120/1.130 *
G Valve Face Angle	45°	45°	45°	45°
H Valve Stem Diameter	.3105/.3110	.3090/.3095	.3105/.3110	.3084/.3091

*2.125" on all K341 and K321 engines with spec suffix "D" and later.

2. Install the starter pulley, lock washer, and retaining nut. Tighten the retaining nut to the specified torque.

BREAKER POINTS

1. Install the pushrod.
2. Position the breaker points and fasten them with the two screws.
3. Place the cover gasket into position and attach the magneto lead.
4. Set the gap and install the cover.

CARBURETOR

Insert a new gasket and assemble the carburetor to the intake port with the two attaching screws.

GOVERNOR ARM AND LINKAGE

1. Insert the carburetor linkage in the throttle arm.
2. Connect the governor arm to the carburetor linkage and slide the governor arm into the governor shaft.

3. Position the governor spring in the speed control disc on the K141, K161, and K181.
4. Before tightening the clamp bolt, turn the shaft counterclockwise with pliers as far as it will go; pull the arm as far as it will go to the left (away from the carburetor), tighten the nut, and check for freedom of movement. Adjust the governor.

BLOWER HOUSING AND FUEL TANK

Install the head baffle, cylinder baffle, and the blower housing, in that order. The smaller cap screws are used on the bottom of the crankcase. Install the fuel tank and connect the fuel line.

RUN-IN PROCEDURE

1. Fill the crankcase with a non-detergent oil and run it under load for 5 hours to break it in.
2. Drain oil and refill crankcase with the recommended detergent type oil. Non-detergent oil must not be used except for break-in.

Engine Rebuilding Specifications

Specifications	K141 2⁷/₈ Bore	K141 2¹⁵/₁₆ Bore	K161 2⁷/₈ Bore	K161 2¹⁵/₁₆ Bore	K181	K241	K301	K321	K341
Displacement									
Cubic Inches	16.22	16.9	16.22	16.9	18.6	23.9	29.07	31.27	35.89
Cubic Centimeter	265.8	276.99	265.8	276.99	304.8	391.65	476.37	528.46	588.24
Horsepower (Max RPM)	6.25	6.25	7.0	7.0	8.0	10.0	12.0	14.0	16.0
Cylinder Bore									
New Diameter	2.875	2.9375	2.875	2.9375	2.9375	3.251	3.375	3.500	3.750
Maximum Wear Diameter	2.878	2.9405	2.878	2.9405	2.9405	3.2545	3.3785	3.503	3.753
Maximum Taper	.0025	.0025	.0025	.0025	.0025	.0015	.0015	.0015	.0015
Maximum out of Round	.005	.005	.005	.005	.005	.005	.005	.005	.005
Crankshaft									
End Play (Free)	.002 .023	.002 .023	.002 .023	.002 .023	.002 .023	.003 .020	.003 .020	.003 .020	.003 .020
Crankpin									
New Diameter	1.186	1.186	1.186	1.186	1.186	1.500	1.500	1.500	1.500
Maximum Out of Round	.0005	.0005	.0005	.0005	.0005	.0005	.0005	.0005	.0005
Maximum Taper	.001	.001	.001	.001	.001	.001	.001	.001	.001
Camshaft									
Run Clearance on Pin	.0005 .003	.0005 .003	.0005 .003	.0005 .003	.0005 .003	.001 .0035	.001 .0035	.001 .0035	.001 .0035
End Play	.005 .010	.005 .010	.005 .010	.005 .010	.005 .010	.005 .010	.005 .010	.005 .010	.005 .010
Connecting Rod									
Big End Maximum Diameter	1.1885	1.1885	1.1885	1.1885	1.1885	1.5025	1.5025	1.5025	1.5025
Rod-Crankpin Max Clear	.0035	.0035	.0035	.0035	.0035	.0035	.0035	.0035	.0035
Small (Pin) End-New Dia	.62565	.62565	.62565	.62565	.62565	.85975	.87585	.87585	.87585
Rod to Pin Clearance	.0006 .0011	.0006 .0011	.0006 .0011	.0006 .0011	.0006 .0011	.0003 .0008	.0003 .0008	.0003 .0008	.0003 .0008
Piston									
Thrust Face-Max Wear Dia*	2.866	2.9305	2.866	2.9305	2.9305	3.2445	3.3625	3.4945	3.7425
Thrust Face*-Bore Clear	.006 .0075	.006 .008	.006 .0075	.006 .008	.006 .008	.0075 .0085	.0065 .0095	.007 .010	.007 .010
Ring-Max Side Clearance	.006	.006	.006	.006	.006	.006	.006	.006	.006
Ring-End Gap in New Bore	.007 .017	.007 .017	.007 .017	.007 .017	.007 .017	.010 .020	.010 .020	.010 .020	.010 .020
Ring-End Gap in Used Bore	.027	.027	.027	.027	.027	.027	.030	.030	.030
Valve-Intake									
Valve-Tappet Cold Clear	.006 .008	.006 .008	.006 .008	.006 .008	.006 .008	.008 .010	.008 .010	.008 .010	.008 .010
Valve Lift (Zero Lash)	.2778	.2778	.2778	.2778	.2778	.324	.324	.324	.324
Stem to Guide Max Wear Clear	.0045	.0045	.0045	.0045	.0045	.0045	.0045	.0045	.0045
Valve-Exhaust									
Valve-Tappet Cold Clear	.015 .017	.015 .017	.015 .017	.015 .017	.015 .017	.017 .020	.017 .020	.017 .020	.017 .020
Valve Lift (Zero Lash)	.2542	.2542	.2542	.2542	.2542	.324	.324	.324	.324
Stem to Guide Max Wear Clear	.006	.006	.006	.006	.006	.0065**	.0065**	.0065**	.0065**
Tappet									
Clearance in Guide	.0005 .002	.0005 .002	.0005 .002	.0005 .002	.0005 .002	.0008 .0023	.0008 .0023	.0008 .0023	.0008 .0023

Engine Rebuilding Specifications

Specifications	K141 2⁷/₈ Bore	K141 2¹⁵/₁₆ Bore	K161 2⁷/₈ Bore	K161 2¹⁵/₁₆ Bore	K181	K241	K301	K321	K341
Ignition									
Spark Plug Gap-Gasoline	025	025	025	025	025	025	025	025	025
Spark Plug Gap-Gas	018	018	018	018	018	018	018	018	018
Spark Plug Gap (Shielded)	020	020	020	020	020	020	020	020	020
Breaker Point Gap	020	020	020	020	020	020	020	020	020
Trigger Air Gap (Breakerless)									
Spark Run ² BTDC	005 010	005 010	005 010	005 010	005 010	005 010	005 010	005 010	005 010
	20°	20°	20°	20°	20°	20°	20°	20°	20°
Spark Retard	3° BTDC*** (ACR-NONE)	ACR ONLY (No Retard)	3° BTDC*** (ACR-NONE)	ACR ONLY (No Retard)	3° BTDC*** (ACR-NONE)	3° ATDC*** (ACR-NONE)	3° ATDC*** (ACR-NONE)	ACR ONLY (No Retard)	ACR ONLY (No Retard)
Torque Values (Also See Page 15.3)									
Spark Plug (ft. lbs.)	18–22	18–22	18–22	18–22	18–22	18–22	18–22	18–22	18–22
Cylinder Head	15–20 ft. lbs.	15–20 ft. lbs.	15–20 ft. lbs.	15–20 ft. lbs.	15–20 ft. lbs.	25–30 ft. lbs.	25–30 ft. lbs.	25–30 ft. lbs.	25–30 ft. lbs.
Connecting Rod	200 in. lbs.	200 in. lbs.	200 in. lbs.	200 in. lbs.	200 in. lbs.	300 in. lbs.	300 in. lbs.	300 in. lbs.	300 in. lbs.
Flywheel Nut	50–60 ft. lbs.	50–60 ft. lbs.	50–60 ft. lbs.	50–60 ft. lbs.	50–60 ft. lbs.	60–70 ft. lbs.	60–70 ft. lbs.	60–70 ft. lbs.	60–70 ft. lbs.

*Measured just below oil ring and at right angles to piston pin

**Measured at top of guide with valve closed

***Engines built before automatic compression release (ACR)

Section 13

Onan

Small Engines—0-6 hp

This section pertains to the AJ series of industrial engines and the engines installed on AK series electric generating plants. Although the engines are not the same size in displacement, they are mechanically very similar. Please note that the procedures given here pertain directly to the AJ series industrial engines. They can be applied to the AK generating plant engines with a few very minor changes. Carburetor specifications for both engines are the same.

Setting the governor on the electrical generating plants is more critical than on the non-generator engines because the engine speed determines the power output of the generator.

At the end of the chapter, specifications for size, clearances, and torque for both engine series are given.

ENGINE IDENTIFICATION

All Onan engines have an identification plate attached to the left side of the cooling shroud (facing the flywheel) on one cylinder engines, and on the right side of the cooling air duct (facing the flywheel) on four cylinder engines. The Model and Specification number shown there may be interpreted as follows:

1. Factory code for general engine identification.
2. Engine specific type:
● S—Manual starting with stub shaft power take-off.
● MS—Electric starting with stub shaft, starter, and generator.
● MSV—Vacu-Flo cooling. Similar to MS, but with cooling air drawn in through a front end duct.
3. Factory code for optional equipment on the engine.
4. Specification letter. Letter advances alphabetically with production modifications.

Serial numbers consist of a two digit month of manufacture indicator and a six digit engine number.

General Engine Specifications

Model	Bore & Stroke (in.)	Displacement (cu. in.)	Horsepower @ RPM
AK	2.5 × 2.5	12.3	3.7 @ 3600
AJ	2.75 × 2.5	14.9	5.5 @ 3600
LK	3.25 × 3.00	24.9	5.0 @ 1800

Onan engine identification plate

Model and Specification Number breakdown

MAINTENANCE
Air Cleaner Service
OIL BATH AIR CLEANER

The oil bath air cleaner uses engine oil to collect dirt. Dirt is deposited in a sump full of oil in the bottom of the unit, and on a screen contained in the upper portion. Dirt sticks to the screen because the incoming air pulls some of the oil out of the sump, deposits it on the screen, and this wetting action keeps the screen sticky. After each 100 hours, disassemble the unit, drain the oil out of the sump, clean the dirt out of the sump with a solvent soaked rag, and refill the sump with clean oil of the type used in the engine. If the air around the engine is very dusty, inspect the air cleaner more frequently. Make sure to change the oil in the sump before dirt contained in the sump reaches the ring which is slightly above the bottom. In extreme cases, it may be necessary to clean the screen with solvent and coat it with clean oil as part of the service.

Oil bath type air cleaner

Oil viscosity chart

MOISTENED FOAM AIR CLEANER

This cleaning medium in this type of cleaner is an oil soaked sponge mounted around a screen. After each 100 hours, wash the sponge in soap and water or solvent, dry it, soak it in engine oil, and then squeeze excess oil out of it before installation.

DRY PAPER AIR CLEANER

Dry paper air cleaners use a replaceable paper element. All that

Dry paper air cleaner with polyurethane pre-cleaner

is required in servicing is disassembly of the unit, replacement of the element, and a brief removal of dust from inside the unit with a rag. Some of these units also employ a polyurethane precleaning element, which is also replaced at time of service, or a dust cup, which is cleaned with a solvent soaked rag and dried. Service standard units every 100 hours; units with precleaner at 500 hours.

Lubrication
OIL AND FUEL RECOMMENDATIONS

Onan gasoline engines may use either regular grade leaded, unleaded, or low lead gasoline. The use of unleaded or low lead fuels is strongly recommended in order to reduce combustion chamber deposits, especially on governor controlled engines which run at a fairly steady load. If you are switching from leaded to unleaded fuel, combustion chamber deposits should be removed before beginning operation on the unleaded fuel because of its lower octane rating and consequent potential for damage due to engine knock.

It is best to use an SE oil, although oils not so well equipped to resist sludge formation at both high and low temperatures may be used in very moderate service—i.e. long operating periods under light loads. Consult the chart for recommended viscosity ranges for the prevailing temperature.

Crankcase Capacity Chart

AK	1.25
AJ	1.75*
LK	2.0

* Portable Model—1.25

TUNE-UP
Spark Plugs

Spark plug should be replaced every 200 operating hours. Set the gap with a wire gauge to 0.025 in. It is best to torque the

13-2

Tune-Up Specifications

Model	Spark Plug Gap (in.)	Ignition Point Gap (in.)	Ignition Timing (Deg. BTDC @ RPM)
AJ	.025	.022	19 @ 1800, 25 @ 3600
AK	.025	.022	19 @ 1800, 25 @ 3600
LK	.025	.020	19

plugs with a torque wrench to avoid cylinder head damage, distorted gap dimension, or leakage. Use new gaskets and torque to 25–30 ft. lbs. Plugs should also be inspected and cleaned with a wire brush once or twice, depending on type of fuel and service, during their service lift. Use new plug gaskets even when reinstalling old plugs.

Breaker Points

Breaker points must be inspected and cleaned up and gapped or replaced as required. Gain access to the points as described below under ''Ignition Timing.'' Inspect the points for burning and pitting. If only slightly worn or damaged, points may be dressed with an abrasive stone and regapped. Otherwise, they must be replaced.

To replace points, disconnect the primary electrical connection, remove the mounting screws and remove the old set. Install the new set with mounting screws slightly loose. Turn the engine over slowly until points reach their maximum gap. Then, using a flat feeler gauge of proper dimension (see the tune-up chart), shift the contact set base plate as necessary, or adjust the Allen screw, to get the proper gap, tighten the mounting screw, and recheck. In some cases, a slight further adjustment may be needed to get the proper timing. Make sure to lube the cam follower with high temperature grease or oil, as necessary, when replacing or servicing points.

Ignition Timing

1. Gain access to the points as follows:
   a. Remove the air housings and blower wheel on Vacu-Flo units, or just the blower housings on pressure-cooled units.
   b. Loosen the flywheel bolt a few turns, and then, while prying outward on the flywheel, tap the flywheel bolt with a hammer.
   c. Remove the breaker cover screw and remove the breaker cover.
2. Set the point gap as described above.
3. Crank the engine slowly by hand in a forward direction until the painted mark on the flywheel and the **TC** mark on the gear cover are aligned. The engine must be on the compression stroke; otherwise, turn it one more revolution.
4. Turn the flywheel slowly and note whether the timing marks are aligned just as the points break. If not, loosen the breaker box mounting screws and shift the breaker box assembly slightly upward if points open too soon, or downward if they open too late. Tighten the breaker box mounting screws.
5. Check the timing with an automotive timing light or with a test lamp run from the primary connection on the breaker box to a good ground on the engine. To check the timing with a test lamp, turn the engine backwards and then slowly approach the point where the timing marks align. The light should go out just as the timing marks align.

SET BREAKER POINT GAP WIDTH AT .017–.024 (0.43 to 0.61 mm) TO GET PROPER 22° BTC TIMING

PLACE A DROP OF LIGHT OIL ON BREAKER ARM PIVOT WHENEVER POINTS ARE INSTALLED OR ADJUSTED

TO ADJUST GAP LOOSEN SCREWS "A" TURN CAM "B"

Setting point gap and timing on battery ignition AJ engines

AJ ENGINES WITH BATTERY IGNITION

1. Timing can be checked either at the timing hole on the front of the blower housing or at a hole on the side of the blower housing. Turn the engine over until it passes the 22 degrees BTDC timing mark so the points will be wide open.
2. Remove the cover on the breaker box and set point gap to 0.020 in. with a flat feeler gauge. Then, install the breaker box cover and check the timing with a timing light while the engine is running.
3. If timing is not correct, reduce point gap to make timing later, or increase it to advance timing. Gap may fall anywhere in the range 0.017–0.024 in. to get correct timing.
4. Repeat the procedure until timing is correct.

Mixture Adjustments

Factory mixture adjustments should not be disturbed unless the engine clearly is running too rich or too lean. Standard settings are: Main Jet – $1\frac{1}{4}$ turns open; Idle Jet – 1 turn open. Do not

THROTTLE STOP SCREW

IDLE FUEL ADJUSTMENT

MAIN FUEL ADJUSTMENT

Locations of idle and main fuel mixture screws

force the needle against the seat by turning it in hard when making these settings. Turn the screw in slowly until it just bottoms very gently, then turn it out the specified amount.

If it is necessary to adjust further, run the engine until it is hot and remove all load. Run the engine at idle speed and adjust the idle mixture screw in and out slowly until highest speed and smoothest running are obtained. Adjust the main jet mixture screw similarly, with the engine running at normal speed without load. Test the response of the engine when accelerating from idle and open the main jet $\frac{1}{4}$ turn more if acceleration response is poor. If the governor "hunts" open and closed, the main jet may be opened as $\frac{1}{4}$ turn more or $\frac{1}{4}$ turn beyond point of highest operating speed without load.

Governor Adjustments

Run the engine or engine/generator under light load until it is hot. On generator sets, engine rpm controls both output frequency and voltage, so both speed and sensitivity settings are critical. Use of an accurate tachometer is the best means of getting good performance without overspeeding the engine.

AK, AJ GENERATORS

Engine speed at full load should be 1800 rpm for four pole generators; 3600 rpm for two pole generators. These speeds should increase to 1900 rpm and 3800 at no load, respectively. On these generators, if speed droop or change in speed is too great, turn the sensitivity adjusting screw outward, then recheck full load speed and adjust as necessary with the speed adjusting nut, turning it clockwise to increase speed. Repeat both adjustments as necessary until full load and no load speeds are both correct. If the unit hunts constantly by opening and closing the throttle in a regular rhythm, decrease sensitivity slightly and then readjust the speed setting.

LK ENGINES

First adjust the speed adjusting nut (clockwise to increase) to the setting called for on the nameplate. If the voltage drop with increase in load is too great, loosen the locknut and turn the sensitivity adjusting screw inward, or toward the shaft, retighten the locknut, and recheck speed. Repeat adjustments as necessary.

Automatic Choke Adjustment

AK, AJ AND LK

1. Loosen the choke cover screws (2) and rotate the cover as necessary to get the following dimension between the inner edge of the choke plate and the wall of the carburetor air horn: At 58°F – $\frac{1}{4}$ in. open; at 66°F – $\frac{1}{2}$ in. open; at 76°F – $\frac{3}{4}$ in. open; at 82°F – fully open.
2. Tighten the choke cover screws.

Valve Adjustment

Adjust the valve clearance with the engine cold and in TDC position with the spark plug about to fire. Adjustment is checked by sliding a flat feeler gauge of the proper dimension (see specifications, noting that clearances are different for exhaust and intake valves) between the tappet and head of the valve stem. If necessary, turn the adjusting nut on the tappet while holding tappet with a second wrench to get a slight pull on the gauge.

Testing Compression

1. Run the engine until hot. Stop and remove spark plugs.
2. Insert the compression gauge in one of the spark plug holes and crank the engine with the starter. Record the reading.
3. Squirt a small amount of SAE 30 oil into the cylinder and repeat the check.
4. Repeat Steps 2–3 for each cylinder. Compression specifications are listed below. If compression is low, but increases substantially when oil is squirted into the cylinder, the compression problem is probably with the pistons, rings, and cylinders. Compression pressures are: (psi).
   - AK, AJ – 95–105
   - LE – 90–110

FUEL SYSTEM

Carburetion

The carburetor is a side draft or horizontal float type with two adjusting screws for the idle needle and the main fuel nozzle needle.

AMBIENT TEMP. (°F)	60	65	70	75	80	85	90	95	100
CHOKE OPENING (Inches)	$\frac{1}{8}$	$\frac{9}{64}$	$\frac{5}{32}$	$\frac{11}{64}$	$\frac{3}{16}$	$\frac{13}{64}$	$\frac{7}{32}$	$\frac{15}{64}$	$\frac{1}{4}$

Thermal-magnetic choke adjustment specifications

BASIC DOWNDRAFT AND SIDEDRAFT CARBURETORS

Removal and Installation

1. Remove air cleaner, fuel line, governor linkage and choke apparatus from carburetor.
2. Remove two carburetor mounting nuts and pull off carburetor. On CC Engines, first remove the intake manifold; then remove the carburetor from the manifold.
3. Reverse above steps to install carburetor on engine.

Overhaul

1. Remove air cleaner adapter and choke from carburetor.
2. Remove main fuel adjustment needle and needle retainer.
3. Remove top of carburetor from carburetor base.
4. Remove carburetor float, lift out float valve and unscrew and remove its seat.
5. Remove no load adjusting needle.
6. Remove throttle plate and throttle shaft.
7. Remove choke plate and choke shaft.
8. Remove nozzle assembly.
9. Soak all components thoroughly in carburetor cleaner, following cleaner manufacturer's instructions. Clean all carbon from carburetor bore, especially in the area of the throttle valve. Blow out passages with compressed air. Avoid using wire to clean out passages.
10. Check adjusting needles and nozzle for damage. If the float is loaded with fuel or damaged, replace it. The float should turn freely on its pin without binding. Invert the carburetor body and measure float level.
11. To adjust float level, bend small lip that needle valve rides on.
12. Check choke and throttle shafts for excessive side play and replace if necessary.
13. Install throttle shaft and valve, using new screws. The bevel on the throttle plate must fit flush with carburetor body. On valve plates marked with a C, install them with the mark on the side toward idle port as viewed from flange end of carburetor. To center throttle valve (Bendix/Zenith Carburetor) back off stop screw, close throttle lever, and seat valve by tapping it with a small screwdriver, then tighten the two throttle plate screws.
14. Install choke shaft and choke plate. Center choke plate in same manner as throttle valve. Always fasten plate in position with new screws.

Exploded view of downdraft carburetor

15. Install main nozzle. Make sure it seats in body casting.
16. Install main fuel adjustment needle and its retainer.
17. Install no load adjusting needle.
18. Install intake valve seat and intake valve.
19. Install float and float pin. Center pin so float bowl doesn't ride against it.
20. Check float level. Adjust if necessary.
21. Install a new body-to-bowl gasket and secure the two sections together.
22. Reinstall choke.
23. Install air horn assembly.
24. Install carburetor on engine.
25. Adjust both main and idle fuel mixture needles, as described in the Tune-Up section.

Mechanical Fuel Pumps

REMOVAL AND INSTALLATION

1. Remove the fuel lines.
2. Remove the two mounting capscrews, and pull the pump off the engine. Remove and discard the gasket.
3. Clean all gasket material from both mounting pad and pump flange. Apply an oil resistant sealer to both sides of a new gasket and to the threads of the attaching bolts.
4. Position the new gasket on the pump flange and position the pump and gasket on the mounting pad. To check their position, make sure the rocker arm rides on the pump cam lobe. Turn the crankshaft until the rocker arm is at the low point of its stroke.
5. Position the pump tightly against the pad and install the mounting bolts, torquing them alternately in several stages to specifications.
6. Connect the fuel lines and then operate the unit to check for leaks.

SERVICE

1. Scribe a mark across the flanges of the pump body and valve housing so these parts can be assembled in original positions.
2. Remove the valve housing from the body of the pump. Tap pump body with a screwdriver to do this.
3. Remove both valves and their gaskets from valve housing.

Exploded view of sidedraft carburetor

Note position of valves in their housing so new valves can be correctly installed.

4. Using a blunt punch, drive rocker arm pin out of pump.

5. Press diaphragm into fuel pump body and then pull rocker arm outward to unhook diaphragm actuating rod from rocker arm link assembly.

6. Remove diaphragm and diaphragm return spring, rocker arm and link assembly, and rocker arm return spring from pump body.

7. Remove diaphragm actuating rod oil seal from pump body.

8. Clean and inspect all fuel pump components and replace all unserviceable parts.

9. Install inlet and outlet valves and their gaskets in their respective positions. Seat valves firmly.

10. Lubricate diaphragm actuating rod.

11. Position fuel pump diaphragm and spring assembly into pump body as shown.

12. Hold diaphragm assembly in pump body and position pump body so mounting flanges faces up. Apply slightly more pressure to lower edge of diaphragm and insert rocker arm link assembly.

13. Hook rocker arm link to diaphragm actuating rod.

14. Install rocker arm return spring and hold it in place by cocking rocker arm slightly.

15. Install the rocker arm pin in the pump body.

16. Position the valve body and pump body so the two previously scribed marks align. Install all screws and lockwashers until they just engage the pump body, being careful not to tear the diaphragm fabric.

17. Alternately and evenly tighten all screws.

Bendix Electric Fuel Pumps

SERVICE

1. Release the bottom cover (1) from the bayonet fittings. Twist the cover by hand to remove it from the pump body.

2. Remove the filter (4), magnet (3) and cover gasket (2) (see appropriate illustration). Wash the filter in cleaning solvent and blow out dirt and cleaning solvent with air pressure. Check the cover gasket and replace if deteriorated. Clean the cover.

3. Remove the retainer spring (5) from plunger tube (11), using thin nose pliers to spread and remove the ends of the retainer from the tube. Then remove the washer (6), O-ring seal (7), cup valve (8), plunger spring (9) and plunger (10) from the tube (11), (see appropriate illustration).

4. Wash all parts in cleaning solvent and blow out with air pressure. If the plunger does not wash clean or if there are any rough spots, gently clean the surface with a crocus cloth. Slosh the pump assembly in cleaning solvent. Blow out the tube with air pressure. Swab the inside of the tube with a cloth wrapped around a stick.

5. Insert the plunger assembly (10) in the tube with the buffer spring end first. Check the fit by slowly raising and lowering

Bendix electric fule pump—removal of retainer, O-ring, cup valve, spring, and plunger

the plunger in the tube. It should move fully without any tendency to stick. If a click cannot be heard, the interrupter assembly is not functioning properly in which case pump should be replaced.

6. To complete the assembly, install the plunger spring (9), cup valve (8), O-ring seal (7) and washer (6) as shown. Compress the spring (9) and assemble the retainer (5) with ends of the retainer in the side holes of tube (11).

7. Place the cover gasket (2) and magnet (3) in the bottom cover (1) and assemble the filter (4) and cover assembly. Twist the cover by hand to hold it in position on the pump housing. Securely tighten the bottom cover.

Onan Electric Fuel Pump

SERVICE

Clean the filters every 100 operating hours. Remove the four phillips screws from the top, and lift off the filter assembly. Clean the two screen-type filters in a safe solvent. Install, ensuring the gasket is in the proper position to prevent leaks.

Governor Repair

1. Draw a sketch of the governor linkage or lay the parts out in position as you disassemble them. Disassemble the linkage. Remove the gear cover.

2. Remove the snap ring that holds the governor cup to the camshaft gear, being ready to catch the flyballs, which will come out as the governor cup is removed.

J SERIES

Typical centrifugal type governor

Bendix electric fuel pump—removal of bottom cover, gasket, magnet, and filter

¾"
(19.05 mm)

WHEN GOVERNOR
IS PROPERLY
ASSEMBLED THE
DIMENSION SHOWN
ON DRAWING WILL
BE AS INDICATED

CAMSHAFT
GEAR

⁷/₃₂"
(5.56 mm)

CENTER PIN

SNAP RING

GOVERNOR CUP

CAMSHAFT

GOVERNOR
FLYBALL

Measuring governor in and out travel

3. Clean all parts thoroughly in a safe solvent. Inspect as follows and replace parts which are found defective:

 a. flyballs for grooves or flat spots.

 b. ball spacers for arms with noticeable wear or damage.

 c. governor cup with a rough or grooved race surface.

 d. governor cup which does not have a free spinning fit on the camshaft center pin, or which is loose and wobbles.

4. To install the governor cup, tilt the engine to make the gear face upward. Space the flyballs at equal distances on the gear and then install the cup and snap ring on the camshaft center pin.

5. On J-series this dimension should be ²⁵/₃₂ in. Hold the cup against the flyballs to make this measurement. If the distance is not correct, pull out the center pin and replace it, pressing the new pin in just the required amount. If the pin extension is ok, grind off the hub of the cup as required. In and out travel distance must be ⁷/₃₂ in. or the engine will race.

ENGINE OVERHAUL

Valves

SERVICE

In order to remove the valves, the cylinder head must be removed. Remove the cylinder head screws and then remove the head from the engine. If the head sticks to the block, hit it lightly with a soft hammer, being careful not to damage any of the cooling fins. Remove the cylinder head gasket and discard it.

Use a conventional type valve spring compressor to compress the valve springs so that the spring retainer can be removed. The retainers are the split, tapered type and will most likely fall out when the valve spring is compressed. After removing the retainers, lift the valve out through the top of the valve guide. Clean the valves of all carbon deposits and inspect them, looking for warpage, worn stems, and burned surfaces that are partially destroyed. Determine whether or not the valve can be reused, whether it can be reground, or if the valve has to be replaced.

Check out the valve stem-to-guide clearance and, if it is too large, the guides must be replaced. They can be removed through the valve chamber. The valve tappets are also replaceable from inside the valve chamber, once the valves have been removed.

In removing the valve guides, first wire brush carbon and other deposits from the top guide surfaces, or the guide bores may be damaged during guide removal. Note that a gasket must be used on the intake valve guides for the LK, engines. Where used, place the gasket on the intake guide and install the intake and exhaust guides from within the valve chamber. Before installing the guides, run a small polishing rod with a crocus cloth through the guide holes to clean out deposits.

VALVE SEAT INSERT REMOVAL AND INSTALLATION

Use a solvent to clean carbon or stuck gasket material from cylinder head and block surfaces. If necessary, follow up using a metal scraper. Inspect the head gasket surfaces for cracks, nicks, or burrs. Check the head for cracks, and replace if any are present. An oil stone may be used to remove burrs or nicks.

Use a straightedge at three angles to check the head for flatness. Try to insert a 0.003 in. flat feeler gauge under the straightedge at any point, and replace the head if it is not flat.

Replace the valve seats if cracked or loose, or excessively worn. On cast iron engines, the seat may be driven out using a knockout tool as shown. The tool must be inserted under the port side of the valve seat with the square end extending over the cylinder bore. With the sharp edge of the tool at the joint between the seat and its recess, strike a sharp blow on the end of the tool with a light hammer. This will crack the insert and permit removal.

—————————— CAUTION ——————————

Since this may shatter the relatively brittle material of the seat, you should wear goggles when performing the procedure.

If the engine has an aluminum block, use a ¾ in. or one inch pipe tap to suit the seat diameter. Place a washer on top of the valve guide for the tap to bottom against. Turn the tap in until the seat begins to turn. As it starts to turn, begin pulling outward on the tap while continuing the turning motion to pull out the seat. Make sure the valve guide is not pushed downward by

VALVE SEAT REMOVAL TOOL

APPLY SHARP BLOW AT THIS POINT

KNOCK-OUT TOOL
USE ONLY ON UNITS
HAVING CAST IRON BLOCK

PORT

Special tool for removing valve seats from engines with cast iron blocks

RIGHT CYLINDER

Cylinder head torque sequences

the tap or, if it is, that it is pressed back into position. Use an oversize replacement seat on aluminum engines.

Clean any carbon or burrs from the insert recess, and install a new seat as follows:

1. Gradually heat the block to 325°F. Place the new seat in dry ice until thoroughly chilled.

2. Insert the pilot of an appropriate special tool in the valve guide hole in the block and quickly and evenly drive in the seat insert. It must seat on the bottom of the recess.

The valve face angle is 44 degrees. The valve seat angle is 45 degrees. the 1 degree interference angle assures a sharp seating surface between the valve and the seat and good sealing characteristics. The valve seat width must be between $\frac{1}{32}$ in. and $\frac{3}{32}$ in. Valves should not be hand lapped if at all possible. This is especially important if stelite valves are used.

To check the valves for a tight seal, make pencil marks around the valve face, then install the valve and rotate it a part of a turn. If the marks are all rubbed off uniformly, then the seal is good.

1. Insert the tappets in the crankcase holes.

2. Install the valves, springs and guides.

3. Using a valve spring compressor, compress each valve spring and insert the valve spring retainer and retainer locks.

4. Set the valve clearance to the specifications listed at the back of the chapter.

5. Install the heads and gaskets to the cylinder block.

6. Tighten the head bolts to the correct torque following the sequence in the appropriate illustration.

7. Install the exhaust manifold, oil lines, spark plugs and carburetor.

Gear Cover

REMOVAL AND INSTALLATION

In order to gain access to the camshaft gear and other internal components of the engine, it is necessary to remove the gear cover. In order to remove the gear cover, the magneto assembly must be removed. Disconnect the spark plug wire at the spark plug and disconnect the stop wire. Remove the attaching screws that hold the magneto assembly to the gear cover and remove the magneto.

When the gear cover is removed, the governor shaft disengages from the governor cup, which is part of the camshaft.

During the installation of the cover, be sure to engage the pin on the cover, with the chamfered hole located in the governor cup. To do this, turn the governor cup so that the hole is located at the top or in the 12 o'clock position. Turn the governor shaft clockwise as far as it will go and hold it there until the cover is installed. Position the cover on the engine and make sure that it fits flush against the engine. Be careful of the gear cover oil seal during installation. Use a new gasket if the old one is damaged.

Timing Gears

REMOVAL AND INSTALLATION

If it becomes necessary to replace either the crankshaft or camshaft gears because of broken teeth, extreme wear, cracks, etc., both gears must be replaced as a pair. Never replace only one of the gears. Both gears are pressed onto their respective shafts.

To remove the crankshaft gear, insert two #10–32 screws into the threaded holes in the gear and tighten the screws alternately a little at a time. The screws will press up against the crankshaft shoulder and force the gear off the end of the crankshaft. On AK engines use a gear puller to remove the crankshaft gear.

Gear cover assembly

Valve and crankcase breather assembly

To remove the camshaft gear, it is necessary to remove the entire camshaft assembly from the engine. First remove the crankshaft gear lock ring and washer. Remove the cylinder head, valve assemblies, fuel pump (if so equipped), and the valve tappets. Remove the governor cup assembly and then remove the camshaft and gear assembly from the engine. The camshaft gear may now be pressed off the camshaft. Do not press on the camshaft center pin as it will be damaged. The governor ball spacer is press fit into the camshaft gear.

When the camshaft gear is replaced on the camshaft, be certain that the gear is properly aligned and the key properly positioned before beginning to press the gear onto the camshaft. Install the governor cup before replacing the camshaft and gear assembly back into position in the engine.

There are two stamped **O** marks, one on each gear, near the gear teeth. When the crankshaft and camshaft gears are meshed, these marks must be exactly opposite each other. When installing the camshaft gear assembly, be sure that the thrust washer that goes behind the camshaft gear is installed. Replace the retaining washer and lockwasher on the crankshaft.

Cylinder Bore, Piston, and Piston Rings

The cylinder can be rebored if it becomes heavily scored or badly worn. If the cylinder bore becomes cracked, replace the cylinder block.

The cylinder can be bored out to 0.010 in., 0.020 in. or 0.030 in. oversize.

NOTE: The cylinder bore on the AK engine can be bored out to 0.040 in. oversize.

There are pistons and rings in the above oversizes available to accommodate an oversized cylinder bore. If the cylinder bore has to be bored out only 0.005 in. to remove the damage in the cylinder, use standard size parts.

Use a ridge reamer to remove the ridge that may be present at the top of the cylinder bore to avoid damaging the piston rings when the piston and connecting rod assembly is removed. Hone the cylinder and create a cross hatch pattern on the cylinder

THESE MARKS MUST COINCIDE WHEN IN-STALLING TIMING GEARS

SCREWS PASS THROUGH GEAR AND BEAR AGAINST CRANK-SHAFT SHOULDER

CRANKSHAFT GEAR

② #10-32 SCREWS

TIGHTEN SCREWS UP ALTERNATELY TO PRESS GEAR OFF EVENLY.

Aligning the timing marks on the timing gears and removing the crankshaft gear

walls if new rings are being installed. Clean the cylinder with SAE 10 engine oil after honing.

Some engines were originally built with 0.005 in. oversize pistons and are so indicated by a letter **E** following the serial number stamped on the identification plate and on the side of the crankcase.

The piston is fitted with two compression rings and one oil control ring. When the piston assembly is removed from the engine, clean off all carbon deposits and open all of the oil return holes in the lower ring groove. Before installing new rings, check the ring gap by installing the rings squarely in the cylinder bore and measuring the gap between the two ends of the rings. If the gap is too small, it is possible to file the ends to obtain the proper size gap.

Tapered type rings are usually marked with the word **TOP** on one side. This side must be installed facing toward the top or closed end of the piston. Position the ring gaps evenly around the circumference of the piston with no ring gap over the piston pin.

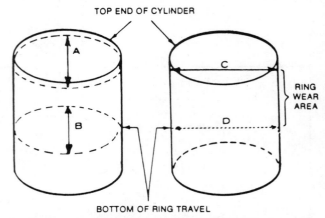

TOP END OF CYLINDER

A

B

C

D

RING WEAR AREA

BOTTOM OF RING TRAVEL

Locations for taking cylinder wear measurements

The piston pin is held in place by two lock rings, one at each end. Make sure that the lock rings are properly installed before installing the piston assembly in the cylinder. Be sure to check the size of the piston, piston pin, piston pin bore, and the size of the cylinder before installing any of these parts back into the engine. Replace any parts that are worn beyond the maximum allowed specification. The piston should be replaced when the side clearance of the top ring reaches 0.008 in., or if there are signs of scuffing, scoring, worn ring lands, fractures, or pre-ignition damage. Measure the piston dimensions at points shown in the illustration.

In fitting rings, install each ring in the cylinder bore, invert the piston and push the ring to the end of the ring travel-about halfway to the end of the bore. With the ring at exactly 90 degrees to the centerline of the bore, measure the end gap with a feeler gauge. Fit rings by choosing the right size for the bore-do not file the ends.

Connecting Rod

Before removing the connecting rod from the crankshaft, mark the cap and rod so they can be installed in exactly the same position from which they are removed.

MEASURE "X" DIAMETER HERE

MEASURE "Y" DIAMETER HERE

Measure piston at points shown

ALIGN OIL HOLE IN
BEARING WITH OIL
HOLE IN CRANK-
SHAFT BEARING
BORE

LINE BORE
ALL BEARINGS
AFTER INSTALLATION

INSTALL WITH THIS
GROOVE AT TOP

CAMSHAFT BEARING

CRANKSHAFT BEARINGS

INSTALL WITH OUT-
SIDE EDGE FLUSH
WITH BOTTOM OF
COUNTERBORED
AREA

EDGE OF BEARING
FLUSH WITH OUT-
SIDE SURFACE

FRONT CAMSHAFT
BEARING

REAR CAMSHAFT
BEARING

Crankshaft main bearings and camshaft bearings for the AJ series engines

If abnormal bearing wear (worn on one side more than the other) is noticed, this would indicate that the connecting rod is bent. It is possible to have the connecting rod straightened, but this should be done at a machine shop or small engine service shop.

Measure all of the bearing surfaces for size, including the piston pin hole and the crankpin bearing. Inspect the bearings for burrs, breaks, pitting and wear. Replace if scored or if the overlay is wiped out. Scratching is also reason for replacement. If bearings look all right, check bearing clearances. Place a piece of Plastigage of the proper width in the bearing cap about ¼ in. off center. Rotate the crankshaft about 30 degrees from bottom center, and install the bearing cap. Torque to specification. Without turning the crankshaft, remove the bearing cap, leaving the flattened Plastigage on the bearing or journal. Compare its widest point with the scale on the Plastigage envelope to determine clearance. If clearance is excessive, replace the connecting rod bearings.

Use a new piston pin to check the pin bushing for wear. A push fit clearance is required. If a new pin will fall through a dry rod bore of its own weight, replace the rod bushing.

To replace the bushings, press them out using a press and proper driving tool. Press in the new bushings so that the ends are flush with the sides of the rod and a ¹⁄₁₆ in. oil groove is formed in the center. If there are oil holes, make sure they are at least half way open. Make sure the pin is a push fit after the bushing is installed.

Replace rod nuts or bolts if the threads are damaged. Replace rods which are nicked or fractured, or which have bores which are out of round more than 0.002 in. Straighten or replace rods which are twisted more than 0.012 in. or out of line more than 0.005 in.

Be sure to reinstall the oil dipper to the connecting rod cap when assembling the piston and connecting rod assembly to the crankshaft.

Cylinder Block
INSPECTION

Check the entire block thoroughly for cracks, and check the cylinder bore for scoring. Measure the bores for out-of-round at points indicated. A is the point where greatest ring wear occurs, and B is the bottom of ring travel. C and D are at the same heights as A and B respectively, but at 90 degrees. Compare (subtract the smaller from the larger) A and B and then compare C and D to determine taper. Cylinder must be rebored for the next oversize piston if taper exceeds 0.005 in. Comparing A to C and B to D indicates out-of-round. If the cylinder is out-of-round 0.002 in., it must be rebored for the next oversize piston.

HONING CYLINDERS TO OVERSIZE DIMENSIONS

1. Anchor the block solidly for either vertical or horizontal honing. Use either a drill press or heavy-duty drill which operates at about 250–450 rpm.

2. Lower the hone into the cylinder until it protrudes ½–¾ in. past the end of the cylinder. Rotate the adjusting nut until the stones just come in contact with the cylinder wall at its narrowest point.

3. Loosen the adjusting nut until the hone can be turned by hand.

4. Connect the drill to the one and start the drill. Move the hone up and down in the cylinder about 40 times per minute. Usually the bottom of the cylinder must be worked out first because it is smaller. When the cylinder takes a uniform diameter, move the hone up and down all the way through the bore. Follow the hone manufacturer's recommendations for wet or dry honing and oiling the hone.

5. Check the diameter of the cylinder regularly during honing. A dial bore gauge is the easiest method but a telescoping gauge can be used. Check the size at six places in the bore; measure twice at the top, middle and bottom at 90 degree angles.

6. When the cylinder is within about 0.002 inch of the desired bore, change to fine stones and finish the bore. The finish should not be smooth but crosshatched as shown. The cross hatch formed by the scratching of the stones should form an angle of 32 degrees. This can be achieved by moving the hone up and down in the cylinder about 40 times per minute.

7. Clean the cylinder block thoroughly with soap, water and clean rags. A clean white rag should not be soiled on the wall after cleaning is complete.

CAUTION

Do not use a solvent or gasoline since they wash the oil from the walls but leave the metal particles.

---

8. Dry the crankcase and coat it with oil.

Crankshaft and Bearings

CRANKSHAFT REMOVAL

1. Remove the lock ring and retainer washer from the crankshaft gear, and pull off gear with a puller.
2. Remove the oil pan, head(s), pistons and rods. See the previous section on piston and ring service.
3. Remove the rear bearing plate. Remove the crankshaft through the rear opening in the crankcase.

INSPECTION

Clean the crankshaft thoroughly, including blowing out all oil passages with compressed air. Check the journals for out-of-round, taper, grooves, and ridges, and place on V-blocks and rotate to measure runout. It should not exceed 0.003 in. Compare all dimensions with specifications: out-of-round should not exceed 0.005 in.; staper should not exceed 0.001 in.; wear should not exceed 0.002 in. If the limits are exceeded, regrind the shaft for the use of undersize bearings.

Replace bearings that are warped, scored, or have been overheated. If such damage is present, check the bearing bores in the block for excessive size. A new rear main bearing seat can be installed by replacing the rear main bearing plate. After doing this, check the main bearing bore alignment on a line boring machine.

The main crankshaft bearings are precision bearings. They are available in the standard size as well as 0.002 in., 0.010 in., 0.020 in. and 0.030 in. undersize. The precision type bearings are NOT to be line reamed.

The bearings are press fit into the cylinder block. Before trying to install the bearing into the cylinder block or bearing plate, heat the plate or block by running hot water over them or placing them in an oven heated to 200°F. This will cause the block or bearing plate to expand and facilitate the installation of the sleeve bearings.

The oil hole in the bearing and the oil hole in the bearing bore must be aligned when the bearings are installed. On pressure lubricated engines, the hole should be opposite the crankshaft. On engines that are splash lubricated, the hole should be upward.

Install rear bearings in the rear bearing plate using a special driver. Install them to $1/64$ in. below the end of the bore. If a special tool is not available and the lock pins must be removed with side cutters or a screw extractor, install new lock pins.

CRANKSHAFT INSTALLATION

1. Oil the bearing surfaces thoroughly. Install the crankshaft from the rear of the crankcase (through the rear bearing plate hole).
2. Put the rear bearing plate gasket in place and lubricate the rear end plate bearing. Slide the thrust washer, with grooves toward the crankshaft and the bearing plate, over the end of the crankshaft.

NOTE: Line up the thrust washer notches with the lock pins before tightening the end plate bolts, or the lock pins and thrust washer will be damaged.

If you have trouble getting the thrust washer to stay in place, it may be lubricated with a light coating of oil.

3. Torque the bearing plate bolts.
4. Heat the timing gear to 350°F. Install a new crankshaft key and drive the gear into position. Install the washer and lock ring.
5. Adjust the crankshaft end play as described below.
6. Complete the reassembly of the engine.

CHECKING END PLAY

Check the end play with the rear bearing plate bolts properly torqued. If end play is excessive, remove the end plate and install a shim between the thrust washer and plate. When installing the plate, line up the notches in the thrust washer and shim with the lock pins. Torque the end plate and recheck end play. If total gasket and shim thickness required is more than 0.015 in., use a steel shim of the proper thickness and two thin gaskets, or gasket compression and consequent loose bolts may result.

Crankshaft Oil Seals

The crankshaft oil seals are installed with the open sides facing toward the inside of the engine. To replace the rear oil seal, the rear bearing plate must be removed. To replace the front oil seal, the front gear cover must be removed. Be careful not to

PRODUCE CROSS HATCH SCRATCHES
FOR FAST RING SEATING

AVOID THIS FINISH

Produce a crosshatch in the bore as shown in the upper part of the illustration

Exploded view of oil pump

Measure end play at location shown

damage the oil seal during installation or to turn back the edge of the seal lip.

LUBRICATION

Onan engines are either splash lubricated or pressure lubricated.

The splash lubrication system consists of an oil dipper attached to the connecting rod cap and various oil passages to catch and channel the oil that is splashed up by the oil dipper.

The pressure lubrication system consists of a gear type oil pump, an oil intake cup, a non-adjustable pressure relief valve, and the various oil passages and channels.

Oil Pump

SERVICE

1. Drain the crankcase oil. Then, remove the gear cover and oil base.

2. Unscrew the intake cup from the oil pump. Loosen the two capscrews holding the pump and remove it.

3. Disassemble the pump by removing the two cap screws holding the cover in position. Inspect for excessive wear of gears and shafts and replace the pump if any parts are badly damaged. Gears may be measured for excess wear by positioning them in the housing, running a straightedge and side of the gear. If either this clearance or the clearance between oil pump teeth is excessive, replace the unit.

4. In reassembling the pump, side clearance between gears and housing can be adjusted by using a thinner gasket. Use the thinnest gasket that permits freedom in operation of the pump.

5. Fully prime the pump with oil before assembly. Mount the pump into position on the engine and adjust for a clearance of 0.005 in. between the pump and crankshaft gears.

6. Install the intake cup on the pump parallel with the bottom of the crankcase.

7. Install the gear cover and oil base, fill the crankcase with oil, and run engine, checking for good oil pressure.

Torque Specifications

ENGINE SERIES		Cylinder Head (Cold)	Conn Rod	Rear Bearing Plate	Flywheel To Crankshaft	Oil Base	Armature Thru Stud / Revolving Armature Units	Spark Plugs
		Use Engine Lubricating Oil As A Thread Lubricant					Do Not Use Any Lubricant On These Threads	
AJ, MAJ	lb-ft	24–26	10–12	20–25	35–40	25–30	25–30	25–30
AK	lb-ft	24–26	10–12	20–25	35–40	25–30	25–30	25–30
LK	lb-ft	29–31	26–28	20–25	35–40	25–30	35–40	25–30

Crankshaft and Connecting Rod Specifications

All measurements are given in inches

Engine Model	Crankshaft				Connecting Rod		
	Main Brg Journal Dia	Main Brg Oil Clearance	Shaft End-Play	Thrust on No.	Journal Diameter	Oil Clearance	Side Clearance
AK	1.6857 – 1.6865	.0030 – .0040	.008 – .0012	Rear	1.3742 – 1.3750	.0015 – .0025	.012 – .035
AJ	1.6857 – 1.6865	.0015 – .0040	.008 – .0012	Rear	1.3742 – 1.3750	.0015 – .0025	.012 – .035
LK	1.9992 – 2.000	.0020 – .0030	.006 – .0012	Rear	1.6252 – 1.6280	.0020 – .0030	.002 – .016

Valve Specifications

Engine Model	Seat Angle (deg)	Face Angle (deg)	Spring Test Pressure (lbs @ in.)	Spring Installed Height (in.)	Guide Diameter (in.)		Stem Diameter (in.)	
					Intake	Exhaust	Intake	Exhaust
AK	45	44	71 – 79 @ 1.313	—	.3110 – .3120	.3110 – .3120	.3095 – .3100	.3090 – .3100
AJ	45	44	71 – 79 @ 1.313	—	.3110 – .3120	.3110 – .3120	.3080 – .3085	.3090 – .3100
LK	45	44	71 – 79 @ 1.375	—	.3440 – .3460	.3440 – .3460	.3425 – .3430	.3410 – .3413

Piston Clearance

Engine Model	Piston to Bore Clearance (in.)
AK	.004 – ① .005
AJ	.006 – ① .008
LK	.0005 – ① .0015

① Measure piston diameter across dimension Y

Ring Gap

All measurements are given in inches

Engine Model	Top Compression	Bottom Compression	Oil Control
AK	.006 – .018	.006 – .018	.006 – .018
AJ	.006 – .024	.006 – .024	.006 – .024
LK	.010 – .023	.010 – .023	.010 – .023

Ring Side Clearance

All measurements are given in inches

Engine Model	Top Compression
AK	.002 – .008
AJ	.002 – .008
LK	.002 – .008

Section 14

Onan

Small Engines — 6-20 hp

ENGINE IDENTIFICATION

All Onan engines have an identification plate attached to the left side of the cooling shroud (facing the flywheel) on one cylinder engines, and on the right side of the cooling air duct (facing the flywheel) on four cylinder engines. The Model and Specification number shown there may be interpreted as follows:

1. Factory code for general engine identification.
2. Engine specific type:
- S — Manual starting with stub shaft power take-off.
- MS — Electric starting with stub shaft, starter, and generator.
- MSV — Vacu-Flo cooling. Similar to MS, but with cooling air drawn in through a front end duct.
3. Factory code for optional equipment on the engine.

4. Specification letter. Letter advances alphabetically with production modifications.

Serial numbers consist of a two digit month of manufacture indicator and a six digit engine number.

General Engine Specifications

Model	Bore & Stroke (In.)	Displacement (cu. in.)	Horsepower @ RPM
LKB	3.25 x 3.00	24.9	8.5 @ 1800
NB	3.562 x 3.00	30.0	12.0 @ 3600
CCK	3.25 x 3.00	49.8	12.9 @ 2700
CCKA	3.25 x 3.00	49.8	16.5 @ 3600
CCKB	3.25 x 3.00	49.8	20.0 @ 3900
BF	3.125 x 2.615	40.3	16.0 @ 3600

DJC - MS / 886 T

```
  1        2       3    4
```

Model and Specification Number breakdown

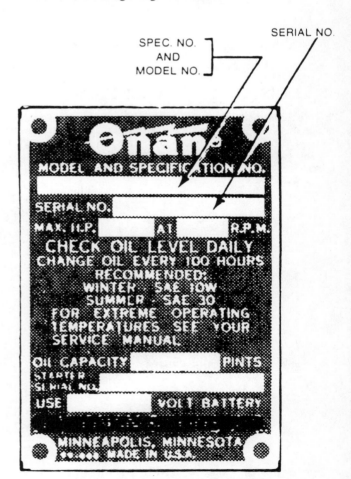

Onan engine identification plate

MAINTENANCE

Air Cleaner Service

OIL BATH AIR CLEANER

The oil bath air cleaner uses engine oil to collect dirt. Dirt is deposited in a sump full of oil in the bottom of the unit, and on a screen contained in the upper portion. Dirt sticks to the screen because the incoming air pulls some of the oil out of the sump, deposits it on the screen, and this setting action keeps the screen sticky. After each 100 hours, disassemble the unit, drain the oil out of the sump, clean the dirt out of the sump with clean oil of the type used in the engine. If the air around the engine is very dusty, inspect the air cleaner more frequently. Make sure to change the oil in the sump before dirt contained in the sump reaches the ring which is slightly above the bottom. In extreme cases, it may be necessary to clean the screen with solvent and coat it with clean oil as part of the service.

FILL WITH OIL TO THIS LEVEL

RENEW OIL BEFORE DIRT REACHES SHELF IN CUP

Oil bath type air cleaner

MOISTENED FOAM AIR CLEANER

This cleaning medium in this type of cleaner is an oil soaked sponge mounted around a screen. After each 100 hours, wash the sponge in soap and water or solvent, dry it, soak it in engine oil, and then squeeze excess oil out of it before installation.

WING NUT

COVER

POLYURETHANE PRE-CLEANER

PAPER ELEMENT

BASE

Dry paper air cleaner with polyurethane precleaner

DRY PAPER AIR CLEANER

Dry paper air cleaners use a replaceable paper element. All that is required in servicing is disassembly of the unit, replacement of the element, and a brief removal of dust from inside the unit with a rag. Some of these units also employ a polyurethane precleaning element, which is also replaced at time of service, or a dust cup, which is cleaned with a solvent soaked rag and dried. Service standard units every 100 hours; units with precleaner at 500 hours.

Lubrication

OIL AND FUEL RECOMMENDATIONS

Onan gasoline engines may use either regular grade leaded, unleaded, or low lead gasoline. The use of unleaded or low lead fuels is strongly recommended in order to reduce combustion chamber deposits, especially on governor controlled engines which run at a fairly steady load. If you are switching from leaded to unleaded fuel, combustion chamber deposits should be removed before beginning operation on the unleaded fuel because of its lower octane rating and consequent potential for damage due to engine knock.

Crankcase Capacity Chart

LKB	2.0
NB	2.0
CCK	4.0
CCKA	3.0
CCKB	4.0
BF	2.0

Oil viscosity chart

It is best to use an SE oil, although oils not so well equipped to resist sludge formation at both high and low temperatures may be used in very moderate service — i.e. long operating periods under light loads. Consult the chart for recommended viscosity ranges for the prevailing temperature.

TUNE-UP

Tune-Up Specifications

Model	Spark Plug Gap (In.)	Ignition Point Gap (In.)	Ignition Timing (Deg. BTDC @ RPM)
LKB	.025	.020	24
NB	.025	.020	22
CCK	.025	.020	19
CCKA	.025	.020	24
CCKB	.025	.020	24
BF	.025	.025	21

Spark Plugs

Spark plug should be replaced every 200 operating hours. Set the gap with a wire gauge to 0.025 in. It is best to torque the plugs with a torque wrench to avoid cylinder head damage, distorted gap dimension, or leakage. Use new gaskets and torque to 15-20 ft. lbs. on NB and BF engines, 25-30 ft. lbs. on other engines. Plugs should also be inspected and cleaned with a wire brush once or twice, depending on type of fuel and service, during their service life. Use new plug gaskets even when reinstalling old plugs.

Breaker Points

Breaker points must be inspected and cleaned up and gapped or replaced as required. On BF engines with side adjust points, they should be replaced every 200 hours. Gain access to the points as described below under "Ignition Timing". Inspect the points for burning and pitting. If only slightly worn or damaged, points may be dressed with an abrasive stone and regapped. Otherwise, they must be replaced.

To replace points, disconnect the primary electrical connection, remove the mounting screws and remove the old set. Install the new set with mounting screws slightly loose. Turn the engine over slowly until points reach their maximum gap. Then, using a flat feeler gauge of proper dimension (see the tune-up chart), shift the contact set base plate as necessary, or adjust the Allen screw, to get the proper gap, tighten the mounting screw, and recheck. In some cases, a slight further adjustment may be needed to get the proper timing. Make sure to lube the cam follower with high temperature grease or oil, as necessary, when replacing or servicing points.

Ignition Timing

CCK, CCKA AND CCKB ENGINES

1. Remove breaker box cover, turn engine until points are wide open, and set point gap to 0.020 in. with a flat feeler gauge.
2. Once point gap is correct, timing is adjusted by shifting the entire breaker box. Note that timing is set at 19 degrees on engines running at 2400 rpm and below, and at 25 degrees on engines running at 2500 and above. On all engines with automatic spark advance, make this preliminary setting at 5 degrees. See Step 4 for the location of the timing marks. A preliminary check of the timing may be made by turning the flywheel backwards and then turning it forward with engine on the compression

Locations of idle and main fuel mixture screws

stroke until the timing marks just line up. If the points open just when the timing marks line up, the timing is o.k. If the timing is too early, shaft the breaker box toward No.1 cylinder to retard the timing; if too late, shift the box in the opposite direction.

3. Tighten the breaker box mounting screws, and check the timing at the suggested rpm as specified in the next step. Shift the breaker box position and recheck until the timing is correct.
4. Timing mark locations and rpm at which timing should be checked are as follows:

Vacu-Flo Engines Without Spark Advance:

Remove the dot button from the top of the blower housing. Run the engine at 1400-1600 rpm. Viewing through the round hole in the blower housing, the TC flywheel mark should line up with the 25 degree mark on the gear cover.

Vacu-Flo Engines With Spark Advance:

Follow the procedure above. After timing at 1400-1600 rpm is set, slow the engine to less than 800 rpm and make sure the timing retards. If not, the timing advance mechanism may need cleaning or repair.

Other CCK Series Engines:

Some types have the timing marks on the gear cover. On these, align the correct mark on the gear cover with the **TC** mark on the flywheel. Other CCK engines have marks on both the gear cover and flywheel. On these, align the **TC** flywheel mark with the correct timing mark on the gear cover, or align the correct timing mark on the flywheel with the **TC** mark on the gear cover. On CCK engines without Vacu-Flo cooling, timing is 24 degrees at over 1100 rpm.

BF ENGINES

Side Adjust Breaker Points

1. Remove two screws and remove the breaker cover.
2. Remove the air intake hose that connects to the blower housing On Power Drawer units, remove the dot button on the blower housing to see the timing marks.
3. Loosen the mounting screw on the points. Rotate the crankshaft clockwise until the timing mark on the gear cover aligns with the mark on the flywheel. Turn another 90 degrees.
4. Insert a screwdriver into the notch on the points and set the gap to 0.025 in. with a flat feeler gauge. Tighten the mounting screw.

5. The points should open when the crankshaft passes the 25 degree mark (26 degrees On Power Drawer engines). Timing may be checked more precisely by connecting a test lamp between the breaker box terminal and a ground on the engine. The lamp should go out just as timing mark lines up. If necessary, change point gap slightly to get timing to occur at the right point. Timing may also be checked on either spark plug with a timing light for greatest accuracy.

Top Adjust Breaker Points

Follow the procedure above for "Side Adjust Breaker Points", but set point gap to 0.021 in. by turning the point gap adjusting screw with an Allen wrench.

NB ENGINES

1. Remove the breaker box cover and crank the engine slowly by hand until the maximum point opening is achieved.
2. Adjust the breaker gap to 0.020 in. with a flat feeler gauge. To adjust the breaker base, loosen the lower mounting screw.
3. Check the timing by turning the engine backward, and then going forward slowly as the 22 degree mark on the flywheel passes the mark on the gear cover. Points should open just as the timing marks line up. If the timing is incorrect, loosen the breaker box mounting screws (located inside the breaker box) and slide the box upward to retard the timing or downward to advance it. Tighten the mounting screws securely and recheck timing.
4. A more accurate check may be made with a test lamp connected from the breaker box terminal to an engine ground. The light should go out when the timing marks align as the engine is turned slowly forward.

Mixture Adjustments

Factory mixture adjustments should not be disturbed unless the engine clearly is running too rich or too lean. Standard settings are: Main Jet — 1¼ turns open; Idle Jet — 1 turn open. Do not force the needle against the seat by turning it in hard when making these settings. Turn the screw in slowly until it just bottoms very gently, then turn it out the specified amount.

If it is necessary to adjust further, run the engine until it is hot and remove all load. Run the engine at idle speed and adjust the idle mixture screw in and out slowly until highest speed and smoothest running are obtained. Adjust the main jet mixture screw similarly, with the engine running at normal speed without load. Test the response of the engine when accelerating from idle and open the main jet ¼ turn more if acceleration response is poor. If the governor "hunts" open and closed, the main jet may be opened as ¼ turn more or ½ turn beyond point of highest operating speed without load.

Governor Adjustments

Run the engine or engine/generator under light load until it is hot. On generator sets, engine rpm controls both output frequency and voltage, so both speed and sensitivity settings are critical. Use of an accurate tachometer is the best mean of getting good performance without overspeeding the engine.

CCK GENERATORS

1. Adjust the carburetor for best mixture with the engine at full load.
2. Adjust the carburetor idle mixture with all load disconnected.
3. Adjust the length of the governor linkage by rotating the ball joint (first loosen the locknut). Adjust, so that with the engine stopped and the governor arm held in the closed position,

the stop screw on the carburetor throttle lever is ¹⁄₃₂ in. from the stop pin. Retighten the locknut.
4. Check the linkage and throttle shaft for binding or looseness and clean up parts or replace them as necessary.
5. Disconnect the booster external spring, and turn the speed adjusting nut to obtain the voltage and speed readings shown on the unit nameplate.
6. Move the governor spring inward to get greater sensitivity until a hunting condition occurs. Then, move it outward (toward the outward end of the governor arm) as necessary to ensure stable operation. Recheck speed adjustment and readjust as necessary.
7. Set the distance between the throttle stop screw and stop pin at ¹⁄₃₂ in.
8. Connect the vacuum booster external spring to the bracket on the governor link. With the unit operating with load, slide the bracket on the governor link just to the position where there is no tension on the external spring. Apply load and carefully watch the cycles. The speed should not drop more than four cycles, and it should recover rapidly. If it is necessary to make the speed booster more or less sensitive, change the cotter pin to another hole in the return spring strap.

CCK INDUSTRIAL VARIABLE SPEED GOVERNOR

1. Run the engine until it is hot and make necessary carburetor adjustments.
2. Adjust the throttle stop screw on the carburetor to a minimum idling speed of 1450 rpm so the governor spring can hold the engine speed at 1500 rpm.
3. Adjust governor spring tension for minimum speed. Shift the lever to the minimum (slow) position and with no load connected, adjust the spring tension for about 1500 rpm with the adjusting nut.
4. Adjust the sensitivity while operating at a minimum speed to attain the smoothest yet quickest no load to full load operation as follows:

To decrease sensitivity and prevent hunting (allow more speed drop from no load to full load operation): Move the governor spring outwards into a different groove or hole in the extension (or on earlier models, turn the sensitivity screw outwards) so that the point of pull by the spring is moved slightly away from the governor shaft.

To increase sensitivity (closer regulation by the governor that permits less speed drop from no load to full load operation): Move the governor spring inward to a different groove or hole in the extension (or on earlier models, turn the sensitivity screw inward) so that the point of pull by spring is moved slightly closer to the governor shaft. Engine speed should not drop more than 100 rpm from noload to full-load.
5. Apply full load and shift the lever until the engine speed reaches the desired maximum speed. Set the screw in the bracket slot to stop lever travel at the desired maximum full load speed position. Approximately 3000 rpm is the recommended maximum full load speed for continuous operation. The speed must be consistent with the load requirements. Adjust it lower if the load must be driven at a lower rpm.

CCK INDUSTRIAL TWO SPEED SOLENOID TYPE GOVERNOR

1. Run the engine until hot and adjust the carburetor for best operation.
2. Adjust the low speed spring tension for the desired low speed (it should be at least 1500 rpm for units that carry load at idle). Increase spring tension to increase speed.
3. Adjust the high speed spring tension by loosening the locknut and turning the plunger on the adjusting stud to give the de-

Locations of CCK series governor adjustments

sired high speed when the plunger is pulled all the way into the solenoid (it should not exceed 3000 rpm).

NOTE: If the plunger does not pull all the way into the solenoid due to excessive tension, electrical damage will occur.

4. Adjust the sensitivity so that no load speed is no more than 100 rpm above full load speed (measure with a tachometer). To increase sensitivity, move the high speed governor spring inward to a different hole of the bracket, or, to decrease it, outward. If, in order to adjust high speed sensitivity, you need to use the hole in the bracket that is occupied by the low speed spring, you can simply move that spring as required. Usually, low speed sensitivity is not critical. If it is necessary to adjust it, proceed as for the high speed spring.

CCKA AND CCKB TRACTOR GOVERNORS

1. Using a tachometer, adjust the throttle stop screw for 1000 rpm on CCKA applications, and to 1200 rpm on CCKB gover-

nors. Readjust the idle mixture as necessary, and then bring the closed throttle speed back to specification.

2. On CCKA units, adjust the nuts on the low speed (smaller) adjusting spring with the control in the "slow" position. Adjust speed to 1200 rpm, and then tighten the nuts against each other to lock.

3. On CCKB units, move the speed control to the "fast" position, and then turn the high speed (larger) adjusting spring adjusting nuts so the engine runs at 3800–3850 rpm. Then tighten the adjusting nuts against each other to lock, making sure rpm does not exceed 3850.

BF TRACTOR GOVERNOR

1. Disconnect the throttle linkage from the governor arm. Then, hold both linkage and governor arm towards the carburetor, and note which hole in the arm is closer to the position of the linkage.

2. Connect the linkage to the hole which more nearly lines up with the position of the link.

Adjustment points of two speed solenoid type governor

3. If it is necessary to increase the sensitivity, move the spring loop into the governor arm hole nearest to the governor. If it is necessary to decrease sensitivity, go in the opposite direction.

4. Adjust the low speed with the adjusting screw on the control wire bracket.

BF Power Drawer

1. Run the unit for 15 minutes under light load to warm up.

2. Adjust the length of the linkage by rotating the ball joint on the governor arm. With the engine stopped and slight tension on the governor spring, adjust so that the stop on the carburetor lever just touches the carburetor bowl.

Voltage and Speed Regulation Limits for BF Power Drawer Governor

Voltage Chart for Checking Governor Regulator	120 Volt 1 Phase 2 Wire
Maximum No-Load Voltage	126
Minimum Full-Load Voltage	110
Speed Chart for Checking Governor Regulation	
Maximum No-Load Speed (RPM)	1890
Hertz (Current Frequency)	63
Minimum Full-Load Speed (RPM)	1770
Hertz	59

3. Adjust with an accurate tachometer and voltmeter. Check voltage and speed at no load and at full load, and compare with specifications in the charts.

4. If it is necessary to change sensitivity, shift the spring toward the outer end of the governor arm to decrease it, or toward the inner end to increase it.

5. After adjusting and checking sensitivity, adjust speed by tightening or loosening the speed adjusting nut at the end of the spring.

Automatic Choke Adjustment

CCK

1. Loosen the choke cover screws (2) and rotate the cover as necessary to get the following dimension between the inner edge of the choke plate and the wall of the carburetor air horn: At 58°F — $\frac{1}{4}$ in. open; at 66°F — $\frac{1}{2}$ in. open; at 76°F — $\frac{3}{4}$ in. open; at 82°F — fully open.

2. Tighten the choke cover screws.

THERMAL-MAGNETIC CHOKE

NOTE: Make sure the engine has been off for at least one hour.

1. Loosen the screw which secures the choke body.

2. Turn the choke body clockwise to enrichen the setting, or counterclockwise to lean it out until measurement of the dimension between the choke and carburetor air horn conforms to specifications shown in the illustration. Tighten the choke body screw.

ELECTRIC SOLENOID CHOKE

1. With the engine cold, disconnect the linkage to the carburetor choke shaft. Rotate the choke lever in the closed direction

AMBIENT TEMP. (°F)	60	65	70	75	80	85	90	95	100
CHOKE OPENING (Inches)	1/8	9/64	5/32	11/64	3/16	13/64	7/32	15/64	1/4

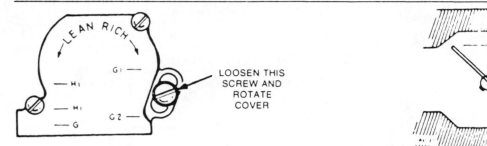

Thermal-magnetic choke adjustment specifications

until the hole in the shaft is aligned with the notch in the shaft bearing.

2. Insert a $\frac{1}{16}$ in. diameter rod through the shaft hole, engaging the rod in the notch of the mounting flange to lock the shaft in place.

3. Loosen the choke lever clamp screw enough to permit moving the lever on the shaft. Remove the air cleaner and verify that the linkage to the choke lever is properly in place. Adjust the choke assembly lever so the choke is from just closed to not more than $\frac{1}{16}$ in. open.

4. Tighten the choke lever clamp screw and remove the locking rod from the shaft.

5. Press downward on the choke lever to the limit of its travel and make sure the choke opens completely. If not, adjust the position of the choke shaft lever as necessary.

6. Make sure that when the engine is hot, the choke is wide open.

BF POWER DRAWER CHOKE

1. Remove the clip and bushing, and loosen the choke lever clamp screw.

2. With the lever fully forward (or, away from carburetor), adjust so the choke valve is completely closed.

3. Tighten the clamp screw, and replace the bushing and clip.

Valve Adjustment

Adjust the valve clearance with the engine cold and in TDC position with the spark plug about to fire. Adjustment is checked by sliding a flat feeler gauge of the proper dimension (see specifications, noting that clearances are different for exhaust and intake valves) between the tappet and head of the valve stem. If necessary, turn the adjusting nut on the tappet while holding tappet with a second wrench to get a slight pull on the gauge.

TESTING COMPRESSION

1. Run the engine until hot. Stop and remove spark plugs.

2. Insert the compression gauge in one of the spark plug holes and crank the engine with the starter. Record the reading.

3. Squirt a small amount of SAE 30 oil into the cylinder and repeat the check.

4. Repeat Steps 2–3 for each cylinder. Compression specifications are listed below. If compression is low, but increases substantially when oil is squirted into the cylinder, the compression problem is probably with the pistons, rings, and cylinders. Compression pressures are: (psi)

- LKB – 100–120
- NB – 105–115   • CCKA, CCKB – 100–120
- CCK – 90–110   • BF – 110–120

FUEL SYSTEM

Carburetion

The carburetor is a side draft or horizontal float type with two adjusting screws for the idle needle and the main fuel nozzle needle.

BASIC DOWNDRAFT AND SIDEDRAFT CARBURETORS

Removal and Installation

1. Remove air cleaner, fuel line, governor linkage and choke apparatus from carburetor.

2. Remove two carburetor mounting nuts and pull off carburetor. On CC Engines, first remove the intake manifold; then remove the carburetor from the manifold.

3. Reverse above steps to install carburetor on engine.

Overhaul

1. Remove air cleaner adapter and choke from carburetor.

2. Remove main fuel adjustment needle and needle retainer.

3. Remove top of carburetor from carburetor base.

Exploded view of downdraft carburetor

Exploded view of sidedraft carburetor

4. Remove carburetor float, lift out float valve and unscrew and remove its seat.

5. Remove no load adjusting needle.

6. Remove throttle plate and throttle shaft.

7. Remove choke plate and choke shaft.

8. Remove nozzle assembly.

9. Soak all components thoroughly in carburetor cleaner, following cleaner manufacturer's instructions. Clean all carbon from carburetor bore, especially in the area of the throttle valve. Blow out passages with compressed air. Avoid using wire to clean out passages.

10. Check adjusting needles and nozzle for damage. If the float is loaded with fuel or damaged, replace it. The float should turn freely on its pin without binding. Invert the carburetor body and measure float level.

11. To adjust float level, bend small lip that needle valve rides on.

12. Check choke and throttle shafts for excessive side play and replace if necessary.

13. Install throttle shaft and valve, using new screws. The bevel on the throttle plate must fit flush with carburetor body. On valve plates marked with a **C**, install them with the mark on the side toward idle port as viewed from flange end of carburetor. lkTo center throttle valve (Bendix/Zenith Carburetor) back off stop screw, close throttle lever, and seat valve by tapping it with a small screwdriver, then tighten the two throttle plate screws.

14. Install choke shaft and choke plate. Center choke plate in same manner as throttle valve. Always fasten plate in position with new screws.

15. Install main nozzle. Make sure it seats in body casting.

16. Install main fuel adjustment needle and its retainer.

17. Install no load adjusting needle.

18. Install intake valve seat and intake valve.

19. Install float and float pin. Center pin so float bowl doesn't ride against it.

20. Check float level. Adjust if necessary.

21. Install a new body-to-bowl gasket and secure the two sections together.

22. Reinstall choke.

23. Install air horn assembly.

24. Install carburetor on engine.

25. Adjust both main and idle fuel mixture needles, as described in the Tune-Up section.

BF ENGINE CARBURETOR

Removal and Installation

1. Remove air cleaner and hose.

2. Disconnect governor and throttle linkage, choke control and fuel line from carburetor.

3. Remove four intake manifold capscrews and lift complete manifold assembly from engine.

4. Remove carburetor from intake manifold.

5. Installation is the reverse of the removal procedure.

Adjusting the float level—BF engine carburetor

Overhaul

Generally follow the procedures applying to the downdraft and sidedraft carburetors as described above. Refer to the two items below for needle and seat replacement and float adjustment.

Replacing Needle and Seat

1. Remove four screws from top of carburetor and lift off float assembly.

2. Invert float assembly as shown.

3. Push out pin that holds float to cover.

4. Remove float and set aside in a clean place. Pull out needle and spring.

5. Remove valve seat and replace with a new one, making sure to use a new gasket.

6. Install new bowl gasket.

7. Clip new needle to float assembly with spring clip. Install float.

Float Adjustment

1. Invert float assembly and casting.

2. With float resting lightly against needle and seat, there should be $1/8$ in. clearance between bowl cover gasket and free end of float.

3. If it is necessary to reset float level, bend float tangs near pin to obtain a $1/8$ in. clearance, as shown.

BF POWER DRAWER CARBURETORS

Removal and Installation

1. Disconnect fuel inlet hose, crankcase breather hose and air inlet hose.

2. Disconnect governor, throttle linkage, and choke control.

3. Remove two hold-down nuts and lift carburetor from intake manifold.

4. Installation is the reverse of the removal procedure.

CHOKE SHAFT
AND LEVER

CHOKE FLY

SLEEVE
ASSEMBLY

NOTE:
ON SOME MODELS ENGINES
FUEL PUMP IS MOUNTED
DIRECTLY TO CARBURETOR

FLOAT VALVE
ASSEMBLY

FLOAT AND LEVER
ASSEMBLY

MAIN ADJUSTING
NEEDLE

CARBURETOR BODY

IDLE ADJUSTING
NEEDLE

THROTTLE STOP
SCREW

THROTTLE SHAFT
AND LEVER

FUEL PUMP

GASKET

THROTTLE FLY

CARBURETOR
INLET PLATE

FUEL
INLET

FUEL LINE

Exploded view of BF engine carburetor

THROTTLE SHAFT

THROTTLE VALVE
RETAINER SCREW

CHOKE SHAFT

THROTTLE
VALVE

THROTTLE
STOP
SCREW

CHOKE VALVE

CHOKE VALVE
RETAINER SCREW

IDLE ADJUSTMENT
SCREW

CARBURETOR BODY

GASKET

NOZZLE

VALVE SEAT

NEEDLE

FLOAT

FLOAT PIN

BODY-TO-BOWL
GASKET

BOWL SCREEN

FLOAT BOWL

GASKET

BOWL RETAINER

MAIN (POWER)
ADJUSTMENT NEEDLE

Exploded view of BF power drawer carburetor

Exploded view of CCK mechanical fuel pump

Overhaul

Generally follow the procedures applying to the downdraft and sidedraft carburetors as described above, referring to the exploded view. Refer to the two items below for needle and seat replacement and float level adjustment.

Replacing Needle and Seat

1. Remove $7/16$ in. hex at base of fuel bowl and lift bowl from carburetor.
2. Push out pin that holds float to carburetor body.
3. Remove float and set aside in a clean place. Pull out needle and using a large screwdriver remove needle valve seat.
4. Install new valve seat and needle and replace float.

Float Adjustment

1. Invert float and casting.
2. With float resting lightly against needle and seat, there should be 0.07–0.11 in. clearance between base of float and carburetor casting.
3. If it is necessary to reset float level, remove float from carburetor and bend float tang near pin to obtain correct float level.

CAUTION

Do not bend the float when installed; doing so may cause deformation of needle or seat.

---

4. Check float carefully for signs of leakage. Repair or replace float if damaged or filled with gasoline.
5. Before assembling carburetor, remove filter screen from float bowl and clean both screen and base of float bowl.
6. Install new gaskets when reassembling.

Mechanical Fuel Pumps

REMOVAL AND INSTALLATION

1. Remove the fuel lines.
2. Remove the two mounting capscrews, and pull the pump off the engine. Remove and discard the gasket.
3. Clean all gasket material from both mounting pad and pump flange. Apply an oil resistant sealer to both sides of a new gasket and to the threads of the attaching bolts.
4. Position the new gasket on the pump flange and position the pump and gasket on the mounting pad. To check their position, make sure the rocker arm rides on the pump cam lobe. Turn the crankshaft until the rocker arm is at the low point of its stroke.
5. Position the pump tightly against the pad and install the mounting bolts, torquing them alternately in several stages to specifications.
6. Connect the fuel lines and then operate the unit to check for leaks.

SERVICE

1. Scribe a mark across the flanges of the pump body and valve housing so these parts can be assembled in original positions.
2. Remove the valve housing from the body of the pump. Tap pump body with a screwdriver to do this.
3. Remove both valves and their gaskets from valve housing. Note position of valves in their housing so new valves can be correctly installed.
4. Using a blunt punch, drive rocker arm pin out of pump.
5. Press diaphragm into fuel pump body and then pull rocker arm outward to unhook diaphragm actuating rod from rocker arm link assembly.
6. Remove diaphragm and diaphragm return spring, rocker arm and link assembly, and rocker arm return spring from pump body.
7. Remove diaphragm actuating rod oil seal from pump body.
8. Clean and inspect all fuel pump components and replace all unserviceable parts.
9. Install inlet and outlet valves and their gaskets in their respective positions. Seat valves firmly.
10. Lubricate diaphragm actuating rod.
11. Position fuel pump diaphragm and spring assembly into pump body as shown.
12. Hold diaphragm assembly in pump body and position pump body so mounting flanges faces up. Apply slightly more pressure to lower edge of diaphragm and insert rocker arm link assembly.
13. Hook rocker arm link to diaphragm actuating rod.
14. Install rocker arm return spring and hold it in place by cocking rocker arm slightly.
15. Install the rocker arm pin in the pump body.
16. Position the valve body and pump body so the two previously scribed marks align. Install all screws and lockwashers until they just engage the pump body, being careful not to tear the diaphragm fabric.
17. Alternately and evenly tighten all screws.

Exploded view of BF pulsating diaphragm fuel pump

BF Engine Pulsating Diaphragm Fuel Pump

SERVICE

1. Disconnect vacuum and fuel lines. Inspect lines for cracks and replace as necessary.

NOTE: On some engines, the pump is mounted on the side of the engine and has an obvious fuel discharge line. On other applications, the pump is an integral part of the carburetor and only a suction line is present.

2. Scribe two lines (one each on opposite ends of the pump) across the pump parts. This will ensure correct alignment of the pump parts with each other and the carburetor when the pump is reassembled.
3. Remove the fuel pump attaching screws.
4. Holding the pump carefully, pull the sections of the pump apart. The diaphragm, plunger, return spring and plate, pump body and gaskets will now be loose.
5. Check parts for wear and damage. Replace them with new parts where necessary.
6. Unclog the pump air bleed hole in the pump base to allow unrestricted movement of pump diaphragm.
7. Replace the gaskets and reassemble the pump.

NOTE: That all parts must be perfectly aligned or there will be leakage and a consequent fire hazard.

8. Install the pump according to the marks scribed above. Reconnect the fuel lines, with the clamps rightly in their proper positions.

Bendix electric fuel pump—removal of bottom cover, gasket, magnet, and filter

Bendix electric fule pump—removal of retainer, O-ring, cup valve, spring, and plunger

Bendix Electric Fuel Pumps

SERVICE

1. Release the bottom cover (1) from the bayonet fittings. Twist the cover by hand to remove it from the pump body.
2. Remove the filter (4), magnet (3) and cover gasket (2) (see appropriate illustration). Wash the filter in cleaning solvent and blow out dirt and cleaning solvent with air pressure. Check the cover gasket and replace if deteriorated. Clean the cover.
3. Remove the retainer spring (5) from plunger tube (11), using thin nose pliers to spread and remove the ends of the retainer from the tube. Then remove the washer (6), O-ring seal (7), cup valve (8), plunger spring (9) and plunger (10) from the tube (11), (see appropriate illustration).
4. Wash all parts in cleaning solvent and blow out with air pressure. If the plunger does not wash clean or if there are any rough spots, gently clean the surface with a crocus cloth. Slosh the pump assembly in cleaning solvent. Blow out the tube with air pressure. Swab the inside of the tube with a cloth wrapped around a stick.
5. Insert the plunger assembly (10) in the tube with the buffer spring end first. Check the fit by slowly raising and lowering the plunger in the tube. It should move fully without any tendency to stick. If a click cannot be heard, the interrupter assembly is not functioning properly in which case pump should be replaced.
6. To complete the assembly, install the plunger spring (9), cup valve (8), O-ring seal (7) and washer (6) as shown. Compress the spring (9) and assemble the retainer (5) with ends of the retainer in the side holes of tube (11).

7. Place the cover gasket (2) and magnet (3) in the bottom cover (1) and assemble the filter (4) and cover assembly. Twist the cover by hand to hold it in position on the pump housing. Securely tighten the bottom cover.

Onan Electric Fuel Pump

SERVICE

Clean the filters every 100 operating hours. Remove the four phillips screws from the top, and lift off the filter assembly. Clean the two screen-type filters in a safe solvent. Install, ensuring the gasket is in the proper position to prevent leaks.

Governor Repair

1. Draw a sketch of the governor linkage or lay the parts out in position as you disassemble them. Disassemble the linkage. Remove the gear cover.

Typical centrifugal type governor

Measuring governor in and out travel

2. Remove the snap ring that holds the governor cup to the camshaft gear, being ready to catch the flyballs, which will come out as the governor cup is removed.

3. Clean all parts thoroughly in a safe solvent. Inspect as follows and replace parts which are found defective:
    a. flyballs for grooves or flat spots.
    b. ball spacers for arms with noticeable wear or damage.
    c. governor cup with a rough or grooved race surface.
    d. governor cup which does not have a free spinning fit on the camshaft center pin, or which is loose and wobbles.

4. To install the governor cup, tilt the engine to make the gear face upward. Space the flyballs at equal distances on the gear and then install the cup and snap ring on the camshaft center pin.

5. On CCK engines, measure the distance the center pin extends outward — it should be $3/4$ in. On J-series this dimension should be $25/32$ in. Hold the cup against the flyballs to make this measurement. If the distance is not correct, pull out the center pin and replace it, pressing the new pin in just the required amount. If the pin extension is ok, grind off the hub of the cup as required. In and out travel distance must be $7/32$ in. or the engine will race.

ENGINE OVERHAUL

Valves

SERVICE

In order to remove the valves, the cylinder head must be removed. Remove the cylinder head screws and then remove the head from the engine. If the head sticks to the block, hit it lightly with a soft hammer, being careful not to damage any of the cooling fins. Remove the cylinder head gasket and discard it.

Use a conventional type valve spring compressor to compress the valve springs so that the spring retainer can be removed. The retainers are the split, tapered type and will most likely fall out when the valve spring is compressed. After removing the retainers, lift the valve out through the top of the valve guide. Clean the valves of all carbon deposits and inspect them, looking for warpage, worn stems, and burned surfaces that are partially destroyed. Determine whether or not the valve can be reused, whether it can be reground, or if the valve has to be replaced.

Check out the valve stem-to-guide clearance and, if it is too large, the guides must be replaced. They can be removed through the valve chamber. The valve tappets are also replaceable from inside the valve chamber, once the valves have been removed.

In removing the valve guides, first wire brush carbon and other deposits from the top guide surfaces, or the guide bores may be damaged during guide removal. Note that a gasket must be used on the intake valve guides for the LKB and BF, engines. Where used, place the gasket on the intake guide and install the intake and exhaust guides from within the valve chamber. Before installing the guides, run a small polishing rod with a crocus cloth through the guide holes to clean out deposits.

VALVE SEAT INSERT REMOVAL AND INSTALLATION

Use a solvent to clean carbon or stuck gasket material from cylinder head and block surfaces. If necessary, follow up using a metal scraper. Inspect the head gasket surfaces for cracks, nicks, or burrs. Check the head for cracks, and replace if any are present. An oil stone may be used to remove burrs or nicks.

Use a straight-edge at three angles to check the head for flatness. Try to insert a 0.003 in. flat feeler gauge under the straight-edge at any point, and replace the head if it is not flat.

Special tool for removing valve seats from engines with cast iron blocks

Replace the valve seats if cracked or loose, or excessively worn. On cast iron engines, the seat may be driven out using a knockout tool as shown. The tool must be inserted under the port side of the valve seat with the square end extending over the cylinder bore. With the sharp edge of the tool at the joint between the seat and its recess, strike a sharp blow on the end of the tool with a light hammer. This will crack the insert and permit removal.

CAUTION

Since this may shatter the relatively brittle material of the seat, you should wear goggles when performing the procedure.

If the engine has an aluminum block, use a ¾ in. or one inch pipe tap to suit the seat diameter. Place a washer on top of the valve guide for the tap to bottom against. Turn the tap in until the seat begins to turn. As it starts to turn, begin pulling outward on the tap while continuing the turning motion to pull out the seat. Make sure the valve guide is not pushed downward by the tap or, if it is, that it is pressed back into position. Use an oversize replacement seat on aluminum engines.

Clean any carbon or burrs from the insert recess, and install a new seat as follows:

1. Gradually heat the block to 325°F. Place the new seat in dry ice until thoroughly chilled.

2. Insert the pilot of an appropriate special tool in the valve guide hole in the block and quickly and evenly drive in the seat insert. It must seat on the bottom of the recess.

The valve face angle is 44 degrees. The valve seat angle is 45 degrees. The 1 degree interference angle assures a sharp seating surface between the valve and the seat and good sealing characteristics. The valve seat width must be between ¹⁄₃₂ in. and ³⁄₆₄ in. Valves should not be hand lapped if at all possible. This is especially important if stellite valves are used.

To check the valves for a tight seal, make pencil marks around the valve face, then install the valve and rotate it a part of a turn. If the marks are all rubbed off uniformly, then the seal is good.

1. Insert the tappets in the crankcase holes.

2. Install the valves, springs and guides.

3. Using a valve spring compressor, compress each valve spring and insert the valve spring retainer and retainer locks.

4. Set the valve clearance to the specifications listed at the back of the chapter.

5. Install the heads and gaskets to the cylinder block.

6. Tighten the head bolts to the correct torque following the sequence in the appropriate illustration.

7. Install the exhaust manifold, oil lines, spark plugs and carburetor.

Gear Cover

REMOVAL AND INSTALLATION

In order to gain access to the camshaft gear and other internal components of the engine, it is necessary to remove the gear cover. In order to remove the gear cover, the magneto assembly must be removed. Disconnect the spark plug wire at the spark plug and disconnect the stop wire. Remove the attaching screws that hold the magneto assembly to the gear cover and remove the magneto.

When the gear cover is removed, the governor shaft disengages from the governor cup, which is part of the camshaft.

During the installation of the cover, be sure to engage the pin on the cover, with the chamfered hole located in the governor cup. To do this, turn the governor cup so that the hole is located at the top or in the 12 o'clock position. Turn the governor shaft clockwise as far as it will go and hold it there until the cover is installed. Position the cover on the engine and make sure that it fits flush against the engine. Be careful of the gear cover oil seal during installation. Use a new gasket if the old one is damaged.

Timing Gears

REMOVAL AND INSTALLATION

If it becomes necessary to replace either the crankshaft or camshaft gears because of broken teeth, extreme wear, cracks, etc., both gears must be replaced as a pair. Never replace only one of the gears. Both gears are pressed onto their respective shafts.

To remove the crankshaft gear, insert two #10–32 screws into the threaded holes in the gear and tighten the screws alternately a little at a time. The screws will press up against the

Cylinder head torque sequences

Valve and crankcase breather assembly

crankshaft shoulder and force the gear off the end of the crankshaft. On AK engines use a gear puller to remove the crankshaft gear.

To remove the camshaft gear, it is necessary to remove the entire camshaft assembly from the engine. First remove the crankshaft gear lock ring and washer. Remove the cylinder head, valve assemblies, fuel pump (if so equipped), and the valve tappets. Remove the governor cup assembly and then remove the camshaft and gear assembly from the engine. The camshaft gear may now be pressed off the camshaft. Do not press on the camshaft center pin as it will be damaged. The governor ball spacer is press fit into the camshaft gear.

When the camshaft gear is replaced on the camshaft, be certain that the gear is properly aligned and the key properly positioned before beginning to press the gear onto the camshaft.

Install the governor cup before replacing the camshaft and gear assembly back into position in the engine.

There are two stamped **O** marks, one on each gear, near the gear teeth. When the crankshaft and camshaft gears are meshed, these marks must be exactly opposite each other. When

installing the camshaft gear assembly, be sure that the thrust washer that goes behind the camshaft gear is installed. Replace the retaining washer and lockwasher on the crankshaft.

Cylinder Bore, Piston, and Piston Rings

The cylinder can be rebored if it becomes heavily scored or badly worn. If the cylinder bore becomes cracked, replace the cylinder block.

The cylinder can be bored out to 0.010 in., 0.020 in., or 0.030 in. oversize. There are pistons and rings in the above oversizes available to accommodate an oversized cylinder bore. If the cylinder bore has to be bored out only 0.005 in. to remove the damage in the cylinder, use standard size parts.

Use a ridge reamer to remove the ridge that may be present at the top of the cylinder bore to avoid damaging the piston rings when the piston and connecting rod assembly is removed. Hone the cylinder and create a cross hatch pattern on the cylinder

Gear cover assembly

IF FEELER WILL ENTER HOLE ½" BALL HAS FALLEN OUT.

walls if new rings are being installed. Clean the cylinder with SAE 10 engine oil after honing.

Some engines were originally built with 0.005 in. oversize pistons and are so indicated by a letter **E** following the serial number stamped on the identification plate and on the side of the crankcase.

The piston is fitted with two compression rings and one oil control ring. When the piston assembly is removed from the engine, clean off all carbon deposits and open all of the oil return holes in the lower ring groove. Before installing new rings, check the ring gap by installing the rings squarely in the cylinder bore and measuring the gap between the two ends of the rings. If the gap is too small, it is possible to file the ends to obtain the proper size gap.

Tapered type rings are usually marked with the word **TOP** on one side. This side must be installed facing toward the top or closed end of the piston. Position the ring gaps evenly around the circumference of the piston with no ring gap over the piston pin.

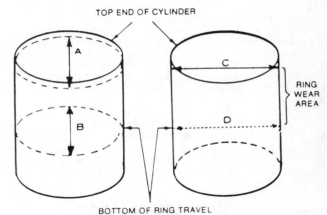

Locations for taking cylinder wear measurements

Aligning the timing marks on the timing gears and removing the crankshaft gear

Measure piston at points shown

The piston pin is held in place by two lock rings, one at each end. Make sure that the lock rings are properly installed before installing the piston assembly in the cylinder. Be sure to check the size of the piston, piston pin, piston pin bore, and the size of the cylinder before installing any of these parts back into the engine. Replace any parts that are worn beyond the maximum allowed specification. The piston should be replaced when the side clearance of the top ring reaches 0.008 in., or if there are signs of scuffing, scoring, worn ring lands, fractures or pre-ignition damage. Measure the piston dimensions at points shown in the illustration.

In fitting rings, install each ring in the cylinder bore, invert the piston and push the ring to the end of the ring travel-about halfway to the end of the bore. With the ring at exactly 90 degrees to the centerline of the bore, measure the end gap with a feeler gauge. Fit rings by choosing the right size for the bore-do not file the ends.

Connecting Rod

Before removing the connecting rod from the crankshaft, mark the cap and rod so they can be installed in exactly the same position from which they are removed.

If abnormal bearing wear (worn on one side more than the other) is noticed, this would indicate that the connecting rod is bent. It is possible to have the connecting rod straightened, but this should be done at a machine shop or small engine service shop.

Measure all of the bearing surfaces for size, including the piston pin hole and the crankpin bearing. Inspect the bearings for burrs, breaks, pitting and wear. Replace if scored or if the overlay is wiped out. Scratching is also reason for replacement. If bearings look all right, check bearing clearances. Place a piece of Plastigage of the proper width in the bearing cap about 1/4 in. off center. Rotate the crankshaft about 30 degrees from bottom center, and install the bearing cap. Torque to specification. Without turning the crankshaft, remove the bearing cap, leaving the flattened Plastigage on the bearing or journal. Compare its widest point with the scale on the Plastigage envelope to determine clearance. If clearance is excessive, replace the connecting rod bearings.

Use a new piston pin to check the pin bushing for wear. A push fit clearance is required. If a new pin will fall through a dry rod bore of its own weight, replace the rod bushing.

To replace the bushings, press them out using a press and proper driving tool. Press in the new bushings so that the ends are flush with the sides of the rod and a 1/16 in. oil groove is formed in the center. If there are oil holes, make sure they are at least half way open. Make sure the pin is a push fit after the bushing is installed.

Replace rod nuts or bolts if the threads are damaged. Replace rods which are nicked or fractured, or which have bores which are out of round more than 0.002 in. Straighten or replace rods which are twisted more than 0.012 in. or out of line more than 0.005 in.

Be sure to reinstall the oil dipper to the connecting rod cap when assembling the piston and connecting rod assembly to the crankshaft.

Cylinder Block
INSPECTION

Check the entire block thoroughly for cracks, and check the cylinder bore for scoring. Measure the bores for out-of-round at points indicated. A is the point where greatest ring wear occurs, and B is the bottom of ring travel. C and D are at the same heights as A and B respectively, but at 90 degrees. Compare (subtract the smaller from the larger) A and B and then compare C and D to determine taper. Cylinder must be rebored for the next oversize piston if taper exceeds 0.005 in. Comparing A to C and B to D indicates out-of-round. If the cylinder is out-of-round 0.002 in., it must be rebored for the next oversize piston.

PRODUCE CROSS HATCH SCRATCHES
FOR FAST RING SEATING

AVOID THIS FINISH

Produce a crosshatch in the bore as shown in the upper part of the illustration

HONING CYLINDERS TO OVERSIZE DIMENSIONS

1. Anchor the block solidly for either vertical or horizontal honing. Use either a drill press or heavy-duty drill which operates at about 250–450 rpm.

2. Lower the hone into the cylinder until it protrudes $\frac{1}{2}$–$\frac{3}{4}$ in. past the end of the cylinder. Rotate the adjusting nut until the stones just come in contact with the cylinder wall at its narrowest point.

3. Loosen the adjusting nut until the hone can be turned by hand.

4. Connect the drill to the one and start the drill. Move the hone up and down in the cylinder about 40 times per minute. Usually the bottom of the cylinder must be worked out first because it is smaller. When the cylinder takes a uniform diameter, move the hone up and down all the way through the bore. Follow the hone manufacturer's recommendations for wet or dry honing and oiling the hone.

5. Check the diameter of the cylinder regularly during honing. A dial bore gauge is the easiest method but a telescoping gauge can be used. Check the size at six places in the bore; measure twice at the top, middle and bottom at 90 degree angles.

6. When the cylinder is within about 0.002 inch of the desired bore, change to fine stones and finish the bore. The finish should not be smooth but crosshatched as shown. The crosshatch formed by the scratching of the stones should form an angle of 32 degrees. This can be achieved by moving the hone up and down in the cylinder about 40 times per minute.

7. Clean the cylinder block thoroughly with soap, water and clean rags. A clean white rag should not be soiled on the wall after cleaning is complete.

—————— CAUTION ——————

Do not use a solvent or gasoline since they wash the oil from the walls but leave the metal particles.

8. Dry the crankcase and coat it with oil.

Crankshaft and Bearings

CRANKSHAFT REMOVAL

1. Remove the lock ring and retainer washer from the crankshaft gear, and pull off gear with a puller.

2. Remove the oil pan, head(s), pistons and rods. See the previous section on piston and ring service.

3. Remove the rear bearing plate. Remove the crankshaft through the rear opening in the crankcase.

INSPECTION

Clean the crankshaft thoroughly, including blowing out all oil passages with compressed air. Check the journals for out-of-round, taper, grooves, and ridges, and place on V-blocks and rotate to measure runout. It should not exceed 0.003 in. Compare all dimensions with specifications: out-of-round should not exceed 0.005 in.; taper should should not exceed 0.001 in.; wear should not exceed 0.002 in. If the limits are exceeded, regrind the shaft for the use of undersize bearings.

Replace bearings that are warped, scored, or have been overheated. If such damage is present, check the bearing bores in the block for excessive size. A new rear main bearing seat can be installed by replacing the rear main bearing plate. After doing this, check the main bearing bore alignment on a line boring machine.

The main crankshaft bearings are precision bearings. They are available in the standard size as well as 0.002 in., 0.010 in., 0.020 in., and 0.030 in. undersize. The precision type bearings are not to be line reamed.

The bearings are press fit into the cylinder block. Before trying to install the bearing into the cylinder block or bearing plate, heat the plate or block by running hot water over them or placing them in an oven heated to 200°F. This will cause the block or bearing plate to expand and facilitate the installation of the sleeve bearings.

The oil hole in the bearing and the oil hole in the bearing bore must be aligned when the bearings are installed. On pressure lubricated engines, the hole should be opposite the crankshaft. On engines that are splash lubricated, the hole should be upward. Install rear bearings in the rear bearing plate using a special driver. Install them to $\frac{1}{64}$ in. below the end of the bore. If a special tool is not available and the lock pins must be removed with the side cutters or a screw extractor, install new lock pins.

CRANKSHAFT INSTALLATION

1. Oil the bearing surfaces thoroughly. Install the crankshaft from the rear of the crankcase (through the rear bearing plate hole).

2. Put the rear bearing plate gasket in place and lubricate the rear end plate bearing. Slide the thrust washer, with grooves toward the crankshaft and the bearing plate, over the end of the crankshaft.

NOTE: Line up the thrust washer notches with the lock pins before tightening the end plate bolts, or the lock pins and thrust washer will be damaged.

If you have trouble getting the thrust washer to stay in place, it may be lubricated with a light coating of oil.

3. Torque the bearing plate bolts.

4. Heat the timing gear to 350°F. Install a new crankshaft key and drive the gear into position. Install the washer and lock ring.

5. Adjust the crankshaft end play as described below.

6. Complete the reassembly of the engine.

CHECKING END PLAY

Check the end play with the rear bearing plate bolts properly torque.

REAR BEARING PLATE

MEASURE ENDPLAY HERE
(Refer to Table of Clearances)

Measure end play at location shown

If end play is excessive, remove the end plate and install a shim between the thrust washer and plate. When installing the plate, line up the notches in the thrust washer and shim with the lock pins. Torque the end plate and recheck end play. If total gasket and shim thickness required is more than 0.015 in., use a steel shim of the proper thickness and two thin gaskets, or gasket compression and consequent loose bolts may result.

Crankshaft Oil Seals

The crankshaft oil seals are installed with the open sides facing toward the inside of the engine. To replace the rear oil seal, the rear bearing plate must be removed. To replace the front oil seal, the front gear cover must be removed. Be careful not to damage the oil seal during installation or to turn back the edge of the seal lip.

LUBRICATION

Onan engines are either splash lubricated or pressure lubricated.

The splash lubrication system consists of an oil dipper attached to the connecting rod cap and various oil passages to catch and channel the oil that is splashed up by the oil dipper.

The pressure lubrication system consists of a gear type oil pump, an oil intake cup, a non-adjustable pressure relief valve, and the various oil passages and channels.

Oil Pump

SERVICE

1. Drain the crankcase oil. Then, remove the gear cover and oil base.
2. Unscrew the intake cup from the oil pump. Loosen the two capscrews holding the pump and remove it.
3. Disassemble the pump by removing the two cap screws holding the cover in position. Inspect for excessive wear of gears and shafts and replace the pump if any parts are badly damaged. Gears may be measured for excess wear by positioning them in the housing, running a straightedge across the sides of the housing, and inserting a flat feeler gauge between the straightedge

Exploded view of oil pump

and side of the gear. If either this clearance or the clearance between oil pump teeth is excessive, replace the unit.

4. In reassembling the pump, side clearance between gears and housing can be adjusted by using a thinner gasket. Use the thinnest gasket that permits freedom in operation of the pump.
5. Fully prime the pump with oil before assembly. Mount the pump into position on the engine and adjust for a clearance of 0.005 in. between the pump and crankshaft gears.
6. Install the intake cup on the pump parallel with the bottom of the crankcase.
7. Install the gear cover and oil base, fill the crankcase with oil, and run engine, checking for good oil pressure.

Crankshaft and Connecting Rod Specifications
All measurements are given in inches

Engine Model	Crankshaft				Connecting Rod		
	Main Brg Journal Dia	Main Brg Oil Clearance	Shaft End-Play	Thrust on No.	Journal Diameter	Oil Clearance	Side Clearance
LKB	1.9992–2.0000	.0020–.0030	.006–.0012	Rear	1.6252–1.6260	.0020–.0030	.002–.016
NB	1.9992–2.0000	.0025–.0038	.006–.0012	Rear	1.6252–1.6260	.0020–.0030	.002–.016
CCK	1.9992–2.0000	.0025–.0038	.006–.0012	Rear	1.6250–1.6252	.0005–.0023	.002–.016
CCKA	1.9992–2.0000	.0025–.0038	.006–.0012	Rear	1.6250–1.6252	.0005–.0023	.002–.016
CCKB	1.9992–2.0000	.0025–.0038	.006–.0012	Rear	1.6250–1.6252	.0005–.0023	.002–.016
BF	1.9992–2.0000	.0025–.0038	.006–.0012	Rear	1.6250–1.6252	.0020–.0033	.002–.016

Valve Specifications

Engine Model	Seat Angle (deg)	Face Angle (deg)	Spring Test Pressure (lbs @ in.)	Spring Installed Height (in.)	Guide Diameter (in.)		Stem Diameter (in.)	
					Intake	Exhaust	Intake	Exhaust
LKB	45	44	71 – 79 @ 1.375	—	.3440 – .3460	.3440 – .3460	.3425 – .3430	.3410 – .3415
NB	45	44	71 – 79 @ 1.375	—	.3440 – .3460	.3440 – .3460	.3425 – .3430	.3410 – .3415
CCK	45	44	71 – 79 @ 1.375	—	.3440 – .3460	.3440 – .3460	.3425 – .3430	.3410 – .3415
CCKA	45	44	71 – 79 @ 1.375	—	.3440 – .3460	.3440 – .3460	.3425 – .3430	.3410 – .3415
CCKB	45	44	71 – 79 @ 1.375	—	.3440 – .3460	.3440 – .3460	.3425 – .3430	.3410 – .3415
BF	45	44	71 – 79 @ 1.375	—	.3440 – .3460	.3440 – .3460	.3425 – .3430	.3410 – .3415

Piston Clearance

Engine Model	Piston to Bore Clearance (in.)
LKB	.0005 – ① .0015
NB	.0025 – ② .0045
CCK	.0015 – ② .0035
CCKA	.0015 – ② .0035
CCKB	.0015 – ② .0035
BF	.0010 – ② .0030

① Measure piston diameter across dimension Y
② Measure piston diameter across dimension X

Ring Gap

All measurements are given in inches

Engine Model	Top Compression	Bottom Compression	Oil Control
LKB	.010 – .023	.010 – .023	.010 – .023
NB	.013 – .023	.013 – .023	.013 – .023
CCK	.010 – .023	.010 – .023	.010 – .023
CCKA	.010 – .023	.010 – .023	.010 – .023
CCKB	.010 – .023	.010 – .023	.010 – .023
BK	.010 – .020	.010 – .020	.010 – .020

Ring Side Clearance

All measurements are given in inches

Engine Model	Top Compression
LKB	002 – 008
NB	002 – 008
CCK	002 – 008

Ring Side Clearance

All measurements are given in inches

Engine Model	Top Compression
CCKA	.002 – 008
CCKB	.002 – 008
BF	002 – 004

Torque Specifications

Engine Series		Use Engine Lubricating Oil As A Thread Lubricant										Do Not Use Any Lubricant On These Threads			
												Armature Thru Stud			
		Cylinder Head (Cold)	Conn Rod	Rear Bearing Plate	Main Bearing (4 Cyl)	Flywheel To Crankshaft	Oil Base	Exhaust Manifold (Tighten Evenly)	Intake Manifold	Damper Flywheel Assy Nut (4 Cyl)	Rocker Arm Stud in Head	Revolving Armature Units	Revolving Field Units	Spark Plugs	Injection Nozzle
CCK, CCKA, CCKB, MCCK, RCCK	lb-ft	29–31	[1]	20–25	—	35–40	43–48	—	15–20	—	—	35–40	—	25–30	—
BF, BG, BFA, BGA	lb-ft	14–16	14–16	25–27	—	35–40	18–23	6–10	—	—	—	45–50	—	15–20	—
B43M, M48M	lb-ft	16–18	14–18	25–27	—	35–40	18–23	9–11	6–10	—	—	—	—	15–20	—
NB	lb-ft	29–31	[1]	30–35	—	30–35 [2]	38–43	—	—	—	—	35–40	—	15–20	—
NH	lb-ft	22–25	27–29	25–27	—	30–35 [2]	18–23	—	—	—	—	35–40	—	15–20	—
NHA, NHB, NHC, NHAV, NHBV, NHCV, NHP, NHPV, N52M	lb-ft	17–19 [5]	27–29	20–23	—	35–40	18–23	10–12	18–20	—	—	45–50	—	15–20	—
JA	lb-ft	28–30 [3]	27–29	40–45 [4]	—	65–70	32–38	13–15	13–15	—	25–30	30–40	—	25–30	—
JB	lb-ft	28–30 [3]	27–29	40–45 [4]	—	65–70	45–50	13–15	13–15	—	25–30	—	55–60	25–30	—
JC	lb-ft	28–30 [3]	27–29	40–45 [4]	97–102	65–70	45–50	13–15	13–15	—	25–30	—	55–60	25–30	—
MJA	lb-ft	44–46 [3]	27–29	40–45 [4]	—	65–70	32–38	13–15	13–15	—	35–40	30–40	—	25–30	—
MJB	lb-ft	44–46 [3]	27–29	40–45 [4]	—	65–70	45–50	13–15	13–15	—	35–40	—	55–60	25–30	—
MJC	lb-ft	44–46 [3]	27–29	40–45 [4]	97–102	65–70	45–50	13–15	13–15	17–21	35–40	—	55–60	25–30	—
MDJA	lb-ft	44–46 [3]	27–29	40–45 [4]	—	65–70	32–38	13–15	13–15	—	35–40	30–40	—	—	20–21
DJA	lb-ft	37–40 [3]	27–29	40–45 [4]	—	65–70	32–38	13–15	13–15	—	35–40	30–40	—	—	20–21
MDJB	lb-ft	44–46 [3]	27–29	40–45 [4]	—	65–70	45–50	13–15	13–15	—	35–40	—	55–60	—	20–21
DJB, DJE	lb-ft	37–40 [3]	27–29	40–45 [4]	—	65–70	45–50	13–15	13–15	—	35–40	—	55–60	—	20–21
MDJE	lb-ft	44–46 [3]	27–29	40–45 [4]	—	65–70	45–50	13–15	13–15	—	35–40	—	55–60	—	20–21
MDJC	lb-ft	44–46 [3]	27–29	40–45 [4]	97–102	65–70	45–50	13–15	13–15	17–21	35–40	—	55–60	—	20–21
DJC	lb-ft	37–40 [3]	27–29	40–45 [4]	97–102	65–70	45–50	13–15	13–15	17–21	35–40	—	55–60	—	20–21
MDJF	lb-ft	44–46 [3]	27–29	40–45 [4]	97–102	65–70	45–50	13–15	13–15	17–21	35–40	—	55–60	—	20–21
RDJE, RDJEA	lb-ft	44–46 [3]	27–29	40–45 [4]	—	65–70	45–50	13–15	13–15	—	35–40	—	—	—	20–21
RJC	lb-ft	44–46 [3]	27–29	40–45 [4]	97–102	65–70	45–50	13–15	13–15	17–21	35–40	—	55–60	—	20–21
RDJC	lb-ft	44–46 [3]	27–29	40–45 [4]	97–102	65–70	45–50	13–15	13–15	17–21	35–40	—	55–60	—	20–21
RDJF	lb-ft	44–46 [3]	27–29	40–45 [4]	97–102	65–70	45–50	13–15	13–15	17–21	35–40	—	55–60	—	20–21

[1] Aluminum rods 24–26 lb-ft (33–35 N·m); forged rods 27–29 (37–39 N·m).
[2] Zinc or aluminum wheel. Cast iron wheel 40–45 lb-ft (54–61 N·m).
[3] Use NEVER-SEEZE or equivalent when torquing to this value.
[4] Use LOCTITE when torquing bolts.
[5] When using compression washers torque should be 13–15 lbs.

Section 15

Tecumseh-Lauson Light Frame 4-Stroke

Small Engines — 0-6 hp

ENGINE IDENTIFICATION

Lauson 4 cycle engines are identified by a model number stamped on a nameplate. The nameplate is located on the crankcase of vertical shaft models and on the blower housing of horizontal shaft models.

A typical model number appears on the illustration showing the location of the nameplate for vertical crankshaft engines. This number is interpreted as follows:

- V — vertical shaft engine
- 60 — 6.0 horsepower
- 70360J — the specification number. The last three numbers (360) indicate that this particular engine is a variation on the basic model line.
- 2361J — serial number
- 2 — year of manufacture
- 361 — the calendar day of manufacture
- J — line and shift location at the factory.

Vertical engine identification

MAINTENANCE

Air Cleaner Service

Service all the oil/foam polyurethane and oil bath air cleaner elements in the same manner as the Briggs and Stratton components. See Chapter Four.

The Tecumseh treated paper element type air cleaner consists

General Engine Specifications Vertical Crankshaft Engines

Model	Bore & Stroke	Displacement	Horsepower
LAV25	2.3125 × 1.8438	7.75	2½
LAV30	2.3125 × 1.8438	7.75	3
TVS75	2.3125 × 1.8438	7.75	3
LV35	2.5000 × 1.8438	9.06	3½
LAV35	2.5000 × 1.8438	9.06	3½
TVS90	2.5000 × 1.8438	9.06	3½
LAV40	2.6250 × 1.9375	10.5	4
TVS105	2.6250 × 1.9375	10.5	4
V40, V40B	2.5000 × 2.2500	11.04	4
VH40	2.5000 × 2.2500	11.04	4
LAV50	2.812 × 1.9375	12.0	5
TVS120	2.812 × 1.9375	12.0	5
V50	2.625 × 2.2500	12.17	5
VH50	2.625 × 2.2500	12.17	5

Horizontal engine identification

of a pleated paper element encased in a metal housing and must be replaced as a unit. A flexible tubing and hose clamps connect the remotely mounted air filter to the carburetor.

Clean the element by lightly tapping it. Do not distort the case. When excessive carburetor adjustment or loss of power results, inspect the air filter to see if it is clogged. Replacing a severely restricted air filter should show an immediate performance improvement.

Check the oil level in the oil bath type air cleaners regularly to make sure the level is correct. To add oil, unscrew the wingnut, pull off the filter element and add oil along the side of the filter until the level is correct. Use the same type and viscosity oil used in the engine.

If the filter is dirty, remove it and wash it in solvent. Also remove the filter bowl, drain the oil, and wash the filter in solvent. Refill the bowl with clean oil after putting it into position on the air horn.

A plain paper element is also used. It should be removed every 10 hours, or more often if the air is dusty. Tap or blow out the dirt from the inside with low pressure air. This type should be replaced at 50 hours. If clogged sooner, it may be washed in soap and water and rinsed by flushing from the inside until the water is clear. Blow dry with low pressure compressed air.

To service the KLEEN-AIRE system, remove the element, wash it in soap and mild detergent, pat dry, and then coat with oil. Squeeze the oil to distribute it evenly and remove the excess. Make sure all mounting surfaces are tight to prevent leakage.

Lubrication

OIL AND FUEL RECOMMENDATIONS

Use fresh (less than one month old) gasoline, of "Regular" grade. Unleaded fuel is preferred, but leaded fuel is acceptable.

Use oil having MS, SC, SD, or SE classification. Use these viscosities for aluminum engines:
- Summer – above 32°F: S.A.E. 30 (S.A.E. 10W30 or 10W40 are acceptable substitutes).
- Winter – Below 32°F: S.A.E. 5W30 (S.A.E. 10W is an acceptable substitute). (Including Snow King Snow Blower Engines)

General Engine Specifications
Horizontal Crankshaft Engines

Model	Bore & Stroke (in.)	Displacement (cu in.)	Horsepower
H25	2.3125 × 1.8438	7.75	2½
H30	2.3125 × 1.8438	7.75	3
H35	2.5000 × 1.8438	9.06	3½
H40	2.5000 × 2.2500	11.04	4
HH40	2.5000 × 2.2500	11.04	4
HS40	2.6250 × 1.9375	10.5	4
H50	2.6250 × 2.2500	12.17	5
HH50	2.6250 × 2.2500	12.17	5
HS50	2.8120 × 1.9375	12.0	5

- Winter – Below 0°F only: S.A.E. 10W diluted with 10% kerosene is an acceptable substitute. (Including Snow King Snow Blower Engines)

Use these viscosities for cast iron engines:
- Summer – Above 32°F: S.A.E. 30
- Winter – Above 32°F: S.A.E. 10W

TUNE-UP
Tune-Up Specifications

The following basic specifications apply to all the engines covered in this section:
- Spark Plug Gap: 0.030 in.
- Ignition Point Gap: 0.020 in.
- Valve Clearance: 0.010 in. for both intake and exhaust

For timing dimension, which varies from engine to engine, see the complete specifications at the rear of this section.

Spark Plug Service

Spark plugs should be removed, cleaned, and adjusted periodically. Check the electrode gap with a wire feeler gauge and adjust the gap. Replace the plugs if the electrodes are pitted and burned or the porcelain is cracked. Refer to the Tecumseh master parts manual for the correct replacement number. Apply a little graphite grease to the threads to prevent sticking. Be sure the cleaned plugs are free of all foreign material.

Breaker Points

ADJUSTMENT

1. Disconnect the fuel line from the carburetor.
2. Remove the mounting screws, fuel tank, and shroud to provide access to the flywheel.
3. Remove the flywheel with either a puller (over 3.5 hp) or by using a screwdriver to pry underneath the flywheel while tapping the top lightly with a soft hammer.
4. Remove the dust cover and gasket from the magneto and crank the engine over until the breaker points of the magneto are fully opened.

5. Check the condition of the points and replace them if they are burned or pitted.

6. Check the point gap with a feeler gauge. Adjust them, if necessary, as per the directions on the dust cover. Refer to the specifications chart at the end of this chapter for point gap.

REPLACEMENT

1. Gain access to the points and inspect them as described above. If the points are badly pitted, follow the remaining steps to replace them.

2. Remove the nuts that hold the electrical leads to the screw on the movable breaker point spring. Remove the movable breaker point from stud.

3. Remove the screw and stationary breaker point. Put a new stationary breaker point on the breaker plate; install the screw, but do not tighten. This point must be moved to make the proper air gap when the points are adjusted.

4. Position a new movable breaker point on the stud.

5. Adjust the breaker point gap with a flat feeler gauge and tighten the screw.

6. Check the new point contact pattern and remove all grease, finger-prints, and dirt from contact surfaces.

7. Adjust the timing as described below.

Ignition Timing

ADJUSTMENT

1. Remove the cylinder head bolts, and move the head (with gasket in place) so that the spark plug hole is centered over the piston.

2. Using a ruler (through the spark plug hole) or special plunger type tool, carefully turn the engine back and forth until the piston is at exactly Top Dead Center. Tighten the thumbscrew on the tool.

3. Find the timing dimension for your engine in the specifications at the rear of the manual. Then, back off the position of the piston until it is about halfway down in the bore. Lower the ruler (or loosen the thumbscrew and lower the plunger, if using the special tool) exactly the required amount (the amount of the timing dimension). Then, hold the ruler in place (or tighten the special tool thumbscrew) and, finally, carefully rotate the engine forward until the piston just touches the ruler or tool plunger.

4. Install a timing light or place a very thin piece of cellophane between the contact points. Loosen and rotate the stator just until the timing light shows a change in current flow or the cellophane pulls out of contact gap easily. Then, tighten stator bolts to specified torque.

5. Install the leads, point cover, flywheel, and shrouding.

Treated paper element air cleaner (used on Craftsman engines)

Exploded view of oil bath type air cleaner

SOLID STATE IGNITION SYSTEM CHECKOUT

The only on-engine check which can be made to determine whether the ignition system is working, is to separate the high tension lead from the spark plug and check for spark. If there is a spark, then the unit is alright and the spark plug should be replaced. No spark indicates that some other part needs replacing.

Check the individual components as follows:

• High Tension Lead: Inspect for cracks or indications of arcing. Replace the transformer if the condition of the lead is questionable.

• Low Tension Leads: Check all leads for shorts. Check the ignition cut-off lead to see that the unit is not grounded. Repair the leads, if possible, or replace them.

• Pulse Transformer: Replace and test for spark.

• Magneto: Replace and test for spark. Time the magneto by turning it counterclockwise as far as it will go and then tighten the retaining screws.

• Flywheel: Check the magnets for strength. With the flywheel off the engine, it should attract a screwdriver that is held 1 in. from the magnetic surface on the inside of the flywheel. Be sure that the key locks the flywheel to the crankshaft.

Chart of Initial Carburetor Adjustments

Adjustment	For Engines Built Prior to 1977	For Engines Built After 1977
Main Adjustment	V50-60-1¼ H50-60-1¼	Same
Idle Adjustment	V50-60-70-1 Turn H50-60-70-1 Turn	Same
Idle Speed (Top of Carburetor) Regulating screw		Back out screw, then turn in until screw just touches throttle lever and continue 1 turn more (if idle RPM is given set final idle speed with a tachometer)

Carburetor Mixture Adjustments

1. If the carburetor has been overhauled, or the engine won't start, make initial mixture screw adjustments as specified in the chart.

2. Start the engine and allow it to warm up to normal running temperature. With the engine running at maximum recommended rpm, loosen the main adjustment screw until engine rpm drops off, then tighten the screw until the engine starts to cut out. Note the number of turns from one extreme to the other. Loosen the screw to a point midway between the extremes.

Schematic of the operation of an air vane governor

DIRECTION OF AIR FLOW
AIR VANE
PIVOT PIN
PIN
UPPER THROTTLE SHAFT LEVER
GOVERNOR SPRING
THROTTLE SHAFT
SPRING BRACKET
LOWER THROTTLE SHAFT LEVER

THROTTLE
LINK
THROTTLE OPEN
THROTTLE CLOSED
CONTROL LEVER
GOVERNOR LEVER
GOVERNOR TENSION SPRING
GOVERNOR SHAFT
GOVERNOR ROD
GOVERNOR GEAR ASSEMBLY
SPOOL
COUNTERWEIGHT

Ⓐ GOVERNED POSITION
Ⓑ NON-GOVERNED POSITION
◆ DIRECTION OF ADJUSTMENT—THIS SCHEMATIC

Schematic of mechanical governor operation

NOTE: Some carburetors have fixed jets. If there is no main adjusting screw and receptacle, no adjustment is needed.

3. After the main system is adjusted, move the speed control lever to the idle position and follow the same procedure for adjusting the idle system.

4. Test the engine by running it under a normal load. The engine should respond to load pickup immediately. An engine that "dies" is too lean. An engine which ran roughly before picking up the load is adjusted too rich.

Governor

ADJUSTMENT

Air Vane Type

1. Operate the engine with the governor adjusting lever or panel control set to the highest possible speed position and check the speed. If the speed is not within the recommended limits, the governed speed should be adjusted.

2. Loosen the locknut on the high speed limit adjusting screw and turn the adjusting screw out to increase the top engine speed.

Mechanical Type

1. Set the control lever to the idle position so that no spring tension affects the adjustment.

2. Loosen the screw so that the governor lever is loose in the clamp.

3. Rotate both the lever and the clamp to move the throttle to the full open position (away from the idle speed regulating screw).

4. Tighten the screw when no end-play exists in the direction of open throttle.

5. Move the throttle lever to the full speed setting and check to see that the control linkage opens the throttle.

Compression Check

1. Run the engine until warm to lubricate and seal the cylinder.

2. Remove the spark plug and install a compression gauge. Turn the engine over with the pull starter or electric starter.

3. Compression on new engines is 80 psi. If the reading is below 60 psi., repeat the test after removing the gauge and squirting about a teaspoonful of engine oil through the spark plug hole. If the compression improves temporarily following this, the problem is probably with the cylinder, piston, and rings. Otherwise, the valves require service.

FUEL SYSTEM

Carburetor

NOTE: Four-cycle Tecumseh engines use float or diaphragm type carburetors.

REMOVAL AND INSTALLATION

1. Drain the fuel tank. Remove the air cleaner and disconnect the carburetor fuel lines.

2. If necessary, remove any shrouding or control panels to provide access to carburetor.

3. Disconnect the choke or throttle control wires at the carburetor.

4. Remove the cap screws, or nuts and lockwashers that hold the carburetor to the engine; remove the carburetor.

5. Secure the carburetor on to engine.

6. Install the shrouding or control panels. Connect the choke and throttle control wires.

7. Position the control panel to carburetor. Connect the carburetor fuel lines.

8. Install the air cleaner.

9. Adjust the carburetor as described above.

GENERAL OVERHAUL INSTRUCTIONS

1. Carefully disassemble the carburetor removing all non-metallic parts, i.e.; gaskets, viton seats and needles, O-rings, fuel pump valves, etc.

The choke position on a float feed carburetor. The closed choke plate restricts air, creating a richened mixture by drawing in a greater proportion of fuel.

NOTE: Nylon check balls used in some diaphragm carburetor models may or may not be serviceable. Check to be sure of serviceability before attempting removal.

2. Clean all metallic parts with solvent.

NOTE: Nylon can be damaged if subjected to harsh cleaners for prolonged periods.

3. The large O-rings sealing the fuel bowl to the carburetor body must be in good condition to prevent leakage. If the O-ring leaks, interfering with the atmospheric pressure in the float bowl, the engine will run rich. Foreign material can enter through the leaking area and cause blocking of the metering orifices. This O-ring should be replaced after the carburetor has been disassembled for repair. Lubricate the new O-ring with a small amount of oil to allow the fuel bowl to slide onto the O-ring properly. Hold the carburetor body in an inverted position

The idle operation of a float feed carburetor. The throttle plate closes, restricting the flow of fuel and air, forcing the engine to run on a reduced volume of fuel and air.

The intermediate operation of a float feed carburetor. The throttle plate "cracks" open to reduce the restriction and the engine runs on an increased volume of fuel and air.

The choke position of the diaphragm carburetor. The closed choke plate restricts the amount of air, creating a richened mixture by drawing in a greater proportion of fuel.

FUEL ▮
AIR ▮
MIXTURE ▮

The high speed operation of a float feed carburetor. The air venturi replaces the throttle plate as the restricting device and the engine runs on its greatest volume of fuel and air.

The idling operation of a diaphragm carburetor. The throttle plate closes, restricting the flow of fuel and air, forcing the engine to run on a reduced volume of fuel and air.

Appearance of good and bad mixture adjusting screws

and place the O-ring on the carburetor body and then position the fuel bowl.

4. The small O-rings used on the carburetor adjustment screws must be in good condition or a leak will develop and cause improper adjustment of carburetor.

5. Check all adjusting screws for wear. The illustration shows a worn screw and a good screw. Replace screws that are worn.

6. Check the carburetor inlet needle and seat for wear, scoring, or other damage. Replace defective parts.

7. Check the carburetor float for dents, leaks, worn hinge or other damage.

8. Check the carburetor body for cracks, clogged passages, and worn bushings. Clean clogged air passages with clean, dry compressed air.

9. Check the diaphragms on diaphragm carburetors for cracks, punctures, distortion, or deterioration.

Appearance of good and bad inlet needles

The intermediate operation of a diaphragm carburetor. The throttle plate "cracks" to decrease the restriction and the engine runs on an increased volume of fuel and air.

FUEL   AIR   MIXTURE �▬▬▬

The high speed operation of a diaphragm carburetor. The air ventui replaces the throttle plate as the restricting device and the engine runs on its greatest volume of fuel and air.

10. Check all shafts and pivot pins for wear on the bearing surfaces, distortion, or other damage.

NOTE: Each time a carburetor is disassembled, it is good practice to install a repair kit.

11. Where there is excessive vibration, a damper spring may be used to assist in holding the float against the inlet needle thus minimizing the flooding condition. Two types of springs are available; the float shaft (hinge pin) type and the inlet needle mounted type.

12. Float shaft spring positioning:

a. The spring is slipped over the shaft.

b. The rectangular shaped spring end is hooked onto the float tab.

c. The shorter angled spring end is placed onto the float bowl gasket support.

13. Note that on late model carburetors, the spring clip fastened to the inlet needle has been revised to provide a damping effect. The clip fastens to the needle and is hooked over the float tab.

FLOAT FEED CARBURETOR

Note the following points when rebuilding these carburetors.

Float feed carburetor identification number

External view of a Tecumseh float type carburetor

Removal

Remove the carburetor from the engine. It is easier to remove the intake manifold and carburetor assembly from the engine, disconnect the governor linkage, fuel line, and grounding wire and then disassemble the carburetor from the intake manifold on a work bench. Be sure to note the positions of the governor and the throttle linkage to facilitate reassembly.

Throttle

1. Examine the throttle lever and plate prior to disassembly. Replace any worn parts.

2. Remove the screw in the center of the throttle plate and pull out the throttle shaft lever assembly.

3. When reassembling, it is important that the lines on the

Service hints for float feed carburetors

Loosen screw until it just clears throttle lever, then turn screw in one turn.

Do not attempt to remove blow air through passage.

Check shaft for binding position opening to bottom of air horn.

Blow air through passage. Do not remove restrictor if present.

Proper installation is important. See appropriate chapter for procedures.

Replace.

Must hook over float tab.

Check float for leaks or dents. Clean bowl and adjust float level position gasket or gaskets.

If the carburetor is used on a 20° slant engine, the engine must be in its normal 20° slanted position for adjustment.

Check needle for damage and "O" ring for cracks. Clean all passages in nut with compressed air.

IDLE SPEED ADJUSTMENT

MAIN NOZZLE

CHOKE SHAFT AND LEVER

CHOKE PLATE

HIGH SPEED AIR BLEED

*INLET FITTING

*INLET NEEDLE AND SEAT

*FLOAT BOWL GASKET

INLET NEEDLE CLIP (If Present)

IDLE FUEL TRANSFER PASSAGE AND ANNULAR GROOVE

FLOAT SHAFT

FLOAT

FLOAT BOWL

*GASKET

*NUT AND MAIN ADJUSTMENT SEAT

*MAIN ADJUSTMENT SCREW AND "O" RING SEAL

(DO NOT REMOVE)

Blow air through passage.

Check shaft for looseness or binding. Shutter must be positioned with detent reference marks on top parallel with shaft and to the right or 3 o'clock position.

Check spring for return action and binding.

Remove idle adjustment screw. Check needle tip and condition of "O" ring. Remove welch plug and blow out all passages.

(DO NOT REMOVE PLUG)

Remove float bowl fastener and cup plug drop out metering pin and clean all passages with air. Use a small tap to catch cup, twist and remove. Pin should move freely within the passage. CAUTION: Do not install idle adjustment screw with carburetor upside down as pin will obstruct movement of adjustment screw causing damage.

IDLE AND INTERMEDIATE AIR BLEED

THROTTLE SHAFT AND LEVER

THROTTLE SHUTTER

DETENT REFERENCE MARK

THROTTLE SHAFT RETURN SPRING

IDLE AND INTERMEDIATE FUEL ORIFICES

IDLE AND INTERMEDIATE FUEL CHAMBER (COVERED WITH WELCH PLUG)

IDLE AND INTERMEDIATE FUEL MIXTURE PASSAGE

*IDLE ADJUSTMENT SCREW AND "O" RING

ATMOSPHERIC VENT

SOFT BAFFLE PLUG

*FLOAT BOWL DRAIN

IDLE AND INTERMEDIATE FUEL TRANSFER PASSAGE

METERING ROD-OR PIN IN FUEL TRANSFER PASSAGE

BALL PLUG

CUP PLUG

IDLE AND INTERMEDIATE FUEL TRANSFER PASSAGE

IDLE AND MAIN FUEL PICK UP ORIFICE

throttle plate are facing out when in the closed position. Position the throttle plates with the two lines at 12 and 3 o'clock. The throttle shaft must be held in tight to the bottom bearing to prevent the throttle plate from riding on the throttle bore of the body which would cause excessive throttle plate wear and governor hunting.

Choke

Examine the choke lever and shaft at the bearing points and holes into which the linkage is fastened and replace any worn parts. The choke plate is inserted into the air horn of the carburetor in such a way that the flat surface of the choke is toward the fuel bowl.

Idle Adjusting Screw

Remove the idle screw from the carburetor body and examine the point for damage to the seating surface on the taper. If damaged, replace the idle adjusting needle. Tension is maintained on the screw with a coil spring and sealed with an O-ring. Examine and replace the O-ring if it is worn or damaged.

High Speed Adjusting Jet

Remove the screw and examine the taper. If the taper is damaged at the area where it seats, replace the screw and fuel bowl retainer nut as an assembly.

The fuel bowl retainer nut contains the seat for the screw. Examine the sealing O-ring on the high speed adjusting screw. Replace the O-ring if it indicates wear or cuts. During the reassembly of the high speed adjusting screw, position the coil spring on the adjusting screw, followed by the small brass washer and the O-ring seal.

Fuel Bowl

To remove the fuel bowl, remove the retaining nut and fiber washer. Replace the nut if it is cracked or worn.

The retaining nut contains the transfer passage through which fuel is delivered to the high speed and idle fuel system of the carburetor. It is the large hole next to the hex nut end of the fitting. If a problem occurs with the idle system of the carburetor, examine the small fuel passage in the annular groove in the retaining nut. This passage must be clean for the proper transfer of fuel into the idle metering system.

The fuel bowl should be examined for rust and dirt. Thoroughly clean it before installing it. If it is impossible to properly clean the fuel bowl, replace it.

Check the drain valve for leakage. Replace the rubber gasket on the inside of the drain valve if it leaks.

Examine the large O-ring that seals the fuel bowl to the carburetor body. If it is worn or cracked, replace it with a new one, making sure the same type is used (square or round).

Float

1. Remove the float from the carburetor body by pulling out the float axle with a pair of needle nose pliers. The inlet needle will be lifted off the seat because it is attached to the float with an anchoring clip.
2. Examine the float for damage and holes. Check the float hinge for wear and replace it if worn.
3. The float level is checked by positioning a #4 (0.209 in.) twist drill across the rim between the center leg and the unmachined surface of the index pad, parallel to the float axle pin. If the index pad is machined, the float setting should be made with a #9 (0.180–0.200 in.) twist drill.
4. Remove the float to make an adjustment. Bend the tab on the float hinge to correct the float setting.

NOTE: Direct compressed air in the opposite direction of normal flow of air or fuel (reverse taper) to dislodge foreign matter.

USE NO. 4 DRILL ON RING

RIM

Adjusting the float level

Inlet Needle and Seat

1. The inlet needle sits on a rubber seat in the carburetor body instead of the usual metal fitting.
2. Remove it, place a few drops of heavy engine oil on the seat, and pry it out with a short piece of hooked wire.
3. The grooved side of the seat is inserted first. Lubricate the cavity with oil and use a flat faced punch to press the inlet seat into place.
4. Examine the inlet needle for wear and rounding off of the corners. If this condition does exist, replace the inlet needle.

Fuel Inlet Fitting

1. The inlet fitting is removed by twisting and pulling at the same time.
2. Use sealer when reinstalling the fitting. Insert the tip of the fitting into the carburetor body. Press the fitting in until the shoulder contacts the carburetor. Only use inlet fittings without screens.

Carburetor Body

1. Check the carburetor body for wear and damage.
2. If excessive dirt has accumulated in the atmospheric vent cavity, try cleaning it with carburetor solvent or compressed air. Remove the welch plug only as a last resort.

EXAMINE FLOAT HINGE FOR WEAR AT BEARING SURFACES

INLET NEEDLE EXAMINE CORNERS REPLACE IF ROUNDED

FLOAT

INLET NEEDLE CLIP

REMOVE FLOAT TO MAKE ADJUSTMENTS. BEND THIS TAB TO ADJUST

Float, inlet needle, and clip

NOTE: The carburetor body contains a pressed-in main nozzle tube at a specific depth and position within the venturi. Do not attempt to remove the main nozzle. Any change in nozzle positioning will adversely affect the metering quality and will require carburetor replacement.

3. Clean the accelerating well around the main nozzle with compressed air and carburetor cleaning solvents.

4. The carburetor body contains two cup plugs, neither of which should be removed. A cup plug located near the inlet seat cavity, high up on the carburetor body, seals off the idle bleed. This is a straight passage drilled into the carburetor throat. Do

NO CHOKE

NO MAIN MIXTURE ADJUSTMENT

NO IDLE MIXTURE ADJUSTMENT

Auto-Magic carburetor

AIR BLEED

CARBURETOR BODY

CARBURETOR BODY FUEL SUPPLY ORIFICE TO BOWL NUT

GASKET NOT ON ALL MODELS

BOWL

MAIN JET IN BOWL NUT

GASKET

Variation on non-adjustable float feed carburetor

not remove this plug. Another cup plug is located in the base where the fuel bowl nut seals the idle fuel passage. Do not remove this plug or the metering rod.

5. A small ball plug located on the side of the idle fuel passage seals this passage. Do not remove this ball plug.

6. The welch plug on the side of the carburetor body, just above the idle adjusting screw, seals the idle fuel chamber. This plug can be removed for cleaning of the idle fuel mixture passage and the primary and secondary idle fuel discharge ports. Do not use any tools that might change the size of the discharge ports, such as wire or pins.

Resilient Tip Needle

Replace the inlet needle. Do not attempt to remove or replace the seat in the carburetor body.

Viton Seat

Using a 10–24 or 10–32 tap, turn the tap into the brass seat fitting until it grasps the seat firmly. Clamp the tap shank into a vise and tap the carburetor body with a soft hammer until the seat slides out of the body.

To replace the viton seat, position the replacement over the receptical with the soft rubber like seat toward the body. Use a flat punch and a small hammer to drive the seat into the body until it bottoms on the shoulder.

TECUMSEH AUTOMATIC NON-ADJUSTABLE FLOAT FEED CARBURETOR

This carburetor has neither a choke plate nor idle and main mixture adjusting screws. There is no running adjustment. The float adjustment is the standard Tecumseh float setting of 0.210 in. (#4 drill).

Cleaning

Remove all non-metallic parts and clean them using a procedure similar to that for the other carburetors. Never use wires through any of the drilled holes. Do not remove the baffling welch plug unless it is certain there is a blockage under the plug. There are no blind passageways in this carburetor.

Some engines use a variation on the Automatic Nonadjustable carburetor which has a different bowl hold-on nut and main jet orifices. There are two main jet orifices and a deeper fuel reserve cavity, but service procedures are the same.

WALBRO AND TILLOTSON FLOAT FEED CARBURETORS

Procedures are similar to those for the Tecumseh float carburetor with the exceptions noted below.

Main Nozzle

The main nozzle in Wallbro carburetors is cross drilled after it is installed in the carburetor. Once removed, it cannot be reinstalled, since it is impossible to properly realign the cross drilled holes. Grooved service replacement main nozzles are available which allow alignment of these holes.

Float Shaft Spring

Carefully position the float shaft spring on models so equipped. The spring dampens float action when properly assembled. Use needlenosed pliers to hook the end of the spring over the float hinge and then insert the pin as far as possible before lifting the spring from the hinge into position. Leaving the spring out or improper installation will cause unbalanced float action and result in a touchy adjustment.

Walbro float feed carburetor

Tillotson Model E float carburetor

Float Adjustment

1. To check the float adjustment, invert the assembled float carburetor body. Check the clearance between the body and the float, opposite the hinge. Clearance should be $\frac{1}{8}$ in. $\pm$ $\frac{1}{64}$ in.

2. To adjust the float level, remove the float shaft and float. Bend the lip of the float tang to correct the measurement.

3. Assemble the parts and recheck the adjustment.

TILLOTSON E FLOAT FEED TYPE CARBURETOR

The following adjustments are different for this carburetor.

Running Adjustment

1. Start the engine and allow it to warm up to operating temperatures. Make sure the choke is fully opened after the engine is warmed up.

2. Run the engine at a constant speed while slowly turning the main adjustment screw in until the engine begins to lose speed; then slowly back it out about $\frac{1}{8}$–$\frac{1}{4}$ of a turn until maximum speed and power is obtained (4000 rpm). This is the correct power adjustment.

3. Close the throttle and cause the engine to idle slightly faster than normal by turning the idle speed regulating screw in. Then turn the idle adjustment speed screw in until the engine begins to lose speed; then turn it back $\frac{1}{4}$–$\frac{1}{2}$ of a turn until the engine idles smoothly. Adjust the idle speed regulating screw until the desired idling speed is acquired.

4. Alternately open and close the throttle a few times for an acceleration test. If stalling occurs at idle speeds, repeat the adjustment procedures to get the proper idle speed.

Float Level Adjustment

1. Remove the carburetor float bowl cover and float mechanism assembly.

2. Remove the float bowl cover gasket and, with the complete assembly in an upside down position and the float lever tang resting on the seated inlet needle, a measurement of $1\frac{5}{64}$ in. should be maintained from the free end flat rim, or edge of the cover, to the toe of the float. Measurement can be checked with a standard straight rule or depth gauge.

3. If it is necessary to raise or lower the float lever setting, remove the float lever pin and the float, then carefully bend the float lever tang up or down as required to obtain the correct measurement.

DIAPHRAGM CARBURETORS

Diaphragm carburetors have a rubber-like diaphragm that is exposed to crankcase pressure on one side and to atmospheric pressure on the other side. As the crakcase pressure decreases, the diaphragm moves against the inlet needle allowing the inlet needle to move from its seat which permits fuel to flow through the inlet valve to maintain the correct fuel level in the fuel chamber.

An advantage of this type of system over the float system, is that the engine can be operated in any position.

*INDICATES PARTS THAT CAN BE DAMAGED BY HARSH CARBURETOR CLEANERS

Exploded view of a Tecumseh diaphragm carburetor

CHOKE SHAFT AND LEVER — Check shaft for binding. Position shutter opening towards inlet fitting side of air horn.

CHOKE SHUTTER

MAIN NOZZLE — Blow air through passage

•INLET FITTING SCREEN — Part of inlet fitting. If fuel is restricted, clean or replace fitting.

•INLET FITTING — Bulb primer models have Viton• one way valve in or behind fitting.

•INLET SEAT GASKET — Remove and replace.

•INLET NEEDLE SEAT AND SPRING ASSEMBLY — Proper installation of assembly is important. See appropriate chapter for procedures.

•DIAPHRAGM GASKET — Gasket and diaphragm sequence may be reversed on some models. Head of rivet must touch inlet needle. Rivet is hooked into inlet needle control lever on some models. See appropriate chapter.

•DIAPHRAGM

ATMOSPHERIC VENT HOLE — Hole must be clean. On models with bulb primer, vent hole is very small and is located off center.

IDLE SPEED ADJUSTMENT SCREW — Loosen screw until it just clears throttle lever, then turn screw in one turn.

THROTTLE SHAFT AND LEVER — DETENT REFERENCE MARK ON THROTTLE SHUTTER — Place detent reference mark to proper location. See chapter appropriate to specific carburetor.

THROTTLE SHUTTER

THROTTLE SHAFT RETURN SPRING — Check spring for return action and binding.

IDLE, INTERMEDIATE AND AIR BLEED ORIFICES — Remove welch plug and blow air through air passages.

•IDLE MIXTURE ADJUSTMENT SCREW AND "O" RING (If Present)

•MAIN MIXTURE ADJUSTMENT SCREW AND "O" RING (If Present) — Remove adjustment screw. To adjust 20 slant engines, the engine must be mounted in its normal 20° slant position.

WELCH PLUG (If Present) — Check ball is not serviceable on some models. See appropriate chapter for specific model.

•MAIN NOZZLE CHECK BALL (If Present)

• NON METALLIC ITEMS – CAN BE DAMAGED BY HARSH CARBURETOR CLEANERS

Service hints for the diaphragm carburetors

NOTE: In rebuilding, use carburetor cleaner only on metal parts, except for the main nozzle in the main body.

Throttle Plate

Install the throttle plate with the short line that is stamped in the plate toward the top of the carburetor, parallel with the throttle shaft, and facing out when the throttle is closed.

Choke Plate

Install the choke plate with the flat side of the choke toward the fuel inlet side of the carburetor. The mark faces in and is parallel to the choke shaft.

Idle Mixture Adjustment Screw

There is a neoprene O-ring on the needle. Never soak the O-ring in carburetor solvent. Idle and main mixture screws vary in size and design, so make sure that you have the correct replacement.

Idle Fuel Chamber

The welch plug can be removed if the carburetor is extremely dirty.

Diaphragms

Diaphragms are serviced and replaced by removing the four retaining screws from the cover. With the cover removed, the diaphragm and gasket may be serviced. Never soak the diaphragm in carburetor solvent. Replace the diaphragm if it is cracked or torn. Be sure there are no wrinkles in the diaphragm when it is replaced. The diaphragm rivet head is always placed facing the inlet needle valve.

Inlet Needle and Seat

The inlet seat is removed by using either a slotted screwdriver (early type) or a $9/32$ in. socket. The inlet needle is spring loaded, so be careful when removing it.

Fuel Inlet Fitting

All of the diaphragm carburetors have an integral strainer in the inlet fitting. To clean it, either reverse flush it or use compressed air after removing the inlet needle and seat. If the strainer is lacquered or otherwise unable to be cleaned, replace the fitting.

CRAFTSMAN FUEL SYSTEMS

Changes In Late Model Carburetors

The newest Craftsman carburetors incorporate the following changes:

    a. The cable form of control is replaced by a control knob.

    b. The fuel pickup is longer and has a collar machined into it which must be installed tight against the carburetor body.

    c. The fuel pickup screen is pressed onto the ends of the fill tubes on both models, but the measured depth has changed.

    d. The cross-drilled passages have been eliminated, as has the O-ring on the body. There are no cup plugs.

    e. The fuel tank and reservoir tube have been revised-the reservoir tube being larger.

Disassembly and Service

1. Remove the air cleaner assembly and remove the four screws on the top of the carburetor body to separate the fuel tank from the carburetor.

2. Remove the O-ring from between the carburetor and the fuel tank. Examine it for cracks and damage and replace it if necessary.

3. Carefully remove the reservoir tube from the fuel tank.

Details of pick-up tube used on revised Craftsman carburetor

Observe the end of the tube that rested on the bottom of the fuel tank. It should be slotted.

4. Remove the control valve by turning the valve clockwise until the flange is clear of the retaining boss. Pull the valve straight out and examine the O-ring seal for damage or wear. If possible, use a new O-ring when reassembling.

5. Examine the fuel pick up tube. There are no valves or ball checks that may become inoperative. These parts can normally be cleaned with carburetor solvent. If it is found that the passage cannot be cleared, the fuel pick up tubes can be replaced. Carefully remove the old ones so as not to enlarge the opening in the carburetor body. If the pickup tube and screen must be replaced, follow the directions shown in the illustration for the type of carburetor (early or late model) on which you are working.

Craftsman fuel tank mounted carburetor

LATE MODEL

NO BOWDEN WIRE CONTROL

MANUAL CONTROL KNOB

POSITIONING SPRING

HIGH SPEED NOTCH

FUEL PICK-UP TUBE PRESS TUBE TO BOTTOM ON DRILL SHANK

1-15/32"

PRESS SCREEN TO PROPER DEPTH AFTER PICK-UP TUBE IS SET

BOWDEN CONTROL

FUEL PICK-UP TUBE (SET COLLAR AGAINST CASTING BODY)

1-27/64"

PRESS SCREEN TO PROPER DEPTH AFTER PICK-UP TUBE IS SET

EARLY MODEL

Details of manual control knob and several other changes incorporated in new Craftsman carburetors

6. Assemble the carburetor in reverse order of disassembly. Use new O-rings and gaskets. When assembling reservoir tube, hold the carburetor upside down and place the reservoir tube over the pickup tube with the slotted end up. Make sure the intake manifold gasket is correctly positioned-it can be assembled blocking the intake passage partially.

Adjustments

1. Move the carburetor control valve to the high speed position. The mark on the face of the valve should be in alignment with the retaining boss on the carburetor body.

2. Move the operator's control on the equipment to the high speed position.

FUEL PICK-UP TUBE

CUP PLUG

"O" RING

GASKET

Positioning of the reservoir tube in the fuel tank

3. Insert the bowden wire into the hole of the control valve. Clamp the bowden wire sheath to the carburetor body.

NOTE: If the engine was disassembled and the camshaft removed, be sure that the timing marks on the camshaft gear and the keyway in the crankshaft gear are aligned when reinserting the camshaft. Then lift the camshaft enough to advance the camshaft gear timing mark to the right (clockwise) ONE tooth, as viewed from the power take-off end of the crankshaft.

CRAFTSMAN FLOAT TYPE CARBURETORS

Craftsman float type carburetors are serviced in the smae manner as the other Tecumseh float type carburetors. The throttle control valve has three positions: stop, run, and start.

When the control valve is removed, replace the O-ring. If the engine runs sluggishly, consider the possibility of a leaky O-ring.

To remove the fuel pickup tube, clamp the tube in a vise and then twist the carburetor body. Check the small jet in the air horn while the tube is out. In replacement, position the tube squarely and then press in on the collar until the collar seats.

In assembly, install the gasket and bolt through the bowl, position the centering spacer onto the bolt, and then attach the parts to the carburetor body.

The camshaft timing mark must be advanced one tooth in relation to crankshaft gear timing mark, as shown in the illustra-

3. PRESS IN TUBE TO BOTTOM ON DRILL SHANK

4. DO NOT REMOVE DRILL PRESS ON STRAINER 1-27/64" to 1-7/16"

6. INSERT NEW CUP PLUG

1. REMOVE CUP PLUG

2. INSERT SHANK END OF 1/8" DRILL

5. REMOVE DRILL

EARLY MODEL

1 PRESS TUBE TO SEAT COLLAR AGAINST CASTING

2 PRESS STRAINER TO 1-15/32" DEPTH

LATE MODEL

Pick-up tube replacement

The fuel pick-up tube and O-ring on the early model Craftsman carburetor

tion above. However, when this type carburetor is used with a float bowl reservoir and variable governor adjustment, time it as for other Tecumseh engines — with the camshaft and crankshaft timing marks aligned.

Governor

The mechanical governor is located inside the mounting flange. See engine disassembly instructions, below. To disassemble the governor, see the illustration, and: remove the retaining ring, pull off the spool, remove the second retaining ring, and then pull off the gear assembly and retainer washer.

Check for wear on all moving surfaces, but especially gear teeth, the inside diameter of the gear where it rides on the shaft, and the flyweights where they work against the spool.

If the governor shaft must be replaced, it should be started into the boss with a few taps using a soft hammer, and then pressed in with a press or vise. The shaft can be installed by positioning a wooden block on top and tapping the upper surface of the block, but the use of a vise or press is much preferred.

The shaft must be pressed in until just the required length is exposed, as measured from the top of the shaft boss to the upper end of the shaft. See the chart.

The governor is installed in reverse of the removal procedure. Connect the linkage and then adjust as described in the Tune-Up section.

FOR CRAFTSMAN CARBURETOR ONLY—
AND ONLY IF ENGINE HAS NO INTERNAL
FLYWEIGHT GOVERNOR

Timing an engine with the Craftsman fuel tank mounted carburetor

Start, run and stop throttle positions of a Craftsman float type carburetor

Exploded view of the Craftsman float type carburetor

Measuring governor shaft installed dimension

Exploded view of standard mechanical governor

Timing marks

ENGINE OVERHAUL

Timing Gears

Correctly matched camshaft gear and crankshaft gear timing marks are necessary for the engine to perform properly.

On all camshafts the timing mark is located in line with the cneter of the hobbing hole (small hole in the face of the gear). If no line is visible, use the cneter of the hobbing hole to align with the crankshaft gear marked tooth.

On crankshafts where the gear is held on by a key, the timing mark is the tooth in line with the keyway.

On crankshafts where the gear is pressed onto the crankshaft, a tooth is bevelled to serve as the timing mark.

On engines with a ball bearing on the power take-off end of the crankshaft, look for a bevelled tooth which serves as the crankshaft gear timing mark.

If the engine uses a Craftsman type carburetor, the camshaft timing mark must be advanced clockwise one tooth ahead of the matching timing mark on the crankshaft, the exception being the Craftsman variable governed fuel systems.

Shaft Installed Dimension/ Engine Model Chart

Engine Model	"A" Exposed Shaft Length (see figure)
LAV30-50 H25-35 HS40-50	$1^{5}/_{16}$"
V50	$1^{19}/_{32}$"
H50 HH40	$1^{7}/_{16}$"

Timing marks

CRAFTSMAN CARBURETOR ONLY

Timing marks

NOTE: If one of the timing gears, either the crank-shaft gear or the camshaft gear, is damaged and has to be replaced, both gears should be replaced.

Crankshaft

INSPECTION

Inspect the crankshaft for worn or crossed threads that can't be redressed; worn, scratched, or damaged bearing surfaces; misalignments; flats on the bearing surfaces. Replace the shaft if any of these problems are in evidence — do not try to straighten a bent shaft.

In replacement, be sure to lubricate the bearing surfaces and use oil seal protectors. If the camshaft gear requires replacement, replace the crankshaft gear, also.

LAV35–LAV50 and H35–HS50 crankshafts have a press fit gear. If the camshaft gear requires replacement on these engines, the crankshaft must be replaced, as the gear connot be replaced separately.

Pistons

When removing the pistons, clean the carbon from the upper cylinder bore and head. The piston and pin must be replaced in matched pairs.

Piston-to-rod relationship for LAV50 and HS50 engines

VM70, V80, VM80, HM70, H80, HM80, VM100 & HM100 PISTON AND CONNECTING ROD ASSEMBLY

Piston-to-rod relationship for engines shown

Check piston dimensions

A ridge reamer must be used to remove the ridge at the top of the cylinder bore on some engines.

Clean the carbon from the piston ring groove. A broken ring can be used for this operation.

Some engines have oversize pistons which can be identified by the oversize engraved on the piston top.

There is a definite piston-to-connecting rod-to-crankshaft arrangement which must be maintained when assembling these parts. If the piston is assembled in the bore 180 degrees out of position, it will cause immediate binding of the parts.

Timing marks for the LAV40 and HS40 engines

Ring arrangement and dimensions

Squaring the ring in the bore

Checking the ring gap

Connecting rod assembly for the LAV40 and HS40 engine

Piston Rings

Always replace the piston rings in sets. Ring gaps must be staggered. When using new rings, wipe the cylinder wall with fine emery cloth to deglaze the wall. Make sure the cylinder wall is thoroughly cleaned after deglazing. Check the ring gap by placing the ring squarely in the center of the area in which the rings travel and measuring the gap with a feeler gauge. Do not spread the rings too wide when assembling them to the pistons. Use a ring leader to install the rings on the piston.

The top compression ring has an inside chamfer. This chamfer must go UP. If the second ring has a chamfer, it must also face UP. If there is a notch on the outside diameter of the ring, it must face DOWN.

Check the ring gap on the old ring to determine if the ring should be replaced. Check the ring gap on the new ring to determine if the cylinder should be rebored to take oversize parts.

NOTE: Make sure that the ring gap is measured with the ring fitted squarely in the worn part of the cylinder where the ring usually rides up and down on the piston.

Connecting Rods

Be sure that the match marks align when assembling the connecting rods to the crankshaft. Use new self-locking nuts. Whenever locking tabs are included, be sure that the tabs lock

the nuts securely. NEVER try to straighten a bent crankshaft or connecting rod. Replace them if necessary. When replacing either the piston, rod, crankshaft, or camshaft, liberally lubricate all bearings with engine oil before assembly.

The following engines have offset connecting rods: LAV40, LAV50, HS40, HS50. The LAV40, LAV50, HS40, HS50 engines have the caps fitted from opposite to the camshaft side of the engine.

On engine with Durlock rod bolts, torque the bolts as follows: LAV25–50, H25, 35, HS40–50, 110 in. lbs. V50, H50, VH50, HH50, 150 inch lbs.

Connecting rod assembly for the V70, V80, H70 and H80 engines

Alignment of timing marks for camshaft removal on VM80 and VM100 engines

Connecting rod and piston assembly for the V80 and H80 engines

Check the dimensions of the camshaft

NOTE: Early type caps can be distorted if the cap is not held to the crankpin while threading the bolts tight. Undue force should not be used.

Later rods have serrations which prevent distortion during tightening. They also have match marks which must face out when assembling the rod. On the LAV50 and the HS50 engines, the piston must be fitted to the rod with the arrow on the top of the piston pointing to the right and the match marks on the rod facing you when the piston is pinned to the rod.

Camshaft

Before removing the camshaft, align the timing marks to relieve the pressure on the valve lifters, on most engines.

In installation, align the gears for this type of camshaft as they were right before removal. After installation, turn the crankshaft gear clockwise in order to check for proper alignment of timing marks.

Clean the camshaft in solvent, then blow the oil passages dry with compressed air. Replace the camshaft if it shows wear of evidence of scoring. Check the cam dimensions against those in the chart.

If the engine has a mechanical fuel pump, it may have to be removed to properly reinstall the camshaft. If the engine is equipped with the Insta-matic Ezee-Start Compression Release, and any of the parts have to be replaced due to wear or damage, the entire camshaft must be replaced. Be sure that the oil pump (if so equipped) barrel chamber is toward the fillet of the camshaft gear when assembled.

NOTE: If a damaged gear is replaced, the crankshaft gear should also be replaced.

Valve Springs

The valve springs should be replaced whenever an engine is overhauled. Check the free length of the springs. Comparing one spring with the other can be a quick check to notice any differences. If a difference is noticed, carefully measure the free length, compression length, and strength of each spring. See the specifications chart at the end of this section.

Some valve springs use dampening coils—coils that are wound closer together than most of the coils of the spring.

Where these are present, the spring must be mounted so the dampening coils are on the stationary (upper) end of the spring.

Valve Lifters

The stems of the valves serve as the lifters. On the 4 hp light frame models, the lifter stems are of different lengths. Because this engine is a cross port model, the shorter intake valve lifter goes nearest the mounting flange.

The valve lifters are identical on standard port engines. However once a wear pattern is established, they should not be interchanged.

Valve Grinding and Replacement

Valves and valve seats can be removed and reground with a minimum of engine disassembly.

Remove the valves as follows:

1. Raise the lower valve spring caps while holding the valve heads tightly against the valve seat to remove the valve spring retaining pin. This is best achieved by using a valve spring compressor. Remove the valves, springs, and caps from the crankcase.

2. Clean all parts with a solvent and remove all carbon from the valves.

Dimensions of the valve face

3. Replace distorted or damaged valves. If the valves are in usable condition, grind the valve faces in a valve refacing maching and to the angle given in the specifications chart at the end of this section. Replace the valves if the faces are ground to less than $\frac{1}{32}$ in.

4. Whenever new or reground valves are installed, lap in the valves with lapping compound to insure an air-tight fit.

NOTE: There are valves available with oversize stems.

5. Valve grinding changes the valve lifter clearance. After grinding the valves check the valve lifter clearance as follows:

   a. Rotate the crankshaft until the piston is set at the TDC position of the compression stroke.

   b. Insert the valves in their guides and hold the valves firmly on their seats.

   c. Check for a clearance of 0.010 in. between each valve stem and valve lifter with a feeler gauge.

   d. Grind the valve stem in a v alve resurfacing machine set to grind a perfectly square face with the proper clearance.

6. Install valves as follows on Early Models:

   a. Position the valve spring and upper and lower valve spring caps under the valve guides for the valve to be installed.

   b. Install the valves in the guides, making sure that the valve marked **EX** is inserted in the exhaust port. The valve stem must pass through the valve spring and the valve spring caps.

   c. Insert the blade of a screwdriver under the lower valve spring cap and pry the spring up.

   d. Insert the valve pin through the hole in the valve stem with a long nosed pliers. Make sure the valve pin is properly seated under the lower valve spring cap.

Install the valves as follows on Later Models:

   a. Position the valve caps and spring in the valve compartment.

   b. Install the valves in guides with the valve marked **EX** in exhaust port. The valve stem must pass through the upper valve cap and spring. The lower cap should sit around the valve lifter exposed end.

   c. Compress the valve spring so that the shank is exposed. DO NOT TRY TO LIFT THE LOWER CAP WITH THE SPRING.

   d. Lift the lower valve cap over the valve stem shank and center the cap in the smaller diameter hole.

   e. Release the valve spring tension to lock the cap in place.

REBORING THE CYLINDER

1. First, decide whether to rebore for 0.010 in. or 0.020 in.

2. lkUse any standard commercial hone of suitable size. Chuck the hone in the drill press with the spindle speed of about 600 rpm.

3. Start with coarse stones and center the cylinder under the press spindle. Lower the hone so the lower end of the stones contact the lowest point in the cylinder bore.

4. Rotate the adjusting nut so that the stones touch the cylinder wall and then begin honing at the bottom of the cylinder. Move the hone up and down at a rate of 50 strokes a minute to avoid cutting ridges in the cylinder wall. Every fourth or fifth stroke, move the hone far enough to extend the stones 1 in. beyond the top and bottom of the cylinder bore.

5. Check the bore size and straightness every thirty or forty strokes. If the stones collect metal, clean them with a wire brush each time the hone is removed.

Check the dimensions of the valve seats, guides, and the cylinder

6. Hone with coarse stones until the cylinder bore is within 0.002 in. of the desired finish size. Replace the coarse stones with burnishing stones and continue until the bore is to within 0.0005 in. of the desired size.

7. Remove the burnishing stones and install finishing stones to polish the cylinder to the final size.

8. Clean the cylinder with solvent and dry it thoroughly.

9. Replace the piston and piston rings with the correct oversize parts.

REBORING VALVE GUIDES

The valve guides are permanently installed in the cylinder. However, if the guides wear, they can be rebored to accommodate a $\frac{1}{32}$ in. oversize valve stem. Rebore the valve guides in the following manner:

1. Ream the valve guides with a standard straight shanked hand reamer or a low speed drill press. Refer to the specifications chart at the end of this section for the correct valve stem guide diameter.

2. Redrill the upper and lower valve spring caps to accommodate the oversize valve stem.

3. Reassemble the engine, installing valves with the correct oversize stems in the valve guides.

REGRINDING VALVE SEATS

The valve seats need regrinding only if they are pitted or scored. If there are no pits or scores, lapping in the valves will provide a proper valve seat. Valve seats are not replaceable. Regrind the valve seats as follows:

1. Use a grinding stone or a reseater set to provide the proper angle and seal face dimensions.

2. If the seat is over $\frac{3}{32}$ in. wide after grinding, use a 15 degree stone or cutter to narrow the face to the proper dimensions.

3. Inspect the seats to make sure that the cutter or stone has been held squarely to the valve seat and that the same dimension has been held around the entire circumference of the seat.

4. Lap the valves to the reground seats.

Cylinder head tightening sequence for 8 hp models

Cylinder head tightening sequence for all engines except 8 hp models

Torquing Cylinder Head

Torque the cylinder head to 200 inch lbs. in 4 equal stages of 50 inch lbs. Follow the sequence shown in the appropriate illustration for each tightening stage.

Bearing Service

LIGHTWEIGHT ALUMINUM BEARING REPLACEMENT

The aluminum bearing must be cut out using the rough cut reamer and the procedures shown in steps 1A and 4A. Follow illustrated steps 1A, 2, 3, 4A, 5 and 6 to install bronze bushings in place of the aluminum bearings.

LONG LIFE AND CAST IRON ENGINES WITHOUT BALL BEARINGS

The worn bronze bushing must be driven out before the new bushing can be installed. Follow illustrated steps 1B, 2, 3, 4B, 5 and 6, to replace the main bearings on these units.

LONG LIFE AND CAST IRON ENGINES WITH BALL BEARING ON THE P.T.O. (POWER TAKE-OFF) SIDE OF THE CRANKSHAFT

The side cover containing the ball bearing must be removed and a substitute cover with either a new bronze bushing or aluminum bearing must be used instead. Follow illustrated steps 1B, 2 and 3 only.

SPECIAL TOOLS

The task of main bearing replacement is made easier by using one of two Tecumseh main bearing tool kits. Kit 670161 is used to replace main bearings on the lightweight engines except HS,

LAV40 and 50 models. Kit 670165 is used to replace main bearings on the medium weight engines except HS, LAV40 and 50 models.

GENERAL NOTES ON BUSHING REPLACEMENT

1. Your fingers and all parts must be kept very clean when replacing bushings.

2. On splash lubricated horizontal engines, the oil hole in the bushing is to be lined up with the oil hole that leads into the slot in the original bearing.

3. In the event it is necessary to replace the mounting flange or cylinder cover, the magneto end bearing must be rebushed. The P.kT.O. bearing should also be rebushed to assure proper alignment.

4. Oil should be used to finish-ream the bushings. In the event the crankshaft does not rotate freely repeat the finish-reaming operation without oil.

5. Kerosene should be used as a cutting lubricant while rough-reaming.

6. Be sure that the dowel pins are in the cylinder block when assembling the mounting flange or cylinder cover. Use all bolts to hold the assembly together.

INSTALLING BRONZE BUSHING P.T.O. END

1. DISASSEMBLE
2. POSITION BEARING SUPPORT TOOL
3. PRESS IN BUSHING
4. ALIGN OIL HOLE
5. DRIVE BUSHING UNTIL IT BOTTOMS ON BEARING SUPPORT

BUSHING DRIVER

BUSHING

ALIGN OIL HOLES AND SLOTS

FLANGE OR SIDE COVER

BEARING SUPPORT SMALL END UP

FINISH REAMING P T O. BUSHING

1. REASSEMBLE CYLINDER AND COVER
2. INSERT REAMER THROUGH MAGNETO BEARING
3. FINISH REAM P T O. BUSHING
4. WHEN FINISH REAMER IS THROUGH BUSHING STOP, DO NOT ALLOW ROUGH REAMER TO ENTER NEWLY INSTALLED MAGNETO END BUSHING
5. WASH ALL PARTS CAREFULLY TO REMOVE ALL CHIPS. CHECK OIL HOLES AND PASSAGES, BLOW OUT WITH COMPRESSED AIR
6. REASSEMBLE ENGINE

APPLY OIL TO PREVENT DAMAGE TO BUSHING

USE LIGHT OIL FOR FINISH REAMING. IF CRANKSHAFT IS TIGHT REPEAT WITHOUT OIL

P.T.O. BUSHING

OIL SEAL REMOVED

SNAP RING

On H20-HS50 H.P. horizontal crankshaft engines, remove the oil seal and snap ring to remove the cylinder cover

7. Remove the reamer by rotating it in the same direction as it is turned during the reaming operation. DO NOT TURN THE REAMER BACKWARDS.

BALL BEARING SERVICE — H20 THROUGH HS50 H.P. HORIZONTAL CRANKSHAFT ENGINES

1. Remove the crankshaft P.T.O. end oil seal. Drive an awl or similar tool into the metal seal body and pry out.

2. Use snap ring pliers to remove the snap ring.

3. Reassembly is in reverse order. Secure the cylinder cover, install the snap ring and oil seal. Protect the oil seal to prevent damage during installation.

FLAT

FLAT

TURN CLOCKWISE TO LOCK

TURN COUNTER-CLOCKWISE TO UNLOCK

EXTERIOR OF COVER

H40-HM100 H.P. horizontal crankshaft engines —locking and unlocking the bearing from outside

ROUGH REAMING WORN ALUMINUM BEARING
(MAGNETO SIDE) FOR ALUMINUM BEARING ONLY

HANDLE

USE KEROSENE
ON ROUGH
CUT REAMER

1 DISASSEMBLE ENGINE, CLEAN
AND REMOVE OIL SEALS
2 ASSEMBLE P.T.O. BEARING
TO CYLINDER
(MOUNTING FLANGE 'V' MODELS
CYLINDER COVER 'H' MODELS)

USE ALL
BOLTS
TO HOLD
ASSEMBLY
IN ALIGN-
MENT. DO
NOT USE
GASKET

DOWEL PINS IN
CYLINDER MUST BE
IN PLACE FOR
PROPER ALIGNMENT

MAGNETO
END
BEARING

FINISH REAMER

REAMER GUIDES
IN P.T.O.
BEARING

DISASSEMBLE AFTER
ROUGH REAMING
MAGNETO END

FINISH REAMING NEW MAGNETO BUSHING

ROUGH CUT REAMER

1 REASSEMBLE CYLINDER TO
COVER
2 INSERT REAMER THROUGH
P.T.O. BEARING
3 FINISH REAM MAGNETO
END BUSHING

USE LIGHT OIL ON
FINISH REAMER
(NOTE: IF BUSH-
ING IS TIGHT ON
CRANKSHAFT,
REPAT REAM-
ING WITHOUT
OIL)

REAMER
INSERTED
THROUGH
P.T.O.
BEARING

WORN
BUSHING

NEW BUSHING

FINISH
REAMER

WORN BUSHING REMOVAL (MAGNETO END)
FOR BRONZE BUSHING ONLY

BUSHING DRIVER

CYLINDER (HORIZON-
TAL OR VERTICAL
CRANKSHAFT ENGINE)

1 DISASSEMBLE AND CLEAN ALL
PARTS
2 POSITION BEARING SUPPORT
TOOL WITH LARGE END UP
3 CAREFULLY DRIVE WORN BEAR-
ING OUT OF CYLINDER AND
SIDE COVER OR FLANGE
(EXCEPT BALL
BEARING P.T.O.)

WORN
BRONZE
BUSHING
(MAGNETO
END)

BEARING
SUPPORT
(SMALL END)

USE END WITH LARGER
HOLE TO SUPPORT
BEARING WHEN
REMOVING BUSHING

ROUGH REAMING WORN ALUMINUM BEARING
(P.T.O. END) FOR ALUMINUM BEARING ONLY

REAMER CUTTING
OUT P.T.O.
BEARING

1 AFTER REAMING MAGNETO
END BUSHING BEGIN TO
REAM P.T.O. BUSHING

NEW MAGNETO
END BUSHING
FINISH REAMED

INSTALLING NEW BRONZE BUSHING
(MAGNETO SIDE)

BUSHING
DRIVER

ALIGN
OIL HOLE

1 POSITION BEARING SUP-
PORT TOOL. SMALL END
UP
2 PRESS IN BUSHING
3 ALIGN OIL HOLES
4 DRIVE BUSHING UNTIL IT
BOTTOMS ON BEARING
SUPPORT

NEW
BUSHING

BEARING
SUPPORT
(LARGE END)

POSITION BEARING SUP-
PORT TOOL WITH SMALL-
ER DIAMETER HOLE
FOR INSTALLING BUSHING

WORN BUSHING REMOVAL (P.T.O. END)
FOR BRONZE BUSHING ONLY

BUSHING
DRIVER

FLANGE OR
SIDE COVER

WORN
BUSHING

BEARING SUPPORT
LARGE END UP

BALL BEARING SERVICE – H40, 50 HORIZONTAL CRANKSHAFT ENGINES

1. Prior to attempting removal of the cylinder cover, observe the area around the crankshaft P.T.O. oil seal. Compare it with the illustration, and if there are bearing locks, follow instructions below:

 a. Remove the locking nuts using the proper socket wrench. Note fiber washer located under nut; this must be reinstalled. Lift side cover from cylinder after removing the side cover bolts.

 b. Install the bearing retainer bolts, fiber washer and locking nuts in the proper sequence in the cover.

2. Also note the following points:

 a. On some engines, a locking type retainer bolt is used. To release the bolt, merely loosen the locking nut and turn the retainer bolt counterclockwise to the unlocked position with needle nose pliers to permit the side cover to be removed. Note that the flats on the retainer bolts must be turned so they face the crankshaft to be relocked upon installation. Don't force them! Torque the locking nuts only to 15–22 inch lbs.

 b. The ball bearing used in horizontal crankshaft engines has a restricted fit. The bearing is heated and put onto the cold crankshaft. As the bearing cools it grasps the crankshaft tightly and must be removed cold. Remove the ball bearing with a bearing splitter (separator) and a puller. The bearing may be heated by placing it into a container with a sufficient amount of oil to cover the bearing. The bearing should not rest on the bottom of the container. Suspend the bearing on a wire or set the bearing onto a spacer block of wood or wire mesh. Heat the oil and bearing carefully until the oil smokes, quickly remove the bearing and slide it onto the crankshaft.

 c. The bearing must seat tightly against the thrust washer which in turn rests tightly against the crankshaft gear.

 d. When a ball bearing is used it is not possible to see the keyway in the crankshaft gear which is normally used for timing. Because of this, one tooth of the crankshaft gear is chamfered. This chamfered tooth of the crankshaft gear is positioned opposite the timing mark on the camshaft gear. The use of a ball bearing requires the removal of the crankshaft when it is necessary to remove the camshaft. When replacing the crankshaft and camshaft, mate the timing marks and insert it into the cylinder block as an assembly.

LUBRICATION

Barrel and Plunger Oil Pump System

This system is driven by an eccentric on the camshaft. Oil is drawn through the hollow camshaft from the oil sump on its in-

A—INTAKE STROKE

B—COMPRESSION STROKE

C—CUTAWAY VIEW

Operation of the barrel type oil pump

FLAT

FLAT MUST FACE OUT

Installation of the barrel type oil pump

take stroke. The passage from the sump through the camshaft is aligned with the pump opening. As the camshaft continues rotation (pressure stroke), the plunger force the oil out. The other port in the camshaft is aligned with the pump, and directs oil out of the top of the camshaft.

At the top of the camshaft, oil is forced through a crankshaft passage to the top main bearing groove which is aligned with the drilled crankshaft passage. Oil is directed through this passage to the crankshaft connecting rod journal and then spills from the connecting rod to lubricate the cylinder walls. Splash is used to lubricate the other parts of the engine.

A pressure relief port in the crank case relieves excessive pressures when the oil viscosity is extremely heavy due to cold temperatures, or when the system is plugged or damaged. Normal pressure is 7 psi.

SERVICE

Remove the mounting flange or the cylinder cover, whichever is applicable. Remove the barrel and plunger assembly and separate the parts.

Clean the pump parts in solvent and inspect the pump plunger and barrel for rough spots or wear. If the pump plunger is scored or worn, replace the entire pump.

Before reassembling the pump parts, lubricate all of the parts in engine oil. Manually operate the pump to make sure the plunger slides freely in the barrel.

Lubricate all the parts and position the barrel on the camshaft eccentric. If the oil pump has a chamfer only on one side, that side must be placed toward the camshaft gear. The flat goes away from the gear, thus out to work against the flange oil pickup hole.

Install the mounting flange. Be sure the plunger ball seats in the recess in the flange before fastening it to the cylinder.

SPRAY MIST LUBRICATION

Late model LAV40, LAV30, and LAV35 engines have a spray mist lubrication system. This system is the same as the barrel and plunger oil pump system except that (1) the pressure relief port is changed to a calibrated spray mist orifice and (2) the crankshaft is not rifle-drilled from the top main to the crank pin. Lubrication is sprayed to the narrow rod cap area through the spray mist hole.

SPLASH LUBRICATION

Some engines utilize the splash type lubrication system. The oil dipper, on some engines, is cast onto the lower connecting rod bearing cap. It is important that the proper parts are used to ensure the longest engine life.

H40 THRU H60
CONNECTING ROD
FOR SPLASH LUBRICATION

BE SURE MATCH MARKS
ARE TOGETHER AND
ARE "OUT" DURING
REASSEMBLY

CAST ON OIL DIPPER

Splash type lubrication connecting rod

GEAR TYPE OIL PUMP SYSTEM

The gear type lubrication pump is a crankshaft driven, positive displacement pump. It pumps oil from the oil sump in the engine base to the camshaft, through the drilled camshaft passage to the top main bearing, through the drilled crankshaft, to the connecting rod journal on the crankshaft.

Spillage from the connecting rod lubricates the cylinder walls and normal splash lubricates the other internal working parts. There is a pressure relief valve in the system.

MOUNTING FLANGE

LOCKWASHER

SCREW

COVER

DISPLACEMENT MEMBER

DRIVE GEAR

GASKET

A. DISASSEMBLED

Disassembled view of the gear type lubrication system

Service

Disassemble the pump as follows: remove the screws, lockwashers, cover, gear, and displacement member.

Wash all of the parts in solvent. Inspect the oil pump ddive gear and displacement member for worn or broken teeth, scoring, or other damage. Inspect the shaft hole in the drive gear for wear. Replace the entire pump if cracks, wear, or scoring is evident.

To replace the oil pump, position the oil pump displacement member and oil pump gear on the shaft, then flood all the parts with oil for priming during the initial starting of the engine.

The gasket provides clearance for the drive gear. With a feeler gauge, determine the clearance between the cover and the oil pump gear. The clearance desired is 0.006–0.007 in. Use gaskets, which are available in a variety of sizes, to obtain the correct clearance. Position the oil pump cover and secure it with the screws and lockwashers.

Craftsman Engines Cross Reference Chart

Craftsman Engine Models	See Column	Craftsman Engine Models	See Column	Craftsman Engine Models	See Column
143.50040	7	143.151012 – 143.153032	1	143.171212 – 143.173042	1
143.50045		143.154012 – 143.154142	2	143.174012 – 143.174292	2
		143.155012 – 143.155062	9	143.175012 – 143.175072	9
143.131022 – 143.131102	1	143.156022		143.177012 – 143.177072	7
143.135012 – 143.135112	9	143.157012 – 143.157032	7		
143.137012	7			143.181042 – 143.183042	1
143.137032		143.161012 – 143.163062	1	143.184012 – 143.184212	2
		143.164012 – 143.164202	2	143.184232 – 143.184252	4
143.141012 – 143.143032	1	143.165012 – 143.165052	9	143.184262 – 143.184402	2
143.145012 – 143.145072	9	143.167012 – 143.167042	7	143.185012 – 143.185052	9
143.146022				143.186062	
143.146022		143.171012 – 143.171172	1	143.187022 – 143.187102	3
143.147012 – 143.147032	7	143.171202	2		

Craftsman Engines Cross Reference Chart (cont.)

Craftsman Engine Models	See Column	Craftsman Engine Models	See Column	Craftsman Engine Models	See Column
143.191012–143.191052	1	143.224432	4	143.254302	2
143.194012–143.194052	2	143.225012	13	143.254312	
143.194062	4	143.225022		143.254322	4
143.194072–143.194092	2	143.225032–143.225052	9	143.254332	2
143.194102	4	143.225062	13	143.254342	4
143.194112–143.194142	2	143.225072		143.254352	
143.195012	9	143.225082–143.225102	9	143.254362	2
143.195022		143.227012–143.227072	12	143.254372	4
143.197012	3			143.254382	
143.197022	5	143.233012	1	143.254392	2
143.197032		143.233032		143.254402	4
143.197042–143.197072	3	143.233042		143.254412	
143.197082	5	143.234022–143.234052	2	143.254432	2
		143.234062–143.234092	4	143.254442	4
143.201032–143.203012	1	143.234102–143.234162	2	143.254452	2
143.204022	4	143.234192		143.254462	4
143.204032–143.204052	2	143.234202		143.254472	2
143.204062	4	143.234212–143.234232	4	143.254482	
143.204072–143.204092	2	143.234242–143.234262	2	143.254492	4
143.204102	4	143.235012	13	143.25450⎯–143.254532	2
143.204132		143.235022		143.2550⎯–143.255112	6
143.204142–143.204192	2	143.235032	6	143.257012–143.257072	3
143.204202	4	143.235042	13		
143.205022	9	143.235052		143.264012–143.264042	2
143.207012–143.207052	3	143.235062	9	143.264052–143.264082	4
143.207062	5	143.235072	6	143.264092	2
143.207072	3	143.237012	12	143.264102	4
143.207082	5	143.237022		143.264232–143.264342	2
		143.237032	5	143.264352–143.264372	4
143.213012–143.213042	1	143.237042	3	143.264382	2
143.214012–143.214032	2			143.264392–143.264412	4
143.214042–143.214072	4	143.244032	2	143.264422	2
143.214082–143.214252	2	143.244042	4	143.264432–143.264482	4
143.214262–143.214282	4	143.244052		143.264492	2
143.214292	2	143.244062		143.264502	
143.214302		143.244072–143.244112	2	143.264512	4
143.214312	4	143.244122–143.244142	4	143.264522	2
143.214322		143.244202	2	143.264542	
143.214332	2	143.244212	4	143.264562–143.264672	4
143.214342		143.244222	2	143.264682	2
143.214352	4	143.244232		143.265012–143.265192	6
143.216042–143.216062	3	143.244242	4	143.267012–143.267042	3
143.217012–143.217032	5	143.244252	4		
143.217042–143.217072	3	143.244262–143.244282	2	143.274022–143.274072	4
143.217092	5	143.244292–143.244332	4	143.274092–143.274132	2
143.217102	3	143.245012	6	143.274142	4
		143.245042	9	143.274152	
143.223012–143.223052	1	143.245052–143.245072	13	143.274162–143.274182	2
143.224012	2	132.245082		143.274192–143.274242	4
143.224022		143.245092	6	143.274252	2
143.224032	4	143.245102–143.245132	13	143.274262	4
143.224062	2	143.245142	6	143.274272–143.274322	2
143.224072	4	143.245152		143.274402–143.274482	4
143.224092–143.224132	2	143.245162	13	143.275012–143.275052	6
143.224142	1	143.245172	6	143.276022	4
143.224162–143.224222	2	143.245182		143.277012	3
143.224232	4	143.245192	13	143.277022	
143.224242					
143.224252–143.224282	2	143.254012–143.254052	2	143.284012	2
143.224292	4	143.254062	4	143.284022	1
143.224302		143.254072–143.254122	2	143.284032	2
143.224312–143.224342	2	143.254142–143.254192	4	143.284042	4
143.224352	4	143.254212	2	143.284052	2
143.224362		143.254222		143.284062	
143.224372–143.224422	2	143.254232–143.254292	4	143.284072	4

Craftsman Engines Cross Reference Chart (cont.)

Craftsman Engine Models	See Column	Craftsman Engine Models	See Column	Craftsman Engine Models	See Column
143.284082	2	143.545012 – 143.545042	9	143.614062 – 143.614162	2
143.284092		143.547012 – 143.547032	7	143.615012 – 143.615092	9
143.284102	4			143.617012 – 143.617182	3
143.284112	2	143.551012	1		
143.284142		143.551032		143.621012 – 143.621092	1
143.284152		143.551052 – 143.551192	1	143.624012 – 143.624112	2
143.284162		143.554012 – 143.554082	2	143.625012 – 143.625132	9
143.284182		143.555012 – 143.555052	9	143.627012 – 143.627042	3
143.284212	4	143.557012 – 143.557082	7		
143.284312	2			143.631012 – 143.631092	1
143.284322		143.565022	9	143.634012	2
143.284332	4	143.566212	9	143.634032	
143.284342		143.567012 – 143.567042	7	143.635012	9
143.284352				143.635022	
143.284382	4	143.571002 – 143.571122	1	143.635032	6
143.284402	2	143.571152	2	143.635052	9
143.284412		143.571162	1	143.637012	3
143.284432	4	143.571172			
143.284362		143.574022 – 143.574102	2	143.641012 – 143.641062	1
143.284392	2	143.575012 – 143.575042	9	143.641072	2
143.284442				143.644012 – 143.644082	2
143.284482		143.581002 – 143.581102	1	143.645012 – 143.645032	6
143.284452	4	143.584012 – 143.584142	2	143.647012 – 143.647062	3
143.284472		143.585012 – 143.585042	9		
143.285012	6	143.587012 – 143.587042	3	143.651012 – 143.651072	1
143.285022				143.654022 – 143.654322	2
143.285032		143.591012 – 143.591142	1	143.655012	6
		143.594022 – 143.594082	2	143.655032	
143.521081	9	143.594092	2	143.657012 – 143.657052	3
143.525021	9	143.594102			
143.526011		143.595012	9	143.661012 – 143.661062	1
		143.595042		143.664012 – 143.664332	2
143.531052	1	143.597012 – 143.597032	3	143.665012 – 143.665082	6
143.531082				143.667012	3
143.531122		143.601022 – 143.601062	1	143.667022	
143.531132		143.604012	2	143.667032	6
143.531142	2	143.604022	4	143.667042 – 143.667082	3
143.531152	1	143.604032	2		
143.531172		143.604042		143.674012	2
143.531182		143.604052	4	143.675012	6
143.534012 – 143.534072	2	143.604062	2	143.675022	
143.535012 – 143.535062	9	143.604072		143.675032	9
143.537012	7	143.605012	9	143.675042	6
		143.605022		143.677012	3
143.541012	1	143.605052		143.677022	
143.541042 – 143.541062	1	143.607012 – 143.607032	3		
143.541112 – 143.541152	1	143.607042 – 143.607062	3	143.686072	9
143.541172 – 143.541202	1			143.687012	3
143.541222	1	143.611012 – 143.611112	1		
143.541282 – 143.541302	1	143.614012 – 143.614032	4	143.694126	4
143.544012 – 143.544042	2	143.614042	2	143.694132	2
		143.614052	4		

Torque Specifications

Model/Part	Inch Pounds	Ft. Pounds
Cylinder Head Bolts	160–200	13–16
Connecting Rod Bolts	65–75	5.5–6
TVS75, 90 & 105, 2.5 thru 4 H.P. (Durlok Rod Bolts)	95–110	7.9–9.1
4–5 H.P. Small Frame	80–95	6.6–7.9
ECH90, ECV100, TNT100	75–80	6.2–6.7
TVS120, 5 H.P. Small Frame (Durlok Rod Bolts)	110–130	9.1–10.8
5–6 H.P. Medium Frame	86–110	7.1–9.1
5–6 H.P. Medium Frame (Durlok Rod Bolts)	130–150	10.8–12.5
ECV105, ECV110, ECV120, TNT120	80–95	6.6–7.9
7, 8 & 10 Medium Frame	106–130	8.8–10.8
7, 8 & 10 Medium Frame (Durlok Rod Bolts)	150–170	12.5–14.1
Cylinder Cover or Flange-to-Cylinder	65–110	5.5–9
Cylinder Cover 5–7 H.P. Medium Frame, H Models	100–140	8.3–11.6
Flywheel Nut	360–396	30–33
Spark Plug	180–360	15–30
Magneto Stator to Cylinder	40–90	3.3–7.5
Starter to Blower Housing or Cylinder	40–60	3.5–5
Housing Baffle to Cylinder	48–72	4–6
Breather Cover (Top Mount ECV)	40–50	3.3–4.1
Breather Cover	20–26	1.7–2.1
Intake Pipe to Cylinder	72–96	6–8
Carburetor to Intake Pipe	48–72	4–6
Air Cleaner to Carburetor (Plastic)	8–12	1
Tank Plate to Bracket (Plastic)	100–144	9–12
Tank to Housing	45–65	3.7–5
Muffler Bolts to Cylinder		
1–5 H.P. Small Frame	30–45	2.5–3.5
4–5 H.P. Medium Frame	90–150	8–12
6:1 Gear Reduction Cover to Housing	100–144	8.5–12
Gear Reduction Cover to Housing	65–110	5–9
Oil Drain Plug		
1/8 — 27	35–50	1.1–4.1
1/4 — 18	65–85	4.5–7
3/8 — 18	80–100	6.6–9
5/8 — 18	90–150	7.5–12.5
1/2 — 14	80–100	6.6–9

Torque Specifications

Model/Part	Inch Pounds	Ft. Pounds
Ball Bearing Retainer 2.5		
2.5–5 H.P. Small frame	45–60	3.7–5
5–10 H.P. Medium frame	15–22	1.5
Craftsman Exclusive Fuel System to Cylinder	72–96	6–8
Electric Starter-to-Cylinder	50–60	4–5

Engine Specifications

Reference Column	1	2	3	4	5	6	7	8	9	10	11	12
Displacement	7.75	9.06	10.5	10.0	10.5	12.0	11.04	12.17	11.5	12.0	10.0	12.0
Stroke	$1^{27}/_{32}$″	$1^{27}/_{32}$″	$1^{15}/_{16}$″	$1^{27}/_{32}$″	$1^{15}/_{16}$″	$1^{15}/_{16}$″	2¼″	2¼″	$1^{15}/_{16}$″	$1^{15}/_{16}$″	$1^{27}/_{32}$″	$1^{15}/_{16}$″
Bore	2.3125 2.3135	2.5000 2.5010	2.625 2.626	2.625 2.626	2.625 2.626	2.812 2.813	2.5000 2.5010	2.625 2.626	2.750 2.751	2.812 2.813	2.625 2.626	2.812 2.813
Timing Dimension Before Top Dead Center for Vertical Engines	V.060 .070	V.065	V.035	.035	.035	V.040 .060	V.050	H.050	V.035	V.035	V.035	V.035
Timing Dimension Before Top Dead Center for Horizontal Engines	H.060 .070	H.030 .040	H.035			H.055	H.050	H.050				
Point Setting	.020	.020	.020	.020	.020	.020	.020	.020	.020	.020	.020	.020
Spark Plug Gap	.030	.030	.030	.030	.030	.030	.030	.030	.030	.030	.030	.030
Valve Clearance	.010 Both	.010 Both	.010 Both	.010 Both	.010 Both	.010 Both	.010 Both	.010 Both	.010 Both	.010 Both	.010 Both	.010 Both
Valve Seat Angle	46°	46°	46°	46°	46°	46°	46°	46°	46°	46°	46°	46°
Valve Spring Free Length	1.135″	1.135″	1.135″	1.135″	1.135″	1.135″	1.562″	1.462″	1.135″	1.135″	1.135″	1.135″
Valve Guides Over-Size Dimensions	.2805 .2815	.2805 .2815	.2805 .2815	.2805 .2815	.2807 .2817	In. .280 Ex. .278	.3432 .3442	.343 .344	.2805 .2815	.2805 .2815	.2805 .2815	.2805 .2815
Valve Seat Width	.035 .045	.035 .045	.035 .045	.035 .045	.035 .045	.035 .045	.042 .052	.042 .052	.035 .045	.035 .045	.035 .045	.035 .045
Crankshaft End Play	.005 .027	.005 .027	.005 .027	.005 .027	.005 .027	.005 .027	.005 .027	.005 .027	.005 .027	.005 .027	.005 .027	.005 .027
Crankpin Journal Diameter	.8610 .8615	.8610 .8615	.9995 1.0000	.8610 .8615	.9995 1.0000	.9995 1.0000	1.0615 1.0620	1.0615 1.0620	.9995 1.0000	.9995 1.0000	.8610 .8615	.9995 1.0000
Cylinder Main Bearing Dia.	.8755 .8760	.8755 .8760	1.0005 1.0010	.8755 .8760	1.0005 1.0010	1.0005 1.0010	1.0005 1.0010	1.0005 1.0010	1.0005 1.0010	1.0005 1.0010	.8755 .8760	1.0005 1.0010
Cylinder Cover Main Bearing Dia.	.8755 .8760	.8755 .8760	1.0005 1.0010	.8755 .8760	1.2010 1.2020	1.0005 1.0010	1.0005 1.0010	1.0005 1.0010	1.0005 1.0010	1.0005 1.0010	.8755 .8760	1.0005 1.0010
Conn. Rod. Dia. Crank Bearing	.8620 .8625	.8620 .8625	1.0005 1.0010	.8620 .8625	1.0005 1.0010	1.0005 1.0010	1.0630 1.0635	1.0630 1.0635	1.0005 1.0010	1.0005 1.0010	.8620 .8625	1.0005 1.0010
Piston Diameter	2.3090 2.3095	2.4950 2.4955	2.6260 2.6205	2.6200 2.6205	2.604 2.608	2.8070 2.8075	2.492 2.4945	2.6210 2.6215	2.7450 2.7455	2.8070 2.8075	2.6200 2.6205	2.8070 2.8075
Piston Pin Diameter	.5629 .5631	.5629 .5631	.5629 .5631	.5629 .5631	.5631 .5635	.5629 .5631	.6248 .6250	.6248 .6250	.5629 .5631	.5629 .5631	.5629 .5631	.5629 .5631
Width of Comp. Ring Groove	.0955 .0977	.0955 .0975	.0925 .0935	.0955 .0975	.0955 .0975	.0955 .0975	.0955 .0975	.0955 .0975	.0795 .0815	.0955 .0975	.0955 .0975	.0955 .0975
Width of Oil Ring Grooves	.125 .127	.125 .127	.156 .158	.156 .158	.156 .158	.156 .158	.156 .158	.156 .158	.1565 .1585	.1565 .1585	.1565 .1585	.1565 .1585

TECUMSEH-LAUSON LIGHT FRAME 4-STROKE
SMALL ENGINES – 0–6 HP

Engine Specifications

Reference Column	1	2	3	4	5	6	7	8	9	10	11	12
(Top) Comp. Side Clearance of Ring Groove	.002 / .005	.002 / .003	.002 / .004	.002 / .005	.002 / .005	.003 / .004	.002 / .003	.002 / .004	.002 / .004	.003 / .004	.003 / .0045	.0028 / .0039
(Bot.) Oil			.001 / .004	.001 / .004	.001 / .004	.002 / .003		.002 / .004	.001 / .002	.001 / .002	.0010 / .0030	.0018 / .0038
Ring End Gap	.007 / .020	.007 / .020	.007 / .020	.007 / .020	.007 / .020	.007 / .020	.007 / .020	.007 / .020	.007 / .020	.007 / .020	.007 / .020	.007 / .020
Top Piston Land Clearance	.0015 / 0145	.015 / .018	.0165 / .0215	.017 / .022	.017 / .022	.017 / .022	.015 / .018	.017 / .020	.024 / .027	.018 / .021	.017 / .022	.017 / .022
Piston Skirt Clearance	.0025 / .0040	.0045 / .0060	.0045 / .0060	.0045 / .0060	.0050 / .0065	.0045 / .0060	.0055 / .0070	.0035 / .0050	.0045 / .0060	.0045 / .0060	.0045 / .0060	.0045 / .0060
Camshaft Bearing Dia.	.4975 / .4980	.4975 / .4980	.4975 / .4980	.4975 / .4980	.505 / .513	.4975 / .4980	.6230 / .6235	.6230 / .6235	.4975 / .4980	.4975 / .4980	.4975 / .4980	.4975 / .4980
Dia. of Crankshaft Mag. Main Brg.	.8735 / .8740	.8735 / .8740	.9985 / .9990	.8735 / .8740	.9985 / .9990	.9985 / .9990	.9985 / .9990	.9985 / .9990	.9990 / .9995	.9990 / .9995	.8735 / .8740	.9985 / .9990
Dia. of Crankshaft P.T.O. Main Brg.	.8735 / .8740	.8735 / .8740	.9985 / .9990	.8735 / .8740	.9985 / .9990	.9985 / .9990	.9985 / .9990	.9985 / .9990	.9985 / .9990	.9985 / .9990	.8735 / .8740	.9985 / .9990

A. For VM80 & HM80 engines only - Displacement is 19.41"
B. For VM80 & HM80 engines only - Bore is 3.125" (3⅛"). 3.126"

Note C. For VM80 & HM80 engines only - Piston Diameter is 3.1205" 3.1195"

Cross Reference for Vertical Crankshaft Engines

Size	Model	Column
2½ HP	LAV25	1
3HP	LAV30	1
	TVS75	1
3½ HP	LV35	2
	LAV35	2
	TVS90	2
4 HP	LAV40	3
	TVS105	3
	V40 thru V40B	7
	VH40	7
10.0 CI	ECV100	4
	TNT100	14
10.5 CI	ECV105	5
11.0 CI	ECV110	12
5 HP	LAV50	6
	TVS120	6
	V50	9
	VH50	9
12.0 CI	ECV120	13
	TNT120	15

Cross Reference Chart for Horizontal Crankshaft Engines

Size	Model	Column
2½ HP	H25	1
3 HP	H30	1
3½ HP	H35	2
9.0 CI	ECH90	2
4 HP	H40	7
	HH40	7
	HS40	3
5 HP	H50	9
	HH50	9
	HS50	6

Section 16
Tecumseh-Lauson Light Frame 4-Stroke

Small Engines — 6-20 hp

ENGINE IDENTIFICATION

Lauson 4 cycle engines are identified by a model number stamped on a nameplate. The nameplate is located on the crankcase of vertical shaft models and on the blower housing of horizontal shaft models.

A typical model number appears on the illustration showing the location of the nameplate for vertical crankshaft engines. This number is interpreted as follows:

- V — vertical shaft engine
- 60 — 6.0 horsepower
- 70360J — the specification number. The last three numbers (360) indicate that this particular engine is a variation on the basic model line.
- 2361J — serial number
- 2 — year of manufacture
- 361 — the calendar day of manufacture
- J — line and shift location at the factory

Vertical engine identification

MAINTENANCE

Air Cleaner Service

Service all the oil/foam polyurethane and oil bath air cleaner elements in the same manner as the Briggs and Stratton components. See Chapter Four.

The Tecumseh treated paper element type air cleaner consists of a pleated paper element encased in a metal housing and must be replaced as a unit. A flexible tubing and hose clamps connect the remotely mounted air filter to the carburetor.

General Engine Specifications Vertical Crankshaft Engines

Model	Bore & Stroke	Displacement	Horsepower
V60	2.625 × 2.5000	13.53	6
VH60	2.625 × 2.5000	13.53	6
V70	2.750 × 2.5313	15.0	7
VH70	2.750 × 2.5313	15.0	7
VM70	2.750 × 2.5313	15.0	7
V80	3.062 × 2.5313	18.65	8
VM80	3.125 × 2.5313	19.41	8
VM100	3.187 × 2.5313	17.16	8
ECV100	2.625 × 1.8438	10.0	—
TNT100	2.625 × 1.8434	20.2	—
ECV105	2.625 × 1.9375	10.5	—
ECV110	2.750 × 1.9375	11.5	—
ECV120	2.812 × 1.9375	12.0	—
TNT120	2.812 × 1.9375	12.0	—

Horizontal engine identification

General Engine Specifications
Horizontal Crankshaft Engines

Model	Bore & Stroke (in.)	Displacement (cu in.)	Horse-power
H60	2.6250 × 2.5000	13.53	6
HH60	2.6250 × 2.5000	13.53	6
H70	2.7500 × 2.5313	15.0	7
HH70	2.7500 × 2.5313	15.0	7
HM70	2.9375 × 2.5313	17.16	7
H80	3.0620 × 2.5313	18.65	8
HM80	3.0620 × 2.5313	18.65	8
HM100	3.1870 × 2.5313	20.2	10
ECH90	2.5000 × 1.8438	9.06	—

Clean the element by lightly tapping it. Do not distort the case. When excessive carburetor adjustment or loss of power results, inspect the air filter to see if it is clogged. Replacing a severely restricted air filter should show an immediate performance improvement.

Check the oil level in the oil bath type air cleaners regularly to make sure the level is correct. To add oil, unscrew the wingnut, pull off the filter element and add oil along the side of the filter until the level is correct. Use the same type and viscosity oil used in the engine.

If the filter is dirty, remove it and wash it in solvent. Also remove the filter bowl, drain the oil, and wash the filter in solvent. Refill the bowl with clean oil after putting it into position on the air horn.

A plain paper element is also used. It should be removed every 10 hours, or more often if the air is dusty. Tap or blow out the dirt form the inside with low pressure air. This type should be replaced at 50 hours. If clogged sooner, it may be washed in soap and water and rinsed by flushing from the inside until the water is clear. Blow dry with low pressure compressed air.

To service the KLEEN-AIRE system, move the element, wash it in soap and mild detergent, pat dry, and then coat with oil. squeeze the oil to distribute it evenly and remove the excess. Make sure all mounting surfaces are tight to prevent leakage.

Treated paper element air cleaner (used on Craftsman engines)

Exploded view of oil bath type air cleaner

Lubrication

OIL AND FUEL RECOMMENDATIONS

Use fresh (less than one month old) gasoline, of "Regular" grade. Unleaded fuel is preferred, but leader fuel is acceptable.

Use oil having MS, SC, SD, or SE classification. Use these viscosities for aluminum engines:

- Summer – above 32°F: S.A.E. 30 (S.A.E. 10W30 or 10W40 are acceptable substitutes)
- Winter – Below 32°F: S.A.E. 5W30 (S.A.E. 10W is an acceptable substitute). (Including Snow King Snow Blower Engines)
- Winter – Below 0°F only: S.A.E. 10W diluted with 10% kerosene is an acceptable substitute. (Including Snow King Snow Blower Engines)

Use these viscosities for cast iron engines:
- Summer – Above 32°F: S.A.E. 30
- Winter – Above 32°F: S.A.E. 10W

TUNE-UP

Tune-Up Specifications

The follow basic specifications apply to all the engines covered in this section:

- Spark Plug Gap: 0.030 in.
- Ignition Point Gap: 0.020 in.
- Valve Clearance: 0.010 in. for both intake and exhaust.

For timing dimension, which varies from engine to engine, see the complete specifications at the rear of this section.

Spark Plug Service

Spark plugs should be removed, cleaned, and adjusted periodically. Check the electrode gap with a wire feeler gauge and adjust the gap. Replace the plugs if the electrodes are pitted and burned or the porcelain is cracked. Refer to the Tecumseh master parts manual for the correct replacement number. Apply a little graphite grease to the threads to prevent sticking. Be sure the cleaned plugs are free of all foreign material.

Breaker Points

ADJUSTMENT

1. Disconnect the fuel line from the carburetor.
2. Remove the mounting screws, fuel tank, and shroud to provide access to the flywheel.
3. Remove the flywheel with either a puller (over 3.5 hp) or by using a screwdriver to pry underneath the flywheel while tapping the top lightly with a soft hammer.
4. Remove the dust cover and gasket from the magneto and crank the engine over until the breaker points of the magneto are fully opened.
5. Check the condition of the points and replace them if they are burned or pitted.
6. Check the point gap with a feeler gauge. Adjust them, if necessary, as per the directions on the dust cover. Refer to the specifications chart at the end of this chapter for point gap.

REPLACEMENT

1. Gain access to the points and inspect them as described above. If the points are badly pitted, follow the remaining steps to replace them.
2. Remove the nuts that hold the electrical leads to the screw on the movable breaker point spring. Remove the movable breaker point from stud.

3. Remove the screw and stationary breaker point. Put a new stationary breaker point on the breaker plate; install the screw, but do not tighten. This point must be moved to make the proper air gap when the points are adjusted.
4. Position a new movable breaker point on the stud.
5. Adjust the breaker point gap with a flat feeler gauge and tighten the screw.
6. Check the new point contact pattern and remove all grease, finger-prints, and dirt from contact surfaces.
7. Adjust the timing as described below.

Ignition Timing

ADJUSTMENT

1. Remove the cylinder head bolts, and move the head (with gasket in place) so that the spark plug hole is centered over the piston.
2. Using a ruler (through the spark plug hole) or special plunger type tool, carefully turn the engine back and forth until the piston is at exactly Top Dead Center. Tighten the thumbscrew on the tool.
3. Find the timing dimension for your engine in the specifications at the rear of the manual. Then, back off the position of the piston until it is about halfway down in the bore. Lower the ruler (or loosen the thumbscrew and lower the plunger, if using the special tool) exactly the required amount (the amount of the timing dimension). Then, hold the ruler in place (or tighten the special tool thumbscrew) and, finally carefully rotate the engine forward until the piston just touches the ruler or tool plunger.
4. Install a timing light or place a very thin piece of cellophane between the contact points. Loosen and rotate the stator just until the timing light shows a change in current flow or the cellophane pulls out of contact gap easily. Then, tighten stator bolts to specified torque.
5. Install the leads, point cover, flywheel, and shrouding.

SOLID STATE IGNITION SYSTEM CHECKOUT

The only on-engine check which can be made to determine whether the ignition system is working, is to separate the high tension lead from the spark plug and check for spark. If there is a spark, then the unit is allright and the spark plug should be replaced. No spark indicates that some other part needs replacing.

Check the individual components as follows:

- High Tension Lead: Inspect for cracks or indications of arcing. Replace the transformer if the condition of the lead is questionable.
- Low Tension Leads: Check all leads for shorts. Check the ignition cut-off lead to see that the unit is not grounded. Repair the leads, if possible, or replace them.
- Pulse Transformer: Replace and test for spark.
- Magneto: Replace and test for spark. Time the magneto by turning it counterclockwise as far as it will go and then tighten the retaining screws.
- Flywheel: Check the magnets for strength. With the flywheel off the engine, it should attract a screwdriver that is held 1 in. from the magnetic surface on the inside of the flywheel. Be sure that the key locks the flywheel to the crankshaft.

Carburetor Mixture Adjustments

1. If the carburetor has been overhauled, or the engine won't start, make initial mixture screw adjustments as specified in the chart.
2. Start the engine and allow it to warm up to normal running temperature. With the engine running at maximum rec-

ommended rpm, loosen the main adjustment screw until engine rpm drops off, then tighten the screw until the engine starts to cut out. Note the number of turns from one extreme to the other. Loosen the screw to a point midway between the extremes.

NOTE: Some carburetors have fixed jets. If there is no main adjusting screw and receptacle, no adjustment is needed.

3. After the main system is adjusted, move the speed control lever to the idle position and follow the same procedure for adjusting the idle system.

4. Test the engine by running it under a normal load. The engine should respond to load pickup immediately. An engine that "dies" is too lean. An engine which ran roughly before picking up the load is adjusted too rich.

Chart of Initial Carburetor Adjustments

Adjustment	For Engines Built Prior to 1977	For Engines Built After 1977
Main Adjustment Up to 7 HP	V-60-70-1¼ H-60-70-1¼	Same
Main Adjustment VM70-80-100 & HM70-80-100	1¼	1½
Idle Adjustment Up to 7 HP	V-60-70-1 Turn H-60-70-1 Turn	Same
Idle Adjustment VM70-80-100 & HM70-80-100	1½	1¼
Idle Speed (Top of Carburetor) Regulating screw	Back out screw, then turn in until screw just touches throttle lever and continue 1 turn more (if idle RPM is given set final idle speed with a tachometer)	

Governor

ADJUSTMENT

Air Vane Type

1. Operate the engine with the governor adjusting lever or panel control set to the highest possible speed position and check the speed. If the speed is not within the recommended limits, the governed speed should be adjusted.

2. Loosen the locknut on the high speed limit adjusting screw and turn the adjusting screw out to increase the top engine speed.

Mechanical Type

1. Set the control lever to the idle position so that no spring tension affects the adjustment.

2. Loosen the screw so that the governor lever is loose in the clamp.

Schematic of the operation of an air vane governor

Ⓐ GOVERNED POSITION
Ⓑ NON-GOVERNED POSITION
♦ DIRECTION OF ADJUSTMENT—THIS SCHEMATIC

Schematic of mechanical governor operation

3. Rotate both the lever and the clamp to move the throttle to the full open position (away from the idle speed regulating screw).

4. Tighten the screw when no end-play exists in the direction of open throttle.

5. Move the throttle lever to the full speed setting and check to see that the control linkage opens the throttle.

Compression Check

1. Run the engine until warm to lubricate and seal the cylinder.

2. Remove the spark plug and install a compression gauge. Turn the engine over with the pull starter or electric starter.

3. Compression on new engines is 80 psi. If the reading is below 60 psi., repeat the test after removing the gauge and squirting about a teaspoonful of engine oil through the spark plug hole. If the compression improves temporarily following this, the problem is probably with the cylinder, piston, and rings. Otherwise, the valves require service.

FUEL SYSTEM

Carburetor

NOTE: Four-cycle Tecumseh engines use float or diaphragm type carburetors.

REMOVAL AND INSTALLATION

1. Drain the fuel tank. Remove the air cleaner and disconnect the carburetor fuel lines.

2. If necessary, remove any shrouding or control panels to provide access to carburetor.

3. Disconnect the choke or throttle control wires at the carburetor.

4. Remove the cap screws, or nuts and lockwashers that hold the carburetor to the engine; remove the carburetor.

5. Secure the carburetor on to engine.

6. Install the shrouding or control panels. Connect the choke and throttle control wires.

7. Position the control panel to carburetor. Connect the carburetor fuel lines.

8. Install the air cleaner.

9. Adjust the carburetor as described above.

GENERAL OVERHAUL INSTRUCTIONS

1. Carefully disassemble the carburetor removing all non-metallic parts, i.e.; gaskets, viton seats and needles, O-rings, fuel pump valves, etc.

The idle operation of a float feed carburetor. The throttle plate closes, restricting the flow of fuel and air, forcing the engine to run on a reduced volume of fuel and air.

NOTE: Nylon check balls used in some diaphragm carburetor models may or may not be serviceable. Check to be sure of serviceability before attempting removal.

2. Clean all metallic parts with solvent.

NOTE: Nylon can be damaged if subjected to harsh cleaners for prolonged periods.

3. The large O-rings sealing the fuel bowl to the carburetor body must be in good condition to prevent leakage. If the O-ring leaks, interfering with the atmospheric pressure in the float bowl, the engine will run rich. Foreign material can enter through the leaking area and cause blocking of the metering orifices. This O-ring should be replaced after the carburetor has been disassembled for repair. Lubricate the new O-ring with a small amount of oil to allow the fuel bowl to slide onto the O-ring properly. Hold the carburetor body in an inverted position

The high speed operation of a float feed carburetor. The air venturi replaces the throttle plate as the restricting device and the engine runs on its greatest volume of fuel and air.

The intermediate operation of a float feed carburetor. The throttle plate "cracks" open to reduce the restriction and the engine runs on an increased volume of fuel and air.

and place the O-ring on the carburetor body and then position the fuel bowl.

4. The small O-rings used on the carburetor adjustment screws must be in good condition or a leak will develop and cause improper adjustment of carburetor.

5. Check all adjusting screws for wear. The illustration shows a worn screw and a good screw. Replace screws that are worn.

6. Check the carburetor inlet needle and seat for wear, scoring, or other damage. Replace defective parts.

7. Check the carburetor float for dents, leaks, worn hinge or other damage.

8. Check the carburetor body for cracks, clogged passages, and worn bushings. Clean clogged air passages with clean, dry compressed air.

9. Check the diaphragms on diaphragm carburetors for cracks, punctures, distortion, or deterioration.

10. Check all shafts and pivot pins for wear on the bearing surfaces, distortion, or other damage.

NOTE: Each time a carburetor is disassembled, it is good practice to install a repair kit.

Appearance of good and bad mixture adjusting screws

Appearance of good and bad inlet needles

The choke position of the diaphragm carburetor. The closed choke plate restricts the amount of air, creating a richened mixture by drawing in a greater proportion of fuel.

The idling operation of a diaphragm carburetor. The throttle plate closes, restricting the flow of fuel and air, forcing the engine to run on a reduced volume of fuel and air.

The high speed operation of a diaphragm carburetor. The air ventui replaces the throttle plate as the restricting device and the engine runs on its greatest volume of fuel and air.

The intermediate operation of a diaphragm carburetor. The throttle plate "cracks" to decrease the restriction and the engine runs on an increased volume of fuel and air.

11. Where there is excessive vibration, a damper spring may be used to assist in holding the float against the inlet needle thus minimizing the flooding condition. Two types of springs are available; the float shaft (hinge pin) type and the inlet needle mounted type.

12. Float shaft spring positioning:
   a. The spring is slipped over the shaft.
   b. The rectangular shaped spring end is hooked onto the float tab.
   c. The shorter angled spring end is placed onto the float bowl gasket support.

13. Note that on late model carburetors, the spring clip fastened to the inlet needle has been revised to provide a damping effect. The clip fastens to the needle and is hooked over the float tab.

FLOAT FEED CARBURETOR

Note the following points when rebuilding these carburetors.

Removal

Remove the carburetor from the engine. It is easier to remove the intake manifold and carburetor assembly from the engine, disconnect the governor linkage, fuel line, and grounding wire and then disassemble the carburetor from the intake manifold on a work bench. Be sure to note the positions of the governor and the throttle linkage to facilitate reassembly.

Throttle

1. Examine the throttle lever and plate prior to disassembly. Replace any worn parts.
2. Remove the screw in the center of the throttle plate and pull out the throttle shaft lever assembly.
3. When reassembling, it is important that the lines on the throttle plate are facing out when in the closed position. Position the throttle plates with the two lines at 12 and 3 o'clock. The throttle shaft must be held in tight to the bottom bearing to prevent the throttle plate from riding on the throttle bore of the body which would cause excessive throttle plate wear and governor hunting.

Choke

Examine the choke lever and shaft at the bearing points and holes into which the linkage is fastened and replace any worn

Float feed carburetor identification number

parts. The choke plate is inserted into the air horn of the carburetor in such a way that the flat surface of the choke is toward the fuel bowl.

Idle Adjusting Screw

Remove the idle screw from the carburetor body and examine the point for damage to the seating surface on the taper. If damaged, replace the idle adjusting needle. Tension on the screw is maintained with a coil spring and sealed with an O-ring. Examine and replace the O-ring if it is worn or damaged.

High Speed Adjusting Jet

Remove the screw and examine the taper. If the taper is damaged at the area where it seats, replace the screw and fuel bowl retainer nut as an assembly.

External view of a Tecumseh float type carburetor

Service hints for float feed carburetors

Loosen screw until it just clears throttle lever, then turn screw in one turn.

Do not attempt to remove blow air through passage.

Check shaft for binding position opening to bottom of air horn.

Blow air through passage. Do not remove restrictor if present.

Proper installation is important. See appropriate chapter for procedures.

Replace

Must hook over float tab.

Check float for leaks or dents. Clean bowl and adjust float level position gasket or gaskets.

If the carburetor is used on a 20° slant engine, the engine must be in its normal 20° slanted position for adjustment.

Check needle for damage and "O" ring for cracks. Clean all passages in nut with compressed air.

IDLE SPEED ADJUSTMENT

MAIN NOZZLE

CHOKE SHAFT AND LEVER

CHOKE PLATE

HIGH SPEED AIR BLEED

*INLET FITTING

*INLET NEEDLE AND SEAT

*FLOAT BOWL GASKET

INLET NEEDLE CLIP (If Present)

IDLE FUEL TRANSFER PASSAGE AND ANNULAR GROOVE

FLOAT SHAFT

FLOAT

FLOAT BOWL

*GASKET

NUT AND MAIN ADJUSTMENT SEAT

*MAIN ADJUSTMENT SCREW AND "O" RING SEAL

(DO NOT REMOVE)

Blow air through passage.

Check shaft for looseness or binding. Shutter must be positioned with reference marks on top parallel with shaft and to the right or 3 o'clock position.

Check spring for return action and binding.

IDLE AND INTERMEDIATE AIR BLEED

THROTTLE SHAFT AND LEVER

THROTTLE SHUTTER

DETENT REFERENCE MARK

THROTTLE SHAFT RETURN SPRING

Remove idle adjustment screw. Check needle tip and condition of "O" ring. Remove welch plug and blow out all passages.

IDLE AND INTERMEDIATE ORIFICES

IDLE AND INTERMEDIATE FUEL CHAMBER (COVERED WITH WELCH PLUG)

IDLE AND INTERMEDIATE FUEL MIXTURE PASSAGE

*IDLE ADJUSTMENT SCREW AND "O" RING

(DO NOT REMOVE PLUG)

ATMOSPHERIC VENT

SOFT BAFFLE PLUG

*FLOAT BOWL DRAIN

Remove float bowl fastener and cup plug drop out metering pin and clean all passages with air. Use a small tap to catch cup, twist and remove. Pin should move freely within the passage. CAUTION: Do not install idle adjustment screw with carburetor upside down as pin will obstruct movement of adjustment screw causing damage.

IDLE AND INTERMEDIATE FUEL TRANSFER PASSAGE

METERING ROD OR PIN IN FUEL TRANSFER PASSAGE

BALL PLUG

CUP PLUG

IDLE AND INTERMEDIATE FUEL TRANSFER PASSAGE

IDLE AND MAIN FUEL PICK UP ORIFICE

The fuel bowl retainer nut contains the seat for the screw. Examine the sealing O-ring on the high speed adjusting screw. Replace the O-ring if it indicates wear or cuts. During the reassembly of the high speed adjusting screw, position the coil spring on the adjusting screw, followed by the small brass washer and the O-ring seal.

Fuel Bowl

To remove the fuel bowl, remove the retaining nut and fiber washer. Replace the nut if it is cracked or worn.

The retaining nut contains the transfer passage through which fuel is delivered to the high speed and idle fuel system of the carburetor. It is the large hole next to the hex nut end of the fitting. If a problem occurs with the idle system of the carburetor, examine the small fuel passage in the annular groove in the retaining nut. This passage must be clean for the proper transfer of fuel into the idle metering system.

The fuel bowl should be examined for rust and dirt. Thoroughly clean it before installing it. If it is impossible to properly clean the fuel bowl, replace it.

Check the drain valve for leakage. Replace the rubber gasket on the inside of the drain valve if it leaks.

Examine the large O-ring that seals the fuel bowl to the carburetor body. If it is worn or cracked, replace it with a new one, making sure the same type is used (square or round).

Float

1. Remove the float from the carburetor body by pulling out the float axle with a pair of needle nose pliers. The inlet needle will be lifted off the seat because it is attached to the float with an anchoring clip.
2. Examine the float for damage and holes. Check the float hinge for wear and replace it if worn.
3. The float level is checked by positioning a #4 (0.209 in.) twist drill across the rim between the center leg and the unmachined surface of the index pad, parallel to the float axle pin. If the index pad is machined, the float setting should be made with a #9 (0.180–0.200 in.) twist drill.
4. Remove the float to make an adjustment. Bend the tab on the float hinge to correct the float setting.

NOTE: Direct compressed air in the opposite direction of normal flow of air or fuel (reverse taper) to dislodge foreign matter.

Inlet Needle and Seat

1. The inlet needle sits on a rubber seat in the carburetor body instead of the usual metal fitting.
2. Remove it, place a few drops of heavy engine oil on the seat, and pry it out with a short piece of hooked wire.
3. The grooved side of the seat is inserted first. Lubricate the cavity with oil and use a flat faced punch to press the inlet seat into place.
4. Examine the inlet needle for wear and rounding off of the corners. If this condition does exist, replace the inlet needle.

Fuel Inlet Fitting

1. The inlet fitting is removed by twisting and pulling at the same time.
2. Use sealer when reinstalling the fitting. Insert the tip of the fitting into the carburetor body. Press the fitting in until the shoulder contacts the carburetor. Only use inlet fittings without screens.

Carburetor Body

1. Check the carburetor body for wear and damage.
2. If excessive dirt has accumulated in the atmospheric vent cavity, try cleaning it with carburetor solvent or compressed air. Remove the welch plug only as a last resort.

USE NO. 4 DRILL ON RING

RIM

Adjusting the float level

EXAMINE FLOAT HINGE FOR WEAR AT BEARING SURFACES

INLET NEEDLE EXAMINE CORNERS REPLACE IF ROUNDED

FLOAT

INLET NEEDLE CLIP

REMOVE FLOAT TO MAKE ADJUSTMENTS. BEND THIS TAB TO ADJUST

Float, inlet needle, and clip

NOTE: The carburetor body contains a pressed-in main nozzle tube at a specific depth and position within the venturi. Do not attempt to remove the main nozzle. Any change in nozzle positioning will adversely affect the metering quality and will require carburetor replacement.

3. Clean the accelerating well around the main nozzle with compressed air and carburetor cleaning solvents.
4. The carburetor body contains two cup plugs, neither of which should be removed. A cup plug located near the inlet seat cavity, high up on the carburetor body, seals off the idle bleed. This is a straight passage drilled into the carburetor throat. Do not remove this plug. Another cup plug is located in the base where the fuel bowl nut seals the idle fuel passage. Do not remove this plug or the metering rod.
5. A small ball plug located on the side of the idle fuel passage seals this passage. Do not remove this ball plug.
6. The welch plug on the side of the carburetor body, just above the idle adjusting screw, seals the idle fuel chamber. This plug can be removed for cleaning of the idle fuel mixture passage and the primary and secondary idle fuel discharge ports. Do not use any tools that might change the size of the discharge ports, such as wire or pins.

Auto-Magic carburetor

Resilient Tip Needle

Replace the inlet needle. do not attempt to remove or replace the seat in the carburetor body.

Viton Seat

Using a 10–24 or 10–32 tap, turn the tap into the brass seat fitting until it grasps the seat firmly. Clamp the tap shank into a vise and tap the carburetor body with a soft hammer until the seat slides out of the body.

To replace the viton seat, position the replacement over the receptical with the soft rubber like seat toward the body. Use a flat punch and a small hammer to drive the seat into the body until it bottoms on the shoulder.

Variation on non-adjustable float feed carburetor

Walbro float feed carburetor

TECUMSEH AUTOMATIC NON-ADJUSTABLE FLOAT FEED CARBURETOR

This carburetor has neither a choke plate nor idle and main mixture adjusting screws. There is no running adjustment. The float adjustment is the standard Tecumseh float setting of 0.210 in. (#4 drill).

Cleaning

Remove all non-metallic parts and clean them using a procedure similar to that for the other carburetors. Never use wires through any of the drilled holes. Do not remove the baffling welch plug unless it is certain there is a blockage under the plug. There are no blind passageways in this carburetor.

Some engines use a variation on the Automatic Nonadjustable carburetor which has a different bowl hold-on nut and main jet orifices. There are two main jet orifices and a deeper fuel reserve cavity, but service procedures are the same.

WALBRO AND TILLOTSON FLOAT FEED CARBURETORS

Procedures are similar to those for the Tecumseh float carburetor with the exceptions noted below.

Main Nozzle

The main nozzle in Wallbro carburetors is cross drilled after it is installed in the carburetor. Once removed, it cannot be reinstalled, since it is impossible to properly realign the cross drilled holes. Grooved service replacement main nozzles are available which allow alignment of these holes.

Float Shaft Spring

Carefully position the float shaft spring on models so equipped. The spring dampens float action when properly assembled. Use needlenosed pliers to hook the end of the spring over the float hinge and then insert the pin as far as possible before lifting the spring from the hinge into position. Leaving the spring out or improper installation will cause unbalanced float action and result in a touchy adjustment.

Float Adjustment

1. To check the float adjustment, invert the assembled float

carburetor body. Check the clearance between the body and the float, opposite the hinge. Clearance should be $\frac{1}{8}$ in. $\pm$ $\frac{1}{64}$ in.

2. To adjust the float level, remove the float shaft and float. Bend the lip of the float tang to correct the measurement.

3. Assemble the parts and recheck the adjustment.

TILLOTSON E FLOAT FEED TYPE CARBURETOR

The following adjustments are different for this carburetor.

Running Adjustment

1. Start the engine and allow it to warm up to operating temperatures. Make sure the choke is fully opened after the engine is warmed up.

2. Run the engine at a constant speed while slowly turning the main adjustment screw in until the engine begins to lose speed; then slowly back it out about $\frac{1}{8}$–$\frac{1}{4}$ of a turn until maximum speed and power is obtained (4000 rpm). This is the correct power adjustment.

3. Close the throttle and cause the engine to idle slightly faster than normal by turning the idle speed regulating screw in. Then turn the idle adjustment speed screw in until the engine begins to lose speed; then turn it back $\frac{1}{4}$–$\frac{1}{2}$ of a turn until the engine idles smoothly. Adjust the idle speed regulating screw until the desired idling speed is acquired.

4. Alternately open and close the throttle a few times for an acceleration test. If stalling occurs at idle speeds, repeat the adjustment procedures to get the proper idle speed.

Float Level Adjustment

1. Remove the carburetor float bowl cover and float mechanism assembly.

2. Remove the float bowl cover gasket and, with the complete assembly in an upside down position and the float lever tang resting on the seated inlet needle, a measurement of $\frac{15}{64}$ in. should be maintained from the free end flat rim, or edge of the cover, to the toe of the float. Measurement can be checked with a standard straight rule or depth gauge.

3. If it is necessary to raise or lower the float lever setting, remove the float lever pin and the float, then carefully bend the

Tillotson Model E float carburetor

float lever tang up or down as required to obtain the correct measurement.

WALBRO CARBURETORS FOR V80, VM80, H80, AND HM80 ENGINES

Adjustment

The following initial carburetor adjustments are to be used to start the engine. For proper carburetion adjustment, the atmospheric vent must be open. Examine and clean it if necessary.

1. Idle adjustment — $1\frac{1}{4}$ turn from its seat.

2. High speed adjustment — $1\frac{1}{2}$ turn form its seat.

3. Throttle stop screw — 1 turn after contacting the throttle lever.

After the engine reaches normal operating temperature, make the final adjustments for best idle and high speed within the following ranges. Recommended speeds: Idle: 1800–2300 rpm. High speed: 3450–3750 rpm.

Rebuilding Notes

1. The throttle plate is installed with the lettering (if present) facing outward when closed. The throttle plate is installed on the throttle level with the lever in the closed position if there is binding after the plate is in position, loosen the throttle plate and reposition it.

Walbro Carburetor #631635—engine side

Installation of the throttle plate

Walbro Carburetor #631635—Intake side

Installation of the choke plate

Installation of the fuel bowl and the high speed adjustment needle

2. Before removing the fuel bowl nut, remove the high speed adjusting needle. Use a $^7/_{16}$ in. box wrench or socket to remove the fuel bowl nut. When replacing the fuel bowl nut, be sure to position the fiber gasket under the nut and tighten it securely.

3. Examine the high speed needle tip and, if it appears to be worn, replace it. When the high speed jet is replaced, the main nozzle, which includes the jet seat, should also be replaced. The original main nozzle cannot be used. There are special replacement nozzles available.

4. The inlet needle valve is replaceable if it appears to be worn. The inlet valve seat is also replaceable and should be replaced if the needle valve is replaced.

Float Adjustment

1. The Float setting for this carburetor is 0.070–0.110 in. ($^5/_{64}$–$^7/_{64}$ in.).

2. The float is set in the traditional manner, at the opposite end of the float from the float hinge and needle valve.

3. Bend the adjusting tab to adjust the float level.

DIAPHRAGM CARBURETORS

Diaphragm carburetors have a rubber-like diaphragm that is exposed to crankcase pressure on one side and to atmospheric pressure on the other side. As the crankcase pressure decreases, the diaphragm moves against the inlet needle allowing the inlet needle to move from its seat which permits fuel to flow through the inlet valve to maintain the correct fuel level in the fuel chamber.

An advantage of this type of system over the float system, is that the engine can be operated in any position.

NOTE: In rebuilding, use carburetor cleaner only on metal parts, except for the main nozzle in the main body.

*INDICATES PARTS THAT CAN BE DAMAGED BY HARSH CARBURETOR CLEANERS

Exploded view of a Tecumseh diaphragm carburetor

CHECK SHAFT AND LEVER — Check shaft for binding. Position shutter opening towards inlet fitting side of air horn.

CHOKE SHUTTER

MAIN NOZZLE — Blow air through passage

•INLET FITTING SCREEN — Part of inlet fitting. If fuel is restricted, clean or replace fitting.

•INLET FITTING — Bulb primer models have Viton• one way valve in or behind fitting.

•INLET SEAT GASKET — Remove and replace

•INLET NEEDLE SEAT AND SPRING ASSEMBLY — Proper installation of assembly is important. See appropriate chapter for procedures.

•DIAPHRAGM GASKET

•DIAPHRAGM — Gasket and diaphragm sequence may be reversed on some models. Head of rivet must touch inlet needle. Rivet is hooked into inlet needle control lever on some models. See appropriate chapter.

ATMOSPHERIC VENT HOLE — Hole must be clean. On models with bulb primer, vent hole is very small and is located off center.

Service hints for the diaphragm carburetors

IDLE SPEED ADJUSTMENT SCREW — Loosen screw until it just clears throttle lever, then turn screw in one turn.

THROTTLE SHAFT AND LEVER DETENT REFERENCE MARK ON THROTTLE SHUTTER — Place detent reference mark to proper location. See chapter appropriate to specific carburetor.

THROTTLE SHUTTER

THROTTLE SHAFT RETURN SPRING — Check spring for return action and binding.

IDLE, INTERMEDIATE AND AIR BLEED ORIFICES — Remove welch plug and blow air through air passages.

•IDLE MIXTURE ADJUSTMENT SCREW AND "O" RING (If Present)

•MAIN MIXTURE ADJUSTMENT SCREW AND "O" RING (If Present) — Remove adjustment screw. To adjust 20° slant engines, the engine must be mounted in its normal 20° slant position.

•WELCH PLUG (If Present)

•MAIN NOZZLE CHECK BALL (If Present) — Check ball is not serviceable on some models. See appropriate chapter for specific model.

• NON METALLIC ITEMS - CAN BE DAMAGED BY HARSH CARBURETOR CLEANERS

Craftsman fuel tank mounted carburetor

Throttle Plate

Install the throttle plate with the short line that is stamped in the plate toward the top of the carburetor, parallel with the throttle shaft, and facing out when the throttle is closed.

Choke Plate

Install the choke plate with the flat side of the choke toward the fuel inlet side of the carburetor. The mark faces in and is parallel to the choke shaft.

Idle Mixture Adjustment Screw

There is a neoprene O-ring on the needle. Never soak the O-ring in carburetor solvent. Idle and main mixture screws vary in size and design, so make sure that you have the correct replacement.

LATE MODEL

EARLY MODEL

Details of pick-up tube used on revised Craftsman carburetor

Idle Fuel Chamber

The welch plug can be removed if the carburetor is extremely dirty.

Diaphragms

Diaphragms are serviced and replaced by removing the four retaining screws from the cover. With the cover removed, the diaphragm and gasket may be serviced. Never soak the diaphragm in carburetor solvent. Replace the diaphragm if it is cracked or torn. Be sure there are no wrinkles in the diaphragm when it is replaced. The diaphragm rivet head is always placed facing the inlet needle valve.

Inlet Needle and Seat

The inlet seat is removed by using either a slotted screwdriver (early type) or a $9/32$ in. socket. The inlet needle is spring loaded, so be careful when removing it.

Fuel Inlet Fitting

All of the diaphragm carburetors have an integral strainer in the inlet fitting. To clean it, either reverse flush it or use compressed air after removing the inlet needle and seat. If the strainer is lacquered or otherwise unable to be cleaned, replace the fitting.

CRAFTSMAN FUEL SYSTEMS
Changes in Late Model Carburetors

The newest Craftsman carburetors incorporate the following changes:

    a. The cable form of control is replaced by a control knob.

    b. The fuel pickup is longer and has a collar machined into it which must be installed tight against the carburetor body.

    c. The fuel pickup screen is pressed onto the ends of the fill tubes on both models, but the measured depth has changed.

    d. The cross-drilled passages have been eliminated, as has the O-ring on the body. There are no cup plugs.

    e. The fuel tank and reservoir tube have been revised — the reservoir tube being larger.

Disassembly and Service

1. Remove the air cleaner assembly and remove the four screws on the top of the carburetor body to separate the fuel tank from the carburetor.

2. Remove the O-ring from between the carburetor and the fuel tank. Examine it for cracks and damage and replace it if necessary.

3. Carefully remove the reservoir tube from the fuel tank. Observe the end of the tube that rested on the bottom of the fuel tank. It should be slotted.

4. Remove the control valve by turning the valve clockwise until the flange is clear of the retaining boss. Pull the valve straight out and examine the O-ring seal for damage or wear. If possible, use a new O-ring when reassembling.

5. Examine the fuel pick up tube. There are no valves or ball checks that may become in-operative. These parts can normally be cleaned with carburetor solvent. If it is found that the passage cannot be cleared, the fuel pick up tubes can be replaced. Carefully remove the old ones so as not to enlarge the opening in the carburetor body. If the pickup tube and screen must be replaced, follow the directions shown in the illustration for the type of carburetor (early or late model) on which you are working.

6. Assemble the carburetor in reverse order of disassembly. Use new O-rings and gaskets. When assembling reservoir tube, hold the carburetor upside down and place the reservoir tube over the pickup tube with the slotted end up. Make sure the intake manifold gasket is correctly positioned — it can be assembled blocking the intake passage partially.

NO BOWDEN WIRE CONTROL

MANUAL CONTROL KNOB

POSITIONING SPRING

HIGH SPEED NOTCH

FUEL PICK-UP TUBE PRESS TUBE TO BOTTOM ON DRILL SHANK

1-15/32"

PRESS SCREEN TO PROPER DEPTH AFTER PICK-UP TUBE IS SET

LATE MODEL

BOWDEN CONTROL

FUEL PICK-UP TUBE (SET COLLAR AGAINST CASTING BODY)

1-27/64"

PRESS SCREEN TO PROPER DEPTH AFTER PICK-UP TUBE IS SET

EARLY MODEL

Details of manual control knob and several other changes incorporated in new Craftsman carburetors

FUEL TANK

RESERVOIR TUBE PLACE SLOTTED END TO REST ON BOTTOM OF FUEL TANK

FUEL INLET

The fuel pick-up tube and O-ring on the early model Craftsman carburetor

3 PRESS IN TUBE TO BOTTOM ON DRILL SHANK

4 DO NOT REMOVE DRILL PRESS ON STRAINER 1²⁷/₆₄" to 1⁷/₁₆"

6 INSERT NEW CUP PLUG

1 REMOVE CUP PLUG

5 REMOVE DRILL

2 INSERT SHANK END OF ⅛ DRILL

EARLY MODEL

1 PRESS TUBE TO SEAT COLLAR AGAINST CASTING

2 PRESS STRAINER TO 1¹⁵/₃₂ DEPTH

LATE MODEL

Pick-up tube replacement

CRANKSHAFT

GEAR KEYWAY

ADVANCE MARK ONE TOOTH

CAMSHAFT GEAR

SMALL HOBBING HOLE

FOR CRAFTSMAN CARBURETOR ONLY— AND ONLY IF ENGINE HAS NO INTERNAL FLYWEIGHT GOVERNOR

Timing an engine with the Craftsman fuel tank mounted carburetor

Adjustments

1. Move the carburetor control valve to the high speed position. The mark on the face of the valve should be in alignment with the retaining boss on the carburetor body.
2. Move the operator's control on the equipment to the high speed position.
3. Insert the bowden wire into the hole of the control valve. Clamp the bowden wire sheath to the carburetor body.

NOTE: If the engine was disassembled and the camshaft removed, be sure that the timing marks on the camshaft gear and the keyway in the crankshaft gear are aligned when reinserting the camshaft. Then lift the camshaft enough to advance the camshaft gear timing mark to the right (clockwise) ONE tooth, as viewed from the power take-off end of the crankshaft.

Positioning of the reservoir tube in the fuel tank

CRAFTSMAN FLOAT TYPE CARBURETORS

Craftsman float type carburetors are serviced in the same manner as the other Tecumseh float type carburetors. The throttle control valve has three positions: stop, run, and start.

When the control valve is removed, replace the O-ring. If the engine runs sluggishly, consider the possibility of a leaky O-ring.

To remove the fuel pickup tube, clamp the tube in a vise and then twist the carburetor body. Check the small jet in the air horn while the tube is out. In replacement, position the tube squarely and then press in on the collar until the collar seats.

In assembly, install the gasket and bolt through the bowl, position the centering spacer onto the bolt, and then attach the parts to the carburetor body.

The camshaft timing mark must be advanced one tooth in relation to crankshaft gear timing mark, as shown in the illustration above. However, when this type carburetor is used with a float bowl reservoir and variable governor adjustment, time it as for other Tecumseh engines — with the camshaft and crankshaft timing marks aligned.

Exploded view of the Craftsman float type carburetor

Start, run and stop throttle positions of a Craftsman float type carburetor

Measuring governor shaft installed dimension

Governor

The mechanical governor is located inside the mounting flange. See engine disassembly instructions, below. To disassemble the governor, see the illustration, and: remove the retaining ring, pull off the spool, remove the second retaining ring, and then pull off the gear assembly and retainer washer.

Check for wear on all moving surfaces, but especially gear teeth, the inside diameter of the gear where it rides on the shaft, and the flyweights where they work against the spool.

If the governor shaft must be replaced, it should be started into the boss with a few taps using a soft hammer, and then pressed in with a press or vise. The shaft can be installed by positioning a wooden block on top and tapping the upper surface of the block, but the use of a vise or press is much preferred.

The shaft must be pressed in until just the required length is exposed, as measured from the top of the shaft boss to the upper end of the shaft. See the chart.

The governor is installed in reverse of the removal procedure. Connect the linkage and then adjust as described in the Tune-Up section.

Shaft Installed Dimension/ Engine Model Chart

Engine Model	"A" Exposed Shaft Length (see figure)
TNT100-120 ECV100-105-110-120 ECH90 TVS75-90-105-120	$1^5/_{16}''$
V60 VM70-80-100	$1^{19}/_{32}''$
H60 HH60-70	$1^7/_{16}''$
HM70-80-100	$1^{13}/_{32}''$

ENGINE OVERHAUL
Timing Gears

Correctly matched camshaft gear and crankshaft gear timing marks are necessary for the engine to perform properly.

Exploded view of standard mechanical governor

Timing marks

On all camshafts the timing mark is located in line with the center of the hobbing hole (small hole in the face of the gear). If no line is visible, use the center of the hobbing hole to align with the crankshaft gear marked tooth.

On crankshafts where the gear is held on by a key, the timing mark is the tooth in line with the keyway.

On crankshafts where the gear is pressed onto the crankshaft, a tooth is bevelled to serve as the timing mark.

On engines with a ball bearing on the power take-off end of the crankshaft, look for a bevelled tooth which serves as the crankshaft gear timing mark.

If the engine uses a Craftsman type carburetor, the camshaft timing mark must be advanced clockwise one tooth ahead of the matching timing mark on the crankshaft, the exception being the Craftsman variable governed fuel systems.

Timing marks

CRAFTSMAN CARBURETOR ONLY

Timing marks

NOTE: If one of the timing gears, either the crankshaft gear or the camshaft gear, is damaged and has to be replaced, both gears should be replaced.

Crankshaft

INSPECTION

Inspect the crankshaft for worn or crossed threads that can't be redressed; worn, scratched, or damaged bearing surfaces; misalignments; flats on the bearing surfaces. Replace the shaft if any of these problems are in evidence – do not try to straighten a bent shaft.

In replacement, be sure to lubricate the bearing surfaces and use oil seal protectors. If the camshaft gear requires replacement, replace the crankshaft gear, also.

Pistons

When removing the pistons, clean the carbon from the upper cylinder bore and head. The piston and pin must be replaced in matched pairs.

A ridge reamer must be used to remove the ridge at the top of the cylinder bore on some engines.

Clean the carbon from the piston ring groove. A broken ring can be used for this operation.

VM70, V80, VM80, HM70, H80, HM80, VM100 & HM100 PISTON AND CONNECTING ROD ASSEMBLY

Piston-to-rod relationship for engines shown

Check piston dimensions

Ring arrangement and dimensions

Some engines have oversize pistons which can be identified by the oversize engraved on the piston top.

There is a definite piston-to-connecting rod-to-crankshaft arrangement which must be maintained when assembling these parts. If the piston is assembled in the bore 180 degrees out of position, it will cause immediate binding of the parts.

Piston Rings

Always replace the piston rings in sets. Ring gaps must be staggered. When using new rings, wipe the cylinder wall with fine emery cloth to deglaze the wall. Make sure the cylinder wall is thoroughly cleaned after deglazing. Check the ring gap by placing the ring squarely in the center of the area in which the rings travel and measuring the gap with a feeler gauge. Do not spread the rings too wide when assembling them to the pistons. Use a ring leader to install the rings on the piston.

The top compression ring has an inside chamfer. This chamfer must go UP. If the second ring has a chamfer, it must also face UP. If there is a notch on the outside diameter of the ring, it must face DOWN.

Check the ring gap on the old ring to determine if the ring should be replaced. Check the ring gap on the new ring to determine if the cylinder should be rebored to take oversize parts.

NOTE: Make sure that the ring gap is measured with the ring fitted squarely in the worn part of the cylinder where the ring usually rides up and down on the piston.

Connecting Rods

Be sure that the match marks align when assembling the connecting rods to the crankshaft. Use new self-locking nuts. Whenever locking tabs are included, be sure that the tabs lock the nuts securely. NEVER try to straighten a bent crankshaft or connecting rod. Replace them if necessary. When replacing either the piston, rod, crankshaft, or camshaft, liberally lubricate all bearings with engine oil before assembly.

The following engines have offset connecting rods: V70, VH70, VM70, V80, VM80, H70, HH70, HM70, H80, HM80. ECV105, ECV110, ECM120, and TNT120 engines have the caps fitted from opposite to the camshaft side of the engine. The following engines have the cap fitted from the camshaft side: V70,

Connecting rod assembly for the V70, V80, H70 and H80 engines

Connecting rod and piston assembly for the V80 and H80 engines

Squaring the ring in the bore

Checking the ring gap

VH70, VM70, V80, VM80, H70, HH70, HM70, H80, HM80, VM100 and HM100.

On H70, HH70, HM70, H80, HM80, V70, VH70, VM70, V80, VM80, VM100 and HM100 engines, a dipper is stamped into the lockplate. Use a new lockplate whenever the rod cap is removed. On engine with Durlock rod bolts, torque the bolt as follows: TVS75, 90, 105, 120, ECH90, TNT100, 120, ECV100, 105, 110 and 120–110 inch lbs.

V60–80, H60–80, VH60–70, HH60–70, VM70–100 and HM70–100–150 inch lbs.

NOTE: Early type caps can be distorted if the cap is not held to the crank pin while threading the bolts tight. Undue force should not be used.

Later rods have serrations which prevent distortion during tightening. They also have match marks which must face out when assembling the rod. On the V80 and H80 engines, the piston and rod must fit so that the number inside the casting is on the rod side of the rod/cap combination.

Camshaft

Before removing the camshaft, align the timing marks to relieve the pressure on the valve lifters, on most engines.

On VM80 and VM100 engines, while the basic timing is the same, the crankshaft must be rotated so that the timing mark is located 3 teeth further counterclockwise (referring to crankshaft gear rotation). This clears the camshaft of the compression release mechanism for easier removal.

In installation, align the gears for this type of camshaft as they were right before removal. After installation, turn the crankshaft gear clockwise in order to check for proper alignment of timing marks.

Alignment of timing marks for camshaft removal on VM80 and VM100 engines

Clean the camshaft in solvent, then blow the oil passages dry with compressed air. Replace the camshaft if it shows wear of evidence of scoring. Check the cam dimensions against those in the chart.

If the engine has a mechanical fuel pump, it may have to be removed to properly reinstall the camshaft. If the engine is equipped with the Insta-matic Ezee-Start Compression Release, and any of the parts have to be replaced due to wear or damage, the entire camshaft must be replaced. Be sure that the oil pump (if so equipped) barrel chamber is toward the fillet of the camshaft gear when assembled.

NOTE: If a damaged gear is replaced, the crankshaft gear should also be replaced.

Check the dimensions of the camshaft

Valve Springs

The valve springs should be replaced whenever an engine is overhauled. Check the free length of the springs. Comparing one spring with the other can be a quick check to notice any difference. If a difference is noticed, carefully measure the free length, compression length, and strength of each spring. See the specifications chart at the end of this section.

Some valve springs use dampening coils that are wound closer together than most of the coils of the spring. Where these are present, the spring must be mounted so the dampening coils are on the stationary (upper) end of the spring.

Valve Lifters

The stems of the valves serve as the lifters. On the 4 hp light frame models, the lifter stems are of different lengths. Because this engine is a cross port model, the shorter intake valve lifter goes nearest the mounting flange.

The valve lifters are identical on standard port engines. However once a wear pattern is established, they should not be interchanged.

Valve Grinding and Replacement

Valves and valve seats can be removed and reground with a minimum of engine disassembly.

Remove the valves as follows:

1. Raise the lower valve spring caps while holding the valve heads tightly against the valve seat to remove the valve spring retaining pin. This is best achieved by using a valve spring compressor. Remove the valves, springs, and caps from the crankcase.

2. Clean all parts with a solvent and remove all carbon from the valves.

3. Replace distorted or damaged valves. If the valves are in usable condition, grind the valve faces in a valve refacing machine and to the angle given in the specifications chart at the end of this section. Replace the valves if the faces are ground to less than $\frac{1}{32}$ in.

4. Whenever new or reground valves are installed, lap in the valves with lapping compound to insure an air-tight fit.

NOTE: There are valves available with oversize stems.

Dimensions of the valve face

5. Valve grinding changes the valve lifter clearance. After grinding the valves check the valve lifter clearance as follows:

a. Rotate the crankshaft until the piston is set at the TDC position of the compression stroke.

b. Insert the valves in their guides and hold the valves firmly on their seats.

c. Check for a clearance of 0.010 in. between each valve stem and valve lifter with a feeler gauge.

d. Grind the valve stem in a valve resurfacing machine set to grind a perfectly square face with the proper clearance.

6. Install valves as follows on Early Models:

a. Position the valve spring and upper and lower valve spring caps under the valve guides for the valve to be installed.

b. Install the valves in the guides, making sure that the valve marked **EX** is inserted in the exhaust port. The valve stem must pass through the valve spring and the valve spring caps.

c. Insert the blade of a screwdriver under the lower valve spring cap and pry the spring up.

d. Insert the valve pin through the hole in the valve stem with a long nosed pliers. Make sure the valve pin is properly seated under the lower valve spring cap.

Install the valves as follows on Later Models:

a. Position the valve caps and spring in the valve compartment.

b. Install the valves in guides with the valve marked **EX** in exhaust port. The valve stem must pass through the upper valve cap and spring. The lower cap should sit around the valve lifter exposed end.

c. Compress the valve spring so that the shank is exposed. DO NOT TRY TO LIFT THE LOWER CAP WITH THE SPRING.

d. Lift the lower valve cap over the valve stem shank and center the cap in the smaller diameter hole.

e. Release the valve spring tension to lock the cap in place.

REBORING THE CYLINDER

1. First, decide whether to rebore for 0.010 in. or 0.020 in.

2. Use any standard commercial hone of suitable size. Chuck the hone in the drill press with the spindle speed of about 600 rpm.

3. Start with coarse stones and center the cylinder under the press spindle. Lower the hone so the lower end of the stones contact the lowest point in the cylinder bore.

4. Rotate the adjusting nut so that the stones touch the cylinder wall and then begin honing at the bottom of the cylinder. Move the hone up and down at a rate of 50 strokes a minute to avoid cutting ridges in the cylinder wall. Every fourth or fifth stroke, move the hone far enough to extend the stones 1 in. beyond the top and bottom of the cylinder bore.

5. Check the bore size and straightness every thirty or forty strokes. If the stones collect metal, clean them with a wire brush each time the hone is removed.

6. Hone with coarse stones until the cylinder bore is within 0.002 in. of the desired finish size. Replace the coarse stones with burnishing stones and continue until the bore is to within 0.0005 in. of the desired size.

7. Remove the burnishing stones and install finishing stones to polish the cylinder to the final size.

8. Clean the cylinder with solvent and dry it thoroughly.

9. Replace the piston and piston rings with the correct oversize parts.

REBORING VALVE GUIDES

The valve guides are permanently installed in the cylinder.

Check the dimensions of the valve seats, guides, and the cylinder

However if the guides wear, they can be rebored to accommodate a $\frac{1}{32}$ in. oversize valve stem. Rebore the valve guides in the following manner:

1. Ream the valve guides with a standard straight shanked hand reamer or a low speed drill press. Refer to the specifications chart at the end of this section for the correct valve stem guide diameter.

2. Redrill the upper and lower valve spring caps to accommodate the oversize valve stem.

3. Reassemble the engine, installing valves with the correct oversize stems in the valve guides.

REGRINDING VALVE SEATS

The valve seats need regrinding only if they are pitted or scored. If there are no pits or scores, lapping in the valves will provide a proper valve seat. Valve seats are not replaceable. Regrind the valve seats as follows:

1. Use a grinding stone or a reseater set to provide the proper angle and seal face dimensions.

2. If the seat is over $\frac{3}{64}$ in. wide after grinding, use a 15 degree stone or cutter to narrow the face to the proper dimensions.

3. Inspect the seats to make sure that the cutter or stone has been held squarely to the valve seat and that the same dimension has been held around the entire circumference of the seat.

4. Lap the valves to the reground seats.

Torquing Cylinder Head

Torque the cylinder head to 200 inch lbs. in 4 equal stages of 50 inch lbs. Follow the sequence shown in the appropriate illustration for each tightening stage.

Bearing Service

LIGHTWEIGHT ALUMINUM BEARING REPLACEMENT

The aluminum bearing must be cut out using the rough cut reamer and the procedures shown in steps 1A and 4A. Follow illustrated steps 1A, 2, 3, 4A, 5 and 6 to install bronze bushings in place of the aluminum bearings.

Cylinder head tightening sequence for all engines except 8 hp models

Cylinder head tightening sequence for 8 hp models

LONG LIFE AND CAST IRON ENGINES WITHOUT BALL BEARINGS

The worn bronze bushing must be driven out before the new bushing can be installed. Follow illustrated steps 1B, 2, 3, 4B, 5 and 6, to replace the main bearings on these units.

LONG LIFE AND CAST IRON ENGINES WITH BALL BEARING ON THE P.T.O. (POWER TAKE-OFF) SIDE OF THE CRANKSHAFT

The side cover containing the ball bearing must be removed and a substitute cover with either a new bronze bushing or aluminum bearing must be used instead. Follow illustrated steps 1B, 2 and 3 only.

SPECIAL TOOLS

The task of main bearing replacement is made easier by using one of two Tecumseh main bearing tool kits. Kit 670161 is used to replace main bearings on the lightweight engines except HS models. Kit 670165 is used to replace main bearings on the medium weight engines except HS models.

GENERAL NOTES ON BUSHING REPLACEMENT

1. Your fingers and all parts must be kept very clean when replacing bushings.
2. On splash lubricated horizontal engines, the oil hole in the bushing is to be lined up with the oil hole that leads into the slot in the original bearing.
3. In the event it is necessary to replace the mounting flange or cylinder cover, the magneto end bearing must be rebushed. The P.T.O. bearing should also be rebushed to assure proper alignment.
4. Oil should be used to finish-ream the bushings. In the

INSTALLING BRONZE BUSHING P T O END

1 DISASSEMBLE
2 POSITION BEARING SUPPORT TOOL
3 PRESS IN BUSHING
4 ALIGN OIL HOLE
5 DRIVE BUSHING UNTIL IT BOTTOMS ON BEARING SUPPORT

BUSHING DRIVER

BUSHING

ALIGN OIL HOLES AND SLOTS

FLANGE OR SIDE COVER

BEARING SUPPORT SMALL END UP

FINISH REAMING P T O BUSHING

APPLY OIL TO PREVENT DAMAGE TO BUSHING

1 REASSEMBLE CYLINDER AND COVER
2 INSERT REAMER THROUGH MAGNETO BEARING
3 FINISH REAM P T O BUSHING
4 WHEN FINISH REAMER IS THROUGH BUSHING STOP, DO NOT ALLOW ROUGH REAMER TO ENTER NEWLY INSTALLED MAGNETO END BUSHING
5 WASH ALL PARTS CAREFULLY TO REMOVE ALL CHIPS CHECK OIL HOLES AND PASSAGES. BLOW OUT WITH COMPRESSED AIR
6 REASSEMBLE ENGINE

USE LIGHT OIL FOR FINISH REAMING IF CRANKSHAFT IS TIGHT REPEAT WITHOUT OIL

P T O BUSHING

ROUGH REAMING WORN ALUMINUM BEARING
(MAGNETO SIDE) FOR ALUMINUM BEARING ONLY

HANDLE

USE KEROSENE ON ROUGH CUT REAMER

USE ALL BOLTS TO HOLD ASSEMBLY IN ALIGN-MENT DO NOT USE GASKET

1 DISASSEMBLE ENGINE CLEAN AND REMOVE OIL SEALS
2 ASSEMBLE P T O BEARING TO CYLINDER (MOUNTING FLANGE V MODELS CYLINDER COVER H MODELS)

DOWEL PINS IN CYLINDER MUST BE IN PLACE FOR PROPER ALIGNMENT

MAGNETO END BEARING

REAMER GUIDE IN P T O BEARING

FINISH REAMER

DISASSEMBLE AFTER ROUGH REAMING MAGNETO END

FINISH REAMING NEW MAGNETO BUSHING

ROUGH CUT REAMER

USE LIGHT OIL ON FINISH REAMER (NOTE IF BUSH-ING IS TIGHT ON CRANKSHAFT, REPAT REAM-ING WITHOUT OIL)

1 REASSEMBLE CYLINDER TO COVER
2 INSERT REAMER THROUGH P T O BEARING
3 FINISH REAM MAGNETO END BUSHING

REAMER INSERTED THROUGH P T O BEARING

WORN BUSHING

FINISH REAMER

NEW BUSHING

WORN BUSHING REMOVAL (MAGNETO END)
FOR BRONZE BUSHING ONLY

BUSHING DRIVER

CYLINDER (HORIZON-TAL OR VERTICAL CRANKSHAFT ENGINE)

1 DISASSEMBLE AND CLEAN ALL PARTS
2 POSITION BEARING SUPPORT TOOL WITH LARGE END UP
3 CAREFULLY DRIVE WORN BEAR-ING OUT OF CYLINDER AND SIDE COVER OR FLANGE (EXCEPT BALL BEARING P T O)

WORN BRONZE BUSHING (MAGNETO END)

BEARING SUPPORT (SMALL END)

USE END WITH LARGER HOLE TO SUPPORT BEARING WHEN REMOVING BUSHING

ROUGH REAMING WORN ALUMINUM BEARING
(P T O END) FOR ALUMINUM BEARING ONLY

REAMER CUTTING OUT P T O BEARING

1 AFTER REAMING MAGNETO END BUSHING BEGIN TO REAM P T O BUSHING

NEW MAGNETO END BUSHING FINISH REAMED

INSTALLING NEW BRONZE BUSHING
(MAGNETO SIDE)

BUSHING DRIVER

ALIGN OIL HOLE

1 POSITION BEARING SUP-PORT TOOL SMALL END UP
2 PRESS IN BUSHING
3 ALIGN OIL HOLES
4 DRIVE BUSHING UNTIL IT BOTTOMS ON BEARING SUPPORT

NEW BUSHING

BEARING SUPPORT (LARGE END)

POSITION BEARING SUP-PORT TOOL WITH SMALL-ER DIAMETER HOLE FOR INSTALLING BUSHING

WORN BUSHING REMOVAL (P T O END)
FOR BRONZE BUSHING ONLY

BUSHING DRIVER

FLANGE OR SIDE COVER

WORN BUSHING

BEARING SUPPORT LARGE END UP

event the crankshaft does not rotate freely repeat the finish-reaming operation without oil.

5. Kerosene should be used as a cutting lubricant while rough-reaming.

6. Be sure that the dowel pins are in the cylinder block when assembling the mounting flange or cylinder cover. Use all bolts to hold the assembly together.

7. Remove the reamer by rotating it in the same direction as it is turned during the reaming operation. DO NOT TURN THE REAMER BACKWARDS.

REMOVAL AND INSTALLATION OF CRANKSHAFT BUSHING FOR 8 AND 10 HORSEPOWER ENGINES

The illustrations show the mounting flange for a vertical engine. Procedures also apply to the cylinder cover for a horizontal engine. Use tool No. 670247 removal end and arbor press to press bushing from P.T.O. bearing end. Note the position of oil slots in the bushing which must align with the oil slots in the mounting flange.

To install, insert a new bushing on the installation end of tool No. 670247. Position the slots so they properly align with the oil slots in the cover and press the bushing in with an arbor press. After the new bushing is installed, use a light coating of oil and finish reaming with reamer, part No. 670248 (handle 690160). Assemble the P.T.O. mounting flange to the cylinder. Use all bolts with dowel pins to hold the assembly in alignment. Insert the tool through the bushing and cylinder crankshaft magneto end bearing as shown. Rotate the cutting edge clockwise in the P.T.O. bushing. Remove the tool in the same direction of rotation. Do not allow the cutting edge of reamer to touch the magneto end of the cylinder.

Bushing Installation (8 and 10 horsepower engines)

Finish—reaming the bushing (8 and 10 horsepower engines)

Bushing removal (8 and 10 horsepower engines)

BALL BEARING SERVICE H60 THROUGH HM100 H.P. HORIZONTAL CRANKSHAFT ENGINES

1. Prior to attempting removal of the cylinder cover, observe the area around the crankshaft P.T.O. oil seal. Compare it with the illustration, and if there are bearing locks, follow instructions below:

a. Remove the locking nuts using the proper socket wrench. Note fiber washer located under nut: this must be reinstalled. Lift side cover from cylinder after removing the side cover bolts.

b. Install the bearing retainer bolts, fiber washer and locking nuts in the proper sequence in the cover.

2. Also note the following points:

a. On some engines, a locking type retainer bolt is used. To release the bolt, merely loosen the locking nut and turn the retainer bolt counterclockwise to the unlocked position with needle nose pliers to permit the side cover to be removed. Note that the flats on the retainer bolts must be turned so they face the crankshaft to be relocked upon installation. Don't force them! Torque the locking nuts only to 15–22 inch lbs.

b. The ball bearing used in horizontal crankshaft engines has a restricted fit. The bearing is heated and put onto the

H60-HM100 H.P. horizontal crankshaft engines
—locking and unlocking the bearing from outside

Operation of the barrel type oil pump

cold crankshaft. As the bearing cools it grasps the crankshaft tightly and must be removed cold. Remove the ball bearing with a bearing splitter (separator) and a puller. The bearing may be heated by placing it into a container with a sufficient amount of oil to cover the bearing. The bearing should not rest on the bottom of the container. Suspend the bearing on a wire or set the bearing onto a spacer block of wood or wire mesh. Heat the oil and bearing carefully until the oil smokes, quickly remove the bearing and slide it onto the crankshaft.

c. The bearing must seat tightly against the thrust washer which in turn rests tightly against the crankshaft gear.

d. When a ball bearing is used it is not possible to see the keyway in the crankshaft gear which is normally used for timing. Because of this, one tooth of the crankshaft gear is chamfered. This chamfered tooth of the crankshaft gear is positioned opposite the timing mark on the camshaft gear. The use of a ball bearing requires the removal of the crankshaft when it is necessary to remove the camshaft. When replacing the crankshaft and camshaft, mate the timing marks and insert it into the cylinder block as an assembly.

LUBRICATION

Barrel and Plunger Oil Pump System

This system is driven by an eccentric on the camshaft. Oil is drawn through the hollow camshaft from the oil sump on its intake stroke. The passage from the sump through the camshaft

Installation of the barrel type oil pump

is aligned with the pump opening. As the camshaft continues rotation (pressure stroke), the plunger force the oil out. The other port in the camshaft is aligned with the pump, and directs oil out of the top of the camshaft.

At the top of the camshaft, oil is forced through a crankshaft passage to the top main bearing groove which is aligned with the drilled crankshaft passage. Oil is directed through this passage to the crankshaft connecting rod journal and then spills from the connecting rod to lubricate the cylinder walls. Splash is used to lubricate the other parts of the engine.

A pressure relief port in the crankcase relieves excessive pressures when the oil viscosity is extremely heavy due to cold temperatures, or when the system is plugged or damaged. Normal pressure is 7 psi.

SERVICE

Remove the mounting flange or the cylinder cover, whichever is applicable. Remove the barrel and plunger assembly and separate the parts.

Clean the pump parts in solvent and inspect the pump plunger and barrel for rough spots or wear. If the pump plunger is scored or worn, replace the entire pump.

Before reassembling the pump parts, lubricate all of the parts in engine oil. Manually operate the pump to make sure the plunger slides freely in the barrel.

Lubricate all the parts and position the barrel on the camshaft eccentric. If the oil pump has a chamfer only on one side, that side must be placed toward the camshaft gear. The flat goes away from the gear, thus out to work against the flange oil pickup hole.

Install the mounting flange. Be sure the plunger ball seats in the recess in the flange before fastening it to the cylinder.

Splash Lubrication

Some engines utilize the splash type lubrication system. The oil dipper, on some engines, is cast onto the lower connecting rod bearing cap. It is important that the proper parts are used to ensure the longest engine life.

Gear Type Oil Pump System

The gear type lubrication pump is a crankshaft driven, positive displacement pump. It pumps oil from the oil sump in the engine base to the camshaft, through the drilled camshaft passage to the top main bearing, through the drilled crankshaft, to the connecting rod journal on the crankshaft.

Spillage from the connecting rod lubricates the cylinder walls and normal splash lubricates the other internal working parts. There is a pressure relief valve in the system.

H40 THRU H60
CONNECTING ROD
FOR SPLASH LUBRICATION

BE SURE MATCH MARKS
ARE TOGETHER AND
ARE "OUT" DURING
REASSEMBLY

CAST ON OIL DIPPER

Splash type lubrication connecting rod

MOUNTING
FLANGE

LOCKWASHER

SCREW

COVER

DISPLACEMENT
MEMBER

DRIVE GEAR

GASKET

A. DISASSEMBLED

**Disassembled view of the gear type lubrication
system**

SERVICE

Disassemble the pump as follows: remove the screws, lockwashers, cover, gear, and displacement member.

Wash all of the parts in solvent. Inspect the oil pump drive gear and displacement member for worn or broken teeth, scoring, or other damage. Inspect the shaft hole in the drive gear for wear. Replace the entire pump if cracks, wear, or scoring is evident.

To replace the oil pump, position the oil pump displacement member and oil pump gear on the shaft, then flood all the parts with oil for priming during the initial starting of the engine.

The gasket provides clearance for the drive gear. With a feeler gauge, determine the clearance between the cover and the oil pump gear. The clearance desired is 0.006–0.007 in. Use gaskets, which are available in a variety of sizes, to obtain the correct clearance. Position the oil pump cover and secure it with the screws of lockwashers.

Craftsman Engines Cross Reference Chart

Craftsman Engine Models	See Column
143.50040	7
143.50045	
143.131022–143.131102	1
143.135012–143.135112	9
143.136012–143.136052	8
143.137012	7
143.137032	
143.141012–143.143032	1
143.145012–143.145072	9
143.146012	8
143.146022	
143.147012–143.147032	7
143.151012–143.153032	1
143.154012–143.154142	2
143.155012–143.155062	9
143.156012	8
143.156022	
143.157012–143.157032	7
143.161012–143.163062	1
143.164012–143.164202	2
143.165012–143.165052	9
143.166012–143.166052	8
143.167012–143.167042	7
143.171012–143.171172	1
143.171202	2
143.171212–143.173042	1
143.174012–143.174292	2
143.175012–143.175072	9
143.176012–143.176092	8
143.177012–143.177072	7
143.181042–143.183042	1
143.184012–143.184212	2
143.184232–143.184252	4
143.184262–143.184402	2
143.185012–143.185052	9
143.186012	8
143.186022–143.186042	10
143.186052	8
143.186062	
143.186072–143.186112	10
143.186122	8
143.187022–143.187102	3
143.191012–143.191052	1
143.194012–143.194052	2
143.194062	4
143.194072–143.194092	2
143.194102	4
143.194112–143.194142	2
143.195012	9
143.195022	
143.196012–143.196032	8

Craftsman Engine Models	See Column
143.196042–143.196072	10
143.196082	8
143.197012	3
143.197022	5
143.197032	
143.197042–143.197072	3
143.197082	5
143.201032–143.203012	1
143.204022	4
143.204032–143.204052	2
143.204062	4
143.204072–143.204092	2
143.204102	4
143.204132	
143.204142–143.204192	2
143.204202	4
143.205022	9
143.206012	8
143.206022	10
143.206032	8
143.207012–143.207052	3
143.207062	5
143.207072	3
143.207082	5
143.213012–143.213042	1
143.214012–143.214032	2
143.214042–143.214072	4
143.214082–143.214252	2
143.214262–143.214282	4
143.214292	2
143.214302	
143.214312	4
143.214322	
143.214332	2
143.214342	
143.214352	4
143.216012–143.216032	10
143.216042–143.216062	3
143.216072–143.216092	10
143.216122	8
143.216132	11
143.216142	8
143.216152	11
143.216162	
143.216172	
143.216182	8
143.217012–143.217032	5
143.217042–143.217072	3
143.217092	5
143.217102	3
143.223012–143.223052	1
143.224012	2

Craftsman Engine Models	See Column
143.224022	
143.224032	4
143.224062	2
143.224072	4
143.224092–143.224132	2
143.224142	1
143.224162–143.224222	2
143.224232	4
143.224242	
143.224252–143.224282	2
143.224292	4
143.224302	
143.224312–143.224342	2
143.224352	4
143.224362	
143.224372–143.224422	2
143.224432	4
143.225012	13
143.225022	
143.225032–143.225052	9
143.225062	13
143.225072	
143.225082–143.225102	9
143.226012	8
143.226032	
143.226072	10
143.226082	
143.226092–143.226122	11
143.226131–143.226182	8
143.226192	11
143.226202	10
143.226212	
143.226222–143.226262	8
143.226272	10
143.226282	
143.226292	11
143.226302	10
143.226312	11
143.226322	8
143.226332	
143.226342	10
143.226352	11
143.227012–143.227072	12
143.233012	1
143.233032	
143.233042	
143.234022–143.234052	2
143.234062–143.234092	4
143.234102–143.234162	2
143.234192	
143.234202	
143.234212–143.234232	4
143.234242–143.234262	2
143.235012	13

Craftsman Engines Cross Reference Chart (cont.)

Craftsman Engine Models	See Column	Craftsman Engine Models	See Column	Craftsman Engine Models	See Column
143.235022		143.246242	10	143.264422	2
143.235032	6	143.246252	11	143.264432–143.264482	4
143.235042	13	143.246262	10	143.264492	2
143.235052		143.246272–143.247292	11	143.264502	
143.235062	9	143.246302	10	143.264512	4
143.235072	6	143.246312		143.264522	2
143.236012	8	143.246322	11	143.264542	
143.236022–143.236042	11	143.246332		143.264562–143.264672	4
143.236052	8	143.246342	10	143.264682	2
143.236062	11	143.246352	8	143.265012–143.265192	6
143.236072		143.246362	16	143.266012	11
143.236082	8	143.246382		143.266022	
143.236092	10	143.246392	8	143.266032	8
143.236102	8			143.266042	10
143.236112		143.254012–143.254052	2	143.266052	
143.236122	10	143.254062	4	143.266062	8
143.236132	8	143.254072–143.254122	2	143.266082	
143.236142	11	143.254142–143.254192	4	143.266092–143.266132	10
143.236152	8	143.254212	2	143.266142–143.266242	11
143.237012	12	143.254222		143.266252	8
143.237022		143.254232–143.254292	4	143.266262	11
143.237032	5	143.254302	2	143.266272–143.266302	10
143.237042	3	143.254312		143.266312	11
		143.254322	4	143.266322	
143.244032	2	143.254332	2	143.266332	10
143.244042	4	143.254342	4	143.266342	11
143.244052		143.254352		143.266352	10
143.244062		143.254362	2	143.266362	11
143.244072–143.244112	2	143.254372	4	143.266372–143.266412	8
143.244122–143.244142	4	143.254382		143.266422	11
143.244202	2	143.254392	2	143.266432–143.266452	8
143.244212	4	143.254402	4	143.266462	16
143.244222	2	143.254412		143.266472	
143.244232		143.254432	2	143.266482	11
143.244242	4	143.254442	4	143.267012–143.267042	3
143.244252	4	143.254452	2		
143.244262–143.244282	2	143.254462	4	143.274022–143.274072	4
143.244292–143.244332	4	143.254472	2	143.274092–143.274132	2
143.245012	6	143.254482		143.274142	4
143.245042	9	143.254492	4	143.274152	
143.245052–143.245072	13	143.254502–143.254532	2	143.274162–143.274182	2
143.245082		143.255012–143.255112	6	143.274192–143.274242	4
143.245092	6	143.256012	11	143.274252	2
143.245102–143.245132	13	143.256022	8	143.274262	4
143.245142	6	143.256032	10	143.274272–143.274322	2
143.245152		143.256042	11	143.274402–143.274482	4
143.245162	13	143.256052	8	143.275012–143.275052	6
143.245172	6	143.256062	11	143.276022	4
143.245182		143.256072		143.276032	11
143.245192	13	143.256082	8	143.276042	
143.246012	8	143.256092		143.276052	16
143.246022	11	143.256102	10	143.276062–143.276162	11
143.246032		143.256112	11	143.276182	8
143.246042	8	143.256122	8	143.276192	10
143.246052–143.246072	10	143.256132	10	143.276202	8
143.246082	11	143.257012–143.257072	3	143.276222	10
143.246092				143.276242	11
143.246102	10	143.264012–143.264042	2	143.276252	8
143.246112		143.264052–143.264082	4	143.276262	11
143.246122	11	143.264092	2	143.276272	11
143.246132	10	143.264102	4	143.276282	10
143.246142		143.264232–143.264342	2	143.276292	11
143.246152–143.246212	11	143.264352–143.264372	4	143.276302	11
143.246222	10	143.264382	2	143.276322–143.276342	10
143.246232	11	143.264392–143.264412	4	143.276352	11

Craftsman Engines Cross Reference Chart (cont.)

Craftsman Engine Models	See Column	Craftsman Engine Models	See Column	Craftsman Engine Models	See Column
143.276362	16	143.531052	1	143.596022	
143.276372–143.276392	10	143.531082		143.596042	8
143.276402	16	143.531122		143.596052	10
143.276412	8	143.531132		143.596072–143.596122	8
143.276422	10	143.531142	2	143.597012–143.597032	3
143.276432–143.276472	11	143.531152	1		
143.276482	16	143.531172		143.601022–143.601062	1
143.277012	3	143.531182		143.604012	2
143.277022		143.534012–143.534072	2	143.604022	4
		143.535012–143.535062	9	143.604032	2
143.284012	2	143.536012–143.536062	8	143.604042	
143.284022	1	143.537012	7	143.604052	4
143.284032	2			143.604062	2
143.284042	4	143.541012	1	143.604072	
143.284052	2	143.541042–143.541062	1	143.605012	9
143.284062		143.541112–143.541152	1	143.605022	
143.284072	4	143.541172–143.541202	1	143.605052	
143.284082	2	143.541222	1	143.606012–143.606052	10
143.284092		143.541282–143.541302	1	143.606092	8
143.284102	4	143.544012–143.544042	2	143.606102	10
143.284112	2	143.545012–143.545042	9	143.607012–143.607032	3
143.284142		143.546012–143.546022	8	143.607042–143.607062	3
143.284152		143.547012–143.547032	7		
143.284162				143.611012–143.611112	1
143.284182		143.551012	1	143.614012–143.614032	4
143.284212	4	143.551032		143.614042	2
143.284312	2	143.551052–143.551192	1	143.614052	4
143.284322		143.554012–143.554082	2	143.614062–143.614162	2
143.284332	4	143.555012–143.555052	9	143.615012–143.615092	9
143.284342		143.556012–143.556282	8	143.616012	10
143.284352		143.557012–143.557082	7	143.616022–143.616112	8
143.284372	16			143.616122	10
143.284382	4	143.565022	9	143.616132	8
143.284402	2	143.566002–143.566202	8	143.616142	
143.284412		143.566212	9	143.617012–143.617182	3
143.284432	2	143.566222–143.566252	8		
143.284362		143.567012–143.567042	7	143.621012–143.621092	1
143.284392	2			143.624012–143.624112	2
143.284442		143.571002–143.571122	1	143.625012–143.625132	9
143.284482		143.571152	2	143.626012	10
143.284452	4	143.571162	1	143.626022	8
143.284472		143.571172		143.626032	10
143.285012	6	143.574022–143.574102	2	143.626042	8
143.285022		143.575012–143.575042	9	143.626052–143.626122	10
143.285032		143.576002–143.576202	8	143.626132	8
143.286102	16			143.626142	10
143.286022	17	143.581002–143.581102	1	143.626152	
143.286032	10	143.584012–143.584142	2	143.626162	8
143.286072–143.286092	11	143.585012–143.585042	9	143.626172	10
143.286112		143.586012–143.586042	8	143.626182	8
143.286122		143.586052–143.586062	10	143.626192	10
143.286132	10	143.586072–143.586082	8	143.626202	8
143.286142	11	143.586112–143.586142	10	143.626212	10
143.286152		143.586152	8	143.626222–143.626262	8
143.286162		143.586162	10	143.626282	11
143.286172		143.586172–143.586242	8	143.626292	10
		143.586252	10	143.626302	8
143.505010	8	143.586262–143.586282	8	143.626312	10
143.505011		143.587012–143.587042	3	143.626322	
143.521081	9			143.627012–143.627042	3
143.525021	9	143.591012–143.591142	1		
143.526011		143.594022–143.594082	2	143.631012–143.631092	1
143.526021	8	143.594092	2	143.634012	2
143.526031	8	143.594102		143.634032	
		143.595012	9	143.635012	9
		143.595042		143.635022	
		143.596012	10		

Craftsman Engines Cross Reference Chart (cont.)

Craftsman Engine Models	See Column	Craftsman Engine Models	See Column	Craftsman Engine Models	See Column
143.635032	6	143.656202	8	143.674012	2
143.635052	9	143.656212	11	143.675012	6
143.636012	11	143.656222		143.675022	
143.636022		143.656232	10	143.675032	9
143.636032	10	143.656242	11	143.675042	6
143.636042	11	143.656252	8	143.676012	11
143.636052	8	143.656262	10	143.676022	
143.636062	10	143.656272		143.676032	10
143.636072	11	143.656282	11	143.676042	11
143.637012	3	143.657012 – 143.657052	3	143.676052	
				143.676062	16
143.641012 – 143.641062	1	143.661012 – 143.661062	1	143.676072	
143.641072	2	143.664012 – 143.664332	2	143.676102	10
143.644012 – 143.644082	2	143.665012 – 143.665082	6	143.676112	8
143.645012 – 143.645032	6	143.666012	10	143.676122	10
143.646012 – 143.646032	10	143.666022		143.676132	8
143.646042	11	143.666032	11	143.676142	11
143.646052		143.666042 – 143.666072	10	143.676152	16
143.646072 – 143.646102	10	143.666082	11	143.676162	
143.646112	8	143.666092		143.676172	10
143.646122	10	143.666102 – 143.666142	8	143.676182	11
143.646132		143.666152	11	143.676192	10
143.646142	11	143.666162		143.676202	11
143.646152	10	143.666172	8	143.676212	16
143.646162	11	143.666202		143.676222	11
143.646172	10	143.666222	10	143.676232	8
143.646182		143.666232	11	143.676242	8
143.646192	8	143.666242	8	143.676252	11
143.646202	10	143.666252	10	143.676262	16
143.646212 – 143.646232	11	143.666272	8	143.677012	3
143.647012 – 143.647062	3	143.666282	10	143.677022	
		143.666292	8		
143.651012 – 143.651072	1	143.666302	10		
143.654022 – 143.654322	2	143.666312		143.686012	17
143.655012	6	143.666322	11	143.686022	
143.655032		143.666332	16	143.686032	11
143.656012 – 143.656052	8	143.666342	10	143.686042	
143.656062	10	143.666352	11	143.686052	
143.656082	11	143.666362	16	143.686062	10
143.656092	8	143.666372	8	143.686072	9
143.656102	10	143.666382	10	143.687012	3
143.656112	8	143.667012	3		
143.656122 – 143.656152	10	143.667022		143.694126	4
143.656162 – 143.656182	8	143.667032	6	143.694132	2
143.656192	10	143.667042 – 143.667082	3	143.694134	11

Torque Specifications

Model/Part	Inch Pounds	Ft. Pounds
Cylinder Head Bolts	160–200	13–16
Connecting Rod Bolts	65–75	5.5–6
ECH90, ECV100, TNT100	75–80	6.2–6.7
TVS120, 5 H.P. Small Frame (Durlok Rod Bolts)	110–130	9.1–10.8
6 H.P. Medium Frame	86–110	7.1–9.1
6 H.P. Medium Frame (Durlok Rod Bolts)	130–150	10.8–12.5
ECV105, ECV110, ECV120, TNT120	80–95	6.6–7.9
7, 8 & 10 Medium Frame	106–130	8.8–10.8
7, 8 & 10 Medium Frame (Durlok Rod Bolts)	150–170	12.5–14.1
Cylinder Cover or Flange-to-Cylinder	65–110	5.5–9
Cylinder Cover 6–7 H.P. Medium Frame, H Models	100–140	8.3–11.6
Flywheel Nut	360–396	30–33
Spark Plug	180–360	15–30
Magneto Stator to Cylinder	40–90	3.3–7.5
Starter to Blower Housing or Cylinder	40–60	3.5–5
Housing Baffle to Cylinder	48–72	4–6
Breather Cover (Top Mount ECV)	40–50	3.3–4.1
Breather Cover	20–26	1.7–2.1
Intake Pipe to Cylinder	72–96	6–8
Carburetor to Intake Pipe	48–72	4–6
Air Cleaner to Carburetor (Plastic)	8–12	1
Tank Plate to Bracket (Plastic)	100–144	9–12
Tank to Housing	45–65	3.7–5
Muffler Bolts to Cylinder		
1–5 H.P. Small Frame	30–45	2.5–3.5
4–5 H.P. Medium Frame	90–150	8–12
6 : 1 Gear Reduction Cover to Housing	100–144	8.5–12
Gear Reduction Cover to Housing	65–110	5–9
Oil Drain Plug		
1/8 — 27	35–50	1.1–4.1
1/4 — 18	65–85	4.5–7
3/8 — 18	80–100	6.6–9
5/8 — 18	90–150	7.5–12.5
1/2 — 14	80–100	6.6–9
Ball Bearing Retainer 2.5		
2.5–5 H.P. Small Frame	45–60	3.7–5
5–10 H.P. Medium Frame	15–22	1.5
Craftsman Exclusive Fuel System to Cylinder	72–96	6–8
Electric Starter-to-Cylinder	50–60	4–5

Cross Reference for Vertical Crankshaft Engines

Size	Model	Column
6 HP	V60	8
	VH60	8
7 HP	V70	10
	VH70	10
	VM70	17
8 HP	V80	11
	VM80	11
10 HP	VM100	16

Cross Reference Chart for Horizontal Crankshaft Engines

Size	Model	Column
6 HP	H60	8
	HH60	8
7 HP	H70	10
	HH70	10
	HM70	17
8 HP	H80	11
	HM80	11
10 HP	HM100	16

Engine Specifications

Reference Column	1	2	3	4	5	6	7	8	9
Displacement	7.75	9.06	10.5	10.0	10.5	12.0	11.04	13.53	12.17
Stroke	1²⁷/₃₂″	1²⁷/₃₂″	1¹⁵/₁₆″	1²⁷/₃₂″	1¹⁵/₁₆″	1¹⁵/₁₆″	2¹/₄″	2¹/₂″	2¹/₄″
Bore	2.3125 2.3135	2.5000 2.5010	2.625 2.626	2.625 2.626	2.625 2.626	2.812 2.813	2.5000 2.5010	2.625 2.626	2.625 2.626
Timing Dimension Before Top Dead Center for Vertical Engines	V.060 .070	V.065	V.035	.035	.035	V.040 .060	V.050	V.050	H.050
Timing Dimension Before Top Dead Center for Horizontal Engines	H.060 .070	H.030 .040	H.035			H.055	H.050	H.050	H.050
Point Setting	.020	.020	.020	.020	.020	.020	.020	.020	.020
Spark Plug Gap	.030	.030	.030	.030	.030	.030	.030	.030	.030
Valve Clearance	.010 Both	.010 Both	.010 Both	.010 Both	.010 Both	.010 Both	.010 Both	.010 Both	.010 Both
Valve Seat Angle	46°	46°	46°	46°	46°	46°	46°	46°	46°
Valve Spring Free Length	1.135″	1.135″	1.135″	1.135″	1.135″	1.135″	1.562″	1.462″	1.462″
Valve Guides Over-Size Dimensions	.2805 .2815	.2805 .2815	.2805 .2815	.2805 .2815	.2807 .2817	In. .280 Ex. .278	.3432 .3442	.3432 .3442	.343 .344
Valve Seat Width	.035 .045	.035 .045	.035 .045	.035 .045	.035 .045	.035 .045	.042 .052	.042 .052	.042 .052
Crankshaft End Play	.005 .027	.005 .027	.005 .027	.005 .027	.005 .027	.005 .027	.005 .027	.005 .027	.005 .027
Crankpin Journal Diameter	.8610 .8615	.8610 .8615	.9995 1.0000	.8610 .8615	.9995 1.0000	.9995 1.0000	1.0615 1.0620	1.0615 1.0620	1.0615 1.0620
Cylinder Main Bearing Dia.	.8755 .8760	.8755 .8760	1.0005 1.0010	.8755 .8760	1.0005 1.0010	1.0005 1.0010	1.0005 1.0010	1.0005 1.0010	1.0005 1.0010
Cylinder Cover Main Bearing Dia.	.8755 .8760	.8755 .8760	1.0005 1.0010	.8755 .8760	1.2010 1.2020	1.0005 1.0010	1.0005 1.0010	1.0005 1.0010	1.0005 1.0010
Conn. Rod. Dia. Crank Bearing	.8620 .8625	.8620 .8625	1.0005 1.0010	.8620 .8625	1.0005 1.0010	1.0005 1.0010	1.0630 1.0635	1.0630 1.0635	1.0630 1.0635
Piston Diameter	2.3090 2.3095	2.4950 2.4955	2.6200 2.6205	2.6200 2.6205	2.604 2.608	2.8070 2.8075	2.492 2.4945	2.6210 2.6215	2.6210 2.6215
Piston Pin Diameter	.5629 .5631	.5629 .5631	.5629 .5631	.5629 .5631	.5631 .5635	.5629 .5631	.6248 .6250	.6248 .6250	.6248 .6250
Width of Comp. Ring Groove	.0955 .0977	.0955 .0975	.0925 .0935	.0955 .0975	.0955 .0975	.0955 .0975	.0955 .0975	.0955 .0975	.0955 .0975
Width of Oil Ring Groove	.125 .127	.125 .127	.156 .158	.156 .158	.156 .158	.156 .158	.156 .158	.156 .158	.156 .158

Engine Specifications (cont.)

Reference Column		1	2	3	4	5	6	7	8	9
Side Clearance of Ring Groove	(Top) Comp.	.002 .005	.002 .003	.002 .004	.002 .005	.002 .005	.003 .004	.002 .003	.002 .004	.002 .004
	(Bot.) Oil			.001 .004	.001 .004	.001 .004	.002 .003		.002 .004	.002 .004
Ring End Gap		.007 .020	.007 .020	.007 .020	.007 .020	.007 .020	.007 .020	.007 .020	.007 .020	.007 .020
Top Piston Land Clearance		.0015 .0145	.015 .018	.0165 .0215	.017 .022	.017 .022	.017 .022	.015 .018	.017 .020	.017 .020
Piston Skirt Clearance		.0025 .0040	.0045 .0060	.0045 .0060	.0045 .0060	.0050 .0065	.0045 .0060	.0055 .0070	.0035 .0050	.0035 .0050
Camshaft Bearing Dia.		.4975 .4980	.4975 .4980	.4975 .4980	.4975 .4980	.505 .513	.4975 .4980	.6230 .6235	.6230 .6235	.6230 .6235
Dia. of Crankshaft Mag. Main Brg.		.8735 .8740	.8735 .8740	.9985 .9990	.8735 .8740	.9985 .9990	.9985 .9990	.9985 .9990	.9985 .9990	.9985 .9990
Dia. of Crankshaft P.T.O. Main Brg.		.8735 .8740	.8735 .8740	.9985 .9990	.8735 .8740	.9985 .9990	.9985 .9990	.9985 .9990	.9985 .9990	.9985 .9990

A. For VM80 & HM80 engines only—Displacement is 19.41".
B. For VM80 & HM80 engines only—Bore is 3.125"/3.126" (3⅛").
C. For VM80 & HM80 engines only—Piston Diameter is 3.1205"/3.1195".

Engine Specifications

Reference Column	10	11	12	13	14	15	16	17
Displacement	15.0	18.65 See Note A	11.5	12.0	10.0	12.0	20.2	17.16
Stroke	2¹⁷/₃₂"	2¹⁷/₃₂" See Note B	1¹⁵/₁₆"	1¹⁵/₁₆"	1²⁷/₃₂"	1¹⁵/₁₆"	2¹⁷/₃₂"	2¹⁷/₃₂"
Bore	2.750 2.751	3.062 3.063	2.750 2.751	2.812 2.813	2.625 2.626	2.812 2.813	3.187 3.188	2.9375 2.9385
Timing Dimension Before Top Dead Center for Vertical Engines	V.050	V.070	V.035	V.035	V.035	V.035	V.070	V.070

Engine Specifications (cont.)

Reference Column		10	11	12	13	14	15	16	17
Timing Dimension Before Top Dead Center for Horizontal Engines		H.050	H.070					H.070	H.070
Point Setting		.020	.020	.020	.020	.020	.020	.020	.020
Spark Plug Gap		.030	.030	.030	.030	.030	.030	.030	.030
Valve Clearance		.010 Both	.010 Both	.010 Both	.010 Both	.010 Both	.010 Both	.010 Both	.010 Both
Valve Seat Angle		46°	46°	46°	46°	46°	46°	46°	46°
Valve Spring Free Length		1.462″	1.462″	1.135″	1.135″	1.135″	1.135″	1.462″	1.462″
Valve Guides Over-Size Dimensions		.3432 .3442	.3432 .3442	.2805 .2815	.2805 .2815	.2805 .2815	.2805 .2815	.3432 .3442	.3432 .3442
Valve Seat Width		.042 .052	.042 .052	.035 .045	.035 .045	.035 .045	.035 .045	.042 .052	.042 .052
Crankshaft End Play		.005 .027	.005 .027	.005 .027	.005 .027	.005 .027	.005 .027	.005 .027	.005 .027
Crankpin Journal Diameter		1.1865 1.1870	1.1865 1.1870	.9995 1.0000	.9995 1.0000	.8610 .8615	.9995 1.0000	1.1865 1.1870	1.1865 1.1870
Cylinder Main Bearing Dia.		1.0005 1.0010	1.0005 1.0010	1.0005 1.0010	1.0005 1.0010	.8755 .8760	1.0005 1.0010	1.0005 1.0010	1.0005 1.0010
Cylinder Cover Main Bearing Dia.		1.0005 1.0010	1.1890 1.1895	1.0005 1.0010	1.0005 1.0010	.8755 .8760	1.0005 1.0010	1.1890 1.1895	1.1890 1.1895
Conn. Rod. Dia. Crank Bearing		1.1880 1.1885	1.1880 1.1855	1.0005 1.0010	1.0005 1.0010	.8620 .8625	1.0005 1.0010	1.1880 1.1885	1.1880 1.1885
Piston Diameter		2.7450 2.7455	3.0575 3.0585 See Note C	2.7450 2.7455	2.8070 2.8075	2.6200 2.6205	2.8070 2.8075	3.1817 3.1842	2.9325 2.9335
Piston Pin Diameter		.6248 .6250	.6248 .6250	.5629 .5631	.5629 .5631	.5629 .5631	.5629 .5631	.6248 .6250	.6248 .6250
Width of Comp. Ring Groove		.0795 .0805	.0955 .0975	.0795 .0815	.0955 .0975	.0955 .0975	.0955 .0975	.0955 .0975	.0975 .0955
Width of Oil Ring Groove		.188 .189	.188 .190	.1565 .1585	.1565 .1585	.1565 .1585	.1565 .1585	.188 .190	.188 .190
Side Clearance of Ring Groove	(Top) Comp.	.002 .003	.003 .004	.002 .004	.003 .004	.003 .0045	.0028 .0039	.0020 .0050	.0028 .0051
	(Bot.) Oil	.001 .003	.002 .003	.001 .002	.001 .002	.0010 .0030	.0018 .0038	.001 .004	.0018 .0029

Engine Specifications (cont.)

Reference Column	10	11	12	13	14	15	16	17
Ring End Gap	.007 .020	.007 .020	.007 .020	.007 .020	.007 .020	.007 .020	.007 .020	.007 .020
Top Piston Land Clearance	.023 .028	.031 .034	.024 .027	.018 .021	.017 .022	.017 .022	.029 .034	.030 .035
Piston Skirt Clearance	.0045 .0060	.0035 .0055	.0045 .0060	.0045 .0060	.0045 .0060	.0045 .0060	.0028 .0063	.004 .006
Camshaft Bearing Dia.	.6230 .6235	.6230 .6235	.4975 .4980	.4975 .4980	.4975 .4980	.4975 .4980	.6230 .6235	.6230 .6235
Dia. of Crankshaft Mag. Main Brg.	.9985 .9990	.9985 .9990	.9990 .9995	.9990 .9995	.8735 .8740	.9985 .9990	.9985 .9990	.9985 .9990
Dia. of Crankshaft P.T.O. Main Brg.	.9985 .9990	1.1870 1.1875	.9985 .9990	.9985 .9990	.8735 .8740	.9985 .9990	1.1870 1.1875	1.1870 1.1875

Section 17

Tecumseh-Lauson 2-Stroke

Small Engines—0-6 hp

ENGINE IDENTIFICATION

The identification taps may be located at a variety of places on the engine. The type number is the most important number since it must be included with any correspondence about a particular engine.

Early engines listed the type number as a suffix of the serial number. For example on number 123456789 P 234, 234 is the type number. In the number 123456789 H 104–02B; 104–02B is the type number. In either case the type number is important.

If you use short block to repair the engine, be sure that you transfer the serial number and type number tag to the new short block.

On the newer engines, reference is sometimes made to the model number. The model number tells the number of cylinders, the design (vertical or horizontal) and the cubic inch displacement.

GENERAL ENGINE SPECIFICATIONS

A detailed listing of the great number of Power Products two-stroke engine models is provided in the Type No./Column No. cross reference chart at the back of this section.

MAINTENANCE

Air Cleaners

The instructions below detail the procedures involved in cleaning the various types of elements. See the illustrations for exploded views to aid disassembly and assembly.

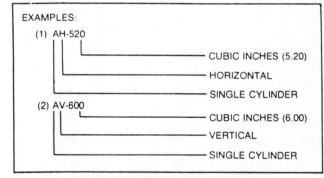

EXAMPLES:

(1) AH-520
- CUBIC INCHES (5.20)
- HORIZONTAL
- SINGLE CYLINDER

(2) AV-600
- CUBIC INCHES (6.00)
- VERTICAL
- SINGLE CYLINDER

Model number interpretation

A. NAMEPLATE ON AIR SHROUD
B. MODEL & TYPE NUMBER PLATE
C. METAL TAG ON CRANKCASE
D. STAMPED ON CRANKCASE
E. STAMPED ON CYLINDER FLANGE
F. STAMPED ON STARTER PULLEY

Location of identification numbers on 2 cycle engines

SCREEN MUST BE POSITIONED WITH EDGES AWAY FROM ELEMENT

POLYURETHANE ELEMENT

COVER

Exploded view of polyurethane air cleaner

POLYURETHANE AIR CLEANER

1. Wash the element in a solvent or detergent and water solution by squeezing similar to a sponge.
2. Clean the air cleaner housing and cover with the same solution. Dry thoroughly.
3. Dry the element by squeezing or with compressed air if available.
4. Apply a generous quantity of oil to the element sides and open ends. Squeeze vigorously to distribute oil and to remove excess oil.

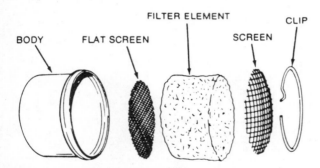

BODY   FLAT SCREEN   FILTER ELEMENT   SCREEN   CLIP

Exploded view of aluminum foil type air cleaner

PAPER ELEMENT

Exploded view of paper element type air cleaner

NOTE

THIS GASKET USED ONLY ON POWER PRODUCTS 641 SERIES AND UP WITH A #450226 ELEMENT

GASKET

Exploded view of felt type air cleaner

ALUMINUM FOIL AIR CLEANER

1. Dip the aluminum foil filter in solvent. Flush out all dirt particles.
2. Shake out the filter thoroughly to remove all solvent, then dip the filter element in oil. Allow the oil to drain from the filter. Clean the screens and filter body.

NOTE: The concave screen and retainer cover or ring are not used on later models. They are replaced with a clip which rolls into the groove in the lip of the body.

FELT TYPE AIR CLEANERS

1. To clean felt air cleaners, merely blow compressed air through the element in the reverse direction to normal air flow. Felt elements may also be washed in nonflammable solvents or soapy water. Blow dry with compressed air.

NOTE: Power Products type numbers 641 and up, use a gasket between the element and the base. The gasket is used only with this element; earlier versions did not have a gasket.

FIBER ELEMENT AIR CLEANER

1. Remove the filter and place the cover in a normal position on the filter. With the filter element down to semi-seal it, blow compressed air through the cover hole to reverse air flow, forcing dirt particles out.
2. Clean the cover mounting bracket with a damp cloth.

DRY PAPER AIR CLEANER

1. Tap the element on a workbench or any solid object to dislodge larger particles of dirt.
2. Wash the element in soap and water. Rinse from the inside until it is thoroughly flushed and the water coming through is free of soap.
3. Allow the element to dry completely or use low pressure compressed air blown from the inside to speed the process.
4. Inspect the element for cracks or holes, and replace if necessary.

Lubrication

OIL AND FUEL RECOMMENDATIONS

Power Products 2 cycle engines are mist-lubricated by oil mixed with the gasoline. For the best performance, use regular grade, leaded fuel, with 2 cycle or outboard oil rated SAE 30 or SAE 40.

Regular grade unleaded fuel is an acceptable substitute. The terms 2 cycle or outboard are used by various manufacturers to designate oil they have designed for use in 2 cycle engines. Multiple weight oil such as all season 10W–30, are not recommended.

If you have to mix the gas and oil when the temperature is below 35°F, heat up the oil first, then mix it with the gas. Oil will not mix with gas when the temperature is approaching freezing. However, if you use oil that has been warmed first it will not be affected by low temperatures.

The proportion of oil to fuel is absolutely critical to two-stroke operation. If too little oil is used, overheating and damage to engine parts will occur (this can even result from running the engine too lean). If excessive oil is used, spark plug fouling, smoke in the exhaust, and even misfire can occur. Mix carefully and precisely. Follow Power Products recommendations for your particular engine, and disregard fuel container labels.

Fuel/oil mix must be clean and fresh. Fuel deteriorates enough to form troublesome gum and varnish after more than a month. Dirt in the fuel can cause clogging of carburetor passages and even engine wear.

TUNE-UP

Specifications

All spark plugs are gapped at 0.035 in. Because of the great number of individual models of Power Products engines that exist, an individual chart of Tune-Up specifications is impractical. Refer to the charts at the end of this section for breaker point gap and timing dimension specifications.

Spark Plugs

Spark plugs should be removed and cleaned of deposits frequently, especially in two-stroke engines because they burn the lubricating oil right with the fuel. Carefully inspect the plug for severely eroded electrodes or a cracked insulator, and replace the plug if either condition exists or if deposits cannot be adequately removed. Set the gap to 0.035 in. with a wire type feeler gauge, and install the plug, torquing it to 18–22 ft. lbs.

Make sure to replace the plug with one of the same type and heat range. If the plug is fouled, poor quality or old fuel, a rich mixture, or the wrong fuel/oil mix may be at fault. Also, make sure the engine's exhaust ports are not clogged.

Breaker Points

REMOVAL AND INSTALLATION

1. First remove the flywheel as described below:

a. Remove the screws, engine shroud, and starter. Determine the direction of rotation of the flywheel nut by looking at the threads. Then, place a box wrench on the nut and tap with a soft hammer in the proper direction.

b. The flywheel is removed with a special puller or a special knock off tool. The knock off tool must not be used on 660, 670, or 1500 ball bearing models. To use the knock off tool, screw it onto the crankshaft until it is within 1/16 in. of the flywheel. Hold the flywheel firmly and rap the top of the puller sharply with a hammer to jar it loose. Pull the flywheel off.

c. If a puller is being used, the flywheel will have three cored holes into which a set of self-tapping screws are turned. The handle which operates the bolt at the center of the puller's collar is then turned to pull the flywheel off the crankshaft. If this is not adequate to do the job, heat the center of the alloy flywheel with a butane torch to expand it before turning the puller handle.

2. Remove the nuts that hold the electrical leads to the screw

POINT GAP

Adjusting point gap

on the movable breaker point spring. Remove the movable breaker point from the stud.

3. Remove the screw and stationary breaker point. Put a new stationary breaker point on breaker plate; install the screw, but do not tighten it fully.

4. Position a new movable breaker point on the post.

5. Check that the new points contact each other properly and remove all grease, fingerprints, and dirt from the points.

ADJUSTMENT

1. If necessary (as when checking the gap of old points), loosen the screw which mounts the stationary breaker contact. Rotate the crankshaft until the contact cam follower rests right on the highest point of the cam.

2. Using a flat feeler gauge of the dimension shown under "point gap" in the charts at the rear of this section, check the dimension of the gap and, if incorrect, move the breaker base in the appropriate direction by wedging a screwdriver between the dimples on the base plate and the notch in the breaker plate. When gap is correct, tighten the screw. Recheck the gap and, if necessary, reset it.

Adjusting Ignition Timing

1. Remove the spark plug and install a special timing tool or thin ruler. With the tool lockscrew loose or the ruler riding on the piston, rotate the crankshaft back and forth to find Top Dead Center. Tighten the tool lockscrew or use a straightedge across the cylinder head, if you're using a ruler, to measure Top Center. Look up the timing dimension in the specifications at the back of this section.

JUST AS POINTS OPEN TIGHTEN STATOR BOLTS. BE SURE POINTS ARE CLEAN AND ALIGNED

TURN STATOR INTO CRANKSHAFT ROTATION

IF CONTINUITY LIGHT IS USED, REMOVE WIRES

Adjusting ignition timing

2. Turn the crankshaft backwards so the piston descends. Then, reset the position of the special tool downward the amount of the dimension and tighten the lockscrew, or move the ruler down that amount.

3. Very carefully bring the piston upward by turning the crankshaft until the top of the piston just touches the tool or ruler.

4. Install a piece of cellophane between the contact surfaces. Loosen the ignition stator lockscrews and turn the stator until the cellophane is clamped tightly between the contact surfaces. Turn the stator until the cellophane can just be pulled out (as contacts start to open), and then tighten the stator lockscrews.

Magneto Armature Air Gap

There are two types of magnetos. Compare appearance of your engine with each illustration to determine whether it looks like the type which employs a gap of 0.005–0.008 in. or the type with a gap of 0.015 in.

1. Loosen the two screws which hold the laminations and coil to the block. Turn the flywheel around so the magnets line up directly with the ends of the laminations. Pull the laminations/coil assembly upward.

2. Insert an air gap gauge (Power Products Part #670216) or an equivalent (0.005–0.008 in.) gauge between the flywheel and laminations on either side. On units with a gap of 0.015, use two of the above numbered parts or a 0.015 in. gauge. Allow the attraction of the magnets to pull the coil/laminations assembly toward the flywheel.

3. On the magneto type with a 0.005–0.008 in. dimension, use Loctite Grade A on the screws and torque to 35–45 inch lbs. On the other type magneto, torque the screws to 20–30 inch lbs. Recheck the gap and readjust, if necessary.

Mixture Adjustment

1. If the engine will not start or the carburetor has recently been disassembled, both idle and main mixture screws may be turned in very gently until they just bottom, and then turned out exactly one turn. In this case, back out the idle speed screw until throttle is free, then turn screw in until it just contacts the throttle. lkTurn it one turn more exactly.

2. Allow the engine to warm up to normal running temperature. With the engine running at maximum recommended rpm, loosen the main metering screw until the engine rolls, then tighten the screw until the engine starts to cut out. Note the number of turns from one extreme to the other. Loosen the screw to a point midway between the extremes.

3. Set the throttle to idle speed and repeat Step 2 for that adjusting screw.

NOTE: Some carburetors do not have a main mixture adjustment. Others employ a drilled mixture screw which provides about the right mixture when fully

AIR GAP DIMENSION .005 .008

MAGNETS

Adjusting magneto armature air gap. Note magnets on flywheel

screwed in. In some weather conditions, operation may be improved by setting this screw just slightly off the seat for smoother running.

Governor Adjustments

POWER TAKEOFF END MECHANICAL GOVERNOR

1. To adjust the governor, remove the outboard bearing housing. Use a $^3/_{32}$ in. allen wrench to loosen the set screw.

2. Squeeze the top and bottom governor rings, fully compressing the governor spring.

3. Hold the upper arm of the bell crank parallel to the crankshaft and insert a $^3/_{32}$ in. allen wrench between the upper ring and the bell crank.

A

FULL THROTTLE POSITION

$^3/_{32}$ INCH ALLEN WRENCH

B

CLOSED THROTTLE POSITION (PARALLEL)

BELL CRANK

$^3/_{32}$ INCH ALLEN WRENCH

GOVERNOR SPRING COMPRESSED

TIGHTEN SETSCREW

Adjusting the PTO end mounted governor

.015"

Adjusting magneto armature air gap. Note wider gap

4. Slide the governor assembly onto the crankshaft so that the allen wrench just touches the bell crank.

5. Tighten the set screw to secure the governor to the crankshaft.

6. Install the bearing adapter, mount the engine, and check the engine speed with a tachometer. It should be about 3200–3400 rpm.

NOTE: Never attempt to adjust the governor by bending the bellcrank or the link.

7. If the speed is not correct, readjust the governor by moving the assembly toward the crankcase to increase speed, and away from the crankcase to decreased speed.

MECHANICAL FLYWHEEL TYPE GOVERNOR ADJUSTMENT

As engine speed increases, the links are thrown outward, compressing the link springs. The links apply a thrust against the slide ring, moving it upward and compressing the governor spring. As the slide ring moves away from the thrust block of the bellcrank assembly, the throttle spring causes the thrust block to maintain engagement and close the throttle slightly.

As the throttle closes and engine speed decreases, force on the slide ring decreases so that it moves downward, pivots the bellcrank outward to overcome the force of the throttle spring, and opens the throttle to speed up the engine. In this manner, the operating speed of the engine is stabilized to the adjusted governor setting.

1. To adjust the governor, loosen the bracket screw and slide the governor bellcrank assembly toward or away from the flywheel. Move the bellcrank toward the flywheel to increase speed and away from the flywheel to decrease speed.

2. Tighten the screw to secure the bracket.

3. Make minor speed adjustments by bending the throttle link at the bend in the center of the link.

NOTE: Do not lubricate the governor assembly or the governor bellcrank assembly of flywheel mounted governors.

ADJUSTING 2 CYCLE AIR VANE GOVERNORS

1. Loosen the self-locking nut that holds the governor spring bracket to the engine crankcase.

Flywheel mounted governor assembly (cutaway view)

2. Adjust the spring bracket to increase or decrease the governor spring tension. Increasing spring tension increases speed and decreasing spring tension decreases speed.

3. After adjusting, the spring bracket should not be closer than $\frac{1}{16}$ in. to the crankcase.

4. Tighten the self-locking nut.

FUEL SYSTEM

Carburetor

REMOVAL AND INSTALLATION

1. Remove the air cleaner. Drain the fuel tank. Disconnect the carburetor fuel lines.

2. If necessary, remove any shrouding or control panels to gain access to the carburetor.

3. Disconnect the choke or throttle control wires at the carburetor.

4. Remove the cap screws, or nuts and lockwashers and remove the carburetor from the engine.

5. To install, reverse the removal procedure, using new gaskets.

OVERHAUL

Tecumseh two-and four-stroke engines employ a common series of carburetors. Refer to the Four-Stroke Tecumseh Engine section for specific carburetor overhaul procedures.

Diagram of an air vane governor

Position detent reference marks of throttle shutter as shown

Idle Governor

SERVICE

1. Remove the shutter fastener and allow the shutter to drop out of the air horn.
2. Note location of the spring end in the disc-shaped throttle lever. The spring should be placed into the same hole during reassembly.
3. Remove the retainer clip and lift out the throttle shaft.
4. Replace all worn parts and reassemble in reverse order.

NOTE: Note the position of the throttle shutter, as shown. The reference marks must be positioned as shown when shutter is installed.

Fuel Pump

SERVICE

Float Type Carburetor With Integral Pump

1. If the engine runs, but roughly, make both carburetor mixture adjustments.

Integral type fuel pump

2. Make sure the fuel supply is adequate and the tank is in the proper position.
3. Make sure the fuel tank valve is open.
4. Make sure the pick-up tube is not cracked.
5. Remove the carburetor and make sure the pulsation passage is properly aligned.
6. Check for air leaks at the gasket surface.
7. Remove the cover and check the condition of the inlet and outlet flaps—if curled, replace the flap leaf.

ENGINE OVERHAUL

Disassembly

SPLIT CRANKCASE ENGINES

1. Remove the shroud and fuel tank if so equipped.
2. Remove the flywheel and the ignition stator.
3. Remove the carburetor and governor linkage. Carefully note the position of the carburetor wire links and springs for reinstallation.

Checking the reed valve clearance

4. Lift off the reed plate and gasket if present and inspect them. They should not bend away from the sealing surface plate more than 0.01 in.
5. Remove the spark plug and inspect it.
6. Remove the muffler. Be sure that the muffler and the exhaust ports are not clogged with carbon. Clean them if necessary.
7. Remove the transfer port cover and check for a good seal.
8. Remove the cylinder head, if so equipped.

NOTE: Some models utilize a locking compound on the cylinder head screws. Removing the screws on such engines can be difficult. This is especially true with screws having slotted head for a straight screwdriver blade. The screws can be removed if heat is applied to the head of the screw with an electric soldering iron.

9. On engines having a governor mounted on the power take-off end of the crankshaft, remove the screws that hold the outboard bearing housing to the crankcase. Clean the PTO end of the crankshaft and remove the outboard bearing housing and bearing. Loosen the set screw that holds the governor assembly to the crankshaft, slide the entire governor assembly from the crankshaft. Remove the screw that holds the governor bellcrank bracket to the crankcase. Remove the governor bellcrank and bracket.
10. Make match marks on the cylinder and crankcase. Remove the four nuts and lockwashers that hold the cylinder to the crankcase.
11. Remove the cylinder by pulling it straight out from the crankcase.
12. To separate the two crankcase halves, remove all of the screws that hold the crankcase halves together.
13. With the crankcase in a vertical position, grasp the top half of the crankcase and hold it firmly. Strike the top end of the crankcase with a rawhide mallet, while holding the assembly over a bench to prevent damage to parts when they fall. The top half of the crankcase should separate from the remaining assembly. Invert the assembly and repeat the procedure to remove the other casting half from the crankcase on ball bearing units.
14.
15. Each time the crankshaft is removed from the crankcase, seals at the end of the crankcase should be replaced. To replace the seals, use a screwdriver or an ice pick to remove the seal retainers and remove and discard the old seals. Install the seals in the bores of the crankcase halves. The seals must be inserted into the bearing well with the channel groove toward the internal side of the crankcase. Retain the seal with the retainer. Seat the retainer spring into the spring groove.

UNIBLOCK ENGINES

1. Remove the shroud and fuel tank. Note the condition of the air vane governor, if so equipped.
2. Remove the starter cap and flywheel nut, noting the position of the belleville washer.
3. Remove the flywheel—see "Breaker Points Removal and Installation," above.
4. Remove the head. Save the old head gasket for use when

replacing the piston, but procure a new gasket for use in final assembly.

5. Remove the cylinder block cover plate to gain access to the connecting rod bolts.

6. Note the location of the connecting rod match marks for reassembly.

7. Remove the piston. Remove the ridge first, if necessary, with a ridge reamer. Push the piston and connecting rod through the top of cylinder.

8. Remove the crankshaft from the cylinder block assembly. On engines with crankshaft ball bearings:

a. Remove the four shroud base screws and tap the shroud base so the base and crankshaft can be removed together.

b. To remove bearing with crankshaft from base: USE SAFETY GLASSES AND HEAT RESISTANT GLOVES. Using a propane torch, heat the area on the base around the outside of the bearing until there is enough expansion to remove the base from the bearing on the crankshaft. Now remove and discard the seal retainer ring, seal retainer and seal.

c. To remove the bearing race, remove the retainer ring on the crankshaft with snap ring pliers, and with the use of a bearing splitter or arbor press, remove the ball bearing.

----------------- CAUTION -----------------

Support the crankshaft's top counterweight to prevent bending. Also, bearing is to be pressed on via the inner race only.

On other engines:

When equipped with a sleeve or needle bearing, use a seal protector and lift the crankshaft out of the cylinder. Be careful not to lose the bearing needles.

When equipped with a ball bearing, use a mallet to strike the crankshaft on the P.T.O. end while holding the block in your hand.

9. In assembly, bear the following points in mind:

a. Use a ring compressor to install piston. Be careful not to allow the rings to catch on the recess for the head gasket. Use the old head gasket to take up the space in the recess. Do not force the piston into the cylinder, or damage to rings or piston could occur.

b. To install the ball bearing on the crankshaft, slide the bearing on the crankshaft and fit it on the shaft by tapping using a mallet and tool, part number 670258 or press the ball bearing on the crankshaft with an arbor press. Install the retainer ring.

c. To install the crankshaft with a ball bearing, heat the shroud base to expand the bearing seat and drop the ball bearing into the seat of the base shroud. Allow it to cool. Install a new seal retainer ring, seal retainer, and seal.

10. After the shroud base and flywheel are back in place, adjust the air gap between the coil core and flywheel as described above under "Magnet Air Gap Adjustment."

Connecting Rod Service

1. For engines using solid bronze or aluminum connecting rods, remove the two self-locking cap screws which hold the connecting rod to the crankshaft and remove the rod cap. Note the match marks on the connecting rod and cap. These marks must be reinstalled in the same position to the crankshaft.

2. Engines using steel connecting rods are equipped with needle bearings at both crankshaft and piston pin end. Remove the two set screws that hold the connecting rod and cap to the crankshaft, taking care not to lose the needle bearings during removal.

3. Needle bearings at the piston pin end of steel rods are caged and can be pressed out as an assembly if damaged.

4. Check the connecting rod for cracks or distortion. Check the bearing surfaces for scoring or wear. Bearing diameters should be within the limits indicated in the table of specifications located at the end of this section.

Connecting rod match marks

5. There are two basic arrangements of needles supplied with the connecting rod crankshaft bearing: split rows of needles and a single row of needles. Service needles are supplied with a beeswax coating. The beeswax holds the needles in position.

6. To install the needle bearings, first make sure that the crankshaft bearing journal and the connecting rod are free from oil and dirt.

7. Place the needle bearings with the beeswax onto a cool metallic surface to stiffen the beeswax. Body temperature will melt the wax, so avoid handling.

8. Remove the paper backing on the bearings and wrap the needles around the crankshaft journal. The beeswax will hold the needles onto the journal. Position the needles uniformly onto the crankpin.

NOTE: When installing the split row of needles, wrap each row of needles around the journal and try to seal them together with gentle but firm pressure to keep the bearings from unwinding.

9. Place the connecting rod onto the journal, position the rod cap, and secure it with the capscrews. Tighten the screw to the proper specifications.

10. Force solvent (lacquer thinner) into the needles just installed to remove the beeswax, then force 30W oil into the needles for proper lubrication.

Piston and Rings Service

1. Clean all carbon from the piston and ring grooves.

2. Check the piston for scoring or other damage.

3. Check the fit of the piston in the cylinder bore. Move the piston from side-to-side to check clearance. If the clearance is not greater than 0.0003 in. and the cylinder is not scored or damaged, then the piston need not be replaced.

4. Check the piston ring side clearance to make sure it is within the limits recommended.

5. Check the piston rings for wear by inserting them into the cylinder about 1/2 in. from the top of the cylinder. Check at various places to make sure that the gap between the ends of the ring does not exceed the dimensions recommended in the specifications table at the end of this section. Bore wear can be checked in the same way, except that a new ring is used to measure the end gap.

6. If replacement rings have a bevelled or chamfered edge, install them with the bevel up toward the top of the piston. Not all engines use bevelled rings. The two rings installed on the piston are identical.

7. When installed, the offset piston used on the AV600 and the AV520 engines must have the **V** stamped in the piston head

A. SPLIT ROWS OF NEEDLE BEARINGS

B. SINGLE ROW OF NEEDLE BEARINGS

Needle bearing arrangement. Double rows of bearings are placed with the tapered edges facing out

(some have hash marks) facing toward the right as the engine is viewed from the top or piston side of the engine.

NOTE: Some AV520 and AV600 engines do not have offset pistons. Only offset pistons will have the "V" or the hash marks on the piston head. Domed pistons must be installed so that the slope of the piston is toward the exhaust port.

Crankshaft Service

1. Use a micrometer to check the bearing journals for out-of-roundness. The main bearing journals should not be more than 0.0005 in. out-of-round. Connecting rod journals should not be more than 0.001 in. out-of-round. Replace a crankshaft that is not within these limits.

NOTE: Do not attempt to regrind the crankshaft since undersize parts are not available.

2. Check the tapered portion of the crankshaft (magneto end), keyways, and threads. Damaged threads may be restored with a thread die. If the taper of the shaft is rusty, it indicates that the engine has been operating with a loose flywheel. Clean the rust off the taper and check for wear. If the taper or keyway is worn, replace the crankshaft.

Checking the crankshaft for out-of-roundness

3. Check all of the bearing journal diameters. They should be within the limits indicated in the specifications table at the end of this section.

4. Check the crankshaft for bends by placing it between two pivot points. Position the dial indicator feeler on the crankshaft bearing surface and rotate the shaft. The crankshaft should not be more than 0.002-0.004 in. out-of-round.

Bearing Service

1. Do not remove the bearings unless they are worn or noisy. Check the operation of the bearings by rotating the bearing cones with your fingers to check for roughness, binding, or any other signs of unsatisfactory operation. If the bearings do not operate smoothly, remove them.

2. To remove the bearings from the crankcase, the crankcase must be heated. Use a hot plate to heat the crankcase to no more than 400°F. Place a $\frac{1}{8}$ in. steel plate over the hot plate to prevent overheating. At this temperature, the bearings should drop out with a little tapping of the crankcase.

3. The replacement bearing is left at room temperature and dropped into the heated crankcase. Make sure that the new bearing is seated to the maximum depth of the cavity.

NOTE: Do not use an open flame to heat the crankcase halves and do not heat the crankcase halves to mmore than 400°F. Uneven heating with an open flame or excessive temperature will distort the case.

4. The needle bearings will fall out of the bearing cage with very little urging. Needles can be reinstalled easily by using a small amount of all-purpose grease to hold the bearings in place.

5. Cage bearings are removed and replaced in the same manner as ball bearings.

6. Sleeve bearings cannot be replaced. Both crankcase halves must be discarded if a bearing is worn excessively.

Assembly

1. The gasket surface where the crankcase halves join must be thoroughly clean before reassembly. Do not buff or use a file or any other abrasive that might damage the mating surfaces.

NOTE: Crankcase halves are matched. If one needs to be replaced, then both must be replaced.

2. Place the PTO half of the crankcase onto the PTO end of the crankshaft. Use seal protectors where necessary.

3. Apply a thin coating of sealing compound to the contact surface of one of the crankcase halves.

4. Position one crankcase half on the other. The fit should be such that some pressure is required to bring the two halves together. If this is not the case, either the crankcase halves and/or the crankshaft must be replaced.

5. Secure the halves with the screws provided, tightening the screws alternately and evenly. Before tightening the screws, check the union of the crankcase halves on the cylinder mounting side. The halves should be flat and smooth at the union to provide a good mounting face for the cylinder. If necessary, realign the halves before tightening the screws.

6. The sleeve tool should be placed into the crankcase bore from the direction opposite the crankshaft. Insert the tapered end of the crankshaft through the half of the crankcase to which the magneto stator is mounted. Remove the seal tools after installing the crankcase halves.

7. Stagger the ring ends on the piston and check for the correct positioning on domed piston models.

8. Place the cylinder gasket on the crankcase end.

9. Place the piston into the cylinder using the chamfer provided on the bottom edge of the cylinder to compress the rings.

10. Secure the cylinder to the crankcase assembly.

Type Number-to-Letter Cross Reference Chart

Type No.	Column Letter	Type No.	Column Letter	Type No.	Column Letter
1 thru 44	A	611	F	1001	D
46 thru 68	B	614–01 thru 614–04	E	1002 thru 1002D	G
69 thru 77	D	614–05 thru 614–06A	F	1003	C
78 thru 80	C	615–01 thru 615–01A	E	1004 thru 1004A	H
81 thru 83	D	615–04	F	1005 thru 1005A	C
84 thru 85	C	615–05 thru 615–10A	E	1006	H
86 thru 87	B	615–11 thru 615–18	F	1007 thru 1008B	C
89	C	615–19 thru 615–27A	E	1009 thru 1010B	G
91	D	615–28	F	1011 thru 1019E	H
93	D	615–29 thru 615–39	E	1020	G
94 thru 98	H	616–01 thru 616–11	E	1021	C
99	C	616–12 and 616–14	F	1022 thru 1022C	D
		616–16 thru 616–40	E	1023 thru 1026A	H
201 thru 208	A	617A thru 617–01	E	1027 thru 1027B	C
209 thru 244	B	617–02 thru 617–03A	F	1028 thru 1030C	H
245	D	617–04 thru 617–04A	E	1031B thru 1031C	G
246 thru 248	B	617–05 thru 617–05A	F	1032	C
249 thru 251	D	617–06	E	1033 thru 1033F	G
252	B	618 thru 618–16	E	1034 thru 1034G	H
253 thru 255	D	618–17	F	1035 thru 1036	C
256 thru 261	B	618–18	E	1037 thru 1039C	D
262 thru 265A	D	619 thru 619–03	E	1040 thru 1041A	D
266 thru 267	B	621 thru 621–11	E	1042 thru 1042C	G
268 thru 272	D	621 thru 621–12A	F	1042D thru 1042E	H
273 thru 275	B	621–13 thru 621–15	E	1042F	G
276	D	622 thru 622–07	E	1042G and 1042H	H
277 thru 279	B	623 and 623A	E	1042I thru 1043B	G
280 thru 281	D	623–01	F	1043C thru 1043F	H
282	B	623–02 thru 623–33E	E	1043G thru 1044E	G
283 thru 284	D	623–34	F	1045 thru 1045F	B
285 thru 286	B	623–35 and 623–36	E	1046 thru 1051B	H
287	D	624 thru 630–09	J	1052	G
288 thru 291B	B	632–02A	K	1053 thru 1054C	H
292 thru 294	D	632–04 and 632–05A	K	1055 thru 1056B	G
295 thru 296	B	633 thru 634–09	J	1057	H
297	F	635A and B	K	1058 thru 1058A	G
298 thru 299	B	635–03B	K	1059	D
		635–04B	K	1060 thru 1060C	H
301 thru 375		635–06A thru 635–10	K	1061 and 1062	C
These are twin		636 thru 636–11	J	1063A thru 1063C	H
cylinder units–		637 thru 637–16	J	1064	G
out of production		638 thru 638–100	F	1065 thru 1067C	D
401 thru 402	L	639	M	1068 and 1068A	G
403	M	640	O	1069	H
404 thru 406B	L	641	K	1070 thru 1070B	C
407 thru 407C	M	642	I	1071 thru 1075D	H
408 thru 410B	L	643	J	1076 thru 1084	H
411 thru 411B	M	650	N	1085 and 1085A	G
412 thru 423	L	670	H	1086 thru 1140A	H
424 thru 427	M				
				1141 thru 1142	G
501 thru 509A	E	701 thru 701–1C	G	1143 thru 1145	H
		701–2 thru 701–2D	H	1146 thru 1149A	I
601 thru 601–01	F	701–3 thru 701–3C	G	1149A thru 1153B	H
602 thru 602–02B	F	701–4 thru 701–4C	H	1158 thru 1159	G
603A thru 603–23	E	701–5 thru 701–5C	G	1160 thru 1161A	I
604 thru 604–25	E	701–6 thru 701–6C	H	1162 and 1162A	G
605 thru 605–18	E	701–7 thru 701–7C	G	1163 thru 1163A	H
606 thru 606–05B	E	701–8 thru 701–11	H	1165 thru 1168A	G
606–06 thru 606–07	F	701–12 thru 701–13A	G	1169 thru 1174	H
606–08 thru 606–13	E	701–14 thru 701–17	H	1175 and 1175A	I
608 thru 608–C	E	701–18 thru 701–19	G	1176 thru 1177A	H
610–A thru 610–14	E	701–19A	H	1179 thru 1179A	I
610–15 thru 610–16	F	701–20 thru 701–22A	G	1180 thru 1181	G
610–19 thru 610–20	E	701–24	H	1182 and 1182A	I

TECUMSEH-LAUSON 2-STROKE
SMALL ENGINES — 0–6 HP

Type Number-to-Letter Cross Reference Chart (cont.)

Type No.	Column Letter	Type No.	Column Letter	Type No.	Column Letter
1183 thru 1185	G	1345	J	1441	P
1186A thru 1186C	J	1348 and 1348A	G	1442 thru 1442B	G
1187 thru 1192	G	1350 thru 1350C	I	1443	I
1192A thru 1196B	J			1444 and 1444A	G
1197 thru 1197A	G	1550A	P	1445	I
1198 and 1198A	H			1446 and 1446A	A
1199 and 1199A	J	1351 thru 1351B	K	1447	G
		1352 thru 1352B	I	1448 thru 1450	F
1206 thru 1208B	J	1353	K	1450A thru 1450B	F
1210 thru 1215C	K	1354 thru 1355A	I	1450C thru 1450E	F
1216 thru 1216A	H	1356 and 1356B	K	1453	K
1217 thru 1220A	J	1357 thru 1358	I	1454 and 1454A	A
1221	H	1359 thru 1362B	K	1455 thru 1456	K
1222 thru 1223	J	1363 thru 1369A	G	1459	G
1224 thru 1225A	H	1372 thru 1375A	G	1460 thru 1460F	A
1226 thru 1227B	K	1376 and 1376A	J	1461	K
1228 thru 1229A	J	1377 thru 1378	K	1462	A
1230 thru 1231	H	1379 thru 1379A	H	1463	K
1232	G	1380 thru 1380B	K	1464 thru 1464B	L
1233 thru 1237B	K	1381	G	1465	A
1238 thru 1238C	H	1382	H	1466 thru 1466A	F
1239 thru 1243	J	1383 thru 1383B	K	1467 thru 1468	K
1244 thru 1245A	K	1384 thru 1385	G	1471 thru 1471B	E
1246 thru 1247	J	1386	I	1472 thru 1472C	L
1248	H	1387 thru 1388	K	1473 thru 1473B	A
1249 thru 1251A	J	1389	G	1474	L
1252 thru 1254A	K	1390 thru 1390B	H	1475 thru 1476	A
1255	J	1391	K	1477	G
1256 thru 1262B	K	1392	J	1478	J
1263	J	1393	G	1479	G
1264 thru 1265E	K	1394 thru 1395A	P	1482 and 1482A	F
1266 thru 1267	G	1396	K	1483	F
1269 thru 1270D	K	1397	H	1484 thru 1484D	A
1271 thru 1271B	G	1398 thru 1399	K	1485	G
1272 thru 1275	K			1486	D
1276	H	1400	K	1487	J
1277 thru 1279D	K	1401 thru 1401F	F	1488 thru 1488D	A
1280	J	1402 and 1402B	G	1489 thru 1490B	C
1283 thru 1284D	K	1403A and 1403B	K	1491	L
1286 thru 1286A	H	1404 and 1404A	K	1493 and 1493A	G
1287	K	1405 thru 1406A	H	1494 and 1495A	B
1288	J	1407 thru 1408	K	1496	G
1289 thru 1289A	G	1409A	J	1497	A
1290 thru 1293A	K	1410 thru 1412A	I	1498	E
1294 thru 1295	J	1413 thru 1416	I	1499	F
1296 thru 1298	K	1417 and 1417A	K		
		1418 and 1418A	P	1500	E
1300 thru 1303	K	1419 and 1419A	K	1501A thru 1501E	A
1304 thru 1307	J	1420 and 1420A	H	1503 thru 1503D	L
1308 thru 1316A	K	1421 and 1421A	K	1506 thru 1507	F
1317 thru 1317B	J	1422 thru 1423	P	1508	G
1318 thru 1320B	K	1424	K	1509	C
1321 thru 1322	J	1425	G	1510	L
1323	K	1426 thru 1426B	P	1511	C
1325	G	1427 and 1427A	H	1512 and 1512A	B
1326 thru 1326F	K	1428 thru 1429A	P	1513	L
1327 thru 1327B	H	1430A	G	1515 thru 1516C	C
1328B and 1328C	K	1431	P	1517	E
1329	J	1432 and 1432A	G	1518	D
1330	H	1433	K	1519 thru 1521	A
1331	J	1434 thru 1435A	P	1522	L
1332	J	1436 and 1436A	I	1523	A
1333	I	1437	J	1524	B
1334	I	1439	I	1527	C
1343 thru 1344A	H	1440 thru 1440D	A	1528	A

Type Number-to-Letter Cross Reference Chart (cont.)

Type No.	Column Letter	Type No.	Column Letter	Type No.	Column Letter
1529A and 1529B	C	2045 thru 2046B	F	40069 thru 40075	N
1530 thru 1530B	A	2047 thru 2048	D		
1531 thru 1535B	C	2049 thru 2049A	F	710101 thru 710116	E
1536	L	2050 thru 2050A	D	710124 thru 710130	E
1537	A	2051 thru 2052A	D	710131 thru 710137	E
1538 thru 1541A	L	2053 thru 2055	F	710138 thru 710149	E
1542	E	2056	D	710154	M
1543 thru 1546	A	2057	F	710201 thru 710209	E
1547	C	2058 thru 2058B	D	710210 thru 710218	E
1549	C	2059 thru 2063	F	710219 thru 710227	E
		2064 thru 2064A	D	710228	M
S–1801 thru 1822	F	2065	F	710150	G
1823 and 1824	E	2066	D	710151	H
		2067 thru 2071B	F	710155	L
2001 thru 2003	B			710157	H
2004 thru 2006	D	2200 thru 2201A	D	710229	H
2007 thru 2007B	B	2202 thru 2204	E	710230	N
2008 thru 2008B	D	2205 thru 2205A	D	710234	N
2009	B	2206 thru 2206A	G		
2010 thru 2011	D	2207	D		
2012	F	2208	G	200.183112	F
2013 thru 2014	B	2768	C	200.18322	F
2015 thru 2018	D			200.193132	F
2020	F			200.193142	F
2021	D	40001 thru 40028	N	200.193152	G
2022 thru 2022B	F	40029 thru 40032C	O	200.193162	G
2023 thru 2026	D	40033	N	200.203172	H
2027	B	40034 thru 40045A	O	200.203182	H
2028 thru 2029	D	40046	N	200.203192	H
2030 thru 2031	B	40047 thru 40052	O	200.213112	H
2032 thru 2033A	D	40053 thru 40054A	N	200.213122	H
2034 thru 2035C	F	40056 thru 40060B	O	200.503111	F
2036 thru 2037	B	40061 thru 40062B	N	200.583111	F
2038	F	40063 thru 40064	O	200.593121	F
2039 thru 2044	C	40065 thru 40066	N	200.613111	F
		40067 thru 40068	O		

Specifications for 2 Cycle Engines with Split Crankcases

	A	B	C	D	E	F	G	H
Bore	1.500 1.5005	1.6253 1.6258	1.7503 1.7508	1.7503 1.7508	2.000	2.000	2.000	2.000
Stroke	1.375	1.50	1.50	1.50	1.50	1.50	1.50	1.50
Displacement Cubic Inches	2.43	3.10	3.60	3.60	4.70	4.70	4.70	4.70
Point Gap	.020	.020	.020	.020	.020	.020	.015	.015
Timing B.T.D.C. Before Top Dead Center	1/16" or .0625	5/32" or .1562	5/32" or .1562	1/4" or .250	5/32" or .1562	5/32" or .1562	5/32"	1/4" or .250 11/64" for "super"
Spark Plug Gap	.030	.030	.030	.030	.030	.030	.030	.030
Piston Ring End Gap	.003 .008	.005 .010	.005 .010	.005 .010	.006 .011	.006 .011	.006 .011	.006 .011
Piston Diameter	1.4966 1.4969	1.6216 1.6219	1.7461 1.7464	1.7461 1.7464	1.9948 1.9951	1.9948 1.9951	1.9948 1.9951	1.9949 1.9955

Specifications for 2 Cycle Engines with Split Crankcases (cont.)

	A	B	C	D	E	F	G	H
Piston Ring Groove Width	.095 .096	.095 .096	.095 .096	.095 .096	.095 .096	.095 .096	.095 .096	.095 .096
Piston Ring Width	.093 .0935	.093 .0935	.093 .0935	.093 .0935	.093 .0935	.093 .0935	.093 .0935	.093 .0935
Piston Pin Diameter	.3750 .3751	.3750 .3751	.3750 .3751	.3750 .3751	.3750 .3751	.3750 .3751	.3750 .3751	Early .3750 .3761 Late .4997 .4999
Connecting Rod Diameter Crank Bearing	.6869 .6874	.6869 .6874	.6869 .6874	.6986 .6989 w/o needles	.6869 .6874	.6869 .6874	.6869 .6874	.6935 .6939 w/o needles
Crankshaft Rod Needle Diameter				.0653 .0655				.0653 .0655
Crank Pin Journal Diameter	.6860 .6865	.6860 .6865	.6860 .6865	.5615 .5618	.6860 .6865	.6860 .6865	.6860 .6865	.5615 .5618
Crankshaft P.T.O. Side Main Brg. Dia.	.6689 .6693	.6689 .6693	.6690 .6694	.6689 .6693	.9995 1.0000	.6689 .6693	.9995 1.0000	.9995 1.0000
Crankshaft Magneto Side Main Brg. Dia.	.6689 .6693	.6689 .6693	.6689 .6693	.6690 .6694	.7495 .7500	.6689 .6693	.7495 .7500	.7495 .7500
Crankshaft End Play	.003 .008	.003 .008	.003 .008	.003 .008	.009 .022	.009 .022	.009 .022	.009 .022

Specifications for 2 Cycle Engines with Split Crankcases

I	J	K	L	M	N	O	P	
2.000	2.000	2.093 2.094	2.2505 2.2510	2.2505 2.2510	2.5030 2.5035	2.5030 2.5035	2.000	Bore
1.50	1.625	1.63	2.00	2.00	1.625	1.680	1.50	Stroke
4.70	5.10	5.80	8.00	8.00	7.98	8.25	4.70	Displacement Cubic Inches
.015	.020	.015	.020	.020	.020	.020	.015	Point Gap
11/64" or .175	11/64" or .175	3/32" or .095	1/8" or .125	3/32" or .095	11/64" or .175	11/64" or .175	.90	Timing B.T.D.C. Before Top Dead Center
.035	.030	.030	.030	.030	.030	.030	.035	Spark Plug Gap
.006 .011	.006 .011	.006 .011	.007 .015	.007 .015	.005 .013	.005 .013	.006 .011	Piston Ring End Gap
1.9948 1.9951	1.9951 1.9948	2.0880 2.0883	2.2460 2.2463	2.2460 2.2463	2.4960 2.4963	2.4960 2.4963	1.9948 1.9951	Piston Diameter
.095 .096	.095 .096	T.0655 .0665 L.0645 .0655	.095 .096	.095 .096	T.0655 .0665 L.0645 .0655	T.0655 .0665 L.0645 .0655	.095 .096	Piston Ring Groove Width

Specifications for 2 Cycle Engines with Split Crankcases (cont.)

I	J	K	L	M	N	O	P	
.093 .0935	.093 .0935	.0615 .0625	.093 .0935	.093 .0935	.0615 .0625	.0615 .0625	.093 .0935	Piston Ring Width
.4997 .4999	.3750 .3751	.4997 .4999	.5000 .5001	.3000 .5001	.4997 .4999	.4997 .4999	.4997 .4999	Piston Pin Diameter
.6941 .6944	.6869 .6874	.9407 .9412	1.000 1.0004	1.000 1.0004	.9407 .9412	.9407 .9412	.6941 .6944	Connecting Rod Diameter Crank Bearing
.0653 .0655		.0943 .0945	.0943 .0945	.0943 .0945	.0943 .0945	.0943 .0945	.0653 .0655	Crankshaft Rod Needle Diameter
.6860 .6865	.6860	.7499 .7502	.8096 .8099	.8096 .8099	.7499 .7502	.7499 .7502	.6860 .6865	Crank Pin Journal Diameter
.9995 1.0000	.9995 1.0000	.6990 .6994	.9839 .9842	.9839 .9842	.7871 .7875	.7871 .7875	.9995 1.0000	Crankshaft P.T.O. Side Main Brg. Dia.
.7495 .7500	.7495 .7500	.6990 .6994	.9839 .9842	.9839 .9842	.7498 .7501	.7498 .7501	.7495 .7500	Crankshaft Magneto Side Main Brg. Dia.
.009 .022	.009 .022	.003 .008	.003 .008	.003 .008	.008 .013	.008 .013	.009 .022	Crankshaft End Play

Uniblock Cross Reference Chart

Type No.	Column No.	Type No.	Column No.	Type No.	Column No.
Vertical Crankshaft Engines		**Horizontal Crankshaft Engines**		**Horizontal Crankshaft Engines**	
638 thru 638–100	6	1401 thru 1401F	16	1508	7
		1401G. H	17	1509	3
642–01. A	9A	1401J	27	1510	12
642–02, A, B, C, D	9A	1402 and 1402B	7	1511	3
642–02E. F	9B	1425	7	1512 and 1512A	2
642–03, A, B	9A	1430A	7	1513	12
642–04. A, B, C	9A	1432 and 1432A	7	1515 thru 1516C	3
642–05, A, B	9A	1440 thru 1440D	1	1517	5
642–06. A	9A	1442 thru 1442B	7	1518	4
642–07. A, B	9A	1444 and 1444A	7	1519 thru 1521	1
		1448 thru 1450	16	1522	12
642–07C	9B	1450A thru 1450B	16	1523	1
642–08	9B	1450C thru 1450E	16	1524	2
642–08A, B	9A	1450F	17	1525A	16
642–09 thru 642–14	9A	1454 and 1454A	1	1527	3
642–13A. 14A. 14B	9B	1459	7	1528	1
642–15 thru 642–22	9B	1460 thru 1460F	1	1529A and 1520B	3
642–24 thru 642–30	9C	1462	1	1530 thru 1530B	1
		1464 thru 1464B	12	1531	3
670–01 thru 670–101	8	1465	1	1534A	17
		1466 thru 1466A	16	1535B	3
200–183112	6	1471 thru 1471B	5	1536	12
200–183122	6	1472 thru 1472C	12	1537	1
200–193132	6	1473 thru 1473B	1	1538 thru 1541A	12
200–193142	6	1474	12	1542	5
200–193152	7	1475 thru 1476	1	1543 thru 1546	1
200–193162	7	1479	7	1517	3
200–203172	8	1482 and 1482A	16	1519	3
200–203182	8	1483	16	1551	16
200–203192	8	1484 thru 1484D	3	1552	20

Uniblock Cross Reference Chart

Type No.	Column No.	Type No.	Column No.	Type No.	Column No.
Vertical Crankshaft Engines		**Horizontal Crankshaft Engines**		**Horizontal Crankshaft Engines**	
200–213112	8	1485	7	1553	16
200–213122	8	1486	4	1554 and 1554A	3
200–243112	8	1488 thru 1488D	1	1555 and 1556	16
200–283012	8	1489 thru 1490B	3	1561	19
		1491	12	1572	2
		1493 and 1493A	7	1573	3
		1494 and 1495A	2	1574 thru 1577	23
		1496	7	1575	24
		1497	1	1578	25
		1498	5	1581 thru 1582A	23
		1499	16	1583 thru 1595	26
		1500	5	200–503111	16
		1501A thru 1501E	1	200–583111	16
		1503 thru 1503D	12	200–593121	16
		1506	16	200–613111	16
		1506B	17	200–672102	26
		1507	16	200–682102	26

Torque Specifications for 2-Cycle Engines

Application	Torque	Application	Torque
Carburetor and Reed Plates		Cylinder head to cylinder (635 type engine)	45–50 in. pounds
Carburetor to crankcase, Carburetor to adapter, or Carburetor adapter to crankcase	70–75 in. pounds	Cable clip and transfer port cover to cylinder	25–30 in. pounds
Carburetor to snow blowers cover	30–35 in. pounds	Cable clip to cylinder Stop lever to cylinder	25–30 in. pounds
Carburetor outlet fitting	40 in. pounds	Spark plug	18–22 ft. pounds
Reed and cover plates 639 type engines	50–60 in. pounds with Loctite, type A	Transfer cover cross port engines	25–30 in. pounds
Reed to plate 635 type engines	12–18 in. pounds	**Crankshaft and Connecting Rods**	
Crankcase and Cylinder		Aluminum and bronze rods to rod cap	40–50 in. pounds
Crankcase to crankcase cover	23–30 in. pounds	Steel rod to rod cap	70–80 in. pounds
Crankcase to crankcase cover screws	35–40 in. pounds	Flywheel nut on tapered end of crankshaft: Aluminum hub on iron or steel shaft.	18–25 ft. pounds
Mounting cylinder or carburetor to crankcase studs	50 in. pounds	Steel hub flywheel on iron shaft	18–25 ft. pounds
Base to crankcase	240–250 in. pounds	Steel hub flywheel on steel shaft	30 ft. pounds
Cylinder to crankcase nuts	70–75 in. pounds	**Governor and Bell Crank Parts**	
In cylinder	100–110 in. pounds	Power take-off end governor lower ring to crankshaft setscrew	30–35 in. pounds
Spark plug stop lever and head shroud to head	50–60 in. pounds	Bell crank bracket to crankcase	20–25 in. pounds
Cylinder head to cylinder	30–40 in. pounds	Governor cover to crankcase	50–60 in. pounds
Cylinder head to cylinder	50–60 in. pounds 80–90 in. pounds		

Specifications Chart

	1	2	3	4	5	6	7	8	9A	9B	9C	12
Bore	2.093 2.094	2.093 2.094	2.093 2.094	2.093 2.094	2.093 2.094	2.093 2.094	2.093 2.094	2.093 2.094	2.093 2.094	2.093 2.094	2.093 2.094	2.093 2.094
Stroke	1.250	1.410	1.410	1.410	1.410	1.500	1.500	1.500	1.500	1.500	1.500	1.410
Cu. In. Displacement	4.40	4.80	4.80	4.80	4.80	5.20	5.20	5.20	5.20	5.20	5.20	4.80
Point Gap	.017	.017	.017	.017	.017	.018	.017	.020	.018	.020	.020	.017
Timing B.T.D.C.	.122″	.100″	.135″	.100″	.135″	.100″	.185″	.070″	.110″	.085″ See Note 1	.078″ See Note 2	.135″
Spark Plug Gap	.035	.035	.035	.035	.035	.035	.035	.035	.035	.035	.035	.035
Piston Ring End Gap	.007 .017	.007 .017	.006 .011	.006 .014	.006 .011	.006 .014	.007 .017	.006 .016	.007 .017	.006 .016	.006 .016	.007 .017
Piston Diameter	2.0080 2.0870	2.0880 2.0870	2.0885 2.0875	2.0885 2.0875	2.0885 2.0875	2.0880 2.0870	2.0880 2.0870	2.0880 2.0870	2.0880 2.0870	2.0880 2.0820	2.0880 2.0870	2.0880 2.0870
Piston Ring Groove Width (Top)	.0655 .0665	.0655 .0665	.0655 .0665	.0975 .0985	.0655 .0665	.0975 .0985	.0655 .0665	.0655 .0665	.0655 .0665	.0655 .0665	.0655 .0665	.0655 .0665
Piston Ring Groove Width (Bot.)	.0645 .0655	.0645 .0655	.0645 .0655	.0955 .0965	.0645 .0655	.0955 .0965	.0645 .0655	.0645 .0655	.0645 .0655	.0645 .0655	.0645 .0655	.0645 .0655
Piston Ring Width	.0625 .0615	.0625 .0615	.0625 .0615	.0925 .0935	.0625 .0615	.0925 .0935	.0625 .0615	.0625 .0615	.0625 .0615	.0625 .0615	.0625 .0615	.0625 .0615
Piston Pin Diameter	.4999 .4997	.4999 .4997	.4999 .4997	.3750 .3751	.4999 .4997	.3750 .3751	.4999 .4997	.4999 .4997	.4999 .4997	.4999 .4997	.4999 .4997	.4999 .4997
Connecting Rod Diameter Crank Bearing	—	—	—	.6886 .6879 Dowels	—	—	—	1.0053 1.0023 Liner Dia.	—	1.0053 1.0023 Liner Dia.	1.0053 1.0023 Liner Dia	—
Crankshaft Rod Needle Dia.	.0655 .0653	.0655 .0653	.0655 .0653	—	.0655 .0653	—	.0655 .0653	.0781 .0780	—	.0781 .0780	.0781 .0780	.0655 .0653
Crank Pin Journal Diameter	.5618 .5611	.5621 .5614	.5621 .5614	.6865 .6857	.5618 .5611	.6865 .6857	.5618 .5611	.8450 .8442	.6865 .6857	.8450 .8442	.8450 .8442	.5621 .5614
Crankshaft P.T.O. Side Main Brg. Dia.	.6695 .6691	.6695 .6691	.6695 .6691	.6695 .6691	.6695 .6691	.8750 .8745	.6694 .6690	1.0003 .9998	.8750 .8745	1.0003 .9998	1.0003 .9998	.6695 .6691
Crankshaft Magneto Side Main Brg. Dia.	.6695 .6691	.6695 .6691	.6695 .6691	.6695 .6691	.6695 .6691	.7500 .7495	.6694 .6690	.6695 .6691	.7500 .7495	.7503 .7498	.6695 .6691	.6695 .6691
Crankshaft End Play	None	None	None	None	None	.003 .016	None	None	.003 .016	.003 .016	None	None

NOTE 1: 642-08 14A, 14B B.T.D.C. = .110″
       642-16D, 19A, 20A, 21 22 B.T.D.C. = .078″
NOTE 2: 642-24, 26, 29 B.T.D.C. = .085″
       643-2a, 25, 26 B.T.D.C. = .085″
NOTE 3: 643-13 B.T.D.C. = .095″
NOTE 4: 643-03A, 05A, 13, 14 = .020″

TECUMSEH-LAUSON 2-STROKE
SMALL ENGINES – 0–6 HP

Specifications Chart

	14	16	17	19	20	23	24	25	26	27
Bore	2.093 2.094	2.093 2.094	2.093 2.094	2.093 2.094	2.093 2.094	2.093 2.094	2.093 2.094	2.093 2.094	2.093 2.094	2.093 2.094
Stroke	1.500	1.500	1.500	1.410	1.250	1.500	1.410	1.410	1.500	1.500
Cu. In. Displacement	5.20	5.20	5.20	4.80	4.40	5.20	4.80	4.80	5.20	5.20
Point Gap	.018	.017	.017	.017	.017	.017	.017	.020	.020	.017
Timing B.T.D.C.	.100"	.110"	.110"	.100"	.122"	.110"	.135"	Fixed	.062"	.100"
Spark Plug Gap	.035	.035	.035	.035	.035	.035	.035	.035	.035	.035
Piston Ring End Gap	.006 .014	.006 .016	.006 .016	.007 .017	.007 .017	.006 .016	.007 .017	.007 .017	.006 .016	.006 .016
Piston Diameter	2.0880 2.0870	2.0885 2.0875	2.0890 2.0880	2.0880 2.0870	2.0880 2.0870	2.0880 2.0870	2.0880 2.0870	2.0880 2.0870	2.0880 2.0870	2.0885 2.0875
Piston Ring Groove Width (Top)	.0975 .0985	.0645 .0655	.0645 .0655	.0655 .0665	.0655 .0665	.0655 .0665	.0655 .0665	.0655 .0665	.0655 .0665	.0655 .0665
Piston Ring Groove Width (Bot.)	.0955 .0965	.0645 .0655	.0645 .0655	.0645 .0655	.0645 .0655	.0645 .0655	.0645 .0655	.0645 .0655	.0645 .0655	.0645 .0655
Piston Ring Width	.0925 .0935	.0625 .0615	.0625 .0615	.0625 .0615	.0625 .0615	.0625 .0615	.0625 .0615	.0625 .0615	.0625 .0615	.0625 .0615
Piston Pin Diameter	.3750 .3751	.3750 .3751	.4999 .4997	.4999 .4997	.4999 .4997	.4999 .4997	.4999 .4997	.4999 .4997	.4999 .4997	.4999 .4997
Connecting Rod Diameter Crank Bearing	—	—	—	—	—	.8534 .8504 Liner Dia.	—	—	.8534 .8504 Liner Dia.	.8919 .8924
Crankshaft Rod Needle Dia.	—	—	—	.0655 .0653	.0655 .0653	.0781 .0780	.0655 .0653	.0655 .0653	.0781 .0780	.0781 .0780
Crank Pin Journal Diameter	.6965 .6857	.6868 .6857	.6865 .6857	.5621 .5614	.5618 .5611	.6927 .6919	.5621 .5614	.5621 .5614	.6927 .6919	.6927 .6922
Crankshaft P.T.O. Side Main Brg. Dia.	.8750 .8745	.6695 .6691	1.0003 .9998	.6695 .6691	.6695 .6691	.6695 .6691	.6695 .6691	.6695 .6691	.7503 .7498	.6695 .6691
Crankshaft Magneto Side Main Brg. Dia.	.7500 .7495	.7500 .7495	.7500 .7495	.6695 .6691	.6695 .6691	.7503 .7498	.6695 .6691	.6695 .6691	.6695 .6691	.7503 .7498
Crankshaft End Play	.003 .016	None	None	None	None	None	None	None	None	.003 .016

A. on Engines prior to 660-24: 7505, 7198

Section 18

Tecumseh-Lauson 2-Stroke

Small Engines — 6-20 hp

ENGINE IDENTIFICATION

The identification tags may be located at a variety of places on the engine. The type number is the most important number since it must be included with any correspondence about a particular engine.

Early engines listed the type number as a suffix of the serial number. For example on number 123456789 P 234, 234 is the type number. In the number 123456789 H 104–02B; 104–02B is the type number. In either case the type number is important.

If you use short block to repair the engine, be sure that you transfer the serial number and type number tag to the new short block.

On the newer engines, reference is sometimes made to the model number. The model number tells the number of cylinders, the design (vertical or horizontal) and the cubic inch displacement.

GENERAL ENGINE SPECIFICATIONS

A detailed listing of the great number of Power Products two-stroke engine models is provided in the Type No./Column No. cross reference chart at the back of this section.

MAINTENANCE

Air Cleaners

The instructions below detail the procedures involved in cleaning the various types of elements. See the illustrations for exploded views to aid disassembly and assembly.

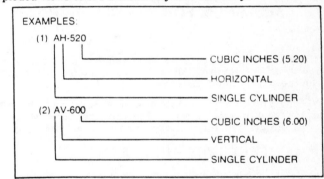

EXAMPLES:

(1) AH-520
- CUBIC INCHES (5.20)
- HORIZONTAL
- SINGLE CYLINDER

(2) AV-600
- CUBIC INCHES (6.00)
- VERTICAL
- SINGLE CYLINDER

Model number interpretation

A. NAMEPLATE ON AIR SHROUD
B. MODEL & TYPE NUMBER PLATE
C. METAL TAG ON CRANKCASE
D. STAMPED ON CRANKCASE
E. STAMPED ON CYLINDER FLANGE
F. STAMPED ON STARTER PULLEY

Location of identification numbers on 2 cycle engines

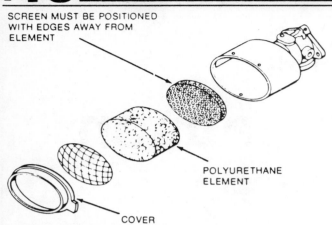

SCREEN MUST BE POSITIONED
WITH EDGES AWAY FROM
ELEMENT

POLYURETHANE
ELEMENT

COVER

Exploded view of polyurethane air cleaner

POLYURETHANE AIR CLEANER

1. Wash the element in a solvent or detergent and water solution by squeezing similar to a sponge.
2. Clean the air cleaner housing and cover with the same solution. Dry thoroughly.
3. Dry the element by squeezing or with compressed air if available.
4. Apply a generous quantity of oil to the element sides and open ends. Squeeze vigorously to distribute oil and to remove excess oil.

ALUMINUM FOIL AIR CLEANER

1. Dip the aluminum foil filter in solvent. Flush out all dirt particles.
2. Shake out the filter thoroughly to remove all solvent, then dip the filter element in oil. Allow the oil to drain from the filter. Clean the screens and filter body.

NOTE: The concave screen and retainer cover or ring are not used on later models. They are replaced with a clip which rolls into the groove in the lip of the body.

FELT TYPE AIR CLEANERS

1. To clean felt air cleaners, merely blow compressed air through the element in the reverse direction to normal air flow. Felt elements may also be washed in non-flammable solvents or soapy water. Blow dry with compressed air.

NOTE: Power Products type numbers 641 and up, use a gasket between the element and the base. The gasket is used only with this element; earlier versions did not have a gasket.

BODY FLAT SCREEN FILTER ELEMENT SCREEN CLIP

Exploded view of aluminum foil type air cleaner

NOTE

THIS GASKET USED ONLY ON POWER PRODUCTS 641 SERIES AND UP WITH A #450226 ELEMENT

GASKET

Exploded view of felt type air cleaner

FIBER ELEMENT AIR CLEANER

1. Remove the filter and place the cover in a normal position on the filter. With the filter element down to semi-seal it, blow compressed air through the cover hole to reverse air flow, forcing dirt particles out.
2. Clean the cover mounting bracket with a damp cloth.

DRY PAPER AIR CLEANER

1. Tap the element on a workbench or any solid object to dislodge larger particles of dirt.
2. Wash the element in soap and water. Rinse from the inside until it is thoroughly flushed and the water coming through is free of soap.
3. Allow the element to dry completely or use low pressure compressed air blown from the inside to speed the process.
4. Inspect the element for cracks or holes, and replace if necessary.

PAPER ELEMENT

Exploded view of paper element type air cleaner

Lubrication

OIL AND FUEL RECOMMENDATIONS

Power Products 2 cycle engines are mist-lubricated by oil mixed with the gasoline. For the best performance, use regular grade, leaded fuel, with 2 cycle or outboard oil rated SAE 30 or SAE 40. Regular grade unleaded fuel is an acceptable substitute. The terms 2 cycle or outboard are used by various manufacturers to designate oil they have designed for use in 2 cycle engines. Multiple weight oil such as all season 10W–30, are not recommended.

If you have to mix the gas and oil when the temperature is below 35°F, heat up the oil first, then mix it with the gas. Oil will not mix with gas when the temperature is approaching freezing. However, if you use oil that has been warmed first it will not be affected by low temperatures.

The proportion of oil to fuel is absolutely critical to two-stroke operation. If too little oil is used, overheating and damage to engine parts will occur (this can even result from running the engine too lean). If excessive oil is used, spark plug fouling, smoke in the exhaust, and even misfire can occur. Mix carefully and precisely. Follow Power Products recommendations for your particular engine, and disregard fuel container labels.

Fuel/oil mix must be clean and fresh. Fuel deteriorates enough to form troublesome gum and varnish after more than a month. Dirt in the fuel can cause clogging of carburetor passages and even engine wear.

TUNE-UP
Specifications

All spark plugs are gapped at 0.035 in. Because of the great number of individual models of Power Products engines that exist, an individual chart of Tune-Up specifications is impractical. Refer to the charts at the end of this section for breaker point gap and timing dimension specifications.

Spark Plugs

Spark plugs should be removed and cleaned of deposits frequently, especially in two-stroke engines because they burn the lubricating oil right with the fuel. Carefully inspect the plug for severely eroded electrodes or a cracked insulator, and replace the plug if either condition exists or if deposits cannot be adequately removed. Set the gap to 0.035 in. with a wire type feeler gauge, and install the plug, torquing it to 18–22 ft. lbs.

Make sure to replace the plug with one of the same type and heat range. If the plug is fouled, poor quality or old fuel, a rich mixture, or the wrong fuel/oil mix may be at fault. Also, make sure the engine's exhaust ports are not clogged.

Breaker Points

REMOVAL AND INSTALLATION

1. First remove the flywheel as described below:
   a. Remove the screws, engine shroud, and starter. Determine the direction of rotation of the flywheel nut by looking at the threads. Then, place a box wrench on the nut and tap with a soft hammer in the proper direction.
   b. The flywheel is removed with a special puller or a special knock off tool. The knock off tool must not be used on 660, 670, or 1500 ball bearing models. To use the knock off tool, screw it onto the crankshaft until it is within 1/16 in. of the flywheel. Hold the flywheel firmly and rap the top of the puller sharply with a hammer to jar it loose. Pull the flywheel off.
   c. If a puller is being used, the flywheel will have three cored holes into which a set of self-tapping screws are turned. The handle which operates the bolt at the center of the puller's collar is then turned to pull the flywheel off the crankshaft. If this is not adequate to do the job, heat the center of the alloy flywheel with a butane torch to expand it before turning the puller handle.
2. Remove the nuts that hold the electrical leads to the screw on the movable breaker point spring. Remove the movable breaker point from the stud.
3. Remove the screw and stationary breaker point. Put a new stationary breaker point on breaker plate; install the screw, but do not tighten it fully.
4. Position a new movable breaker point on the post.

POINT GAP

Adjusting point gap

5. Check that the new points contact each other properly and remove all grease, fingerprints, and dirt from the points.

ADJUSTMENT

1. If necessary (as when checking the gap of old points), loosen the screw which mounts the stationary breaker contact. Rotate the crankshaft until the contact cam follower rests right on the highest point of the cam.
2. Using a flat feeler gauge of the dimension shown under "point gap" in the charts at the rear of this section, check the dimension of the gap and, if incorrect, move the breaker base in the appropriate direction by wedging a screwdriver between the dimples on the base plate and the notch in the breaker plate. When gap is correct, tighten the screw. Recheck the gap and, if necessary, reset it.

Adjusting Ignition Timing

1. Remove the spark plug and install a special timing tool or thin ruler. With the tool lockscrew loose or the ruler riding on the piston, rotate the crankshaft back and forth to find Top Dead Center. Tighten the tool lockscrew or use a straightedge across the cylinder head, if you're using a ruler, to measure Top Center. Look up the timing dimension in the specifications at the back of this section.
2. Turn the crankshaft backwards so the piston descends. Then, reset the position of the special tool downward the amount of the dimension and tighten the lockscrew, or move the ruler down that amount.
3. Very carefully bring the piston upward by turning the crankshaft until the top of the piston just touches the tool or ruler.
4. Install a piece of cellophane between the contact surfaces.

JUST AS POINTS OPEN TIGHTEN STATOR BOLTS. BE SURE POINTS ARE CLEAN AND ALIGNED

TURN STATOR INTO CRANKSHAFT ROTATION

IF CONTINUITY LIGHT IS USED, REMOVE WIRES

Adjusting ignition timing

Loosen the ignition stator lockscrews and turn the stator until the cellophane is clamped tightly between the contact surfaces. Turn the stator until the cellophane can just be pulled out (as contacts start to open), and then tighten the stator lockscrews.

Magneto Armature Air Gap

There are two types of magnetos. Compare appearance of your engine with each illustration to determine whether it looks like the type which employs a gap of 0.005–0.008 in. or the type with a gap of 0.015 in.

1. Loosen the two screws which hold the laminations and coil to the block. Turn the flywheel around so the magnets line up directly with the ends of the laminations. Pull the laminations/coil assembly upward.

2. Insert an air gap gauge (Power Products Part #670216) or an equivalent (0.005–0.008 In.) gauge between the flywheel and laminations on either side. On units with a gap of 0.015 in., use two of the above numbered parts or a 0.015 in. gauge. Allow the attraction of the magnets to pull the coil/laminations assembly toward the flywheel.

Adjusting magneto armature air gap. Note magnets on flywheel

3. On the magneto type with a 0.005–0.008 in. dimension, use Loctite Grade A on the screws and torque to 35–45 inch lbs. On the other type magneto, torque the screws to 20–30 inch lbs. Recheck the gap and readjust, if necessary.

Mixture Adjustment

1. If the engine will not start or the carburetor has recently been disassembled, both idle and main mixture screws may be turned in very gently until they just bottom, and then turned out exactly one turn. In this case, back out the idle speed screw until throttle is free, then turn screw in until it just contacts the throttle. Turn it one turn more exactly.

2. Allow the engine to warm up to normal running temperature. With the engine running at maximum recommended rpm,

Adjusting magneto armature air gap. Note wider gap

loosen the main metering screw until the engine rolls, then tighten the screw until the engine starts to cut out. Note the number of turns from one extreme to the other. Loosen the screw to a point midway between the extremes.

3. Set the throttle to idle speed and repeat Step 2 for that adjusting screw.

NOTE: Some carburetors do not have a main mixture adjustment. Others employ a drilled mixture screw which provides about the right mixture when fully screwed in. In some weather conditions, operation may be improved by setting this screw just slightly off the seat for smoother running.

Governor Adjustments

POWER TAKEOFF AND MECHANICAL GOVERNOR

1. To adjust the governor, remove the outboard bearing housing. Use a $^3/_{32}$ in. allen wrench to loosen the set screw.

2. Squeeze the top and bottom governor rings, fully compressing the governor spring.

3. Hold the upper arm of the bell crank parallel to the crankshaft and insert a $^3/_{32}$ in. allen wrench between the upper ring and the bell crank.

4. Slide the governor assembly onto the crankshaft so that the allen wrench just touches the bell crank.

5. Tighten the set screw to secure the governor to the crankshaft.

6. Install the bearing adapter, mount the engine, and check the engine speed with a tachometer. It should be about 3200–3400 rpm.

Adjusting the PTO end mounted governor

NOTE: Never attempt to adjust the governor by bending the bellcrank or the link.

7. If the speed is not correct, readjust the governor by moving the assembly toward the crankcase to increase speed, and away from the crankcase to decreased speed.

MECHANICAL FLYWHEEL TYPE GOVERNOR ADJUSTMENT

As engine speed increases, the links are thrown outward, compressing the link springs. The links apply a thrust against the slide ring, moving it upward and compressing the governor spring. As the slide ring moves away from the thrust block of the bellcrank assembly, the throttle spring causes the thrust block to maintain engagement and close the throttle slightly.

As the throttle closes and engine speed decreases, force on the slide ring decreases so that it moves downward, pivots the bellcrank outward to overcome the force of the throttle spring, and opens the throttle to speed up the engine. In this manner, the operating speed of the engine is stabilized to the adjusted governor setting.

1. To adjust the governor, loosen the bracket screw and slide the governor bellcrank assembly toward or away from the flywheel. Move the bellcrank toward the flywheel to increase speed and away from the flywheel to decrease speed.
2. Tighten the screw to secure the bracket.
3. Make minor speed adjustments by bending the throttle link at the bend in the center of the link.

NOTE: Do not lubricate the governor assembly or the governor bellcrank assembly of flywheel mounted governors.

ADJUSTING 2 CYCLE AIR VANE GOVERNORS

1. Loosen the self-locking nut that holds the governor spring bracket to the engine crankcase.
2. Adjust the spring bracket to increase or decrease the governor spring tension. Increasing spring tension increases speed and decreasing spring tension decreases speed.
3. After adjusting, the spring bracket should not be closer than $\frac{1}{16}$ in. to the crankcase.
4. Tighten the self-locking nut.

Diagram of an air vane governor

Flywheel mounted governor assembly (cutaway view)

FUEL SYSTEM

Carburetor

REMOVAL AND INSTALLATION

1. Remove the air cleaner. Drain the fuel tank. Disconnect the carburetor fuel lines.
2. If necessary, remove any shrouding or control panels to gain access to the carburetor.
3. Disconnect the choke or throttle control wires at the carburetor.
4. Remove the cap screws, or nuts and lockwashers and remove the carburetor from the engine.
5. To install, reverse the removal procedure, using new gaskets.

OVERHAUL

Tecumseh two-and four-stroke engines employ a common series of carburetors. Refer to the Four-Stroke Tecumseh Engine section for specific carburetor overhaul procedures.

Idle Governor

SERVICE

1. Remove the shutter fastener and allow the shutter to drop out of the air horn.

Position detent reference marks of throttle shutter as shown

2. Note location of the spring end in the disc-shaped throttle lever. The spring should be placed into the same hole during reassembly.

3. Remove the retainer clip and lift out the throttle shaft.

4. Replace all worn parts and reassemble in reverse order.

NOTE: Note the position of the throttle shutter, as shown. The reference marks must be positioned as shown when shutter is installed.

Fuel Pump

SERVICE

Float Type Carburetor With Integral Pump

1. If the engine runs, but roughly, make both carburetor mixture adjustments.

2. Make sure the fuel supply is adequate and the tank is in the proper position.

3. Make sure the fuel tank is open.

4. Make sure the pick-up tube is not cracked.

5. Remove the carburetor and make sure the pulsation passage is properly aligned.

6. Check for air leaks at the gasket surface.

7. Remove the cover and check the condition of inlet and outlet flaps-if curled, replace the flap leaf.

Integral type fuel pump

ENGINE OVERHAUL

Disassembly

SPLIT CRANKCASE ENGINES

1. Remove the shroud and fuel tank if so equipped.

2. Remove the flywheel and the ignition stator.

3. Remove the carburetor and governor linkage. Carefully note the position of the carburetor wire links and springs for reinstallation.

4. Lift off the reed plate and gasket if present and inspect them. They should not bend away from the sealing surface plate more than 0.010 in.

5. Remove the spark plug and inspect it.

6. Remove the muffler. Be sure that the muffler and the exhaust ports are not clogged with carbon. Clean them if necessary.

7. Remove the transfer port cover and check for a good seal.

8. Remove the cylinder head, if so equipped.

Checking the reed valve clearance

NOTE: Some models utilize a locking compound on the cylinder head screws. Removing the screws on such engines can be difficult. This is especially true with screws having slotted head for a straight screwdriver blade. The screws can be removed if heat is applied to the head of the screw with an electric soldering iron.

9. On engines having a governor mounted on the power take-off end of the crankshaft, remove the screws that hold the outboard bearing housing to the crankcase. Clean the PTO end of the crankshaft and remove the outboard bearing housing and bearing. Loosen the set screw that holds the governor assembly to the crankshaft, slide the entire governor assembly from the crankshaft. Remove the screw that holds the governor bellcrank bracket to the crankcase. Remove the governor bellcrank and bracket.

10. Make match marks on the cylinder and crankcase. Remove the four nuts and lockwashers that hold the cylinder to the crankcase.

11. Remove the cylinder by pulling it straight out from the crankcase.

12. To separate the two crankcase halves, remove all of the screws that hold the crankcase halves together.

13. With the crankcase in a vertical position, grasp the top half of the crankcase and hold it firmly. Strike the top end of the crankcase with a rawhide mallet, while holding the assembly over a bench to prevent damage to parts when they fall. The top half of the crankcase should separate from the remaining assembly.

14. Invert the assembly and repeat the procedure to remove the other casting half from the crankcase on ball bearing units.

15. Each time the crankshaft is removed from the crankcase, seals at the end of the crankshaft should be replaced. To replace the seals, use a screwdriver or an ice pick to remove the seal retainers and remove and discard the old seals. Install the seals in the bores of the crankcase halves. The seals must be inserted into the bearing well with the channel groove toward the internal side of the crankcase. Retain the seal with the retainer. Seat the retainer spring into the spring groove.

UNIBLOCK ENGINES

1. Remove the shroud and fuel tank. Note the condition of the air vane governor, if so equipped.

2. Remove the starter cap and flywheel nut, noting the position of the belleville washer.

3. Remove the flywheel—see "Breaker Points Removal and Installation," above.

4. Remove the head. Save the old head gasket for use when replacing the piston, but procure a new gasket for use in final assembly.

5. Remove the cylinder block cover plate to gain access to the connecting rod bolts.

6. Note the location of the connecting rod match marks for reassembly.

7. Remove the piston. Remove the ridge first, if necessary, with a ridge reamer. Push the piston and connecting rod through the top of cylinder.

8. Remove the crankshaft from the cylinder block assembly. On engines with crankshaft ball bearings:

   a. Remove the four shroud base screws and tap the shroud base so the base and crankshaft can be removed together.

   b. To remove bearing with crankshaft from base: USE SAFETY GLASSES AND HEAT RESISTANT GLOVES. Using a propane torch, heat the area on the base around the outside of the bearing until there is enough expansion to remove the base from the bearing on the crankshaft. Now remove and discard the seal retainer ring, seal retainer and seal.

   c. To remove the bearing race, remove the retainer ring on the crankshaft with snap ring pliers, and with the use of a bearing splitter or arbor press, remove the ball bearing.

CAUTION

Support the crankshaft's top counterweight to prevent bending. Also, bearing is to be pressed on via the inner race only.

On other engines:

When equipped with a sleeve or needle bearing, use a seal protector and lift the crankshaft out of the cylinder. Be careful not to lose the bearing needles.

When equipped with a ball bearing, use a mallet to strike the crankshaft on the P.T.O. end while holding the block in your hand.

9. In assembly, bear the following points in mind:

   a. Use a ring compressor to install piston. Be careful not to allow the rings to catch on the recess for the head gasket. Use the old head gasket to take up the space in the recess. Do not force the piston into the cylinder, or damage to rings or piston could occur.

   b. To install the ball bearing on the crankshaft, slide the bearing on the crankshaft and fit it on the shaft by tapping using a mallet and tool, part number 670258 or press the ball bearing on the crankshaft with an arbor press. Install the retainer ring.

   c. To install the crankshaft with a ball bearing, heat the shroud base to expand the bearing seat and drop the ball bearing into the seat of the base shroud. Allow it to cool. Install a new seal retainer ring, seal retainer, and seal.

10. After the shroud base and flywheel are back in place, adjust the air gap between the coil core and flywheel as described above under "Magnet Air Gap Adjustment."

Connecting Rod Service

1. For engines using solid bronze or aluminum connecting rods, remove the two self-locking cap screws which hold the connecting rod to the crankshaft and remove the rod cap. Note the match marks on the connecting rod and cap. These marks must be reinstalled in the same position to the crankshaft.

BRONZE BUSHING

MATCH MARKS

ROD CAP

Connecting rod match marks

2. Engines using steel connecting rods are equipped with needle bearings at both crankshaft and piston pin end. Remove the two set screws that hold the connecting rod and cap to the crankshaft, taking care not to lose the needle bearings during removal.

3. Needle bearings at the piston pin end of steel rods are caged and can be pressed out as an assembly if damaged.

4. Check the connecting rod for cracks or distortion. Check the bearing surfaces for scoring or wear. bearing diameters should be within the limits indicated in the table of specifications located at the end of this section.

5. There are two basic arrangements of needles supplied with the connecting rod crankshaft bearing: split rows of needles and a single row of needles. Service needles are supplied with a beeswax coating. The beeswax holds the needles in position.

6. To install the needle bearings, first make sure that the crankshaft bearing journal and the connecting rod are free from oil and dirt.

7. Place the needle bearings with the beeswax onto a cool metallic surface to stiffen the beeswax. Body temperature will melt the wax, so avoid handling.

8. Remove the paper backing on the bearings and wrap the needles around the crankshaft journal. The beeswax will hold the needles onto the journal. Position the needles uniformly onto the crankpin.

NOTE: When installing the split row of needles, wrap each row of needles around the journal and try to seal them together with gentle but firm pressure to keep the bearings from unwinding.

9. Place the connecting rod onto the journal, position the rod cap, and secure it with the capscrews. Tighten the screw to the proper specifications.

10. Force solvent (lacquer thinner) into the needles just installed to remove the beeswax, then force 30W oil into the needles for proper lubrication.

Piston and Rings Service

1. Clean all carbon from the piston and ring grooves.
2. Check the piston for scoring or other damage.
3. Check the fit of the piston in the cylinder bore. Move the piston from side-to-side to check clearance. If the clearance is not greater than 0.003 in. and the cylinder is not scored or damaged, then the piston need not be replaced.

A. SPLIT ROWS OF NEEDLE BEARINGS

B. SINGLE ROW OF NEEDLE BEARINGS

Needle bearing arrangement. Double rows of bearings are placed with the tapered edges facing out

4. Check the piston ring side clearance to make sure it is within the limits recommended.

5. Check the piston rings for wear by inserting them into the cylinder about ½ in. from the top of the cylinder. Check at various places to make sure that the gap between the ends of the ring does not exceed the dimensions recommended in the specifications table at the end of this section. Bore wear can be checked in the same way, except that a new ring is used to measure the end gap.

6. If replacement rings have a bevelled or chamfered edge, install them with the bevel up toward the top of the piston. Not all engines use bevelled rings. The two rings installed on the piston are identical.

7. When installed, the offset piston used on the AV600 and the AV520 engines must have the **V** stamped in the piston head (some have hash marks) facing toward the right as the engine is viewed from the top or piston side of the engine.

NOTE: Some AV520 and AV600 engines do not have offset pistons. Only offset pistons will have the "V" or the hash marks on the piston head. Domed pistons must be installed so that the slope of the piston is toward the exhaust port.

Crankshaft Service

1. Use a micrometer to check the bearing journals for out-of-roundness. The main bearing journals should not be more than 0.0005 in. out-of-round. Connecting rod journals should not be more than 0.001 in. out-of-round. Replace a crankshaft that is not within these limits.

NOTE: Do not attempt to regrind the crankshaft since undersize parts are not available.

2. Check the tapered portion of the crankshaft (magneto end), keyways, and threads. Damaged threads may be restored with a thread die. If the taper of the shaft is rusty, it indicates that the engine has been operating with a loose flywheel. Clean the rust off the taper and check for wear. If the taper or keyway is worn, replace the crankshaft.

3. Check all of the bearing journal diameters. They should be within the limits indicated in the specifications table at the end of this section.

Checking the crankshaft for out-of-roundness

4. Check the crankshaft for bends by placing it between two pivot points. Position the dial indicator feeler on the crankshaft bearing surface and rotate the shaft. The crankshaft should not be more than 0.002–0.004 in. out-of-round.

Bearing Service

1. Do not remove the bearings unless they are worn or noisy. Check the operation of the bearings by rotating the bearing cones with your fingers to check for roughness, binding, or any other signs of unsatisfactory operation. If the bearings do not operate smoothly, remove them.

2. To remove the bearings from the crankcase, the crankcase must be heated. Use a hot plate to heat the crankcase to no more than 400°F. Place a ⅛ in. steel plate over the hot plate to prevent overheating. At this temperature, the bearings should drop out with a little tapping of the crankcase.

3. The replacement bearing is left at room temperature and dropped into the heated crankcase. Make sure that the new bearing is seated to the maximum depth of the cavity.

NOTE: Do not use an open flame to heat the crankcase halves and do not heat the crankcase halves to more than 400°F. Uneven heating with an open flame or excessive temperature will distort the case.

4. The needle bearings will fall out of the bearing cage with very little urging. Needles can be reinstalled easily by using a small amount of all-purpose grease to hold the bearings in place.

5. Cage bearings are removed and replaced in the same manner as ball bearings.

6. Sleeve bearings cannot be replaced. Both crankcase halves must be discarded if a bearing is worn excessively.

Assembly

1. The gasket surface where the crankcase halves join must be thoroughly clean before reassembly. Do not buff or use a file or any other abrasive that might damage the mating surfaces.

NOTE: Crankcase halves are matched. If one needs to be replaced, then both must be replaced.

2. Place the PTO half of the crankcase onto the PTO end of the crankshaft. Use seal protectors where necessary.

3. Apply a thin coating of sealing compound to the contact surface of one of the crankcase halves.

4. Position one crankcase half on the other. The fit should be such that some pressure is required to bring the two halves together. If this is not the case, either the crankcase halves and/or the crankshaft must be replaced.

5. Secure the halves with the screws provided, tightening the screws alternately and evenly. Before tightening the screws, check the union of the crankcase halves on the cylinder mounting side. The halves should be flat and smooth at the union to provide a good mounting face for the cylinder. If necessary, realign the halves before tightening the screws.

6. The sleeve tool should be placed into the crankcase bore from the direction opposite the crankshaft. Insert the tapered end of the crankshaft through the half of the crankcase to which the magneto stator is mounted. Remove the seal tools after installing the crankcase halves.

7. Stagger the ring ends on the piston and check for the correct positioning on domed piston models.

8. Place the cylinder gasket on the crankcase end.

9. Place the piston into the cylinder using the chamfer provided on the bottom edge of the cylinder to compress the rings.

10. Secure the cylinder to the crankcase assembly.

Type Number-to-Letter Cross Reference Chart

Type No.	Column Letter	Type No.	Column Letter	Type No.	Column Letter
1 thru 44	A	611	F	1001	D
46 thru 68	B	614–01 thru 614–04	E	1002 thru 1002D	G
69 thru 77	D	614–05 thru 614–06A	F	1003	C
78 thru 80	C	615–01 thru 615–01A	E	1004 thru 1004A	H
81 thru 83	D	615–04	F	1005 thru 1005A	C
84 thru 85	C	615–05 thru 615–10A	E	1006	H
86 thru 87	B	615–11 thru 615–18	F	1007 thru 1008B	C
89	C	615–19 thru 615–27A	E	1009 thrù 1010B	G
91	D	615–28	F	1011 thru 1019E	H
93	D	615–29 thru 615–39	E	1020	G
94 thru 98	H	616–01 thru 616–11	E	1021	C
99	C	616–12 and 616–14	F	1022 thru 1022C	D
		616–16 thru 616–40	E	1023 thru 1026A	H
201 thru 208	A	617A thru 617–01	E	1027 thru 1027B	C
209 thru 244	B	617–02 thru 617–03A	F	1028 thru 1030C	H
245	D	617–04 thru 617–04A	E	1031B thru 1031C	G
246 thru 248	B	617–05 thru 617–05A	F	1032	C
249 thru 251	D	617–06	E	1033 thru 1033F	G
252	B	618 thru 618–16	E	1034 thru 1034G	H
253 thru 255	D	618–17	F	1035 thru 1036	C
256 thru 261	B	618–18	E	1037 thru 1039C	D
262 thru 265A	D	619 thru 619–03	E	1040 thru 1041A	D
266 thru 267	B	621 thru 621–11	E	1042 thru 1042C	G
268 thru 272	D	621 thru 621–12A	F	1042D thru 1042E	H
273 thru 275	B	621–13 thru 621–15	E	1042F	G
276	D	622 thru 622–07	E	1042G and 1042H	H
277 thru 279	B	623 and 623A	E	1042I thru 1043B	G
280 thru 281	D	623–01	F	1043C thru 1043F	H
282	B	623–02 thru 623–33E	E	1043G thru 1044E	G
283 thru 284	D	623–34	F	1045 thru 1045F	B
285 thru 286	B	623–35 and 623–36	E	1046 thru 1051B	H
287	D	624 thru 630–09	J	1052	G
288 thru 291B	B	632–02A	K	1053 thru 1054C	H
292 thru 294	D	632–04 and 632–05A	K	1055 thru 1056B	G
295 thru 296	B	633 thru 634–09	J	1057	H
297	F	635A and B	K	1058 thru 1058A	G
298 thru 299	B	635–03B	K	1059	D
		635–04B	K	1060 thru 1060C	H
301 thru 375		635–06A thru 635–10	K	1061 and 1062	C
These are twin		636 thru 636–11	J	1063A thru 1063C	H
cylinder units –		637 thru 637–16	J	1064	G
out of production		638 thru 638–100	F	1065 thru 1067C	D
401 thru 402	L	639	M	1068 and 1068A	G
403	M	640	O	1069	H
404 thru 406B	L	641	K	1070 thru 1070B	C
407 thru 407C	M	642	I	1071 thru 1075D	H
408 thru 410B	L	643	J	1076 thru 1084	H
411 thru 411B	M	650	N	1085 and 1085A	G
412 thru 423	L	670	H	1086 thru 1140A	H
424 thru 427	M				
				1141 thru 1142	G
501 thru 509A	E	701 thru 701–1C	G	1143 thru 1145	H
		701–2 thru 701–2D	H	1146 thru 1149A	I
601 thru 601–01	F	701–3 thru 701–3C	G	1149A thru 1153B	H
602 thru 602–02B	F	701–4 thru 701–4C	H	1158 thru 1159	G
603A thru 603–23	E	701–5 thru 701–5C	G	1160 thru 1161A	I
604 thru 604–25	E	701–6 thru 701–6C	H	1162 and 1162A	G
605 thru 605–18	E	701–7 thru 701–7C	G	1163 thru 1163A	H
606 thru 606–05B	E	701–8 thru 701–11	H	1165 thru 1168A	G
606–06 thru 606–07	F	701–12 thru 701–13A	G	1169 thru 1174	H
606–08 thru 606–13	E	701–14 thru 701–17	H	1175 and 1175A	I
608 thru 608–C	E	701–18 thru 701–19	G	1176 thru 1177A	H
610–A thru 610–14	E	701–19A	H	1179 thru 1179A	I
610–15 thru 610–16	F	701–20 thru 701–22A	G	1180 thru 1181	G
610–19 thru 610–20	E	701–24	H	1182 and 1182A	I

TECUMSEH-LAUSON 2-STROKE
SMALL ENGINES — 6–20 HP

Type Number-to-Letter Cross Reference Chart (cont.)

Type No.	Column Letter	Type No.	Column Letter	Type No.	Column Letter
1183 thru 1185	G	1345	J	1441	P
1186A thru 1186C	J	1348 and 1348A	G	1442 thru 1442B	G
1187 thru 1192	G	1350 thru 1350C	I	1443	I
1192A thru 1196B	J			1444 and 1444A	G
1197 thru 1197A	G	1550A	P	1445	I
1198 and 1198A	H			1446 and 1446A	A
1199 and 1199A	J	1351 thru 1351B	K	1447	G
		1352 thru 1352B	I	1448 thru 1450	F
1206 thru 1208B	J	1353	K	1450A thru 1450B	F
1210 thru 1215C	K	1354 thru 1355A	I	1450C thru 1450E	F
1216 thru 1216A	H	1356 and 1356B	K	1453	K
1217 thru 1220A	J	1357 thru 1358	I	1454 and 1454A	A
1221	H	1359 thru 1362B	K	1455 thru 1456	K
1222 thru 1223	J	1363 thru 1369A	G	1459	G
1224 thru 1225A	H	1372 thru 1375A	G	1460 thru 1460F	A
1226 thru 1227B	K	1376 and 1376A	J	1461	K
1228 thru 1229A	J	1377 thru 1378	K	1462	A
1230 thru 1231	H	1379 thru 1379A	H	1463	K
1232	G	1380 thru 1380B	K	1464 thru 1464B	L
1233 thru 1237B	K	1381	G	1465	A
1238 thru 1238C	H	1382	H	1466 thru 1466A	F
1239 thru 1243	J	1383 thru 1383B	K	1467 thru 1468	K
1244 thru 1245A	K	1384 thru 1385	G	1471 thru 1471B	E
1246 thru 1247	J	1386	I	1472 thru 1472C	L
1248	H	1387 thru 1388	K	1473 thru 1473B	A
1249 thru 1251A	J	1389	G	1474	L
1252 thru 1254A	K	1390 thru 1390B	H	1475 thru 1476	A
1255	J	1391	K	1477	G
1256 thru 1262B	K	1392	J	1478	J
1263	J	1393	G	1479	G
1264 thru 1265E	K	1394 thru 1395A	P	1482 and 1482A	F
1266 thru 1267	G	1396	K	1483	F
1269 thru 1270D	K	1397	H	1484 thru 1484D	A
1271 thru 1271B	G	1398 thru 1399	K	1485	G
1272 thru 1275	K			1486	D
1276	H	1400	K	1487	J
1277 thru 1279D	K	1401 thru 1401F	F	1488 thru 1488D	A
1280	J	1402 and 1402B	G	1489 thru 1490B	C
1283 thru 1284D	K	1403A and 1403B	K	1491	L
1286 thru 1286A	H	1404 and 1404A	K	1493 and 1493A	G
1287	K	1405 thru 1406A	H	1494 and 1495A	B
1288	J	1407 thru 1408	K	1496	G
1289 thru 1289A	G	1409A	J	1497	A
1290 thru 1293A	K	1410 thru 1412A	I	1498	E
1294 thru 1295	J	1413 thru 1416	I	1499	F
1296 thru 1298	K	1417 and 1417A	K		
		1418 and 1418A	P	1500	E
1300 thru 1303	K	1419 and 1419A	K	1501A thru 1501E	A
1304 thru 1307	J	1420 and 1420A	H	1503 thru 1503D	L
1308 thru 1316A	K	1421 and 1421A	K	1506 thru 1507	F
1317 thru 1317B	J	1422 thru 1423	P	1508	G
1318 thru 1320B	K	1424	K	1509	C
1321 thru 1322	J	1425	G	1510	L
1323	K	1426 thru 1426B	P	1511	C
1325	G	1427 and 1427A	H	1512 and 1512A	B
1326 thru 1326F	K	1428 thru 1429A	P	1513	L
1327 thru 1327B	H	1430A	G	1515 thru 1516C	C
1328B and 1328C	K	1431	P	1517	E
1329	J	1432 and 1432A	G	1518	D
1330	H	1433	K	1519 thru 1521	A
1331	J	1434 thru 1435A	P	1522	L
1332	J	1436 and 1436A	I	1523	A
1333	I	1437	J	1524	B
1334	I	1439	I	1527	C
1343 thru 1344A	H	1440 thru 1440D	A	1528	A

Type Number-to-Letter Cross Reference Chart (cont.)

Type No.	Column Letter	Type No.	Column Letter	Type No.	Column Letter
1529A and 1529B	C	2045 thru 2046B	F	40069 thru 40075	N
1530 thru 1530B	A	2047 thru 2048	D		
1531 thru 1535B	C	2049 thru 2049A	F	710101 thru 710116	E
1536	L	2050 thru 2050A	D	710124 thru 710130	E
1537	A	2051 thru 2052A	D	710131 thru 710137	E
1538 thru 1541A	L	2053 thru 2055	F	710138 thru 710149	E
1542	E	2056	D	710154	M
1543 thru 1546	A	2057	F	710201 thru 710209	E
1547	C	2058 thru 2058B	D	710210 thru 710218	E
1549	C	2059 thru 2063	F	710219 thru 710227	E
		2064 thru 2064A	D	710228	M
S–1801 thru 1822	F	2065	F	710150	G
1823 and 1824	E	2066	D	710151	H
		2067 thru 2071B	F	710155	L
2001 thru 2003	B			710157	H
2004 thru 2006	D	2200 thru 2201A	D	710229	H
2007 thru 2007B	B	2202 thru 2204	E	710230	N
2008 thru 2008B	D	2205 thru 2205A	D	710234	N
2009	B	2206 thru 2206A	G		
2010 thru 2011	D	2207	D	200.183112	F
2012	F	2208	G	200.18322	F
2013 thru 2014	B	2768	C	200.193132	F
2015 thru 2018	D			200.193142	F
2020	F	40001 thru 40028	N	200.193152	G
2021	D	40029 thru 40032C	O	200.193162	G
2022 thru 2022B	F	40033	N	200.203172	H
2023 thru 2026	D	40034 thru 40045A	O	200.203182	H
2027	B	40046	N	200.203192	H
2028 thru 2029	D	40047 thru 40052	O	200.213112	H
2030 thru 2031	B	40053 thru 40054A	N	200.213122	H
2032 thru 2033A	D	40056 thru 40060B	O	200.503111	F
2034 thru 2035C	F	40061 thru 40062B	N	200.583111	F
2036 thru 2037	B	40063 thru 40064	O	200.593121	F
2038	F	40065 thru 40066	N	200.613111	F
2039 thru 2044	C	40067 thru 40068	O		

Uniblock Cross Reference Chart

Type No.	Column No.	Type No.	Column No.	Type No.	Column No.
Vertical Crankshaft Engines		**Horizontal Crankshaft Engines**		**Horizontal Crankshaft Engines**	
638 thru 638–100	6	1398 thru 1399	11	1509	3
639 thru 639–13A	13			1510	12
		1400	11	1511	3
		1401 thru 1401F	16	1512 and 1512A	2
		1401G, H	17	1513	12
640–02 thru 640–06B	21	1401J	27	1515 thru 1516C	3
640–07 thru 640–18	22	1402 and 1402B	7	1517	5
641 thru 641–14	11	1425	7	1518	4
642–01, A	9A	1430A	7	1519 thru 1521	1
642–02, A, B, C, D	9A	1432 and 1432A	7	1522	12
642–02E, F	9B	1440 thru 1440D	1	1523	1
642–03, A, B	9A	1442 thru 1442B	7	1524	2
642–04, A, B, C	9A	1444 and 1444A	7	1525A	16
642–05, A, B	9A	1448 thru 1450	16	1527	3
642–06, A	9A	1450A thru 1450B	16	1528	1
642–07, A, B	9A	1450C thru 1450E	16	1529A and 1520B	3
		1450F	17	1530 thru 1530B	1
642–07C	9B	1454 and 1454A	1	1531	3
642–08	9B	1459	7	1534A	17
642–08A, B	9A	1460 thru 1460F	1	1535B	3
642–09 thru 642–14	9A	1462	1	1536	12
642–13A, 14A, 14B	9B	1464 thru 1464B	12	1537	1

Uniblock Cross Reference Chart

Type No.	Column No.	Type No.	Column No.	Type No.	Column No.
Vertical Crankshaft Engines		**Horizontal Crankshaft Engines**		**Horizontal Crankshaft Engines**	
642–15 thru 642–22	9B	1465	1	1538 thru 1541A	12
642–24 thru 642–30	9C	1466 thru 1466A	16	1542	5
643–01, A, 03, A	10A	1471 thru 1471B	5	1543 thru 1546	1
643–03B, C	10B	1472 thru 1472C	12	1517	3
643–04, 05A	10A	1473 thru 1473B	1	1519	3
643–05B	10B	1474	12	1550A	15
643–13, 14	10A	1475 thru 1476	1	1551	16
643–14A, B, C	10B	1479	7	1552	20
643–15	10A	1482 and 1482A	16	1553	16
643–15A thru 643–28	10B	1483	16	1554 and 1554A	3
		1484 thru 1484D	3	1555 and 1556	16
		1485	7	1557 thru 1560	15
650	14	1486	4	1561	19
660–11 thru 660–32	18	1488 thru 1488D	1	1562 thru 1571	15
670–01 thru 670–101	8	1489 thru 1490B	3	1572	2
		1491	12	1573	3
200–183112	6	1493 and 1493A	7	1574 thru 1577	23
200–183122	6	1494 and 1495A	2	1575	24
200–193132	6	1496	7	1578	25
200–193142	6	1497	1	1581 thru 1582A	23
200–193152	7	1498	5	1583 thru 1595	26
200–193162	7	1499	16		
200–203172	8	1500	5		
200–203182	8	1501A thru 1501E	1	200–503111	16
200–203192	8	1503 thru 1503D	12	200–583111	16
200–213112	8	1506	16	200–593121	16
200–213122	8	1506B	17	200–613111	16
200–243112	8	1507	16	200–672102	26
200–283012	8	1508	7	200–682102	26

Specifications for 2 Cycle Engines with Split Crankcases

	A	B	C	D	E	F	G	H
Bore	1.500 1.5005	1.6253 1.6258	1.7503 1.7508	1.7503 1.7508	2.000	2.000	2.000	2.000
Stroke	1.375	1.50	1.50	1.50	1.50	1.50	1.50	1.50
Displacement Cubic Inches	2.43	3.10	3.60	3.60	4.70	4.70	4.70	4.70
Point Gap	.020	.020	.020	.020	.020	.020	.015	.015
Timing B.T.D.C. Before Top Dead Center	1/16" or .0625	5/32" or .1562	5/32" or .1562	1/4" or .250	5/32" or .1562	5/32" or .1562	5/32"	1/4" or .250 11/64" for "super"
Spark Plug Gap	.030	.030	.030	.030	.030	.030	.030	.030
Piston Ring End Gap	.003 .008	.005 .010	.005 .010	.005 .010	.006 .011	.006 .011	.006 .011	.006 .011
Piston Diameter	1.4966 1.4969	1.6216 1.6219	1.7461 1.7464	1.7461 1.7464	1.9948 1.9951	1.9948 1.9951	1.9948 1.9951	1.9949 1.9955
Piston Ring Groove Width	.095 .096	.095 .096	.095 .096	.095 .096	.095 .096	.095 .096	.095 .096	.095 .096
Piston Ring Width	.093 .0935	.093 .0935	.093 .0935	.093 .0935	.093 .0935	.093 .0935	.093 .0935	.093 .0935

Specifications for 2 Cycle Engines with Split Crankcases (cont.)

	A	B	C	D	E	F	G	H
Piston Pin Diameter	.3750 .3751	.3750 .3751	.3750 .3751	.3750 .3751	.3750 .3751	.3750 .3751	.3750 .3751	Early .3750 .3761 Late .4997 .4999
Connecting Rod Diameter Crank Bearing	.6869 .6874	.6869 .6874	.6869 .6874	.6986 .6989 w/o needles	.6869 .6874	.6869 .6874	.6869 .6874	.6935 .6939 w/o needles
Crankshaft Rod Needle Diameter				.0653 .0655				.0653 .0655
Crank Pin Journal Diameter	.6860 .6865	.6860 .6865	.6860 .6865	.5615 .5618	.6860 .6865	.6860 .6865	.6860 .6865	.5615 .5618
Crankshaft P.T.O. Side Main Brg. Dia.	.6689 .6693	.6689 .6693	.6690 .6694	.6689 .6693	.9995 1.0000	.6689 .6693	.9995 1.0000	.9995 1.0000
Crankshaft Magneto Side Main Brg. Dia.	.6689 .6693	.6689 .6693	.6689 .6693	.6690 .6694	.7495 .7500	.6689 .6693	.7495 .7500	.7495 .7500
Crankshaft End Play	.003 .008	.003 .008	.003 .008	.003 .008	.009 .022	.009 .022	.009 .022	.009 .022

Specifications for 2 Cycle Engines with Split Crankcases

I	J	K	L	M	N	O	P	
2.000	2.000	2.093 2.094	2.2505 2.2510	2.2505 2.2510	2.5030 2.5035	2.5030 2.5035	2.000	Bore
1.50	1.625	1.63	2.00	2.00	1.625	1.680	1.50	Stroke
4.70	5.10	5.80	8.00	8.00	7.98	8.25	4.70	Displacement Cubic Inches
.015	.020	.015	.020	.020	.020	.020	.015	Point Gap
11/64" or .175	11/64" or .175	3/32" or .095	1/8" or .125	3/32" or .095	11/64" or .175	11/64" or .175	.90	Timing B.T.D.C. Before Top Dead Center
.035	.030	.030	.030	.030	.030	.030	.035	Spark Plug Gap
.006 .011	.006 .011	.006 .011	.007 .015	.007 .015	.005 .013	.005 .013	.006 .011	Piston Ring End Gap
1.9948 1.9951	1.9951 1.9948	2.0880 2.0883	2.2460 2.2463	2.2460 2.2463	2.4960 2.4963	2.4960 2.4963	1.9948 1.9951	Piston Diameter
.095 .096	.095 .096	T.0655 .0665 L.0645 .0655	.095 .096	.095 .096	T.0655 .0665 L.0645 .0655	T.0655 .0665 L.0645 .0655	.095 .096	Piston Ring Groove Width
.093 .0935	.093 .0935	.0615 .0625	.093 .0935	.093 .0935	.0615 .0625	.0615 .0625	.093 .0935	Piston Ring Width
.4997 .4999	.3750 .3751	.4997 .4999	.5000 .5001	.3000 .5001	.4997 .4999	.4997 .4999	.4997 .4999	Piston Pin Diameter

Specifications for 2 Cycle Engines with Split Crankcases (cont.)

I	J	K	L	M	N	O	P	
.6941 .6944	.6869 .6874	.9407 .9412	1.000 1.0004	1.000 1.0004	.9407 .9412	.9407 .9412	.6941 .6944	Connecting Rod Diameter Crank Bearing
.0653 .0655		.0943 .0945	.0943 .0945	.0943 .0945	.0943 .0945	.0943 .0945	.0653 .0655	Crankshaft Rod Needle Diameter
.6860 .6865	.6860	.7499 .7502	.8096 .8099	.8096 .8099	.7499 .7502	.7499 .7502	.6860 .6865	Crank Pin Journal Diameter
.9995 1.0000	.9995 1.0000	.6990 .6994	.9839 .9842	.9839 .9842	.7871 .7875	.7871 .7875	.9995 1.0000	Crankshaft P.T.O. Side Main Brg. Dia.
.7495 .7500	.7495 .7500	.6990 .6994	.9839 .9842	.9839 .9842	.7498 .7501	.7498 .7501	.7495 .7500	Crankshaft Magneto Side Main Brg. Dia.
.009 .022	.009 .022	.003 .008	.003 .008	.003 .008	.008 .013	.008 .013	.009 .022	Crankshaft End Play

Specifications Chart

	1	2	3	4	5	6	7	8	9A	9B	9C	12
Bore	2.093 2.094	2.093 2.094	2.093 2.094	2.093 2.094	2.093 2.094	2.093 2.094	2.093 2.094	2.093 2.094	2.093 2.094	2.093 2.094	2.093 2.094	2.093 2.094
Stroke	1.250	1.410	1.410	1.410	1.410	1.500	1.500	1.500	1.500	1.500	1.500	1.410
Cu. In. Displacement	4.40	4.80	4.80	4.80	4.80	5.20	5.20	5.20	5.20	5.20	5.20	4.80
Point Gap	.017	.017	.017	.017	.017	.018	.017	.020	.018	.020	.020	.017
Timing B.T.D.C.	.122″	.100″	.135″	.100″	.135″	.100″	.185″	.070″	.110″	.085″ See Note 1	.078″ See Note 2	.135″
Spark Plug Gap	.035	.035	.035	.035	.035	.035	.035	.035	.035	.035	.035	.035
Piston Ring End Gap	.007 .017	.007 .017	.006 .011	.006 .014	.006 .011	.006 .014	.007 .017	.006 .016	.007 .017	.006 .016	.006 .016	.007 .017
Piston Diameter	2.0080 2.0870	2.0880 2.0870	2.0885 2.0875	2.0885 2.0875	2.0885 2.0875	2.0880 2.0870	2.0880 2.0870	2.0880 2.0870	2.0880 2.0870	2.0880 2.0820	2.0880 2.0870	2.0880 2.0870
Piston Ring Groove Width (Top)	.0655 .0665	.0655 .0665	.0655 .0665	.0975 .0985	.0655 .0665	.0975 .0985	.0655 .0665	.0655 .0665	.0655 .0665	.0655 .0665	.0655 .0665	.0655 .0665
Piston Ring Groove Width (Bot.)	.0645 .0655	.0645 .0655	.0645 .0655	.0955 .0965	.0645 .0655	.0955 .0965	.0645 .0655	.0645 .0655	.0645 .0655	.0645 .0655	.0645 .0655	.0645 .0655
Piston Ring Width	.0625 .0615	.0625 .0615	.0625 .0615	.0925 .0935	.0625 .0615	.0925 .0935	.0625 .0615	.0625 .0615	.0625 .0615	.0625 .0615	.0625 .0615	.0625 .0615
Piston Pin Diameter	.4999 .4997	.4999 .4997	.4999 .4997	.3750 .3751	.4999 .4997	.3750 .3751	.4999 .4997	.4999 .4997	.4999 .4997	.4999 .4997	.4999 .4997	.4999 .4997
Connecting Rod Diameter Crank Bearing	—	—	—	.6886 .6879 Dowels	—	—	—	1.0053 1.0023 Liner Dia.	—	1.0053 1.0023 Liner Dia.	1.0053 1.0023 Liner Dia	—
Crankshaft Rod Needle Dia.	.0655 .0653	.0655 .0653	.0655 .0653	—	.0655 .0653	—	.0655 .0653	.0781 .0780	—	.0781 .0780	.0781 .0780	.0655 .0653
Crank Pin Journal Diameter	.5618 .5611	.5621 .5614	.5621 .5614	.6865 .6857	.5618 .5611	.6865 .6857	.5618 .5611	.8450 .8442	.6865 .6857	.8450 .8442	.8450 .8442	.5621 .5614

Specifications Chart

	1	2	3	4	5	6	7	8	9A	9B	9C	12
Crankshaft P.T.O. Side Main Brg. Dia.	.6695 .6691	.6695 .6691	.6695 .6691	.6695 .6691	.6695 .6691	.8750 .8745	.6694 .6690	1.0003 .9998	.8750 .8745	1.0003 .9998	1.0003 .9998	.6695 .6691
Crankshaft Magneto Side Main Brg. Dia.	.6695 .6691	.6695 .6691	.6695 .6691	.6695 .6691	.6695 .6691	.7500 .7495	.6694 .6690	.6695 .6691	.7500 .7495	.7503 .7498	.6695 .6691	.6695 .6691
Crankshaft End Play	None	None	None	None	None	.003 .016	None	None	.003 .016	.003 .016	None	None

NOTE 1: 642-08 14A, 14B B.T.D.C. = .110"
  642-16D, 19A, 20A, 21 22 B.T.D.C. = .078"
NOTE 2: 642-24, 26, 29 B.T.D.C. = .085"
  643-2a, 25, 26 B.T.D.C. = .085"
NOTE 3: 643-13 B.T.D.C. = .095"
NOTE 4: 643-03A, 05A, 13, 14 = .020"

Specifications Chart

		14	16	17	19	20	23	24	25	26	27
Bore		2.093 2.094	2.093 2.094	2.093 2.094	2.093 2.094	2.093 2.094	2.093 2.094	2.093 2.094	2.093 2.094	2.093 2.094	2.093 2.094
Stroke		1.500	1.500	1.500	1.410	1.250	1.500	1.410	1.410	1.500	1.500
Cu. In. Displacement		5.20	5.20	5.20	4.80	4.40	5.20	4.80	4.80	5.20	5.20
Point Gap		.018	.017	.017	.017	.017	.017	.017	.020	.020	.017
Timing B.T.D.C.		.100"	.110"	.110"	.100"	.122"	.110"	.135"	Fixed	.062"	.100"
Spark Plug Gap		.035	.035	.035	.035	.035	.035	.035	.035	.035	.035
Piston Ring End Gap		.006 .014	.006 .016	.006 .016	.007 .017	.007 .017	.006 .016	.007 .017	.007 .017	.006 .016	.006 .016
Piston Diameter		2.0880 2.0870	2.0885 2.0875	2.0890 2.0880	2.0880 2.0870	2.0880 2.0870	2.0880 2.0870	2.0880 2.0870	2.0880 2.0870	2.0880 2.0870	2.0885 2.0875
Piston Ring Groove Width	(Top)	.0975 .0985	.0645 .0655	.0645 .0655	.0655 .0665	.0655 .0665	.0655 .0665	.0655 .0665	.0655 .0665	.0655 .0665	.0655 .0665
	(Bot.)	.0955 .0965	.0645 .0655	.0645 .0655	.0645 .0655	.0645 .0655	.0645 .0655	.0645 .0655	.0645 .0655	.0645 .0655	.0645 .0655
Piston Ring Width		.0925 .0935	.0625 .0615	.0625 .0615	.0625 .0615	.0625 .0615	.0625 .0615	.0625 .0615	.0625 .0615	.0625 .0615	.0625 .0615
Piston Pin Diameter		.3750 .3751	.3750 .3751	.4999 .4997	.4999 .4997	.4999 .4997	.4999 .4997	.4999 .4997	.4999 .4997	.4999 .4997	.4999 .4997
Connecting Rod Diameter Crank Bearing		—	—	—	—	—	.8534 .8504 Liner Dia.	—	—	.8534 .8504 Liner Dia.	.8919 .8924
Crankshaft Rod Needle Dia.		—	—	—	.0655 .0653	.0655 .0653	.0781 .0780	.0655 .0653	.0655 .0653	.0781 .0780	.0781 .0780
Crank Pin Journal Diameter		.6965 .6857	.6868 .6857	.6865 .6857	.5621 .5614	.5618 .5611	.6927 .6919	.5621 .5614	.5621 .5614	.6927 .6919	.6927 .6922
Crankshaft P.T.O. Side Main Brg. Dia.		.8750 .8745	.6695 .6691	1.0003 .9998	.6695 .6691	.6695 .6691	.6695 .6691	.6695 .6691	.6695 .6691	.7503 .7498	.6695 .6691
Crankshaft Magneto Side Main Brg. Dia.		.7500 .7495	.7500 .7495	.7500 .7495	.6695 .6691	.6695 .6691	.7503 .7498	.6695 .6691	.6695 .6691	.6695 .6691	.7503 .7498
Crankshaft End Play		.003 .016	None	None	None	None	None	None	None	None	.003 .016

A. on Engines prior to 660-24: 7505.
  7198

Torque Specifications for 2-Cycle Engines

Application	Torque	Application	Torque
Carburetor and Reed Plates		Cylinder head to cylinder (635 type engine)	45–50 in. pounds
Carburetor to crankcase, Carburetor to adapter, or Carburetor adapter to crankcase	70–75 in. pounds	Cable clip and transfer port cover to cylinder	25–30 in. pounds
Carburetor to snow blowers cover	30–35 in. pounds	Cable clip to cylinder Stop lever to cylinder	25–30 in. pounds
Carburetor outlet fitting	40 in. pounds	Spark plug	18–22 ft. pounds
Reed and cover plates 639 type engines	50–60 in. pounds with Loctite, type A	Transfer cover cross port engines	25–30 in. pounds
Reed to plate 635 type engines	12–18 in. pounds	**Crankshaft and Connecting Rods**	
Crankcase and Cylinder		Aluminum and bronze rods to rod cap	40–50 in. pounds
Crankcase to crankcase cover	23–30 in. pounds	Steel rod to rod cap	70–80 in. pounds
Crankcase to crankcase cover screws	35–40 in. pounds	Flywheel nut on tapered end of crankshaft: Aluminum hub on iron or steel shaft.	18–25 ft. pounds
Mounting cylinder or carburetor to crankcase studs	50 in. pounds	Steel hub flywheel on iron shaft	18–25 ft. pounds
Base to crankcase	240–250 in. pounds	Steel hub flywheel on steel shaft	30 ft. pounds
Cylinder to crankcase nuts	70–75 in. pounds	**Governor and Bell Crank Parts**	
In cylinder	100–110 in. pounds	Power take-off end governor lower ring to crankshaft setscrew	30–35 in. pounds
Spark plug stop lever and head shroud to head	50–60 in. pounds	Bell crank bracket to crankcase	20–25 in. pounds
Cylinder head to cylinder	30–40 in. pounds	Governor cover to crankcase	50–60 in. pounds
Cylinder head to cylinder	50–60 in. pounds 80–90 in. pounds		

Section 19

Wisconsin Robin

Small Engines—0-6 hp

ENGINE IDENTIFICATION

There is an identification plate attached to the upper blower housing which lists the engine model on the left and individual engine serial number on the right. Both of these should be used when requesting parts.

General Engine Specifications

Model	Bore & Stroke (in.)	Displacement (cu in.)	Horsepower @ RPM
EY18W	2.56 x 2.16	11.14	4.6 @ 3,600

MAINTENANCE

Air Cleaner
SERVICE

The air cleaner should be serviced every 50 hours in normal operation, more often (daily) if the air is dusty.

First disassemble the unit as follows:

1. Remove the air cleaner cover, element and element retainer.

2. Disconnect the breather tube from the inspection cover.

3. Remove the two capscrews and remove the air cleaner body and gasket from the carburetor.

4. Wipe all metal parts clean. Wash the element only in either kerosene, or water and a liquid detergent. Do not wash in any other solvents.

5. Wrap the element in a cloth and squeeze it dry.

6. Pour engine oil into the filter until it is saturated, and then squeeze out the excess.

7. Reassemble in reverse of above. Make sure element retainer is installed with arrow up.

Lubrication
OIL AND FUEL RECOMMENDATIONS

Oil

Use only MS or SD oils of the viscosity shown in the chart for both normal operation and breakin. See chart for crankcase capacities.

Oil Viscosity Chart

Season or Temperature	Grade of Oil*
Spring, Summer or Autumn (+120°F to +40°F)	SAE 30
Winter (+40°F to +15°F) (Below +15°F)	SAE 20 SAE 10W-30

* Use oils classified as Service MS or SD

Crankcase Capacity	
Model	Capacity (pts.)
EY18W	1.25

AIR CLEANER BODY

MOUNT ELEMENT RETAINER THUS

BREATHER TUBE

ELEMENT

COVER

LATCH

Exploded view of air cleaner

Fuel

Use only regular grade gasoline with an octane rating of 90 or above. Use only a reputable brand of fuel, and make sure it is free of both dirt and water. Note that low octane fuel may cause the engine to detonate and result in severe damage. Do not use gasoline that is more than one month old.

TUNE-UP

Spark Plug

The spark plug should be checked frequently for fouling, widened or otherwise improper gap, or more serious problems such as eroded electrodes or a cracked insulator. Clean carbon deposits from the electrodes with a wire brush, and set the gap to 0.020–0.025 in. with a wire type feeler gauge. Bend only the side electrode. If electrodes are severely eroded so that the side electrode has lost its sharp, rectangular cross-section or the center electrode is very short, or if there are any cracks in the insulator, the spark plug should be replaced.

Always check the plug gap before installation, even if the plug is new. Install the plug with a new gasket and torque it to 24–27 ft. lbs.

Tune-Up Specifications

Model	Plug Type	Plug Gap (in.)	Point Gap (in.)	Ignition Timing (deg Before Top Center)
EY18W	①	.020–.025	.014	23

① 14 mm. Champion L86,
     AC 44F
     NGK B6HS

Breaker Points

REMOVAL AND INSTALLATION

1. Remove the three mounting screws and remove the starter pulley. Install a 14mm socket wrench onto the flywheel nut and tap the wrench sharply with a soft hammer. If you have a wheel puller, remove the nut. Otherwise, unscrew it until it is flush with the end of the crankshaft.
2. Remove the flywheel with a wheel puller or as follows:
3. Place a large screwdriver between the crankcase and the flywheel. The screwdriver must be in line with the keyway. While wedging the flywheel outward with the screwdriver, strike the outer end of the flywheel nut with a soft (brass, wood or plastic) hammer. This will bring the flywheel off the crankshaft taper.
4. Take off the point cover by removing mounting screws and pulling it off. Unscrew the terminal nut connecting condenser and coil wires to the contact set, and disconnect the wires. Remove the contact lockscrew, and remove the contacts.
5. Position the pin contacts on the crankcase, install the lockscrew, and tighten it just slightly.
6. Adjust the contact gap and time the engine as described below.

SETTING POINT GAP AND IGNITION TIMING

1. If necessary, remove the flywheel and point cover as described in Steps 1–4 above. Loosen the contact lockscrew.

Aligning ignition timing marks

2. Turn the engine over until the contact cam follower is at the very peak of the cam. Using a flat .014 feeler gauge, slide the points back and forth via the adjusting knob until the gauge just slides between the contact surfaces. Tighten the contact lockscrew. Recheck gap and, if necessary, reset it.
4. Connect a timing light between the coil primary ground. Align the **M** timing mark on the rim of the flywheel with the **D** mark on the lower left side of the crankcase by rotating the flywheel. See illustration.
5. Turn the flywheel counterclockwise slowly until the light either goes on or goes off, depending on the type. Turn it very slowly clockwise just until the light reacts again. If the timing marks are aligned, timing is correct. If the **M** mark is below the **D** mark (the **D** is on the crankcase), the breaker gap is too large; if the **M** is above the **D**, the breaker gap is too small. Reset the contact gap as necessary. If timing is off about 2 degrees or 1/8 in. of flywheel rotation, the gap must be changed about 0.001 in.
6. Reset the contact gap and recheck timing until timing is correct.
7. Remove the timing light and install the timing cover and flywheel. In installing the flywheel, slide it into place with key and keyway aligned, and then install retaining nut and tighten it just until lockwasher collapses. Finally, put a wrench on the nut and tap it lightly once or twice, or torque the nut to 44–47 ft. lbs. with a torque wrench.

Magneto Service

1. If the engine is hard to start or runs erratically, first service the spark plug. If this does not cure the problem, remove the flywheel and check for loose or broken ignition wires. Inspect the points. See "Breaker Point Removal and Installation," above. Repair the wires or replace the points and condenser. Set the point gap and timing, as required.
2. Check the spark by removing the plug and grounding it against the engine block with the high tension wire connected. Then, spin the engine. If the spark is erratic or weak, remove the ignition coil by removing the flywheel, feeding the high tension wire through the grommet in the wall of the cylinder block, disconnecting the primary wire, and removing the mounting screws and coil. Replace the coil in reverse of the above.

Governor Adjustment

If the governor lever has been loosened or removed, perform the governor lever adjustment. Otherwise, skip to "Speed Regulation," below.

GOVERNOR LEVER ADJUSTMENT

1. Mount the governor lever with the clamp screw just slightly loose.

Adjusting the governor lever

2. Install both the control rod and spring which connect the governor lever and the throttle lever. Mount the control lever assembly onto the crankcase but do not tighten the wingnut.

3. Connect the governor spring between the holes of both governor lever and control lever as shown in the illustration below.

4. Turn the control lever counterclockwise until the throttle in the carburetor is fully open. Lock the lever in this position by tightening the wingnut.

5. With the clamp screw loose so the governor shaft will turn independently of the governor lever, turn the governor shaft (use a screwdriver in the groove in the end of the shaft) until you can feel the internal vane stop against the flywheel thrust sleeve (this is as far as the shaft can be turned without excessive force). Tighten the lever clamp nut.

SPEED REGULATION

1. Remove the load from the engine, and run it until it is hot.
2. Install the governor spring hooked to holes **1** and **B** as shown above.

Adjusting governor speed

3. Loosen the control lever wing nut so that the lever is free to move. Loosen the locknut on the adjusting screw. Find the desired no-load speed on the chart above that corresponds with the speed at which you want the engine to run when fully loaded.

4. Hold the control lever down so the adjusting screw is against the crankcase stop. Turn the adjusting screw in or out until the required no-load speed is obtained. Tighten the adjusting screw locknut.

5. If the engine is to operate at a fixed speed, tighten the control lever wingnut.

Governor No-Load/Load RPM Chart

Load RPM	No Load RPM EY18W	Spring Holes EY18W
1800	2370	1-B
2000	2515	1-B
2200	2665	1-B
2400	2815	1-B
2600	2975	1-B
2800	3140	1-B
3000	3310	1-B
3200	3485	1-B
3400	3670	1-B
3600	3855	1-B

Carburetor

OVERHAUL

1. Using the illustration as a guide, remove the choke phillips head screws (17), and then take out the choke valve and shaft (18 and 19).

2. Remove the choke shaft retainer spring and ball (30 and 31) to prevent their being lost (these are used on models EY25W and EY27W).

3. Remove the throttle plate phillips head screws (17) and take out the choke valve and shaft (18 and 19), being careful not to damage the edges of the throttle valve.

4. Remove the throttle stop screw and spring (14 and 25).

5. Remove the main jet holder (16) and then take off the float bowl (13).

6. Remove the main jet from the (15) jet holder.

7. Remove the main nozzle (9) from the carburetor body.

8. Remove the idle jet (27) using an appropriate tool to prevent damage to it.

9. Remove the float pin (12), float (11), and needle valve (20).

10. Inspect the float. Replace it if dented, fuel has leaked in, or if the float hinge pin or the tab that limits float travel is worn.

11. Clean all parts thoroughly in a solvent such as Bendix Metalclene or Speeclene, rinse in a cleaning solvent, and blow out all passages with low pressure compressed air in the reverse direction of normal flow. Make sure all carbon deposits have been removed from the throttle bore and idle discharge holes.

NOTE: Never use a drill or wire to clean jets!

Carburetor—exploded view

Reference Number	Part Number	Part Name
1.	EYY790-350	Flywheel Puller
	EY016508500	Bolt, 8 x 50 mm (3 required)
2.	EYY790-282	Valve Spring Compressor
3.	EYY790-324 (for 18W engine)	Valve Guide Puller
	EYY790-524 (for 25W, 27W engines)	Valve Guide Puller
4.	EYH640-118	Seat Cutter (45°)

5. 10, 12, and 14mm open end wrenches.

In addition, various special pullers, a valve spring compressor, and a valve set cutter are required. See the illustration. Wisconsin-Robin part numbers are provided for your use in ordering the tools either from them or from a tool manufacturer who makes equivalent tools which are cross-referenced.

The tools and Wisconsin-Robin part numbers are listed below:

Engine Disassembly, Reassembly, and Overhaul

Engine disassembly instructions are provided below, in proper sequence. Reassembly is primarily a simple reversal of disassembly. Notes on the extra procedures to be performed as the engine is put back together are included with the disassembly instructions for each part of the engine.

FUEL TANK

1. Disconnect the fuel line at the carburetor. Note that in reassembly, it is advisable to use a new fuel line.
2. Remove the fuel tank from the bracket.
3. Remove the tank bracket from the cylinder head.
4. In reassembly, wash the element per "Air Cleaner Service" instructions above, and correctly mount the element retainer as illustrated there.

FLYWHEEL SHROUD

1. Disconnect the coil primary wire from the stop button wire.
2. Remove the flywheel shroud from the cylinder case, and the baffle from the head.
3. Remove the baffle from the cylinder block.

AIR CLEANER

1. Remove the air cleaner cover, element and element retainer.
2. Disconnect the breather tube from the inspection cover.
3. Remove the two capscrews and remove the air cleaner body and gasket from the carburetor.

MUFFLER

Remove the two hex nuts and remove the muffler and gasket from cylinder case.

12. Reassemble, keeping the following points in mind:

  a. Replace the needle valve and seat with a matched valve and seat (as included in a repair kit).

  b. Install the new idle jet and adjusting screw which are included in the repair kit. Make sure the idle jet is tightened firmly.

  c. Install the new main jet, tighten securely, and then install the main jet holder and torque to 5.5 ft. lbs.

  d. When assembling the choke, make sure the flat on the choke valve faces the main air jet.

ENGINE OVERHAUL
Tools

To overhaul this engine, which uses metric fasteners, you should have the following tools:
1. 10mm thin wall socket
2. 10mm standard socket
3. 10mm deep socket
4. 12, 14 and 18mm standard sockets.

Special tools

GOVERNOR LEVER AND CARBURETOR

1. Disconnect the governor spring from the lever and speed control assembly.
2. Remove the governor lever from the shaft, and, at the same time, disconnect the rod and spring from the lever and carburetor.
3. Remove the two nuts and lockwashers, and remove the two nuts and lockwashers, and remove the carburetor, insulating plate and gaskets from the cylinder case.
4. If necessary, the speed control assembly can be removed from the side of the crankcase by removing the wing nut and clip.
5. In reassembly, refer to "Governor Adjustment" above.

STARTING PULLEY AND FLYWHEEL

1. To remove the starting pulley, first remove the three mounting screws.
2. Place a 14mm socket wrench on the flywheel nut and give the wrench a sharp blow with a soft hammer. Remove the nut, spring washer and pulley.
3. Attach puller to flywheel — turn center bolt clockwise until flywheel becomes loose enough to be removed. If puller is not available, screw flywheel nut flush with end of crankshaft to protect shaft threads from being damaged. Place the end of a large screwdriver between the crankcase and flywheel in line with the keyway. Then, strike the end of the flywheel nut with a babbitt or other soft hammer and at the same time wedge outward with the screwdriver.
4. Disconnect the high tension cable from the spark plug, and slip the cable along with the rubber grommet through the hole, to the inside of the crankcase. Then remove the ignition coil along with the attached high tension cable by taking out the two mounting screws, and disconnecting the breaker assembly wire.
5. Remove the contact breaker and condenser by removing the point cover, and removing mounting screws from the cylinder case. Slip the coil primary wire along with the grommet through the hole in the side of the crankcase.
6. In reassembly, refer to "Setting Point Gap and Timing."
7. Securely tighten the flywheel nut after the timing is finalized, but first be sure the woodruff key is in position on the shaft. Do not drive the flywheel onto the taper of the crankshaft and do not overtighten the flywheel nut. Simply turn the nut until the lockwasher collapses. Then, tighten it by placing a wrench on the nut and giving the handle of the wrench 1 or 2 sharp blows with a soft hammer. If a torque wrench is available, tighten to 44–47 ft. lbs.

CYLINDER HEAD AND SPARK PLUG

1. Remove the spark plug from the cylinder head.
2. Loosen the mounting nuts and remove the cylinder head along with the gasket.
3. Clean the carbon from the combustion chamber and the dirt from among the cooling fins. Check the cylinder head mounting face for distortion. If warpage is evident, replace head.
4. In reassembly, use a new cylinder head gasket and spark plug. Torque the head nuts to 22 ft. lbs. Leave the spark plug out temporarily, for each in turning engine over for the remainder of assembly and for timing adjustments. When installing the spark plug, tighten it 24–27 ft. lbs. torque.

INTAKE AND EXHAUST VALVES

1. Remove the valve inspection cover, breather plate and gaskets from the cylindercase.
2. Lift the valve springs, by means of a compressor tool EYY790–282 or equivalent and remove the retainer locks with

Correcting the valve seats

Pulling out valve guides with special tool

long nose pliers. Release the compressor tool and remove the intake valve and exhaust valve along with their respective spring and retainer. A standard Automotive type valve lifter can be used for removing valves, but is not practical for reasembly.

———————— **CAUTION** ————————
Do not damage the gasket surface of the tappet chamber with the compressor tool.

3. Clean carbon and gum deposits from the valves, seats, ports and guides.
4. In reassembly, replace valves that are badly burned, pitted or warped.
5. Correct the valve seat by using a 45 degree seat cutter tool No. EYH640–118 or equivalent as illustrated. The finished seat width should be 0.047–0.059 in. — maximum usable width is 0.098 in.

———————— **CAUTION** ————————
Do not use an electric power driven grinding wheel to correct the valve seats.

6. Valve guides should be replaced when the valve stem clearance becomes excessive. Use a valve guide puller tool EYY790–324 or equivalent. See illustration. Draw the valve guides out and press new guides in using the same puller tool. Refer to the illustration for clearance specifications and proper assembly.
7. After correcting the valve seats and replacing the valve guides, lap the valves in place until a uniform ring will show entirely around the face of the valve. Clean the valves, and wash the block thoroughly with a hot solution of soap and water. Wipe the cylinder walls with clean, lint free rags and light engine oil. Do not assemble the valve springs until the tappet clearance has been checked. See "Tappet Adjustment" below.

VALVE GUIDE

A – VALVE FACE ANGLE	45°
B – SEAT ANGLE	45°
C – GUIDE INSIDE DIAMETER	0.2756 to 0.2770"
D – VALVE STEM DIAMETER	0.273 to 0.274"
MAXIMUM ALLOWABLE CLEARANCE BETWEEN C AND D	0.006"

Valve Specifications

TAPPET ADJUSTMENT

With the tappet in its lowest position, hold valve down and insert a feeler gauge between the valve and tappet stem. The clearance for both intake and exhaust, with the engine cold is: 0.006–0.008 in.

If the clearance is less than it should be, grind the end of valve stem a very little at a time and remeasure. Stems must be ground square and flat.

If the clearance is too large, sink the valve seat with seat cutter tool.

After obtaining the correct clearance, assemble the valve springs and retainers, and secure them in place with the retainer locks. Check the operation of the valves by turning the crankshaft over by hand and remeasure the tappet clearance.

GEAR COVER AND CRANKSHAFT REMOVAL

1. Place a rag under the engine to absorb the remaining oil. Remove the gear cover mounting screws.

2. With a soft hammer, tap at even intervals around the outer surface of the gear cover until it breaks free of the crankcase face. Break the cover free carefully, so as to avoid damaging the oil seal.

3. Inspect the adjusting collar, oil seal, governor shaft and yoke. Replace any parts that are damaged or excessively worn.

4. Remove the flywheel woodruff key.

5. Pull the crankshaft out from open end of crankcase and take care not to damage the oil seal. If necessary, loosen shaft by tapping lightly at the flywheel end with a soft hammer.

NOTE: The "Gear Cover and Crankshaft Installation" procedure is located with camshaft, connecting rod, and

piston service, since that is the normal sequence of doing overhaul work.

CAMSHAFT, TAPPETS AND TIMING MARKS

1. To prevent the tappets from falling out and becoming damaged when the camshaft is removed, turn the crankcase over on its side as shown. Push the tappets inward to clear the cam lobes, and remove the camshaft.

2. Withdraw the tappets and mark them for identification with the hole from which they were removed.

3. In reassembly, put the tappets back in their corresponding guide hole. This will eliminate unnecessary valve stem grinding for obtaining correct tappet clearance.

4. Mount governor sleeve on end of camshaft by holding the shaft in a vertical position with the flyweights hanging down. Align the groove in the flange with the pin in the gear face and install the sleeve on shaft so that its flange fits in the groove between the heavy end and the thrust tabs of all three flyweights. Mount camshaft assembly in crankcase.

5. The timing marks on the camshaft gear and crankshaft gear must be matched up. Mount the camshaft so that the marked tooth on crankshaft gear is between the two marked teeth of the camshaft gear, see the illustration. If the valve timing is off, the engine will not function properly or may not run at all.

TIMING MARK ON CRANKSHAFT GEAR

CAMSHAFT   GOVERNOR SLEEVE   TAPPETS

Postion in which engine should be placed while removing camshaft

CONNECTING ROD AND PISTON

Disassembly

1. Straighten out the bent tabs of the lock plate and remove the bolts from the connecting rod.

2. Remove lock plate, oil dipper and connecting rod cap.

3. Scrape off all carbon deposits that might interfere with the removal of the piston from the upper end of the cylinder. Use a ridge reamer.

4. Turn the crankshaft until the piston is at top, then push the connecting rod and piston assembly upward and out through the top of the cylinder.

5. Remove the piston from connecting rod by taking out one of the snap rings and then removing the piston pin. A new snap ring should be used in reassembly.

Reassembly

1. Use a ring expander tool to prevent the ring from becoming distorted or broken when installing on the piston.

2. If an expander tool is not available, install the rings by placing the open end of the ring on the first land of the piston. Spread the ring only far enough to slip it over the piston and into the correct groove, being careful not to distort the ring.

3. With the expander tool, assemble the bottom ring first and work upward, installing the top ring last.

4. Mount the scraper ring with the scraper edge down, otherwise oil pumping and excessive oil consumption will result. Refer to the illustration for correct placement of rings.

5. Measure the diameter of the piston in the center of the thrust faces at the bottom of the piston skirt, as illustrated.

6. Measure the cylinder bore and inspect it for out-of-round and taper. If the cylinder is scored or worn more than 0.005 in. over standard size, it should be rebored and fitted with an oversize piston and rings. Refer to the chart for the clearance between the piston and cylinder. Size, clearance and wear limits are given in more detail at the end of this chapter.

7. When installing the piston in the cylinder, oil the piston, rings, wrist pin, rod bearings and cylinder wall before assembly. Stagger the piston ring gaps 90 degrees apart around the piston. Use a piston ring compressor.

8. Turn the crankshaft to the bottom of the stroke and tap the piston down until the rod contacts the crankpin. Mount the connecting rod cap so that the cast rib between the face of rod and bolt boss matches up with the cast rib on the connecting rod. Assemble oil dipper to cap. The dipper should be toward the gear cover end of the connecting rod cap if engine is operated on a tilt toward the take off end. Mount the dipper toward the flywheel end if it's tilted in that direction or with a no tilt operation.

Connecting rod and oil dipper assembly

Measure piston diameter as shown

Location and positioning of rings

PISTON TO CYLINDER AT PISTON SKIRT THRUST FACE		EY18W .0016 .0032"
CONNECTING ROD TO CRANK PIN	DIA.	.0021 .0031"
	SIDE	.008 .0235"
PISTON PIN TO CONNECTING ROD		.0004 .0012"
D - CRANKSHAFT PIN DIAMETER		1.0210 1.0215"
W - CRANKSHAFT PIN WIDTH		.9846 .9882"
PISTON RING GAP		.002 .010"
PISTON RINGS - SIDE CLEARANCE IN GROOVES		.0004 .0030"
PISTON PIN TO PISTON		.00035" tight to .00039" loose

STANDARD CRANK PIN DIMENSIONS

Piston, ring, and rod clearance specifications

Install a new rod bolt lock plate. Mount the connecting rod bolts and tighten to the following torque specifications.
● Model EY18W—12.5–14.5 ft. lbs.

Check for free movement of the connecting rod by turning the crankshaft over slowly. If satisfactory, bend the tabs on the lock plate against the hex head flat of the connecting rod bolts.

GEAR COVER AND CRANKSHAFT INSTALLATION

1. In reassembly, inspect the crankcase oil seal and main bearing for possible replacement. Mount the crankshaft with extreme care so as not to damage the lips of the oil seal. Use an oil seal sleeve if available.

2. End Play is regulated by means of the adjusting collar at the gear end of the crankshaft. This should be set immediately before mounting the gear cover as explained below.

NOTE: Crankshaft End Play is regulated by the length of the adjusting collar. The end play should be 0.001–0.009 in., engine cold. The adjusting collar is located between the crankshaft gear and main bearing at the take-off end of the engine. Replacement of the collar is seldom necessary unless the crankshaft or gear cover is replaced. To determine what length adjusting collar to use, refer to the illustration and steps below.

1. With the gear cover removed, tap the end of the crankshaft slightly to insure that the shaft is shouldered against the front end main bearing.

2. Use a depth micrometer and measure the distance between the machined surface of crankcase face and end of crank gear (dimension A)

3. Measure the distance between the machined surface of gear cover and end of main bearing (dimension B).

4. The compressed thickness of gear cover gasket is 0.007 in. (dimension C).

5. Select an adjusting collar that is 0.001–0.009 in. less than the total length of A, B, and C from the chart.

6. Apply oil to the bearing surfaces, gear train and tappets. Also lubricate the lips of the oil seal and add a light film of oil on the gear cover face to hold the gasket in place.

7. Mount the adjusting collar to the crankshaft with the recess toward the crankgear.

8. Assemble the gear cover, being sure that the governor yoke is in a downward position, and be extremely careful not to damage the lips of the oil seal. If available, mount an oil seal sleeve on the crankshaft to prevent damage to the oil seal lips.

--- CAUTION ---

Be sure the timing marks on the crankshaft and camshaft gear remain correctly mated when the end of the camshaft is fitted into the bearing hole of the gear cover.

Adjusting Collar Dimension Chart

Model	Collar Length	Part Number
EY18W	.701 to .709″	EY18D2112a
	.709 to .717″	EY18D2112b
	.717 to .725″	EY18D2112c

9. Tap the gear cover in place with a soft hammer, remove the oil seal sleeve and tighten the gear cover capscrews to 13 ft. lbs. torque.

10. Tap the end of the crankshaft with a soft hammer so that the crankshaft will shoulder against the main bearing at the flywheel end.

11. Tap the crankshaft in the opposite direction (from flywheel end) to seat the adjusting collar against the main bearing at the takeoff end.

12. Attach a dial indicator to one of the $5/64$–24 tapped holes on the face of the crankcase with the indicator plunger resting against the end of the crankshaft and set the dial at 0.

CRANKSHAFT

MAIN BEARING

CRANK GEAR

ADJUSTING COLLAR

Measuring points for dimensions critical to crankshaft end play adjustment

13. Wedge a screwdriver between the flywheel and the crankcase. The movement of the flywheel away from the engine block will register as end play on the indicator dial.

14. If end play is not within the limits of 0.007–0.009 in., use the reading from the indicator dial to determine the new length adjusting collar to use. Replace the collar with one of the proper dimension, if necessary.

FINAL CHECKOUT AND BREAK-IN

After the major moving parts are assembled, turn the engine over using the starter pulley. Make sure it turns without unusual resistance. Check the basic ignition adjustments (point gap and timing), and make preliminary carburetor and governor adjustments. Refill the crankcase and fuel tank.

Break the engine in by running it as specified on the chart below, proceeding from top to bottom. Check the engine frequently for leaks, and make final carburetor and governor adjustments during the sequence.

Break-in Sequence

Load			Speed	Time
No load			2500 rpm	10 minutes
No load			3000 rpm	10 minutes
No load			3600 rpm	10 minutes
1.75 hp	2.5 hp	2.65 hp	3600 rpm	30 minutes
3.5 hp	5.0 hp	5.3 hp	3600 rpm	60 minutes

Clearances and Wear Limits

Description	Model EY18W Tol	Limit	Description	Model EY18W Tol	Limit
Flatness of cylinder head	.004	.008	Piston pin to connecting rod clearance	.0004 to .0012	.0032
Cylinder bore	2.5591 2.5599	2.5655	Con. rod—large and small bore alignment (parallel)	.002	.004
Bore-out of round	.0005	.003	Con. rod—large and small bore centers	3.9370	3.943
Cylindricity (taper)	.0006	.006	Crank pin—out of round	—	.0002
Valve seat width	.047 .059	.098	Cylindricity (taper)	—	.0002
Valve guide bore	.2755 .2770	.280	Crank pin—parallel	—	.00032
Piston diameter at skirt thrust faces (standard size)	2.5567 2.5575	2.5535	Crankshaft journal diameter (take-off end)	.9838 .9842	.982
Piston to cylinder clearance at skirt thrust faces	.0016 to .0032	.008	Crankshaft journal (flywheel end)	.9839 .9843	.982
Ring groove width (top and 2nd ring)	.0985 .0995	.105	Crankshaft end play	.001 .009	—
Ring groove width (all ring)	.1575 .1585	.164	Camshaft (cam rise)	1.2087 1.2165	1.199
Ring width (top and 2nd ring)	.097 .098	.0935	Camshaft (journal diameter)	.5889 .5893	.587
Ring width (oil ring)	.1563 .1570	.1523	Valve spring (free height)	1.4173	1.3582
Piston rings—side clearance in groove	.0005 .0030	.006	Valve spring (squareness)	—	.039
Ring gap (at cylinder skirt)	.002 .010	.040	Valve stem diameter	.273 .274	.270
Pin hole in piston	.5509 .5513	.5523	Valve stem clearance to guide	.0016 .0039	.006
Piston pin diameter	.5509 .5512	.5499	Valve stem—lock-pl. to groove clearance	.0016 .0059	.020
Piston pin to piston fit	.00035 tight to 00039 loose	.0023 loose	Valve—length from groove to stem end	.1575	.0785
Connecting rod (crank pin end)	1.0236 1.0241	1.026	Valve—tappet clearance	.006 .008	.002 .010
Crank pin diameter	1.0210 1.0215	1.019	Tappet—length	1.811	1.801
			Tappet—stem to guide clearance	.0010 .0024	.004
			Ignition timing	23° (±2°)	—

Clearances and Wear Limits

Description	Model EY18W		Description	Model EY18W	
	Tol	Limit		Tol	Limit
Connecting rod to crank pin clearance	.0021 to .0031	.005	Breaker contact opening	.014 (±.001)	—
Connecting rod side clearance	.0080 .0235	.039	Spark plug gap	.020 .025	—
Connecting rod (piston pin end)	.5516 .5520	.5555	Spark plug	14 mm, Champ. L86, AC 44F, NGK B6HS	

Section 20

Wisconsin

Small Engines—6-20 hp

ENGINE IDENTIFICATION

There is a Wisconsin name plate attached to the blower housing of the engine on which is stamped the model number, serial number, and specification number along with the size and rpm rating. The model, serial, and specification number must be given when obtaining replacement parts for any of the engines. Make certain that the identification plate remains with the engine on which it was originally installed.

General Engine Specifications

Model	Bore & Stroke (in.)	Displacement (cu. in.)	Horsepower @ RPM
S-7D	3 x 2⅝	18.6	7.25 @ 3600
S-8D	3⅛ x 2⅝	20.2	8.25 @ 3600
TR-10D	3⅛ x 2⅝	20.2	—
TRA-10D	3⅛ x 2⅞	22.05	10.1 @ 3600
TRA-12D	3½ x 2⅞	27.66	12.0 @ 3600
ALN	2⅝ x 2¾	14.9	6.0 @ 3600
BKN	2⅞ x 2¾	17.8	7.0 @ 3600
AEN	3 x 3¼	23.0	9.2 @ 3600
AENL	3 x 3¼	23.0	9.2 @ 3600
AENS	3 x 3¼	23.0	9.2 @ 3600

MAINTENANCE

Air Cleaner Service

DRY ELEMENT TYPE

If the unit is operated in a very dusty atmosphere, remove the element by unscrewing the wingnut and removing the cover.

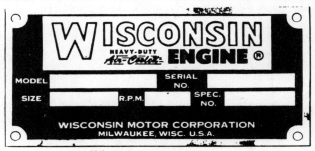

Wisconsin name plate

Shake out accumulated dirt (do not tap) once each day. Under normal operating conditions, the most frequent service required is a weekly washing of the element. Rinse the element under cold water and then dip repeatedly into a solution of mild, non-sudsing detergent and warm water. Rinse in cold water and allow to dry overnight. Avoid freezing temperatures until the element is dry.

Dry type air cleaner

After five washings, or one year of service, whichever comes first, replace the element.

OIL BATH TYPE

All Series Except ACN and BKN

If the engine is run in a very dusty atmosphere, service the cleaner daily according to the instructions printed on the unit.

Ordinarily, remove the wingnut and pull off the cover and filter element and wash it in solvent. Drain oil from the filter bowl and clean it with a clean rag. Fill the filter bowl with engine oil to the line, and reassemble the unit. Install the wing nut.

ACN and BKN

The air cleaner must be serviced frequently: weekly in a clean atmosphere, and as often as daily in a dusty atmosphere. Ordinary service consists of snapping off the spring bail and removing the bowl from the bottom of the unit for service. Clean out the cup and baffle, and then refill with about ¼ pint of engine oil.

The filter element does not ordinarily need service and may be left on the engine. However, if extreme conditions have made it dusty, remove it from the engine bracket and wash it in solvent.

Oil bath type air cleaner

free of moisture and sediment. Remember that fuel of too low an octane rating may cause engine knock and severe damage.

Check the oil level every 8 hours and replenish. Check more frequently when the engine is new. Drain the old oil and replace it every 50 operating hours. Always change the oil when the engine is hot, by removing the crankcase drain plug. Fill crankcase to the level of the filler plug hole.

Crankcase Capacity Chart

S-7D, S-8D, TR-10D, TRA-10D, TRA-12D	1 qt.
AEN, AENL, AENS	3 pts.
ACN, BKN	2 pts.
Clutch Unit Housing—ACN, BKN	½ pt.
Reduction Unit Housing—ACN, BKN	1 pt.

Special type oil bath air cleaner used on series ACN and BKN

AIR CLEANER BODY

BAIL

BAFFLE

OIL LEVEL

AIR CLEANER BOWL

Lubrication

OIL AND FUEL RECOMMENDATIONS

Oils of grades MS, SD, or SE may be used in Wisconsin engines. Viscosity recommendations are as follows:

- Above 40°F — SAE 30
- 15°F ± 40°F — SAE 20–20W
- 0°F ± 15°F — SAE 10W
- Below 0°F — SAE 5W-20

Fuel should be regular grade of 90 octane or above. Fuel should be of known quality to provide adequate protection against gum formation, and adequate assurance that it will be

TUNE-UP

Spark Plugs

The outside of the plug, the electrodes and insulator on the underside should be kept clean. As often as significant deposits form, remove the plug and wire brush deposits away. Set the spark plug gap by bending the side electrode to get a gap of 0.030 in., as measured by a wire type feeler gauge. Clean the threads on the plug and in the cylinder head before installing the plug. Use a new gasket, and torque to 25–30 ft. lbs. If the plug has deposits than cannot be removed, or if the electrodes are badly burned or there is evidence of cracking of the insulator, either inside or outside, replace the plug.

Breaker Points

REMOVAL AND INSTALLATION

1. Remove the breaker box cover.
2. Disconnect the terminal strip by loosening the screw and pulling it off the contact set.
3. Remove the point attaching screws and remove the contact set.

Tune-Up Specifications

Model	Plug Type	Plug Gap (in.)	Point Gap (in.)	Idle Speed
S-7D	AC-C86, Champion D-16J	.030	.020	①
S-8D	AC-C86, Champion D-16J	.030	.020	①
TR-10D	AC-C86, Champion D-16J	.030	.020	②
TRA-10D	AC-C86, Champion D-16J	.030	.020	②
TRA-12D	AC-C86, Champion D-16J	.030	.020	②
ACN	AC-C86, Champion D-16J	.030	.020 ③	④
BKN	AC-C86, Champion D-16J	.030	.020 ③	④
AEN	AC-C86, Champion D-16J	.030	.020 ③	④
AENL	AC-C86, Champion D-16J	.030	.020 ③	④
AENS	AC-C86, Champion D-16J	.030	.020 ③	④

① Throttle screw 2 turns open, or lowest smooth speed
② Throttle screw 1¼ turns open, or lowest smooth speed
③ Applies to battery ignition—with magneto, point gap is .015
④ Lowest smooth speed

4. To install, first position the points, noting that on some types a prong located on the underside of the contact set must fit into a hole in the breaker box. Install the mounting screw or screws just tightly enough to hold the contacts in place.

5. Set the gap and time the engine as described in the procedures below.

SETTING BREAKER GAP

1. If necessary, loosen the breaker mounting screws so the point gap can be changed. Turn the engine flywheel back and forth until the contacts are as far apart as they can be.

2. Place the screwdriver in the adjusting slot and slide a flat feeler gauge of the proper dimension (see tune-up chart) between the contacts.

3. Adjust the gap with the screwdriver until the gauge has a very slight pull when sliding straight through the point gap.

Locations of breaker points, lockscrew, and terminal strip in a typical breaker box

4. On AEN, AENL, AENS, ACN, and BKN, tighten the contact mounting screw. Then, recheck the gap. Reset if necessary. On these engines, timing need not be reset after contact gap adjustment unless the magneto or timer position has been disturbed.

5. On all other engines, leave the contact mounting screw(s) only slightly tight, and proceed to the engine timing procedure below.

Ignition Timing Adjustment

S–7D, S–8D, TR–10D, TRA–10D and TRA–12D

1. Remove the breaker box cover. Disconnect the coil primary wire at the bottom of the breaker box.

2. Line up the flywheel timing mark and the pointer with the engine on compression stroke. The timing mark on the flywheel can be seen through the opening on the right side of the flywheel shroud. The engine is on the compression stroke if the breaker arm push pin is moving as the timing marks approach alignment.

3. Connect a self powered test lamp or timing light between an engine ground and the terminal stud on the bottom of the breaker box. If necessary, slightly loosen the contact set mounting screw so the gap can easily be changed.

4. Close the points slowly with a screwdriver in the adjusting slot, just until the light goes out. Tighten the mounting screw.

5. Turn the flywheel counterclockwise until the light goes on, and then rotate it slowly forward and stop just as light goes out. At this point, the timing marks should be lined up. If necessary, readjust the gap slightly. Widen the gap if the light goes out too early; narrow the gap if the light goes out too soon.

6. Install the breaker cover and reconnect the primary wire to the terminal stud.

Magneto timing marks for ACN, BKN engines

AEN, AENL, AENS, ACN and BKN with Magneto

1. The magneto need not be timed unless it is removed from the engine. On ACN and BKN engines, take off the shroud and remove the timing inspection hole plug. On AEN, AENL and AENS engines, simply remove the plug, which is located near the magneto mounting. Remove the spark plug. Then, turn the engine over until the piston is coming up on compression stroke (air will be expelled from the spark plug hole) and the **D/C** and **X** marked vane on the flywheel lines up with the mark on the vertical centerline of the cooling shroud. On AEN, AENL and AENS engines, remove the plug from the hole in the shroud to see the flywheel marks.

2. When installing the magneto on ACN and BKN engines, mesh the magneto and camshaft gears so the two timing marks line up. They are visible through the inspection hole located to the left side of the flywheel. On AEN, AENL and AENS engines,

Flywheel timing marks can be seen by removing the plug in the fan shroud on AEN, AENL, and AENS engines

mesh the magneto gears so that the **X** marked gear tooth is visible through the inspection hole.

3. Timing may be checked by slowly rotating the engine past the point where the flywheel **D/C** and **X** marked vanes pass the vertical centerline mark. The impulse coupling will snap when the marks are lined up, if the timing is correct.

AEN, AENL, AENS, ACN and BKN Engines with Battery Ignition

1. Set the engine flywheel at the position described in Step 1 of the procedure above, in the same way.

2. If the timer unit has been removed from the engine, turn the timer cam counterclockwise, using the gear on the back of the unit, until the points just begin to open (you will feel increased friction). Then, mount the timer to the engine.

3. Loosen the clamp lever screw which keeps the unit from rotating. On ACN and BKN engines, turn the timing unit clockwise $\frac{3}{64}$ in. as measured on the circumference of the timer body, to get 2 degrees of spark advance. On AEN, AENL and AENS engines, rotate the unit clockwise $\frac{1}{8}$ in. to get 5 degrees of advance.

4. Mark the timing marks with chalk, install the spark plug, and connect a timing light.

5. Start the engine and run it at 1800 rpm or higher, as measured with a tach. On ACN and BKN engines, turn the unit as required to align the flywheel mark and the running advance timing mark, located on the shroud to the left (counterclockwise) of the centerline mark. On AEN, AENL and AENS engines, rotate the unit as necessary to align the marked vane of the flywheel with the running advance timing hole in the shroud. Tighten the clamp screw.

Carburetor Mixture Adjustments

1. If the engine seems to be running very poorly due to improper fuel/air mixture, or if it will not start, make the following preliminary settings. Make the settings by turning the mixture screw in until it seats only very gently, then outward the required number of turns:
- S–7D, S–8D – Turn main jet adjustment out 1–1¼ turns.
- AEN, AENL, AENS – Turn main jet adjustment out 1¼ turns.
- ACN, BKB – Turn main jet adjustment out 1¼ turns.
- TRA–12D, TRA–10D, TR–10D – Turn main jet open 1¼ turns, idle jet open 1 turn.

2. After making preliminary settings, run the engine until hot and check acceleration. If engine stumbles, open the main jet ¼ turn at a time until response is smooth.

3. Slow the engine down to idle speed and adjust idle mixture screw in or out for the smoothest idle.

Governor Adjustment

S–7D, S–8D, TR–10D, TRA–10D AND TRA–12D

1. Loosen the governor lever clamp screw so that the fulcrum shaft can be turned independently of the governor lever and the lever moves to full throttle position. Then, turn the shaft counterclockwise until the internal governor vane stops against the flywheel thrust pin.

2. Tighten the clamp screw. See "Speed Adjustment" below.

ACN, BKN, AENL, AEN AND AENS

1. Disconnect the rod at the governor lever. Move the rod as far as possible toward the carburetor.

2. Move the governor lever as far as possible in the same direction.

3. Hold both parts in this position and turn the rod in or out of the swivel block until the hole in the lever indexes with the end of the rod. On ACN and BKN engines, turn the rod two more turns inward. Install the rod in the lever and install the cotter pin. See "Speed Adjustment" below.

SPEED ADJUSTMENT

The governor lever is provided with a number of holes so that the engine can be operated at different speeds. If the governor spring has been removed from the hole, or if the speed range of the engine is to be changed, the proper hole in the lever must first be selected, and the adjusting screw must then be turned for fine adjustment.

1. Run the engine until hot, and connect a tachometer. Open the throttle control and install the governor spring into each of the holes in the lever to get as close as possible to the desired rpm (holes further away from the fulcrum of the lever give more speed). Once you've found the hole nearest the desired speed, note which hole you are using.

2. Loosen the locknut, disconnect the spring, and turn the screw for more tension to increase speed, or for less tension to decrease it.

Compression Check

No precise method of checking compression is required. However, on engines without compression release, compression may be checked by spinning the engine in the normal direction of rotation and checking for a substantial increase in resistance when the piston begins coming up on the compression stroke.

Generally, when compression is poor, the engine requires disassembly and major work. However, compression can be low because a long period without operation has permitted oil to drain off the cylinder walls. If this is suspected, remove the spark plug and squirt a small quantity of engine oil into the combustion chamber to seal it.

FUEL SYSTEM

Carburetor

DISASSEMBLY AND REASSEMBLY

NOTE: See item below for inspection and cleaning procedure.

Zenith 87 and Wisconsin L–51

1. Remove the three bowl assembly screws (37 & 38) and lockwashers (36) and separate fuel bowl (30) from throttle body (9).

2. Remove the main jet adjustment (34) and fiber washer (33), using a $9/16$ in. open end wrench.

3. Remove the main jet (32) and fiber washer (31), using Zenith Tool No. C161–83 main jet wrench or equivalent.

4. Remove the idle jet (29), using a small screwdriver.

5. Remove the bowl drain plug (35).

6. Remove the float axle (26) by pressing against the end with the blade of a screwdriver.

7. Remove the float (27).

8. Remove the fuel valve needle (25), with your fingers.

9. Remove the fuel bowl to throttle body gasket (28).

10. Remove the main discharge jet (23), using a small screwdriver.

11. Remove the fuel valve seat (25) and fiber washer (24), using Zenith Tool No. C161–85 or equivalent.

12. Remove the idle adjusting needle (11) and spring (10).

Making S-7D and S-8D governor adjustment

13. Install the fuel valve seat (25) and fiber washer (24), using Zenith Tool No. C161–85 or equivalent.

14. Install the main discharge jet (23), using a small screwdriver.

15. Install fuel valve needle (25) in seat (25), followed by float (27). Insert tapered end of float axle (26) into float bracket on side opposite slot and push through the other side. Press float axle (26) into slotted side until the axle is centered in bracket.

Exploded view of Zenith 87 series carburetor

61/64" ±1/32"

REMOVE GASKET

HOLD FLOAT SO THAT
LEVER CONTACTS HEAD
OR PIN WITHOUT
PRESSURE

Adjusting float level—Zenith 87 series carburetor

16. Check position of float assembly for correct measurement to obtain proper fuel level using a depth gauge.

NOTE: Do not bend, twist, or apply pressure on the float body. With bowl cover assembly in an inverted position, viewed from free end of float, the float body must be centered and at right angles to the machined surface. The float setting is measured from the machined surface (no gasket) of float bowl cover to top side of float body at highest point. This measurement should be $\frac{61}{64}$ in., plus or minus $\frac{1}{32}$ in. To increase or decrease distance between float body and machined surface use long nosed pliers and bend lever close to float body. Replace with new float if position is off more than $\frac{1}{16}$ in.

17. Install throttle body to fuel bowl assembly gasket (29) on machined surface of throttle body (9).
18. Install the idle adjusting needle (11) and spring (10).
19. Install the main jet (32) and fiber washer (31), using Zenith Tool No. C161–83 main jet wrench or equivalent.
20. Install the main jet adjusting needle assembly (34) and fiber washer (33), using a $\frac{9}{16}$ in. open end wrench.
21. Install the idle jet (29), using a small screwdriver.
22. Install the bowl drain plug (35).
23. Install the three bowl assembly screws (38) and lockwashers (36) through the fuel bowl and into the throttle body and draw down firmly and evenly.

Zenith 72Y6

1. Remove the three assembly screws (2) that hold the bowl cover to the bowl.
2. Separate the bowl cover assembly (1) from the bowl assembly.
3. Remove the float axle (13) and float (12).
4. Remove the bowl cover gasket (16).
5. Remove the fuel valve needle and seat (15) with the gasket (14). Remove the fuel valve seat. Use tool C161–85 or equivalent.
6. Remove the idle adjusting screw needle (3) and spring (4).
7. Remove the throttle stop screw (5) and spring (6).
8. Hold the bowl cover inverted with the mounting flange to your right. Note that the closed throttle plate slopes down and away from you and that there is a mark stamped on the high side of the throttle plate. It is important that upon reassembly this same relationship is retained.
9. Hold the throttle shaft in the closed position, remove the throttle plate screw and lockwasher (11) and throttle plate (10).

10. Remove the throttle shaft and lever (9).
11. With a screwdriver, or similar tool, remove the throttle shaft dust seal retainer (8) and rubber shaft seal (7).
12. Remove the venturi (17) and idle tube (18) by inverting bowl cover.

NOTE: Do not attempt to remove the main discharge jet. This part is pressed in and is not a serviceable item.

13. Remove the lower main jet plug if a plug is used or remove the main jet adjustment (20) and gasket (21). Then remove the main jet (22). Use tool C161–83 or equivalent.
14. Hold the bowl (19) in a vertical position with the air intake and bottom of bowl next to you and note that the closed choke plate slopes down and away. Observe the marking on the choke plate. It is important that upon reassembly, this same relationship is retained.
15. Remove the choke plate screws and lockwashers (26), choke plate (25), shaft and lever (24).

NOTE: During cleaning, do not soak the float in solvent.

16. Hold the bowl (19) vertically (with air intake upward), and insert the choke shaft and lever with the lever pointing downward.
17. Install the choke plate (25) with the letter "Z" toward the bottom of bowl.
18. Start, but do not tighten both the choke plate screw and the lockwasher (26).
19. Center the choke plate in the air intake bore by lightly tapping the choke plate on the high side. Hold it in this position with a finger and tighten the choke plate screws.
20. Install the main jet (22) using tool C161–83 or equivalent.

NOTE: Before installing the main jet (20) and gasket (21), turn the adjusting needle several turns to the left

Exploded view of Zenith 72Y6 carburetor

(counterclockwise) to avoid damage to the main jet orifice during assembly.

21. Install the main jet plug with a new fiber washer (21).

22. Hold the bowl in the operating position and install the idle jet (18), tube end down.

23. Install the venturi with the key at the lower edge of the venturi, in the matching slot at the choke valve side of the bowl.

24. Assemble a new rubber dust seal (7) against the throttle shaft bearing, with the lips of the seal toward the outside.

25. Install and stake the seal retainer washer (8).

26. Assemble the throttle shaft and lever (9) with the wide open stop lug (narrow lug) on shaft lever in contact with the stop on the casting when the shaft is in the wide open throttle position.

27. Hold the bowl cover inverted, with the mounting flange toward your right.

28. Install the throttle plate (10) with the mark stamped on the throttle plate on the high side of the plate and toward you.

29. Start, but do not tighten both throttle plate screw and lockwasher (11).

30. Gently tap the high side of the throttle plate to center the plate. Hold in this position with a finger and tighten the throttle plate screws.

31. Install the throttle stop screw (5) and spring (6).

32. Install the idle adjusting needle (3) and spring (4).

33. Install the fuel valve seat (15) with a fiber washer (14). Use tool C161–85 or equivalent.

34. Assemble the fuel valve needle and bowl cover gasket (16).

35. Carefully examine the float assembly (12) for evidence of wear or damage. This type of float is not adjustable and wear in any part of the fuel valve and float hinge assembly will raise the fuel level.

36. Install the float and float axle pin (13). Insert the bowl cover and check the float in the closed position. The float setting will be within limits if the float is parallel to the gasket seating surfaces of the bowl cover. Any necessary float correction should be made by replacing worn parts. DO NOT attempt to bend the float bracket.

37. Attach the bowl cover assembly (1) to the bowl using the three assembly screws (2).

Zenith 68–7

Use the detailed exploded view to guide you in disassembling and reassembling the carburetor. Clean and inspect all parts as described in the section below. When the float has been reassembled to the throttle body, invert the throttle body and support the float so that the lever contacts the head of the float pin without pressure. Measure from the surface of the casting (without gasket) to the top surface of the float (which is the bottom surface during normal operation). The distance should be $15/32$ in. plus or minus $1/32$ in. If distance is incorrect, bend the float lever close to the float body with long nose pliers.

CLEANING AND INSPECTION

1. Clean all of the metal parts in a suitable solvent, removing all carbon deposits from the throttle bore and idle discharge passages. To ensure that all dirt is removed, blow compressed air through all passages in the throttle body and fuel bowl in the reverse direction from normal flow.

NOTE: Never use wire or a drill to clean jet orifices or idle port openings.

2. Check the float and make sure that it is not soaked with gasoline. Also check for wear on the float hinge and where the float contacts the inlet needle. Replace the float if any of the above conditions exist.

3. Inspect the main jet adjustment needle and the idle adjusting needle tapered ends to make sure that they are smooth and

Exploded view of Zenith 68-7 carburetor

not grooved from being seated too hard. If there is a groove around the end of the taper, or if it is pitted, replace the needle.

4. Check the fuel inlet valve and seat for wear or damage. Replace the entire assembly as a unit if it doesn't look like new.

5. All gaskets, seals, retainers, and rubber O-rings must be replaced every time the carburetor is overhauled, with the possible exception of the rubber O-rings which can be retained if they are in good condition.

6. Make preliminary adjustments of the main and idle jet adjusting needles before remounting the carburetor on the engine.

Fuel Pump

These instructions refer to overhaul of the LP–62 Series fuel pumps used on some Wisconsin engines. The pump requires rebuilding sometime after 500 hours of operation.

1. Disconnect the fuel lines and, if so equipped, remove the fuel strainer.

2. Scribe a mark across the two halves of the body. Use this mark to positively indicate fuel line inlet and outlet positions. Then, remove the fuel head-to-bracket screws (12), and remove the fuel head (10).

3. Turn the fuel head over, note the positions of the valve assemblies, and then discard them. Clean the fuel head thoroughly in kerosene using a fine wire brush.

4. Hold the head with the diaphragm surface upward, and evenly press in new valve gaskets. Carefully press in new valve assemblies evenly and without any distortion. Make sure each assembly faces in the proper direction — they are check valves.

Fuel pump—exploded view

5. Remove the rocker arm spring (11) from the lower diaphragm section by inserting a screwdriver between coils and prying it out.

6. Hold the mounting bracket (9) in your left hand with the rocker arm toward you and your thumbnail on the end of the link (8). Compress the diaphragm spring (3) by placing the heel of your other hand on the diaphragm (2), and then rotate your hand 90 degrees clockwise to unhook the diaphragm from the link. Remove the diaphragm.

7. Clean the mounting bracket in the same way you cleaned the fuel head.

8. Install the new diaphragm spring onto the bracket (9). Reconnect the new diaphragm to the link by reversing the removal procedure (Step 6). Replace the rocker arm spring (11).

9. Mount the completed mounting bracket assembly (9) onto the engine, using a new gasket (13).

10. Crank the engine over until the diaphragm is laying flat on the mounting bracket. Remount the fuel head (10) with match marks aligned, tightening the screws only three turns. Crank the engine over until the diaphragm is pulled down to its lowest position. Fully tighten screws.

11. Remount the strainer (if so equipped) and install fuel lines to proper connections.

Governor

The governor consists of hardened parts which are only slightly stressed, so repair is rarely necessary. If parts must be replaced,

however, the following procedure is useful in reassembling these heavily sprung parts:

1. Slip the spacer onto the camshaft first. Then, separate the flyweights far enough to permit the thrust sleeve to pass between them. Slide the thrust sleeve back so the flyweights will be closed down between the two flanges of the thrust sleeve.

ENGINE OVERHAUL

The disassembly, inspection, and assembly of each component of the engine is discussed separately, because many times it is not necessary to disassemble the entire engine. The order in which the disassembly procedures are given may be changed to suit the job.

Whenever the engine is either partially or completely disassembled, all of the parts removed should be thoroughly cleaned. Be sure to use new gaskets when reassembling the engine and to lubricate all bearings.

If the engine is to be completely overhauled, remove the engine from the machinery it drives or operates and remove any accessories. If an external component is to be removed, or a minor adjustment made, it may not be necessary to remove the engine from the equipment it powers.

Fuel Tank

Close the fuel tank outlet valve and remove the fuel line. Unscrew the nuts or bolts that retain the tank to the cylinder head bolts and crankcase. The tank and bracket may then be removed as a complete unit. Replace the tank in reverse order, torquing cylinder head in sequence.

Air Cleaner and Carburetor

Unscrew the wing nut and remove the air cleaner. Remove the breather line at the inspection cover, the throttle rod clip at the governor lever, and the fuel line. Unscrew the bolts which hold the carburetor bracket and manifold to the engine and remove the carburetor and air cleaner bracket and the manifold as one. Replace in reverse order.

Starter Sheave and Flywheel Shroud

Remove the starter sheave by removing the three screws and washers which retain it to the flywheel. Remove the top cover and the cylinder side shroud. Disconnect the governor spring and remove the four screws that hold the flywheel shroud to the back plate. The entire flywheel shroud may now be removed.

Assembling the governor

The back plate can be removed, if necessary, only after the flywheel is removed. Reassemble in the reverse order. Apply a $\frac{1}{4}$
in. long bead of #271 Loctite to the thread ends of the capscrews
for mounting the sheet metal starter sheave. Use plain washers
and lockwashers in place of the rubber washers used previously,
and torque to 9–10 ft. lbs.

Rope Starter Sheave

Loosen and remove the rope starter sheave by installing a
wrench on the hexagonal hub of the sheave and striking a sharp
blow in the proper direction. Install in reverse order.

Air Shroud

Remove the cylinder head capscrews and, in cases where so
equipped, the crankcase capscrews, and remove the air shroud.
Usually, the fuel tank must be removed to remove this shroud,
and in some cases, common mounting bolts may be used.

Cylinder Head

Remove the spark plug and unscrew the five cap screws that attach the cylinder head. Remove the cylinder head and gasket.
Clean the carbon from the combustion chamber and all dirt
from the cooling fins. Use a new gasket when installing the
head. If screws of different lengths are used, judge their locations in reassembly from the lengths of the bosses on the head.
 Torque the bolts precisely according to instructions below:
 • ACN and BKN – Torque bolts to 14–18 ft. lbs.
 • TR–10D, TRA–10D, TRA–12D, S–7D and S–8D – torque to
10 ft. lbs. all around; then to 14 ft. lbs; and finally to 18 ft. lbs.
 • AEN, AENL and AENS – Coat screw threads with a mixture of oil and graphite and torque to 32 ft. lbs.

Valves and Valve Seat Inserts

Remove the valve inspection cover which is also the breather assembly. Use a valve spring compressor to compress the valve
springs. On TR and TRA Model engines, be careful not to damage the breather reed in the valve spring compartment in compressing valve springs. Remove the valve spring retainers, the
compressor, and the valve springs; take the valve out from the
top of the cylinder block. Clean all carbon deposits from the
valves, seats, ports, and guides. Inspect the condition of the

**LARGE INSIDE RADIUS
TOWARD INLET VALVE**

**Removal and installation of the cylinder head and
gasket**

valves, stems, guides, and seats, looking for burned, pitted,
scored, or warped surfaces.
 The exhaust valve and seat are made of stellite. A valve rotator is used on the exhaust valve only. Clean the valve rotator
and make sure that it operates properly.
 Both intake and exhaust valves have removeable seat inserts
on AEN, AENL and AENS models. On all other models, only the

**REMOVING SEAT INSERT
WITH PULLER**

**DF-69
INSERT DRIVER**

**DF-66-A
INSERT PULLER**

**Removal and installation of the valve seat inserts
using special tools**

VALVES

**RETAINER
LOCKS**

**Removal and installation of the valves, retainers,
and springs**

exhaust seat insert is replaceable. Valve seats are removed by means of a special puller. After the new seats are installed, they should be ground to the proper angle.

Before grinding the seats or valves, check the valve-to-guide clearance. The illustration shows specifications for TR, TRA and S series engines. For AEN, AENL, AENS and ACH, stem-to-guide clearance is 0.003–0.005 in. initially, and the limit is 0.007 in. Valve and seat angles are 45 degrees for these engines, also. Try replacing the valve to get the proper clearance. If clearance is still excessive, the guides can be pressed out and new ones installed (pressed in). A special tool, Wisconsin DF–72 driver or equivalent is required in installation of new guides. The guide must go in with the internal chamfer downward (towards the camshaft). All guides are pressed in with the top surface flush with the guide boss except for exhaust valve guides on TR and TRA series engines. On these models, the exhaust guide must extend ¹⁄₃₂ in. above the guide boss.

On TR and TRA series engines only, valve guides must be reamed to the dimensions shown in the illustration after they have been pressed into the guide bosses.

The valves should be ground (machined) at an authorized Wisconsin engine service outlet or other qualified machine shop to the specifications shown in the illustration. Then, they must be lapped, using a valve grinding compound by turning them back and forth from above with light downward pressure. Check the effectiveness of the lapping process by putting a dye such as

	Intake	Exhaust
TR and TRA Series Engines:	.006	.015
AEN, AENL, AENS Series Engines:	.008	.016
S-7D, S-8D Engines:	.006	.012
ACN, BKN Engines:	.008	.014

"Prussian Blue" or a similar product on the valve sealing surface and seating the valve. The dye will show the pattern of the effective contact between valve and seat on the seat. The pattern shown must be a wide, uniform ring.

Finally, clean the valves and block with soap and water, rinse and wipe thoroughly, and then apply a coating of light oil to the cylinder walls to prevent rust.

Valve tappet clearance must be checked before the springs and keepers are reassembled except on engines with adjustable tappets. Install the valves into the guides and seat them. Turn the camshaft as necessary until the cam for the valve to be checked points downward (the tappet is at the lowest possible position). If the engine uses a compression release, make sure the tappet is not riding on the compression release spoiler cam. Check the clearance between the head of the valve stem and the tappet. Clearance should be:

On engines with adjustable tappets, loosen the locknut with an open end wrench and turn the adjusting nut with another open end wrench until the gauge fits between the tappet and valve stem and can be pulled between the two with a slight amount of effort. On engines with plain tappets, if the clearance is smaller than specification, so that gauge cannot be inserted without lifting the valve off the seat, remove the valve and grind a small amount off the end of the stem. Recheck the clearance until it is adequate. Make sure the stem end is ground absolutely flat (parallel to the valve face) and that all grinding chips, etc. are removed from the valve stem before installation.

Assemble the springs, spring and spring seats or rotators, compress the springs, and install the retainer locks. Make sure the springs are seated properly in the locator cups. If they are not properly seated, they could cock to one side and cause the valve to stick.

Flywheel

ALL TR, TRA AND S ENGINES

If the flywheel is to be removed, loosen the retaining nut before the gear cover on the opposite end is removed.

NOTE: Do not try to loosen the flywheel after the gear cover is removed. Do not strike the crankshaft when it is not supported by the gear cover.

To remove the flywheel, first straighten the tab of the washer under the flywheel retaining nut. Place the correct size wrench on the flywheel retaining nut and strike the wrench sharply with a hammer to loosen the nut. Do not remove the nut completely, just unscrew it until it is flush with the end of the crankshaft. Turn the crankshaft until the keyway is at 10 o'clock. Pry outward on the flywheel with the outer end of the prybar at the

A – VALVE FACE ANGLE		45°
B – SEAT INSERT ANGLE		45°
C – GUIDE INSIDE DIAMETER		.312 – .313
D – VALVE STEM DIAMETER	INLET	.310 – .311
	EXH.	.309 – .310
MAXIMUM ALLOWABLE CLEARANCE BETWEEN C AND D		.006

Valve and guide measurements for TR and TRA and S-7 and S-8 Series engines

10 o'clock position on the flywheel. At the same time strike the end of the crankshaft with a soft hammer. This will loosen the flywheel from the tapered end of the crankshaft. Loosen the flywheel, but do not remove it at this point. It is necessary for the flywheel to remain on the crankshaft and support it while the gear cover and connecting rod are removed. Remove the flywheel only after the piston and connecting rod are removed.

When reassembling the engine, install the flywheel after the crankshaft is installed. Make sure that the woodruff key is in place before positioning the flywheel onto the crankshaft. Do not drive the flywheel onto the crankshaft by striking it with a hammer. Place a small length of pipe against the hub of the flywheel and tap the end of the pipe with a soft hammer until the flywheel is seated on the crankshaft taper. Assemble the washer and nut to the crankshaft with the tab of the washer inserted into the keyway of the flywheel. Tighten the nut only enough to hold the flywheel in place. Only after the crankshaft endplay has been adjusted is the flywheel nut to be tightened by sharply striking the wrench with a soft hammer. Bend the tab of the washer up against the nut.

AENL, AEN, AENS, ACN AND BKN ENGINES

1. Remove the four air intake screen mounting screws, and remove the screen.
2. Pull outward on the flywheel air fins, and gently tap on the end of the crankshaft with a soft hammer (do not use an ordinary, hard hammer) until the flywheel slides off the crankshaft taper.
3. To install the flywheel, first put the crankshaft key into position in the crankshaft keyway. Then, line up the keyway in the flywheel with the key, and slide the flywheel into position on the crankshaft taper. Finally, position a piece of pipe around the crankshaft and against the hub of the flywheel and strike the end of it sharply with the hammer.

Gear Cover
TR, TRA AND S ENGINES

To remove the gear cover, unscrew the cover cap screws and remove the governor lever. Tap the two dowel pins lightly from the crankcase side to break the cover loose from the crankcase.

NOTE: A steel ball for the end thrust of the camshaft will most likely fall out when the cover is removed. Remove the spring from the end of the camshaft so it won't be lost.

To reassemble the gear cover to the engine, position the spring into the end of the camshaft and mount the governor flyweight assembly. Lubricate the bearings, gears and tappets. Tap the dowel pins into the crankcase until they protrude about 1/8 in. from the mounting flange face. Place a finger full of grease into the hole in the cover to retain the camshaft spring and ball in place. Lubricate the lip of the oil seal with engine oil. Lubricate the gear cover face with a light film of oil to hold the gasket in place. The best means of getting the oil seal onto the crankshaft is to make a tapered installation sleeve such as that shown in the illustration. If such a sleeve is available, or you can make one, install it onto the crankshaft.

Position the governor lever as shown in the illustration. Then, gently locate the cover around the crankshaft. If the seal sleeve is being used, it can simply be pushed into position. If a seal sleeve is not available, press the cover into position very carefully. It may be necessary to hold the crankshaft still and very gently rotate the cover back and forth in order to get it over the crankshaft sealing surface without damaging the seal.

Finally, remove the seal sleeve (if used) and torque the cover capscrews to 8 ft. lbs. Tap the dowel pins into place.

Removing the gear cover

SLEEVE FOR ASSEMBLING GEAR COVER WITH OIL SEAL, ON TO CRANKSHAFT.

Dimensions of a suitable oil seal installation sleeve

Positioning governor lever during gear cover installation

Connecting Rod, Piston and Piston Rings

TR, TRA AND S SERIES ENGINES

Unscrew the two cap bolts which hold the connecting rod cap to the connecting rod. The oil dipper will come off with the cap screws. Tap the ends of the bolts to loosen the connecting rod cap.

Remove all deposits from the cylinder that might hinder the removal of the piston. This is done with a ridge reamer.

Turn the crankshaft until the piston is at the top of the cylinder and push the connecting rod and piston assembly up and out of the engine.

The piston skirt is elliptical in shape. When measuring the piston-to-cylinder wall clearance, you must take the measurement at the bottom of the piston skirt thrust face. The thrust faces of the piston skirt are located at a 90 degree angle from the piston pin hole axis.

Install the piston rings so that the rings gaps are 90 degrees apart around the circumference of the piston. A ring expander tool should be used to remove and install piston rings. If the tool is not available, the rings can be installed by placing the open end of the ring into the appropriate groove and working the ring down over the piston. Install the bottom oil control ring first, the scraper ring second and the compression ring last. Be careful not to bend or distort the rings in any way. A notch mark or the word "top" will be stamped on each ring so as to identify which side of the ring should face the top of the piston. Before installing the piston assembly into the cylinder, oil the rings, cylinder wall, rod bearings, wrist pin and the piston itself. Use a ring compressor to install the piston assembly into the cylinder bore.

PISTON TO CYLINDER AT PISTON SKIRT THRUST FACES		.004 to .0045"
PISTON RING GAP		.010 to .020"
PISTON RING SIDE CLEARANCE IN GROOVES	TOP RING	.002 to .0035"
	2nd RING	.001 to .0025"
	OIL RING	.002 to .0035"
CONNECTING ROD TO CRANK PIN	DIAMETER	.0015 to .0005"
	SIDE	.009 to .016"
PISTON PIN TO CONNECTING ROD		.0002 to .0008"
PISTON PIN TO PISTON		.0000 to .0008" tight

STANDARD CRANK PIN DIMENSIONS

Specifications for S series and TR and TRA engines except piston-to-cylinder specification for TRA-12D; that figure is .0025–.003 in

The piston and rod are mounted with the arrows on the connecting rod bolt boss and on the cap matched up and facing toward the open end of the crankcase. The oil hole in the dipper, which is integral with the cap, will be toward the camshaft side of the engine.

If the cylinder is worn more than 0.005 in. beyond the standard size, you should have it reground at an authorized Wisconsin shop or other reputable machine shop. It might be wise to consult with the machinist as to whether or not the rings should be replaced with a set of chromium rings. Rotate the crankshaft until it is at the bottom of its stroke. Tap the piston down so that the connecting rod seats onto the crankpin. Tighten the cap screws to 18–22 ft. lbs.

Removing the piston and connecting rod assembly

Assembling a piston ring to the piston

Install the connecting rod and piston assembly with cast arrow facing open end of crankcase

ARROW TOWARD OPEN END OF CRANKCASE

PISTON TO CYLINDER AT PISTON SKIRT	MODEL ACN	
	Up to 3000 R.P.M.	.005 to .0055"
	3000 R.P.M. & above	.006 to .0065"
	MODEL BKN	
	Up to 3000 R.P.M.	.0055 to .006"
	3000 R.P.M. & above	.006 to .0065"
PISTON RING GAP		.012 to .022"
PISTON RING SIDE CLEARANCE IN GROOVES	TOP RING	.002 to .0035"
	2nd, 3rd RING	.001 to .0025"
	OIL RING	.0025 to .004"
CONNECTING ROD TO CRANK PIN – SIDE CLEARANCE		.009 to .016"
CONNECTING ROD SHELL BEARING TO CRANK PIN DIA. (VERTICAL)		.0009 to .0032"
CONNECTING ROD BABBITT BEARING TO CRANK PIN		.0007 to .002"
PISTON PIN TO CONNECTING ROD		.0001 to .0007"
PISTON PIN TO PISTON		.0000 to .0008" tight

$\frac{3}{32}$ R.   1.0010 / 1.0003 DIA.   1.005 / 1.000

Piston, ring, and connecting rod specifications for ACN and BKN engines

CYLINDER BORE		3.0005 to 2.9995
PISTON TO CYLINDER AT PISTON SKIRT (THRUST FACE)	CAM-GROUND	SPLIT-SKIRT
	.003 to .0035"	.0045 to .005"
PISTON RING GAP		.010 to .022"
PISTON RING SIDE CLEARANCE IN GROOVES	TOP RING	.002 to .0035"
	2nd, 3rd RING	.001 to .0025"
	OIL RING	.0025 to .004"
PISTON PIN TO CONNECTING ROD BUSHING		.0005 to .0011"
PISTON PIN TO PISTON		.0000 to .0008" tight
CONNECTING ROD TO CRANK PIN – SIDE CLEARANCE		.009 to .018"
CONNECTING ROD SHELL BEARING TO CRANK PIN DIA. (VERTICAL)		.0011 to .0030"
CONNECTING ROD BABBITT BEARING TO CRANK PIN		.0007 to .0020"

Standard Crank Pin Dimensions $\frac{1}{8}$ R.   1.1260 / 1.1255 DIA.   1.255 / 1.250

Piston, ring, and connecting rod specifications for AEN, AENL, and AENS engines

AEN, AENL, AENS, ACN AND BKN ENGINES

Drain the oil from the crankcase, and then place the engine on its side. Remove the base capscrews and washers, and remove the base and gasket. On AEN, AENL and AENS, remove the two capscrews which hold the oil pump to the crankcase and remove it.

Use a ½ in. socket wrench to remove the hex locknuts from the rod bolts. If there are lockwasher tabs, these must be straightened first. Tap the ends of the rod bolts lightly to free the cap, and remove it.

Use a ridge reamer to remove all carbon deposits from the cylinder wall above the piston. Turn the crankshaft until the piston is at the top of the cylinder. Push the rod and piston out through the top of the cylinder from below.

NOTE: Do not let the rod bolts come in contact with the crankpin!

AEN, AENL and AENS engines were originally furnished with babbit cast connecting rod bearings. The shell bearing type rods are now used, and these are interchangeable with the older type rod for service replacement. In reassembling shell bearings, make sure the locating lug for both bearing halves are on the same side of the rod—the side on which numbers are stamped. Fit the bearings according to the specifications shown.

In installing rings, use an expander, or, if none is available, install the rings open end first. Be careful to open the ring only far enough to get in onto the piston. Install the rings so the gaps are 90 degrees apart. Go from bottom to top. Make sure the oil scraper ring is mounted as shown, with the scraper edge down, or severe oil pumping will result.

Piston ring locations for AEN, AENL, AENS, ACN engines. Note position in which oil scraper ring is mounted

If the cylinder is worn more than 0.005 in. beyond the standard size, you should have it reground at a Wisconsin authorized shop or other reputable machine shop. It might be wise to consult with the machinist on whether or not the rings should be replaced with a set of chromium rings.

On the AENL engine, if the split skirt type piston originally used is to be re-used, be sure to install it with the split toward the manifold side of the engine. In the case of cam ground pistons used on AEN, AENL and AENS engines, install the piston with the wide section of the skirt (wide thrust face) toward the fuel tank. Piston-to-cylinder clearance is measured at the center of the thrust face, at the bottom of the skirt.

When installing the piston into the cylinder, oil the rings, piston pin, rod bearings and cylinder wall. Use a ring compressor to hold the rings compressed while sliding the piston into the cylinder. On AEN, AENL and AENS engines, the numbers stamped on the rod and cap must be on the same side and the oil hole in the cap must face toward the oil pump.

On ACN and BKN engines, the arrow cast onto the connecting rod bolt boss must face toward the take-off end of the crankcase and the oil hole in the rod must face the camshaft.

The rod cap must be installed with the cast arrow lining up with the arrow on the rod.

Turn the crankshaft to Bottom Center position, and insert the piston into the cylinder, using a ring compressor until after the rings enter cylinder. Tap the piston down (with the rod hanging straight down) until the rod contacts the crankpin. Install the cap in the proper position as described above, and install the bolts and nuts (use new nuts on AEN, AENL and AENS). Torque to 14–20 ft. lbs. on ACN and BKN engines, 18–20 ft. lbs on AEN, AENL and AENS engines. Fold the lockwasher tabs over hex head and bolt boss, if so equipped.

Install the oil pump on AEN, AENL and AENS engines. Install the engine base using a new gasket. Torque the bolts to 6–8 ft. lbs. on ACN and BKN engines, and to 7–9 ft. lbs. on AEN, AENL and AENS engines.

Camshaft and Valve Tappets

S, TR AND TRA SERIES ENGINES

When removing the camshaft, turn the engine over on its side and push the tappets away from the camshaft so that they will clear the camshaft lobes when the camshaft is removed. the valves must be removed for this operation. After the camshaft is removed, mark the tappets as to location and then remove the valve tappets and inspect them for wear. The tappet stem diameter must be 0.309–0.310 in., and the clearance in the guide hole must be 0.002–0.006 in.

Install tappets into their original guide holes before installing the camshaft. Install the camshaft with the timing mark on the ~amshaft gear located between the two marked teeth on the

Aligning crankshaft gear and camshaft gear timing marks, S, TR and TRA engines

crankshaft gear. Put the camshaft thrust spring into the end of the camshaft before installing the gear cover. Adjust or check the valve tappet clearance as described above.

ACN, BKN, AEN, AENL AND AENS ENGINES

To remove the camshaft, first raise the tappets until they clear the cam lobes. Pry out the expansion plug from the flywheel end of the crankcase. With a drift punch, drive out the camshaft pin from the flywheel end of the crankcase until it emerges from the opposite end of the crankcase. The camshaft should drop down inside the crankcase, with the expansion plug emerging in front of the camshaft pin.

On installation, align the timing marks as shown. Use new expansion plugs and make sure to drive in the camshaft support pin from the takeoff end of the crankcase.

Adjusting valve tappet clearance

Breaker Push Pin and Bushing

TR, TRA AND S SERIES ENGINES

Remove the breaker arm push pin and inspect it for wear. Replace parts as necessary. If you're replacing the pin, install the assist spring, small end toward the groove in the tapered end, from the plain end. The pin goes into the guide hole with the plain end toward the camshaft. If there is excessive clearance between pin and bushing, replace the bushing and then ream it to an inside diameter of 0.2785–0.2790 in. Bushings are pressed in. Loctite may be used to fasten them in place if there is excessive clearance between outside of the bushing and the crankcase.

Aligning crankshaft, camshaft, and magneto timing marks, ACN, BKN, AEN, AENL, and AENS engines

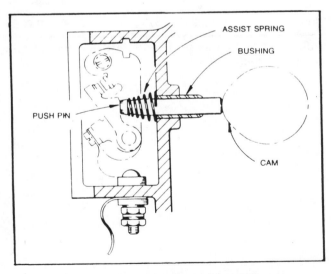

Breaker push pin and bushing

Crankshaft

TR, TRA AND S SERIES ENGINES

The crankshaft is removed after the gear cover has been removed, the connecting rod disconnected and raised up out of the way. Remove the flywheel nut, flywheel, and the woodruff key.

The crankshaft may now be pulled out of the open end of the crankcase. When reinstalling the crankshaft, mount the flywheel after the crankshaft is inserted into the crankcase. The flywheel supports the crankshaft while the connecting rod is attached. The flywheel nut is tightened only enough to hold the flywheel during end-play adjustment.

Stator Plate and End-Play Adjustment

The end-play of the crankshaft is adjusted by the application of various size gaskets behind the stator plate, which doubles in function as the front bearing support and an adaptor for the magneto coil. The stator plate should not be removed from the crankcase unless it has to be replaced.

Removal and installation of the stator plate

To remove the stator plate, remove the four retaining screws and tap the plate from the inside until it falls off. Reassemble the stator plate to the crankcase using new gaskets with the same total thickness as those originally installed. The stator plate mounting screws are to be torqued 8 ft. lbs. on all S, TR, and TRA series engines except the TRA–12D engine; tighten them on TRA–12D engines to 20–22 ft. lbs. End play is checked after the crankshaft, gear cover, and flywheel are mounted. End play is 0.002–0.005 in. on S–7D and S–8D engines, and 0.001–0.004 on TR–10D, TRA–10D, and TRA–12D.

Crankshaft end-play is measured with a dial indicator mounted on the PTO side of the crankshaft and a lever prying behind the flywheel. If new crankshaft roller bearings have been installed, they must be properly seated by tapping the ends of the crankshaft with a lead hammer before measuring the crankshaft end-play.

Crankshaft and End-Play

Remove the four main bearing plate capscrews at the power takeoff end. Pry off the plate and pull the crankshaft out.

On installation, use the same thickness of gaskets initially, and torque the mounting bolts to 10–12 ft. lbs. on ACN and BKN engines, and 20–22 ft. lbs. on AEN, AENL and AENS engines. Check the end play as described at the end of the section above. It should be 0.001–0.003 in. on AEN, AENL and AENS engines, and 0.002–0.005 in. on ACN and BKN engines. Change the gasket thicknesses in order to correct improper end play.

Make sure to align the punch mark on the front face of the crankshaft between two marked teeth of the camshaft gear.

Oil Pump

ACN AND BKN ENGINES

Drain the crankcase, place the engine on its side, and remove the engine base. Carefully note the order of disassembly of the check balls, springs, and other parts.

When assembling the pump, tap the check ball at the bottom of the pump very lightly with a punch and hammer to seat it. After the pump is assembled, fill the engine base with oil and work the pump plunger up and down with a screwdriver in order to check the pump's operation and fill the oil trough. Use a new base gasket and torque the bolts to 6–8 ft. lbs.

AEN, AENL AND AENS ENGINES

Remove the engine base by draining oil placing engine on its side, and then removing the capscrews and washers. Remove the two oil pump mounting capscrews and remove the oil pump. The main pump plunger, springs, and check balls come out the top, once the pump is away from the drive pushrod. The plug on the side of the discharge tube is removed to gain access to the discharge check ball and spring. Wash all parts in a good solvent.

New plunger-to-bore clearance is 0.003–0.005 in. The limit is 0.008 in. The pump should be replaced if the clearance is greater than the limit. Inspect the check ball seat in the bottom of the pump cylinder for wear, pitting, or dirt. Clean or replace the pump as necessary.

On reassembly:

1. Drop the intake check ball into the bottom of the pump cylinder and tap it very lightly in order to seat it. Insert the retainer, spring, and plunger into the bore. Install the discharge check ball and spring into the discharge tube.

2. Fill the engine base and put the oil pump into position. Operate the plunger with your finger to prime the pump and check operation. Install the oil pump mounting bolts.

Priming ACN and BKN oil pump

3. Make sure the oil pump pushrod makes good contact with the plunger and the strainer screen is in good condition and properly mounted.

4. Install the engine base, using a new gasket. Torque the mounting bolts to 7–9 ft. lbs.

Break-In

An overhauled engine should be operated at 1600–1800 rpm with no load for one-half hour. It should be operated at normal operating rpm, but still without load, for an additional four hours.

Valve Seat and Face Angle	Valve-to-Guide Clearance	
	Inlet	Exhaust
45°	0.001–0.003	0.003–0.005

Valve and guide measurements for BKN and all AE series engines.

Section 21

Wisconsin Robin

Small Engines — 6-20 hp

ENGINE IDENTIFICATION

There is an identification plate attached to the upper blower housing which lists the engine model on the left and individual engine serial number on the right. Both of these should be used when requesting parts. The engine models are EY25W and EY27W.

General Engine Specifications

Model	Bore & Stroke (In.)	Displacement (cu In.)	Horsepower @ RPM
EY25W	2.83 x 2.44	15.40	6.5 @ 3,600
EY27W	2.91 x 2.44	16.26	7.5 @ 3,600

MAINTENANCE

Air Cleaner

SERVICE

The air cleaner should be serviced every 50 hours in normal operation, more often (daily) if the air is dusty.

First disassemble the unit as follows:

1. Remove the air cleaner cover, element and element retainer.
2. Disconnect the breather tube from the inspection cover.
3. Remove the two capscrews and remove the air cleaner body and gasket from the carburetor.
4. Wipe all metal parts clean. Wash the element only in either kerosene, or water and a liquid detergent. Do not wash in any other solvents.
5. Wrap the element in a cloth and squeeze it dry.
6. Pour engine oil into the filter until it is saturated, and then squeeze out the excess.
7. Reassemble in reversal of above. Make sure element retainer is installed with arrow up.

AIR CLEANER BODY

MOUNT ELEMENT RETAINER THUS

BREATHER TUBE

ELEMENT

COVER

LATCH

Exploded view of air cleaner

Lubrication

OIL AND FUEL RECOMMENDATIONS

Oil

Use only MS or SD oils of the viscosity shown in the chart for both normal operation and breakin. See chart for crankcase capacities.

Oil Viscosity Chart

Season or Temperature	Grade of Oil*
Spring, Summer or Autumn (+120°F to +40°F)	SAE 30

Oil Viscosity Chart

Season or Temperature	Grade of Oil*
Winter	
(+40°F to +15°F)	SAE 20
(Below +15°F)	SAE 10W-30

*Use oils classified as Service MS or SD

Crankcase Capacity	
Model	Capacity (pts.)
EY25W, EY27W	1.6

Fuel

Use only regular grade gasoline with an octane rating of 90 or above. Use only a reputable brand of fuel, and make sure it is free of both dirt and water. Note that low octane fuel may cause the engine to detonate and result in severe damage. Do not use gasoline that is more than one month old.

TUNE-UP

Spark Plug

The spark plug should be checked frequently for fouling, widened or otherwise improper gap, or more serious problems such as eroded electrodes or a cracked insulator. Clean carbon deposits from the electrodes with a wire brush, and set the gap to 0.020–0.025 in. with a wire type feeler gauge. Bend only the side electrode. If electrodes are severely eroded so that the side electrode has lost its sharp, rectangular cross-section or the center electrode is very short, or if there are any cracks in the insulator, the spark plug should be replaced.

Always check the plug gap before installation, even if the plug is new. Install the plug with a new gasket and torque it to 24–27 ft. lbs.

Tune-Up Specifications

Model	Plug Type	Plug Gap (in.)	Point Gap (in.)	Ignition Timing (deg Before Top Center)
EY25W	①	.020–.025	.014	23
EY27W	①	.020–.025	.014	23

① 14 mm. Champion L86,
   AC 44F
   NGK B6HS

Breaker Points

REMOVAL AND INSTALLATION

1. Remove the three mounting screws and remove the starter pulley. Install a 14mm socket wrench onto the flywheel nut and tap the wrench sharply with a soft hammer. If you have a wheel puller, remove the nut. Otherwise, unscrew it until it is flush with the end of the crankshaft.

2. Remove the flywheel with a wheel puller or as follows:

3. Place a large screwdriver between the crankcase and the flywheel. The screwdriver must be in line with the keyway. While wedging the flywheel outward with the screwdriver, strike the outer end of the flywheel nut with a soft (brass, wood or plastic) hammer. This will bring the flywheel off the crankshaft taper.

4. Take off the point cover by removing mounting screws and pulling it off. Unscrew the terminal nut connecting condenser and coil wires to the contact set, and disconnect the wires. Remove the contact lockscrew, and remove the contacts.

5. Position the pin contacts on the crankcase, install the lockscrew, and tighten it just slightly.

6. Adjust the contact gap and time the engine as described below.

SETTING POINT GAP AND IGNITION TIMING

1. If necessary, remove the flywheel and point cover as described in Steps 1–4 above. Loosen the contact lockscrew.

2. Turn the engine over until the contact cam follower is at the very peak of the cam. Using a flat 0.014 feeler gauge, slide the points back and forth via the adjusting knob until the gauge just slides between the contact surfaces. Tighten the contact lockscrew. Recheck gap and, if necessary, reset it.

3. Connect a timing light between the coil primary ground. Align the **M** timing mark on the rim of the flywheel with the **D** mark on the lower left side of the crankcase by rotating the flywheel. See illustration.

ADVANCE TIMING MARK ON FLYWHEEL RIM

TIMING MARK ON CRANKCASE

Aligning ignition timing marks

4. Turn the flywheel counterclockwise slowly until the light either goes on or goes off, depending on the type. Turn it very slowly clockwise just until the light reacts again. If the timing marks are aligned, timing is correct. If the **M** mark is below the **D** mark (the **D** is on the crankcase), the breaker gap is too large; if the **M** is above the **D**, the breaker gap is too small. Reset the contact gap as necessary. If timing is off about 2 degrees or 1/8 in. of flywheel rotation, the gap must be changed about 0.001 in.

5. Reset the contact gap and recheck timing until timing is correct.

6. Remove the timing light and install the timing cover and flywheel. In installing the flywheel, slide it into place with key and keyway aligned, and then install retaining nut and tighten it just until lockwasher collapses. Finally, put a wrench on the nut and tap it lightly once or twice, or torque the nut to 44–47 ft. lbs. with a torque wrench.

Magneto Service

1. If the engine is hard to start or runs erratically, first service the spark plug. If this does not cure the problem, remove the flywheel and check for loose or broken ignition wires. Inspect the points. See "Breaker Point Removal and Installation," above. Repair the wires or replace the points and condenser. Set the point gap and timing, as required.

2. Check the spark by removing the plug and grounding it against the engine block with the high tension wire connected. Then, spin the engine. If the spark is erratic or weak, remove the ignition coil by removing the flywheel, feeding the high tension wire through the grommet in the wall of the cylinder block, disconnecting the primary wire, and removing the mounting screws and coil. Replace the coil in reverse of the above.

Governor

If the governor lever has been loosened or removed, perform the governor lever adjustment. Otherwise, skip to "Speed Regulation," below.

GOVERNOR LEVER ADJUSTMENT

1. Mount the governor lever with the clamp screw just slightly loose.

2. Install both the control rod and spring which connect the governor lever and the throttle lever. Mount the control lever assembly onto the crankcase but do not tighten the wingnut.

3. Connect the governor spring between the holes of both governor lever and control lever as shown in the illustration below.

Adjusting the governor lever

4. Turn the control lever counterclockwise until the throttle in the carburetor is fully open. Lock the lever in this position by tightening the wingnut.

5. With the clamp screw loose so the governor shaft will turn independently of the governor lever, turn the governor shaft (use a screwdriver in the groove in the end of the shaft) until you can feel the internal vane stop against the flywheel thrust sleeve (this is as far as the shaft can be turned without excessive force). Tighten the lever clamp nut.

SPEED REGULATION

1. Remove the load from the engine, and run it until it is hot.

2. Install the governor spring hooked to holes "1" and "B" as shown above.

3. Loosen the control lever wing nut so that the lever is free to move. Loosen the locknut on the adjusting screw. Find the desired no-load speed on the chart above that corresponds with the speed at which you want the engine to run when fully loaded.

Adjusting governor speed

4. Hold the control lever down so the adjusting screw is against the crankcase stop.
Turn the adjusting screw in or out until the required no-load speed is obtained. Tighten the adjusting screw locknut.

5. if the engine is to operate at a fixed speed, tighten the control lever wingnut.

Governor No-Load/Load RPM Chart

Load RPM	No Load RPM		Spring Holes	
	EY25W	EY27W	EY25W	EY27W
1800	2330	2210	1-B	1-B
2000	2445	2375	1-B	1-B
2200	2595	2500	1-B	1-B
2400	2745	2660	1-B	1-B
2600	2900	2850	1-B	1-B
2800	3065	3020	1-B	1-B
3000	3230	3210	1-B	1-B
3200	3400	3385	1-B	1-B
3400	3580	3590	1-B	1-B
3600	3765	3760	1-B	1-B

Carburetor—exploded view

Carburetor

OVERHAUL

1. Using the illustration as a guide, remove the choke phillips head screws (17), and then take out the choke valve and shaft (18 and 19).

2. Remove the choke shaft retainer spring and ball (30 and 31) to prevent their being lost.

3. Remove the throttle plate phillips head screws (17) and take out the choke valve and shaft (18 and 19), being careful not to damage the edges of the throttle valve.

4. Remove the throttle stop screw and spring (14 and 25).

5. Remove the main jet holder (16) and then take off the float bowl (13).

6. Remove the main jet from the (15) jet holder.

7. Remove the main nozzle (9) from the carburetor body.

8. Remove the idle jet (27) using an appropriate tool to prevent damage to it.

9. Remove the float pin (12), float (11), and needle vale (20).

10. Inspect the float. Replace it if dented, fuel has leaked in, or if the float hinge pin or the tab that limits float travel is worn.

11. Clean all parts thoroughly in a solvent such as Bendix Metalclene or Speeclene, rinse in a cleaning solvent, and blow out all passages with low pressure compressed air in the reverse direction of normal flow. Make sure all carbon deposits have been removed from the throttle bore and idle discharge holes.

NOTE: Never use a drill or wire to clean jets!

12. Reassemble, keeping the following points in mind:
   a. Replace the needle valve and seat with a matched valve and seat (as included in a repair kit).

b. Install the new idle jet and adjusting screw which are included in the repair kit. Make sure the idle jet is tightened firmly.

c. Install the new main jet, tighten securely, and then install the main jet holder and torque to 5.5 ft. lbs.

d. When assembling the choke, make sure the flat on the choke valve faces the main air jet.

ENGINE OVERHAUL

Tools

To overhaul this engine, which uses metric fasteners, you should have the following tools:
- 10mm thin wall socket
- 10mm standard socket
- 10mm deep socket
- 12, 14 and 18mm standard sockets.
- 10, 12, and 14mm open end wrenches.

In addition, various special pullers, a valve spring compressor, and a valve set cutter are required. See the illustration. Wisconsin-Robin part numbers are provided for your use in ordering the tools either from them or from a tool manufacturer who makes equivalent tools which are cross-referenced.

The tools and Wisconsin-Robin part numbers are listed below:

Special tools

Reference Number	Part Number	Part Name
1.	EYY790-350	Flywheel Puller
	EY016508500	Bolt, 8 x 50 mm (3 required)
2.	EYY790-282	Valve Spring Compressor
3.	EYY790-524	Valve Guide Puller
4.	EYH640-118	Seat Cutter (45°)

Engine Disassembly, Reassembly, and Overhaul

Engine disassembly instructions are provided below, in proper sequence. Reassembly is primarily a simple reversal of disassembly. Notes on the extra procedures to be performed as the engine is put back together are included with the disassembly instructions for each part of the engine.

FUEL TANK

1. Disconnect the fuel line at the carburetor. Note that in reassembly, it is advisable to use a new fuel line.
2. Remove the fuel tank from the bracket.
3. Remove the tank bracket from the cylinder head.
4. In reassembly, wash the element per "Air Cleaner Service" instructions above, and correctly mount the element retainer as illustrated there.

FLYWHEEL SHROUD

1. Disconnect the coil primary wire from the stop button wire.
2. Remove the flywheel shroud from the cylinder case, and the baffle from the head.
3. Remove the baffle from the cylinder block.

AIR CLEANER

1. Remove the air cleaner cover, element and element retainer.
2. Disconnect the breather tube from the inspection cover.
3. Remove the two capscrews and remove the air cleaner body and gasket from the carburetor.

MUFFLER

Remove the two hex nuts and remove the muffler and gasket from cylinder case.

GOVERNOR LEVER AND CARBURETOR

1. Disconnect the governor spring from the lever and speed control assembly.
2. Remove the governor lever from the shaft, and, at the same time, disconnect the rod and spring from the lever and carburetor.
3. Remove the tow nuts and lockwashers, and remove the carburetor, insulating plate and gaskets from the cylinder case.
4. If necessary, the speed control assembly can be removed from the side of the crankcase by removing the wing nut and clip.
5. In reassembly, refer to "Governor Adjustment" above.

STARTING PULLEY AND FLYWHEEL

1. To remove the starting pulley, first remove the three mounting screws.
2. Place a 14mm socket wrench on the flywheel nut and give the wrench a sharp blow with a soft hammer. Remove the nut, spring washer and pulley.
3. Attach puller to flywheel—turn center bolt clockwise until flywheel becomes loose enough to be removed. If puller is not available, screw flywheel nut flush with end of crankshaft to protect shaft threads from being damaged. Place the end of a large screwdriver between the crankcase and flywheel in line with the keyway. Then, strike the end of the flywheel nut with a babbitt or other soft hammer and at the same time wedge outward with the screwdriver.
4. Disconnect the high tension cable from the spark plug, and slip the cable along with the rubber grommet through the hole, to the inside of the crankcase. Then remove the ignition coil along with the attached high tension cable by taking out the two mounting screws, and disconnecting the breaker assembly wire.
5. Remove the contact breaker and condenser by removing the point cover, and removing mounting screws from the cylinder case. Slip the coil primary wire along with the grommet through the hole in the side of the crankcase.
6. In reassembly, refer to "Setting Point Gap and Timing."
7. Securely tighten the flywheel nut after the timing is finalized, but first be sure the woodruff key is in position on the shaft. Do not drive the flywheel onto the taper of the crankshaft and do not overtighten the flywheel nut. Simply turn the nut until the lockwasher collapses. Then, tighten it by placing a wrench on the nut and giving the handle of the wrench 1 or 2 sharp blows with a soft hammer. If a torque wrench is available, tighten to 44–47 ft. lbs.

CYLINDER HEAD AND SPARK PLUG

1. Remove the spark plug from the cylinder head.
2. Loosen the mounting nuts and remove the cylinder head along with the gasket.
3. Clean the carbon from the combustion chamber and the dirt from among the cooling fins. Check the cylinder head mounting face for distortion. If warpage is evident, replace head.
4. In reassembly, use a new cylinder head gasket and spark plug. Torque the head nuts to 22 ft. lbs. Leave the spark plug out temporarily, for each in turning engine over for the remainder of assembly and for timing adjustments. When installing the spark plug, tighten it 24–27 ft. lbs. torque.

INTAKE AND EXHAUST VALVES

1. Remove the valve inspection cover, breather plate and gaskets from the cylindercase.
2. Lift the valve springs, by means of a compressor tool EYY790–282 or equivalent and remove the retainer locks with long nose pliers. Release the compressor tool and remove the intake valve and exhaust valve along with their respective spring and retainer. A standard Automotive type valve lifter can be used for removing valves, but is not practical for reassembly.

——————————— CAUTION ———————————
Do not damage the gasket surface of the tappet chamber with the compressor tool.

VALVE SEAT
CUTTER

Correcting the valve seats

Pulling out valve guides with special tool

3. Clean carbon and gum deposits from the valves, seats, ports and guides.

4. In reassembly, replace valves that are badly burned, pitted or warped.

5. Correct the valve seat by using a 45 degree seat cutter tool No. lkEYH640–118 or equivalent as illustrated. The finished seat width should be 0.047–0.059 in. — maximum usable width is 0.098 in.

CAUTION

Do not use an electric power driven grinding wheel to correct the valve seats.

A – VALVE FACE ANGLE	45°
B – SEAT ANGLE	45°
C – GUIDE INSIDE DIAMETER	0.2756 to 0.2770"
D – VALVE STEM DIAMETER	0.273 to 0.274"
MAXIMUM ALLOWABLE CLEARANCE BETWEEN C AND D	0.006"

Valve Specifications

6. Valve guides should be replaced when the valve stem clearance becomes excessive. Use a valve guide puller tool EYY790–324 or equivalent. See illustration. Draw the valve guides out and press new guides in using the same puller tool. Refer to the illustration for clearance specifications and proper assembly.

7. After correcting the valve seats and replacing the valve guides, lap the valves in place until a uniform ring will show entirely around the face of the valve. Clean the valves, and wash the block thoroughly with a hot solution of soap and water. Wipe the cylinder walls with clean, lint free rags and light engine oil. Do not assemble the valve springs until the tappet clearance has been checked. See "Tappet Adjustment" below.

TAPPET ADJUSTMENT

With the tappet in its lowest position, hold valve down and insert a feeler gauge between the valve and tappet stem. The clearance for both intake and exhaust, with the engine cold is 0.006–0.008 in.

Valve Specifications

If the clearance is less than it should be, grind the end of valve stem a very little at a time and remeasure. Stems must be ground square and flat.

If the clearance is too large, sink the valve seat with seat cutter tool.

After obtaining the correct clearance, assemble the valve springs and retainers, and secure them in place with the retainer locks. Check the operation of the valves by turning the crankshaft over by hand and remeasure the tappet clearance.

GEAR COVER AND CRANKSHAFT REMOVAL

1. Place a rag under the engine to absorb the remaining oil. Remove the gear cover mounting screws.

2. With a soft hammer, tap at even intervals around the outer surface of the gear cover until it breaks free of the crankcase face. Break the cover free carefully, so as to avoid damaging the oil seal.

3. Inspect the adjusting collar, oil seal, governor shaft and yoke. Replace any parts that are damaged or excessively worn.

4. Remove the flywheel woodruff key.

5. Pull the crankshaft out from open end of crankcase and take care not to damage the oil seal. If necessary, loosen shaft by tapping lightly at the flywheel end with a soft hammer.

NOTE: The "Gear Cover and Crankshaft Installation" procedure is located with camshaft, connecting rod, and piston service, since that is the normal sequence of doing overhaul work.

CAMSHAFT, TAPPETS AND TIMING MARKS

1. To prevent the tappets from falling out and becoming damaged when the camshaft is removed, turn the crankcase over on its side as shown. Push the tappets inward to clear the cam lobes, and remove the camshaft.

2. Withdraw the tappets and mark them for identification with the hole from which they were removed.

3. In reassembly, put the tappets back in their corresponding guide hole. This will eliminate unnecessary valve stem grinding for obtaining correct tappet clearance.

4. Mount governor sleeve on end of camshaft by holding the shaft in a vertical position with the flyweights hanging down. Align the groove in the flange with the pin in the gear face and install the sleeve on shaft so that its flange fits in the groove between the heavy end and the thrust tabs of all three flyweights.

Postion in which engine should be placed while removing camshaft

Mount camshaft assembly in crankcase.

5. The timing marks on the camshaft gear and crankshaft gear must be matched up. Mount the camshaft so that the marked tooth on crankshaft gear is between the two marked teeth of the camshaft gear, see the illustration. If the valve timing is off, the engine will not function properly or may not run at all.

CONNECTING ROD AND PISTON

Disassembly

1. Straighten out the bent tabs of the lock plate and remove the bolts from the connecting rod.
2. Remove lock plate, oil dipper and connecting rod cap.
3. Scrape off all carbon deposits that might interfere with the removal of the piston from the upper end of the cylinder. Use a ridge reamer.
4. Turn the crankshaft until the piston is at top, then push the connecting rod and piston assembly upward and out through the top of the cylinder.
5. Remove the piston from connecting rod by taking out one of the snap rings and then removing the piston pin. A new snap ring should be used in reassembly.

Measure piston diameter as shown

Reassembly

1. Use a ring expander tool to prevent the ring from becoming distorted or broken when installing on the piston.
2. If an expander tool is not available, install the rings by placing the open end of the ring on the first land of the piston. Spread the ring only far enough to slip it over the piston and into the correct groove, being careful not to distort the ring.
3. With the expander tool, assemble the bottom ring first and work upward, installing the top ring last.
4. Mount the scraper ring with the scraper edge down, otherwise oil pumping and excessive oil consumption will result. Refer to the illustration for correct placement of rings.
5. Measure the diameter of the piston in the center of the thrust faces at the bottom of the piston skirt, as illustrated.
6. Measure the cylinder bore and inspect it for out-of-round and taper. If the cylinder is scored or worn more than 0.005 in. over standard size, it should be rebored and fitted with an oversize piston and rings. Refer to the chart for the clearance between the piston and cylinder. Size, clearance and wear limits are given in more detail at the end of this chapter.
7. When installing the piston in the cylinder, oil the piston, rings, wrist pin, rod bearings and cylinder wall before assembly. Stagger the piston ring gaps 90 degrees apart around the piston. Use a piston ring compressor.

Location and positioning of rings

Connecting rod and oil dipper assembly

8. Turn the crankshaft to the bottom of the stroke and tap the piston down until the rod contacts the crankpin. Mount the connecting rod cap so that the cast rib between the face of rod and bolt boss matches up with the cast rib on the connecting rod. Assemble oil dipper to cap. The dipper should be toward the gear cover end of the connecting rod cap if engine is operated on a tilt toward the take off end. Mount the dipper toward the flywheel end if it's tilted in that direction or with a no tilt operation.

Install a new rod bolt lock plate. Mount the connecting rod bolts and tighten to the following torque specifications.

Models EY25W and EY27W — 14.5–18.0 ft. lbs.

Check for free movement of the connection rod by turning the crankshaft over slowly. If satisfactory, bend the tabs on the lock plate against the hex head flat of the connecting rod bolts.

Measuring points for dimensions critical to crankshaft end play adjustment

		EY18W	EY25W	EY27W
PISTON TO CYLINDER AT PISTON SKIRT THRUST FACE		.0016 .0032"	.0024 .0039"	.0028 .0052"
CONNECTING ROD TO CRANK PIN	DIA.	.0021 .0031"	.0016 .0026"	.0016 .0026"
	SIDE	.008 .0235"	.004 .012"	.004 .012"
PISTON PIN TO CONNECTING ROD		.0004 .0012"	.0006 .0014"	.0006 .0014"
D - CRANKSHAFT PIN DIAMETER		1.0210 1.0215"	1.1003 1.1008"	1.1003 1.1008"
W - CRANKSHAFT PIN WIDTH		.9846 .9882"	1.0630 1.0669"	1.0630 1.0669"
PISTON RING GAP		.002 .010"	.002 .010"	.008 .016"
PISTON RINGS - SIDE CLEARANCE IN GROOVES		.0004 .0030"	.0004 .0030"	.0004 .0022"
PISTON PIN TO PISTON		.00035" tight to .00039" loose		

STANDARD CRANK PIN DIMENSIONS

Piston, ring, and rod clearance specifications

GEAR COVER AND CRANKSHAFT INSTALLATION

1. In reassembly, inspect the crankcase oil seal and main bearing for possible replacement. Mount the crankshaft with extreme care so as not to damage the lips of the oil seal. Use an oil seal sleeve if available.

2. End Play is regulated by means of the adjusting collar at the gear end of the crankshaft. This should be set immediately before mounting the gear cover as explained below.

NOTE: Crankshaft End Play is regulated by the length of the adjusting collar. The end play should be 0.001–0.009 in., engine cold. The adjusting collar is located between the crankshaft gear and main bearing at the take-off end of the engine. Replacement of the collar is seldom necessary unless the crankshaft or gear cover is replaced. To determine what length adjusting collar to use, refer to the illustration and steps below.

1. With the gear cover removed, tap the end of the crankshaft slightly to insure that the shaft is shouldered against the front end main bearing.

2. Use a depth micrometer and measure the distance between the machined surface of crankcase face and end of crank gear (dimension A).

3. Measure the distance between the machined surface of gear cover and end of main bearing (dimension B).

4. The compressed thickness of gear cover gasket is 0.007 in. (dimension C).

5. Select an adjusting collar that is 0.001–0.009 in. less than the total length of A, B, and C from the chart.

6. Apply oil to the bearing surfaces, gear train and tappets. Also lubricate the lips of the oil seal and add a light film of oil on the gear cover face to hold the gasket in place.

7. Mount the adjusting collar to the crankshaft with the recess toward the crank gear.

8. Assemble the gear cover, being sure that the governor yoke is in a downward position, and be extremely careful not to damage the lips of the oil seal. If available, mount on oil seal sleeve on the crankshaft to prevent damage to the oil seal lips.

— CAUTION —

Be sure the timing marks on the crankshaft and camshaft gear remain correctly mated when the end of the camshaft is fitted into the bearing hole of the gear cover.

Adjusting Collar Dimension Chart

Model	Collar Length	Part Number
EY25W	.740 to .748"	EY25W2112a
EY27W	.748 to .756"	EY25W2112b
	.756 to .764"	EY25W2112c

9. Tap the gear cover in place with a soft hammer, remove the oil seal sleeve and tighten the gear cover capscrews to 13 ft. lbs. torque.

10. Tap the end of the crankshaft with a soft hammer so that the crankshaft will shoulder against the main bearing at the flywheel end.

11. Tap the crankshaft in the opposite direction (from flywheel end) to seat the adjusting collar against the main bearing at the takeoff end.

12. Attach a dial indicator to one of the $\frac{5}{16}$–24 tapped holes on the face of the crankcase with the indicator plunger resting against the end of the crankshaft and set the dial at 0.

13. Wedge a screwdriver between the flywheel and the crankcase. The movement of the flywheel away from the engine block will register as end play on the indicator dial.

14. If end play is not within the limits of 0.007–0.009 in., use the reading from the indicator dial to determine the new length adjusting collar to use. Replace the collar with one of the proper dimension, if necessary.

FINAL CHECKOUT AND BREAK-IN

After the major moving parts are assembled, turn the engine over using the starter pulley. Make sure it turns without unusual resistance. Check the basic ignition adjustments (point gap and timing), and make preliminary carburetor and governor adjustments. Refill the crankcase and fuel tank.

Break the engine in by running it as specified on the chart below, proceeding from top to bottom. Check the engine frequently for leaks, and make final carburetor and governor adjustments during the sequence.

Break-in Sequence

Load			Speed	Time
No load			2500 rpm	10 minutes
No load			3000 rpm	10 minutes
No load			3600 rpm	10 minutes
1.75 hp	2.5 hp	2.65 hp	3600 rpm	30 minutes
3.5 hp	5.0 hp	5.3 hp	3600 rpm	60 minutes

Torque Specifications
All figures in ft. lbs.

	Model EY25W	Model EY27W
Cylinder head nuts	24.5 to 27	†
Connecting rod bolts	14.5 to 18	†
Flywheel nut	*	*
Spark plug	*	*
Gear cover screws	*	*

Clearances and Wear Limits

Description	Model EY25W		Model EY27W	
	Tol	Limit	Tol	Limit
Flatness of cylinder head	.002	.006	†	†
Cylinder bore	2.8346 2.8354	2.841	2.9134 2.9141	2.9197
Bore-out of round	*	*	*	*
Cylindricity (taper)	*	*	*	*
Valve seat width	*	*	*	*
Valve guide bore	*	*	*	*
Piston diameter at skirt thrust faces (standard size)	2.8315 2.8323	2.8285	2.9090 2.9105	2.9072
Piston to cylinder clearance at skirt thrust faces	.0024 to .0039	.009	.0028 to .0052	.009
Ring groove width (top and 2nd ring)	*	*	.0787 .0797	.0852
Ring groove width (oil ring)	*	*	*	*
Ring width (top and 2nd ring)	*	*	.0776 .0783	.0741
Ring width (oil ring)	*	*	*	*
Piston rings—side clearance in groove	*	*	.0004 .0022	.006

Clearances and Wear Limits (cont.)

Description	Model EY25W		Model EY27W	
	Tol	Limit	Tol	Limit
Ring gap (at cylinder skirt)	•	•	.008 .016	.040
Pin hole in piston	.6297 .6301	.6311	†	†
Piston pin diameter	.6297 .6300	.6287	†	†
Piston pin to piston fit	•	•	•	•
Connecting rod (crank pin end)	1.1024 1.1029	1.104	†	†
Crank pin diameter	1.1003 1.1008	1.0983	†	†
Connecting rod to crank pin clearance	.0016 to .0026	.005	†	†
Connecting rod side clearance	.004 .012	.039	†	†
Connecting rod (piston pin end)	.6306 .6311	.6345	†	†
Piston pin to connecting rod clearance	.0006 to .0014	.0032	†	†
Con. rod—large and small bore alignment (parallel)	•	•	•	•
Con. rod—large and small bore centers	4.3307	4.3366	†	†
Crank pin—out of round	—	•	—	•
Cylindricity (taper)	—	•	—	•
Crank pin—parallel	—	•	—	•
Crankshaft journal diameter (take-off end)	1.1805 1.1809	1.1785	†	†
Crankshaft journal (flywheel end)	1.1806 1.1810	1.179	†	†
Crankshaft end play	•	—	•	—
Camshaft (cam rise)	•	•	•	•
Camshaft (journal diameter)	•	•	•	•
Valve spring (free height)	•	•	•	•
Valve spring (squareness)	—	•	—	•
Valve stem diameter	•	•	•	•
Valve stem clearance to guide	•	•	•	•
Valve stem—lock-pl. to groove clearance	•	•	•	•
Valve—length from groove to stem end	•	•	•	•
Valve—tappet clearance	•	•	•	•
Tappet—length	2.004	1.994	†	†
Tappet—stem to guide clearance	•	•	•	•
Ignition timing	•	—	•	—
Breaker contact opening	•	—	•	—
Spark plug gap	•	—	•	—
Spark plug	14 mm. Champ. L86, AC 44F, NGK B6HS			

*Model EY18W dimensions and specifications apply to Models EY25W and EY27W
†Model EY25W dimensions and specifications apply to Model EY27W

Mechanics Data

SI METRIC TABLES

The following tables are given in SI (International System) metric units. SI units replace both customary (English) and the older gavimetric units. The use of SI units as a new worldwide standard was set by the International Committee of Weights and Measures in 1960. SI has since been adopted by most countries as their national standard.

These tables are general conversion tables which will allow you to convert customary units, which appear in the text, into SI units.

The following are a list of SI units and the customary units, used in this book, which they replace:

To measure:	Use SI units:	Which replace (customary units):
mass	kilograms (kg)	pounds (lbs)
temperature	Celsius (°C)	Fahrenheit (°F)
length	millimeters (mm)	inches (in.)
force	newtons (N)	pounds force (lbs)
capacities	liters (l)	pints/quarts/gallons (pts/qts/gals)
torque	newton·meters (N·m)	foot pounds (ft lbs)
pressure	kilopascals (kPa)	pounds per square inch (psi)
volume	cubic centimeters (cm³)	cubic inches (cu in.)
power	kilowatts (kW)	horsepower (hp)

If you have had any prior experience with the metric system, you may have noticed units in this chart which are not familiar to you. This is because, in some cases, SI units differ from the older gravimetric units which they replace. For example, newtons (N) replace kilograms (kg) as a force unit, kilopascals (kPa) replace atmospheres or bars as a unit of pressure, and, although the units are the same, the name Celsius replaces centigrade for temperature measurement.

If you are not using the SI tables, have a look at them anyway; you will be seeing a lot more of them in the future.

DRILL SIZES IN DECIMAL EQUIVALENTS

Inch	Decimal	Wire	mm
1/64	.0156		.39
	.0157		.4
	.0160	78	
	.0165		.42
	.0173		.44
	.0177		.45
	.0180	77	
	.0181		.46
	.0189		.48
	.0197		.5
	.0200	76	
	.0210	75	
	.0217		.55
	.0225	74	
	.0236		.6
	.0240	73	
	.0250	72	
	.0256		.65
	.0260	71	
	.0276		.7
	.0280	70	
	.0292	69	
	.0295		.75
	.0310	68	
1/32	.0312		.79
	.0315		.8
	.0320	67	
	.0330	66	
	.0335		.85
	.0350	65	
	.0354		.9
	.0360	64	
	.0370	63	
	.0374		.95
	.0380	62	
	.0390	61	
	.0394		1.0
	.0400	60	
	.0410	59	
	.0413		1.05
	.0420	58	
	.0430	57	
	.0433		1.1
	.0453		1.15
	.0465	56	
3/64	.0469		1.19
	.0472		1.2
	.0492		1.25
	.0512		1.3
	.0520	55	
	.0531		1.35
	.0550	54	
	.0551		1.4
	.0571		1.45
	.0591		1.5
	.0595	53	

Inch	Decimal	Wire	mm
	.0610		1.55
1/16	.0625		1.59
	.0630		1.6
	.0635	52	
	.0650		1.65
	.0669		1.7
	.0670	51	
	.0689		1.75
	.0700	50	
	.0709		1.8
	.0728		1.85
	.0730	49	
	.0748		1.9
	.0760	48	
	.0768		1.95
5/64	.0781		1.98
	.0785	47	
	.0787		2.0
	.0807		2.05
	.0810	46	
	.0820	45	
	.0827		2.1
	.0846		2.15
	.0860	44	
	.0866		2.2
	.0886		2.25
	.0890	43	
	.0906		2.3
	.0925		2.35
	.0935	42	
3/32	.0938		2.38
	.0945		2.4
	.0960	41	
	.0965		2.45
	.0980	40	
	.0981		2.5
	.0995	39	
	.1015	38	
	.1024		2.6
	.1040	37	
	.1063		2.7
	.1065	36	
	.1083		2.75
7/64	.1094		2.77
	.1100	35	
	.1102		2.8
	.1110	34	
	.1130	33	
	.1142		2.9
	.1160	32	
	.1181		3.0
	.1200	31	
	.1220		3.1
1/8	.1250		3.17
	.1260		3.2
	.1280		3.25

Inch	Decimal	Wire	mm
	.1285	30	
	.1299		3.3
	.1339		3.4
	.1360	29	
	.1378		3.5
	.1405	28	
9/64	.1406		3.57
	.1417		3.6
	.1440	27	
	.1457		3.7
	.1470	26	
	.1476		3.75
	.1495	25	
	.1496		3.8
	.1520	24	
	.1535		3.9
	.1540	23	
5/32	.1562		3.96
	.1570	22	
	.1575		4.0
	.1590	21	
	.1610	20	
	.1614		4.1
	.1654		4.2
	.1660	19	
	.1673		4.25
	.1693		4.3
	.1695	18	
11/64	.1719		4.36
	.1730	17	
	.1732		4.4
	.1770	16	
	.1772		4.5
	.1800	15	
	.1811		4.6
	.1820	14	
	.1850	13	
	.1850		4.7
	.1870		4.75
3/16	.1875		4.76
	.1890		4.8
	.1890	12	
	.1910	11	
	.1929		4.9
	.1935	10	
	.1960	9	
	.1969		5.0
	.1990	8	
	.2008		5.1
	.2010	7	
13/64	.2031		5.16
	.2040	6	
	.2047		5.2
	.2055	5	
	.2067		5.25
	.2087		5.3

Inch	Decimal	Wire & Letter	mm
	.2090	4	
	.2126		5.4
	.2130	3	
7/32	.2165		5.5
	.2188		5.55
	.2205		5.6
	.2210	2	
	.2244		5.7
	.2264		5.75
	.2280	1	
	.2283		5.8
	.2323		5.9
	.2340	A	
15/64	.2344		5.95
	.2362		6.0
	.2380	B	
	.2402		6.1
	.2420	C	
	.2441		6.2
	.2460	D	
	.2461		6.25
	.2480		6.3
1/4	.2500	E	6.35
	.2520		6.4
	.2559		6.5
	.2570	F	
	.2598		6.6
	.2610	G	
	.2638		6.7
17/64	.2656		6.74
	.2657		6.75
	.2660	H	
	.2677		6.8
	.2717		6.9
	.2720	I	
	.2756		7.0
	.2770	J	
	.2795		7.1
	.2810	K	
9/32	.2812		7.14
	.2835		7.2
	.2854		7.25
	.2874		7.3
	.2900	L	
	.2913		7.4
	.2950	M	
	.2953		7.5
19/64	.2969		7.54
	.2992		7.6
	.3020	N	
	.3031		7.7
	.3051		7.75
	.3071		7.8
	.3110		7.9
5/16	.3125		7.93
	.3150		8.0

Inch	Decimal	Letter	mm
	.3160	O	
	.3189		8.1
	.3228		8.2
	.3230	P	
	.3248		8.25
	.3268		8.3
21/64	.3281		8.33
	.3307		8.4
	.3320	Q	
	.3346		8.5
	.3386		8.6
	.3390	R	
	.3425		8.7
11/32	.3438		8.73
	.3445		8.75
	.3465		8.8
	.3480	S	
	.3504		8.9
	.3543		9.0
	.3580	T	
	.3583		9.1
23/64	.3594		9.12
	.3622		9.2
	.3642		9.25
	.3661		9.3
	.3680	U	
	.3701		9.4
	.3740		9.5
3/8	.3750		9.52
	.3770	V	
	.3780		9.6
	.3819		9.7
	.3839		9.75
	.3858		9.8
	.3860	W	
	.3898		9.9
25/64	.3906		9.92
	.3937		10.0
	.3970	X	
	.4040	Y	
13/32	.4062		10.31
	.4130	Z	
	.4134		10.5
27/64	.4219		10.71
	.4331		11.0
7/16	.4375		11.11
	.4528		11.5
29/64	.4531		11.51
15/32	.4688		11.90
	.4724		12.0
31/64	.4844		12.30
	.4921		12.5
1/2	.5000		12.70
	.5118		13.0
33/64	.5156		13.09
17/32	.5312		13.49

Inch	Decimal	mm
	.5315	13.5
35/64	.5469	13.89
	.5512	14.0
9/16	.5625	14.28
	.5709	14.5
37/64	.5781	14.68
	.5906	15.0
19/32	.5938	15.08
39/64	.6094	15.47
	.6102	15.5
5/8	.6250	15.87
	.6299	16.0
41/64	.6406	16.27
	.6496	16.5
21/32	.6562	16.66
	.6693	17.0
43/64	.6719	17.06
11/16	.6875	17.46
	.6890	17.5
45/64	.7031	17.85
	.7087	18.0
23/32	.7188	18.25
	.7283	18.5
47/64	.7344	18.65
	.7480	19.0
3/4	.7500	19.05
49/64	.7656	19.44
	.7677	19.5
25/32	.7812	19.84
	.7874	20.0
51/64	.7969	20.24
	.8071	20.5
13/16	.8125	20.63
	.8268	21.0
53/64	.8281	21.03
27/32	.8438	21.43
	.8465	21.5
55/64	.8594	21.82
	.8661	22.0
7/8	.8750	22.22
	.8858	22.5
57/64	.8906	22.62
	.9055	23.0
29/32	.9062	23.01
59/64	.9219	23.41
	.9252	23.5
15/16	.9375	23.81
	.9449	24.0
61/64	.9531	24.2
	.9646	24.5
31/32	.9688	24.6
	.9843	25.0
63/64	.9844	25.0
1	1.0000	25.4

GENERAL CONVERSION TABLE

Multiply By	To Convert	To	—
	Length		—
2.54	Inches	Centimeters	.3937
25.4	Inches	Millimeters	.03937
30.48	Feet	Centimeters	.0328
.304	Feet	Meters	3.28
.914	Yards	Meters	1.094
1.609	Miles	Kilometers	.621
	Volume		
.473	Pints	Liters	2.11
.946	Quarts	Liters	1.06
3.785	Gallons	Liters	.264
.016	Cubic inches	Liters	61.02
16.39	Cubic inches	Cubic cms.	.061
28.3	Cubic feet	Liters	.0353
	Mass (Weight)		
28.35	Ounces	Grams	.035
.4536	Pounds	Kilograms	2.20
	Area		
.645	Square inches	Square cms.	.155
.836	Square yds.	Square meters	1.196
	Force		
4.448	Pounds	Newtons	.225
.138	Ft./lbs.	Kilogram/meters	7.23
1.36	Ft./lbs.	Newton-meters	.737
.112	In./lbs.	Newton-meters	8.844
	Pressure		
.068	Psi	Atmospheres	14.7
6.89	Psi	Kilopascals	.145
	Other		
1.104	Horsepower (DIN)	Horsepower (SAE)	.9861
.746	Horsepower (SAE)	Kilowatts (KW)	1.34
1.60	Mph	Km/h	.625
.425	Mpg	Km/1	2.35
—	To obtain	From	Multiply by

TAP DRILL SIZES

NATIONAL COARSE OR U.S.S.						NATIONAL FINE OR S.A.E.					
Screw & Tap Size	Threads Per Inch	Use Drill Number	Screw & Tap Size	Threads Per Inch	Use Drill Number	Screw & Tap Size	Threads Per Inch	Use Drill Number	Screw & Tap Size	Threads Per Inch	Use Drill Number
No. 5	40	39	1/2	13	27/64	No. 5	44	37	1/2	20	29/64
No. 6	32	36	9/16	12	31/64	No. 6	40	33	9/16	18	33/64
No. 8	32	29	5/8	11	17/32	No. 8	36	29	5/8	18	37/64
No. 10	24	25	3/4	10	21/32	No. 10	32	21	3/4	16	11/16
No. 12	24	17	7/8	9	49/64	No. 12	28	15	7/8	14	13/16
1/4	20	8	1	8	7/8	1/4	28	3	1 1/8	12	1 3/64
5/16	18	F	1 1/8	7	63/64	5/16	24	1	1 1/4	12	1 11/64
3/8	16	5/16	1 1/4	7	1 7/64	3/8	24	Q	1 1/2	12	1 27/64
7/16	14	U	1 1/2	6	1 11/32	7/16	20	W			

ENGLISH TO METRIC CONVERSION: MASS (WEIGHT)

Current **mass** measurement is expressed in pounds and ounces (lbs. & ozs.). The metric unit of mass (or weight) is the kilogram (kg). Even although this table does not show conversion of masses (weights) larger than 15 lbs, it is easy to calculate larger units by following the data immediately below.

To convert ounces (oz.) to grams (g): multiply th number of ozs. by 28
To convert grams (g) to ounces (oz.): multiply the number of grams by .035

To convert pounds (lbs.) to kilograms (kg): multiply the number of lbs. by .45
To convert kilograms (kg) to pounds (lbs.): multiply the number of kilograms by 2.2

lbs	kg	lbs	kg	oz	kg	oz	kg
0.1	0.04	0.9	0.41	0.1	0.003	0.9	0.024
0.2	0.09	1	0.4	0.2	0.005	1	0.03
0.3	0.14	2	0.9	0.3	0.008	2	0.06
0.4	0.18	3	1.4	0.4	0.011	3	0.08
0.5	0.23	4	1.8	0.5	0.014	4	0.11
0.6	0.27	5	2.3	0.6	0.017	5	0.14
0.7	0.32	10	4.5	0.7	0.020	10	0.28
0.8	0.36	15	6.8	0.8	0.023	15	0.42

ENGLISH TO METRIC CONVERSION: TEMPERATURE

To convert Fahrenheit (°F) to Celsius (°C): take number of °F and subtract 32; multiply result by 5; divide result by 9
To convert Celsius (°C) to Fahrenheit (°F): take number of °C and multiply by 9; divide result by 5; add 32 to total

| Fahrenheit (F) | Celsius (C) | | | Fahrenheit (F) | Celsius (C) | | | Fahrenheit (F) | Celsius (C) | | |
°F	°C	°C	°F	°F	°C	°C	°F	°F	°C	°C	°F
−40	−40	−38	−36.4	80	26.7	18	64.4	215	101.7	80	176
−35	−37.2	−36	−32.8	85	29.4	20	68	220	104.4	85	185
−30	−34.4	−34	−29.2	90	32.2	22	71.6	225	107.2	90	194
−25	−31.7	−32	−25.6	95	35.0	24	75.2	230	110.0	95	202
−20	−28.9	−30	−22	100	37.8	26	78.8	235	112.8	100	212
−15	−26.1	−28	−18.4	105	40.6	28	82.4	240	115.6	105	221
−10	−23.3	−26	−14.8	110	43.3	30	86	245	118.3	110	230
−5	−20.6	−24	−11.2	115	46.1	32	89.6	250	121.1	115	239
0	−17.8	−22	−7.6	120	48.9	34	93.2	255	123.9	120	248
1	−17.2	−20	−4	125	51.7	36	96.8	260	126.6	125	257
2	−16.7	−18	−0.4	130	54.4	38	100.4	265	129.4	130	266
3	−16.1	−16	3.2	135	57.2	40	104	270	132.2	135	275
4	−15.6	−14	6.8	140	60.0	42	107.6	275	135.0	140	284
5	−15.0	−12	10.4	145	62.8	44	112.2	280	137.8	145	293
10	−12.2	−10	14	150	65.6	46	114.8	285	140.6	150	302
15	−9.4	−8	17.6	155	68.3	48	118.4	290	143.3	155	311
20	−6.7	−6	21.2	160	71.1	50	122	295	146.1	160	320
25	−3.9	−4	24.8	165	73.9	52	125.6	300	148.9	165	329
30	−1.1	−2	28.4	170	76.7	54	129.2	305	151.7	170	338
35	1.7	0	32	175	79.4	56	132.8	310	154.4	175	347
40	4.4	2	35.6	180	82.2	58	136.4	315	157.2	180	356
45	7.2	4	39.2	185	85.0	60	140	320	160.0	185	365
50	10.0	6	42.8	190	87.8	62	143.6	325	162.8	190	374
55	12.8	8	46.4	195	90.6	64	147.2	330	165.6	195	383
60	15.6	10	50	200	93.3	66	150.8	335	168.3	200	392
65	18.3	12	53.6	205	96.1	68	154.4	340	171.1	205	401
70	21.1	14	57.2	210	98.9	70	158	345	173.9	210	410
75	23.9	16	60.8	212	100.0	75	167	350	176.7	215	414

ENGLISH TO METRIC CONVERSION: LENGTH

To convert inches (ins.) to millimeters (mm): multiply number of inches by 25.4

To convert millimeters (mm) to inches (ins.): multiply number of millimeters by .04

Inches	Decimals	Milli-meters	inches	mm		Inches	Decimals	Milli-meters	inches	mm
1/64	0.051625	0.3969	0.0001	0.00254		33/64	0.515625	13.0969	0.6	15.24
1/32	0.03125	0.7937	0.0002	0.00508		17/32	0.53125	13.4937	0.7	17.78
3/64	0.046875	1.1906	0.0003	0.00762		35/64	0.546875	13.8906	0.8	20.32
1/16	0.0625	1.5875	0.0004	0.01016		9/16	0.5625	14.2875	0.9	22.86
5/64	0.078125	1.9844	0.0005	0.01270		37/64	0.578125	14.6844	1	25.4
3/32	0.09375	2.3812	0.0006	0.01524		19/32	0.59375	15.0812	2	50.8
7/64	0.109375	2.7781	0.0007	0.01778		39/64	0.609375	15.4781	3	76.2
1/8	0.125	3.1750	0.0008	0.02032		5/8	0.625	15.8750	4	101.6
9/64	0.140625	3.5719	0.0009	0.02286		41/64	0.640625	16.2719	5	127.0
5/32	0.15625	3.9687	0.001	0.0254		21/32	0.65625	16.6687	6	152.4
11/64	0.171875	4.3656	0.002	0.0508		43/64	0.671875	17.0656	7	177.8
3/16	0.1875	4.7625	0.003	0.0762		11/16	0.6875	17.4625	8	203.2
13/64	0.203125	5.1594	0.004	0.1016		45/64	0.703125	17.8594	9	228.6
7/32	0.21875	5.5562	0.005	0.1270		23/32	0.71875	18.2562	10	254.0
15/64	0.234375	5.9531	0.006	0.1524		47/64	0.734375	18.6531	11	279.4
1/4	0.25	6.3500	0.007	0.1778		3/4	0.75	19.0500	12	304.8
17/64	0.265625	6.7469	0.008	0.2032		49/64	0.765625	19.4469	13	330.2
9/32	0.28125	7.1437	0.009	0.2286		25/32	0.78125	19.8437	14	355.6
19/64	0.296875	7.5406	0.01	0.254		51/64	0.796875	20.2406	15	381.0
5/16	0.3125	7.9375	0.02	0.508		13/16	0.8125	20.6375	16	406.4
21/64	0.328125	8.3344	0.03	0.762		53/64	0.828125	21.0344	17	431.8
11/32	0.34375	8.7312	0.04	1.016		27/32	0.84375	21.4312	18	457.2
23/64	0.359375	9.1281	0.05	1.270		55/64	0.859375	21.8281	19	482.6
3/8	0.375	9.5250	0.06	1.524		7/8	0.875	22.2250	20	508.0
25/64	0.390625	9.9219	0.07	1.778		57/64	0.890625	22.6219	21	533.4
13/32	0.40625	10.3187	0.08	2.032		29/32	0.90625	23.0187	22	558.8
27/64	0.421875	10.7156	0.09	2.286		59/64	0.921875	23.4156	23	584.2
7/16	0.4375	11.1125	0.1	2.54		15/16	0.9375	23.8125	24	609.6
29/64	0.453125	11.5094	0.2	5.08		61/64	0.953125	24.2094	25	635.0
15/32	0.46875	11.9062	0.3	7.62		31/32	0.96875	24.6062	26	660.4
31/64	0.484375	12.3031	0.4	10.16		63/64	0.984375	25.0031	27	690.6
1/2	0.5	12.7000	0.5	12.70						

ENGLISH TO METRIC CONVERSION: TORQUE

To convert foot-pounds (ft. lbs.) to Newton-meters: multiply the number of ft. lbs. by 1.3

To convert inch-pounds (in. lbs.) to Newton-meters: multiply the number of in. lbs. by .11

in lbs	N·m	in lbs	N·m	in lbs	N·m	in lbs	N·m	in lbs	N·m
0.1	0.01	1	0.11	10	1.13	19	2.15	28	3.16
0.2	0.02	2	0.23	11	1.24	20	2.26	29	3.28
0.3	0.03	3	0.34	12	1.36	21	2.37	30	3.39
0.4	0.04	4	0.45	13	1.47	22	2.49	31	3.50
0.5	0.06	5	0.56	14	1.58	23	2.60	32	3.62
0.6	0.07	6	0.68	15	1.70	24	2.71	33	3.73
0.7	0.08	7	0.78	16	1.81	25	2.82	34	3.84
0.8	0.09	8	0.90	17	1.92	26	2.94	35	3.95
0.9	0.10	9	1.02	18	2.03	27	3.05	36	4.0/

ENGLISH TO METRIC CONVERSION: TORQUE

Torque is now expressed as either foot-pounds (ft./lbs.) or inch-pounds (in./lbs.). The metric measurement unit for torque is the Newton-meter (Nm). This unit—the Nm—will be used for all SI metric torque references, both the present ft./lbs. and in./lbs.

ft lbs	N-m	ft lbs	N-m	ft lbs	N-m	ft lbs	N-m
0.1	0.1	33	44.7	74	100.3	115	155.9
0.2	0.3	34	46.1	75	101.7	116	157.3
0.3	0.4	35	47.4	76	103.0	117	158.6
0.4	0.5	36	48.8	77	104.4	118	160.0
0.5	0.7	37	50.7	78	105.8	119	161.3
0.6	0.8	38	51.5	79	107.1	120	162.7
0.7	1.0	39	52.9	80	108.5	121	164.0
0.8	1.1	40	54.2	81	109.8	122	165.4
0.9	1.2	41	55.6	82	111.2	123	166.8
1	1.3	42	56.9	83	112.5	124	168.1
2	2.7	43	58.3	84	113.9	125	169.5
3	4.1	44	59.7	85	115.2	126	170.8
4	5.4	45	61.0	86	116.6	127	172.2
5	6.8	46	62.4	87	118.0	128	173.5
6	8.1	47	63.7	88	119.3	129	174.9
7	9.5	48	65.1	89	120.7	130	176.2
8	10.8	49	66.4	90	122.0	131	177.6
9	12.2	50	67.8	91	123.4	132	179.0
10	13.6	51	69.2	92	124.7	133	180.3
11	14.9	52	70.5	93	126.1	134	181.7
12	16.3	53	71.9	94	127.4	135	183.0
13	17.6	54	73.2	95	128.8	136	184.4
14	18.9	55	74.6	96	130.2	137	185.7
15	20.3	56	75.9	97	131.5	138	187.1
16	21.7	57	77.3	98	132.9	139	188.5
17	23.0	58	78.6	99	134.2	140	189.8
18	24.4	59	80.0	100	135.6	141	191.2
19	25.8	60	81.4	101	136.9	142	192.5
20	27.1	61	82.7	102	138.3	143	193.9
21	28.5	62	84.1	103	139.6	144	195.2
22	29.8	63	85.4	104	141.0	145	196.6
23	31.2	64	86.8	105	142.4	146	198.0
24	32.5	65	88.1	106	143.7	147	199.3
25	33.9	66	89.5	107	145.1	148	200.7
26	35.2	67	90.8	108	146.4	149	202.0
27	36.6	68	92.2	109	147.8	150	203.4
28	38.0	69	93.6	110	149.1	151	204.7
29	39.3	70	94.9	111	150.5	152	206.1
30	40.7	71	96.3	112	151.8	153	207.4
31	42.0	72	97.6	113	153.2	154	208.8
32	43.4	73	99.0	114	154.6	155	210.2